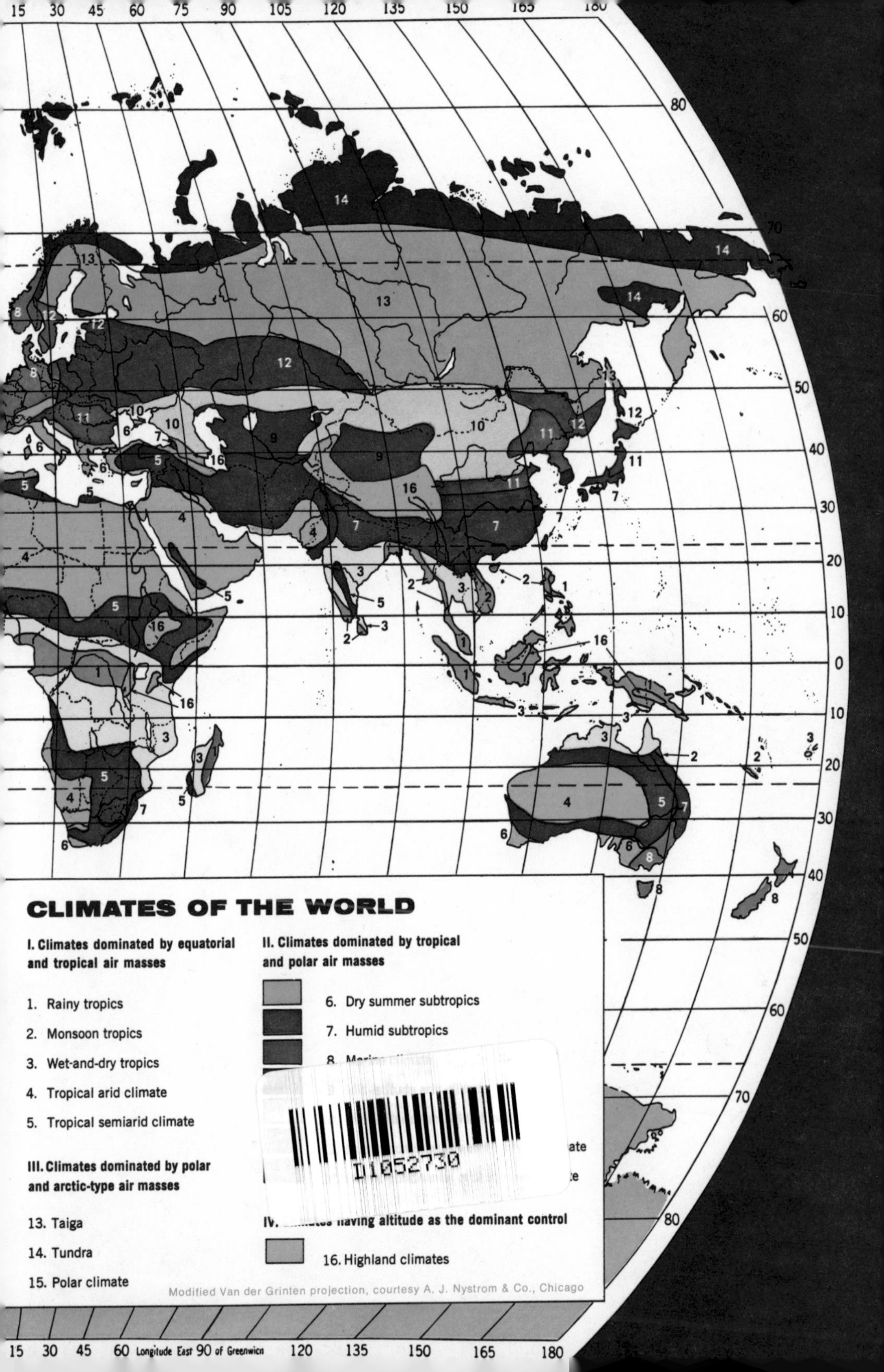
CLIMATES OF THE WORLD
I. Climates dominated by equatorial and tropical air masses
1. Rainy tropics
2. Monsoon tropics
3. Wet-and-dry tropics
4. Tropical arid climate
5. Tropical semiarid climate
II. Climates dominated by tropical and polar air masses
6. Dry summer subtropics
7. Humid subtropics
III. Climates dominated by polar and arctic-type air masses
13. Taiga
14. Tundra
15. Polar climate
having altitude as the dominant control
16. Highland climates
Modified Van der Grinten projection, courtesy A. J. Nystrom & Co., Chicago
Longitude East 90 of Greenwich

GENERAL CLIMATOLOGY

Fourth Edition

HOWARD J. CRITCHFIELD

Western Washington University

PRENTICE-HALL, INC., Englewood Cliffs, New Jersey 07632

Library of Congress Cataloging in Publication Data

Critchfield, Howard J.
General climatology.

Bibliography: p.
Includes index.
1. Climatology. I. Title.
QC981.C73 1983 551.6 82-7536
ISBN 0-13-349217-6 AACR2

Editorial/production supervision and interior design by Maria McKinnon
Cover design by Photo Plus Art
Cover coronagraph of Comet Howard-Koomen-Michels in the vicinity of the sun August 30 and 31, 1979, courtesy of Naval Research Laboratory.
Manufacturing buyer: John Hall

Printed in the United States of America

10 9 8 7 6 5 4 3 2 1

ISBN 0-13-349217-6

PRENTICE-HALL INTERNATIONAL, INC., *London*
PRENTICE-HALL OF AUSTRALIA PTY. LIMITED, *Sydney*
PRENTICE-HALL OF CANADA INC., *Toronto*
PRENTICE-HALL OF INDIA PRIVATE LIMITED, *New Delhi*
PRENTICE-HALL OF JAPAN, INC., *Tokyo*
PRENTICE-HALL OF SOUTHEAST ASIA PTE. LTD., *Singapore*
WHITEHALL BOOKS LIMITED, *Wellington, New Zealand*

Contents

CHAPTER 3 ATMOSPHERIC MOISTURE 42

CHAPTER 4 MOTION IN THE CLIMATE SYSTEM 75

CHAPTER 5 WEATHER DISTURBANCES 107

Preface

The formerly strong emphasis of atmospheric studies on daily *weather* has shifted noticeably toward *climate* during the last half of the twentieth century as we have begun to realize that decisions affecting our future on this planet require a broader perception of time and space. Different world climates not only influence human affairs over the long term, but they also may be changing as a result of our activities. We cannot ignore climate. The prudent alternative is to seek understanding of its causes, its spatial and temporal variations, and its effects as an active element of our environment. The United States Congress has recognized this by passing the National Climate Program Act for the purpose of establishing "a national climate program that will assist the Nation and the world to understand and respond to natural and man-induced climate processes and their implications." (Public Law 95-367, September 17, 1978)

General Climatology considers processes involving energy, moisture, and motion in the earth's climate system. Its objective is to introduce the fundamentals of climatology in a manner that will serve those who have a concern for the global environment, whether their primary interests lie in one of the natural or social sciences, the humanities, or interdisciplinary approaches to problem solving. Its scope, encompassing both the nature of climate and the significance of climate, extends beyond the traditional focus on the atmosphere alone.

Changes in topical emphasis and sequence in this fourth edition accommodate emerging themes in climatological research and applications. A threefold division of the broad field of climatology is retained to allow adjustment to varied educational goals. Part I introduces the exchanges of heat, moisture, and momentum that produce weather and climate. It incorporates basic observational techniques, without which there could be no modern science of climatology. Part II outlines patterns of world climates and their change through time. A new separate section expands the discussion of ocean climates. Part III examines applications of climatology to problems of resource management, food production, energy supplies, health and comfort, and housing. There are new sections on marine resources and aquaculture. A concluding chapter treats modification of climate by human activities. The topics of Part III have been selected to illustrate the methodology and benefits of applied climatology in a wide range of economic, social, and political decisions.

Questions and problems at the end of each chapter are designed to assist review and encourage further study. A selection of books and scientific journals in the bibliography offers initial directions for more detailed investigation of special topics. The text, tables, and figures of this edition employ the international system of units (SI); conversion equivalents are included in the appendix.

I am greatly indebted to the growing number of climatologists whose contributions to an expanding field of knowledge have helped determine the content and organization of this book. Successive revisions have incorporated numerous suggestions by students, teachers, and reviewers. This edition reflects especially valuable comments on the manuscript by Robert B. Batchelder, Boston University, Boston, Massachusetts; Robert A. Muller, Louisiana State University, Baton Rouge, Louisiana; Joe R. Eagleman, The University of Kansas, Lawrence, Kansas; John C. Klink, Miami University, Oxford, Ohio; Phillip J. Smith, Purdue University, Lafayette, Indiana; and Raymond S. Bradley, The University of Massachusetts, Amherst, Massachusetts. As always, errors, inconsistencies, or omissions remain my responsibility.

Howard J. Critchfield
Bellingham, Washington

Part I

THE PHYSICAL ELEMENTS OF WEATHER AND CLIMATE

1

Climate and the Atmosphere

Studies of the planet earth fall into four broad categories that embrace the solid *lithosphere,* water in the liquid *hydrosphere* and frozen *cryosphere,* the mainly gaseous *atmosphere,* and the life forms of the *biosphere.* Although the study of weather and climate focuses on the envelope of gases, continuous interchanges among the "spheres" produce an integrated environment, and no component can be understood adequately without reference to the others.

Weather, the day-to-day state of the atmosphere, consists of short-term variations of energy and mass exchanges within the atmosphere and between the earth and the atmosphere. It results from processes that attempt to equalize differences in the distribution of net radiant energy from the sun. Acting over an extended period of time, these exchange processes accumulate to become *climate.* More than a statistical average, climate is an aggregate of environmental conditions involving heat, moisture, and motion. Any study of climate must consider extremes in addition to means, trends, fluctuations, probabilities, and their variations in time and space.

Climate is an active factor in the physical environment of all living things. Its influences on human welfare range from the immediate effects of weather events to complex responses associated with climatic change. The modern communications media that inform us almost daily of floods, droughts, hurricanes, blizzards, heat waves, or other disasters somewhere in the world also bring news of the resulting property damage, crop failures, famine, or deaths. Dire views of the future promise global heating or cooling, advance or recession of polar ice, changing sea level, expanding deserts, and inevitable world hunger. A growing body of evidence suggests that abuse of the environment may enhance the likelihood of these catastrophes.

For this reason, climatology treats the role of humankind as well as the so-called "natural" factors.

THE SCOPE OF CLIMATOLOGY

Climatology is the science that seeks to describe and explain the nature of climate, why it differs from place to place, and how it is related to other elements of the natural environment and to human activities. The term originated from the Greek words *klima,* referring to the supposed slope of the earth and approximating our concept of latitude, and *logos,* a discourse or study. Climatology is closely allied with *meteorology* (literally, a "discourse on things above"), which treats day-to-day atmospheric conditions and their causes. Often defined as the physics of the atmosphere, meteorology uses the methods of the physical sciences to interpret and explain atmospheric processes; it is equated increasingly with atmospheric science. *Aerology* and *aeronomy* ordinarily denote studies of the upper atmosphere.

Climatology extends the findings of meteorology in space and in time to encompass the entire earth and periods of time as long as observations or scientific inference will permit. Because climatology involves the collection and interpretation of observed data, whether for developing generalizations or for testing theories, it necessitates instrumental and statistical techniques. It is more than statistical meteorology, however. In investigations of spatial patterns and interactions it makes abundant use of the tools of geography, including maps. A close affinity with geography arises from its multiple-factor approach to explanation of phenomena and from a concern with climate as an element of our natural environment.

The study of climate in the following chapters comprises three fundamental subdivisions: physical elements and processes, climatic patterns, and applications. Together they treat the earth's *climate system,* an integrated set of phenomena concerning sun, atmosphere, surface features, life forms, and human activities (see Fig. 1.1). It should be clear that climatology cannot isolate climate or any of its elements from other aspects of the total world environment.

The basic question to be answered in *physical climatology* is: What causes the variations in heat exchange, moisture exchange, and fluid motion from time to time and place to place? That is, why do climates differ? The first step in a scientific approach to the answer is observation. Several observable elements aid the description of weather and climate: intensity and duration of solar radiation, temperature, humidity, evaporation, cloudiness and fog, precipitation, visibility, barometric pressure, and winds. Their occurrence in a particular combination results from transfers of energy and mass in the climate system. The transfer processes are influenced in turn by differences in latitude, altitude, land and water surfaces, mountain barriers, local topography, and such gross atmospheric features as prevailing winds, air masses, and pressure centers. Part I discusses the roles of these factors and processes.

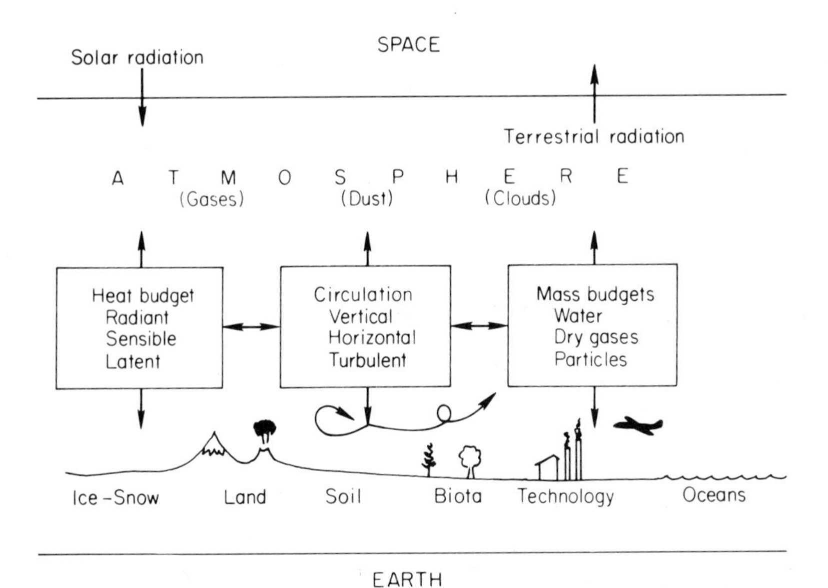

Figure 1.1 The climate system.

Climates exhibit both spatial and temporal variations throughout the world. *Regional climatology* has as its goal the orderly arrangement and explanation of spatial patterns. It includes the identification of significant climatic characteristics and the classification of climatic types, thus providing a link between the physical bases of climate and the investigation of problems in applied climatology. Because it deals with spatial distributions, regional climatology implicates the concept of scale. In keeping with the broad approach of this book, Part II treats the large-scale macroclimatic regions of the world, but in a more detailed analysis medium-scale mesoclimates and the microclimates of small areas also require differentiation and explanation. Indeed, many applications of the principles and methodology of climatology are highly localized. As a result it is common to regard *microclimatology* as a branch of physical climatology or as a group of techniques in applied climatology rather than as a subdivision of regional climatology. Where local climates are closely related to surface conditions, the term *topoclimatology* is sometimes employed to designate studies in the intermediate or mesoscale. The temporal scale of climate involves regional and global trends, which are the bases for projecting future probabilities. A fuller understanding of the nature and causes of change could lead to explanations of the past and forecasts of climate.

The third major division, *applied climatology,* explores the relation of climate to other phenomena and considers its potential effects on human welfare, finally confronting the possibility of modifying climates to meet human needs. It is this phase of general climatology that emphasizes the interdependence of sciences and the unity of human knowledge as well as the utility of climatic data and information.

Not only has there been growing cooperation among the traditional sciences in the investigation of climate itself, but also new combinations have begun to evolve. Bioclimatology, agroclimatology, medical climatology, building climatology, and urban climatology are examples. The proliferation of specialized studies clearly illustrates the wide scope and variety of climate-related phenomena.

NATURE AND ORIGIN OF THE ATMOSPHERE

The atmosphere is a blanket of gases and suspended liquids and solids that entirely envelops the earth, extending outward several thousand kilometers to a zone characterized more by magnetic fields and ionized particles than by the familiar air near the surface. ''Pure'' air is colorless, odorless, tasteless, and cannot be felt except when in motion. It is mobile, compressible, and expansible; it transmits compression waves and experiences tides. Transparent to many forms of radiation, it can absorb others. Although air is not nearly as dense as either land or water, it has weight and exerts pressure, but because it is compressible its density decreases with altitude. About half of the total mass of the atmosphere lies below 5,500 m, and nearly 99 percent lies within 30 km of the earth's surface. Without the atmosphere life would be impossible, and there would be no clouds, winds, or storms—no weather. Among its other functions, the atmosphere acts as a great canopy to protect the earth's surface from the full range of solar effects by day and prevent excessive loss of heat by night.

The more plausible explanations for the creation of the atmosphere are based in part on studies of the atmospheres of other planets. The gaseous elements that comprise the earth's lower atmosphere appear to be scarce elsewhere in the solar system. In the early stages of formation of the planet from ''cosmic gases,'' solar and gravitational effects probably resulted in accretion of some gases and the escape of others. The gases of the present atmosphere are not a direct residue of the earliest form of the planet; rather, they are the evolutionary products of volcanic eruptions, hot springs, chemical breakdown of solid matter, and contributions from the biosphere, including photosynthesis and human activities. Evidence points to stabilization of the atmosphere in approximately its present form by the Cambrian Period, nearly 600 million years ago.

Interactions among land, water, air, and life forms constantly use and renew the atmosphere. For example, the weathering of rocks, burning of fuel, decay of plants, and breathing by animals entail chemical exchanges of oxygen and carbon dioxide. Nitrogen follows a complex cycle through bacterial activity in the soil, animal tissue, organic processes in decay, and return to the air. Plants, animals, bacteria, and chemical reactions in soil and water all help to maintain an intricate balance among land, water, life, and the air.

COMPOSITION OF THE ATMOSPHERE

Air is a mixture of gases, most of which act independently of the others. Simple experiments in physics and chemistry demonstrate that air is neither a single gaseous element nor a chemical compound. Continuous churning and diffusion of the major constituents in the lower atmosphere maintain essentially the same proportions at different times and places, so that the early concept of air as a single element is neither surprising nor entirely illogical.

Four gases—nitrogen, oxygen, argon, and carbon dioxide—account for more than 99 percent of dry air (Table 1.1). Nitrogen alone constitutes nearly four-fifths by volume and oxygen one-fifth. Argon is chemically inert, as are neon, helium, krypton, and xenon. Oxygen combines chemically with many other elements and is necessary for combustion and animal metabolism. Its availability in a free state is due to continuous replenishment by photosynthesis in plants. Carbon dioxide is a product of combustion and is exhaled by animals; as a source of carbon in plant photosynthesis it is essential for both vegetation and animal life. Because carbon dioxide absorbs part of the long-wave radiation from the earth, thereby affecting the global energy budget, increasing amounts in the atmosphere may be a factor in climatic change. Nitrogen does not combine readily with other elements, but it is a constituent of many organic compounds. Another of its main effects in the atmosphere is to dilute oxygen and thus regulate combustion and oxidation.

Ozone, carbon monoxide, oxides of nitrogen and sulfur, hydrocarbons, and other trace gases, including products of biological and technological activity, have complex effects as pollutants and possible agents of climatic change. Ozone is produced by the recombination of oxygen under the influence of ultraviolet radiation at high altitudes to form three-atom molecules (O_3). Its capacity to absorb ultraviolet radiation limits the amount which reaches the earth's surface. Slight changes of ozone concentration in the upper atmosphere could have harmful effects in the biosphere. Chemical reactions between ozone and nitrogen oxides, chlorofluorocarbons, or other trace gases are suspected causes of reduced ozone density. Ozone also is formed by lightning discharges and by the action of sunlight on mixtures of

TABLE 1.1

Principal Gases of Dry Air in the Lower Atmosphere

Gas	*Percent by volume*	*Gas*	*Percent by volume*
Nitrogen (N_2)	78.084	Helium (He)	0.0005
Oxygen (O_2)	20.9476	Methane (CH_4)	0.0002
Argon (Ar)	0.934	Krypton (Kr)	0.00011
Carbon dioxide (CO_2)	0.0314	Hydrogen (H_2)	0.00005
Neon (Ne)	0.0018	Xenon (Xe)	0.0000087

hydrocarbons and oxides of nitrogen near the surface. Some stratospheric ozone is transferred to the surface by vertical circulation. Ozone is a powerful oxidizing agent and therefore a threat to life and property when concentrations become excessive. Carbon monoxide, sulfur dioxide, and nitrogen dioxide are among other gases which occur in varying quantities in the lower atmosphere and have polluting effects far out of proportion to their minute percentages.

Thus far we have considered the major gases of dry air; of more direct importance in weather and climate is water vapor, which has characteristics that set it apart from the other gases. Variations in the proportion of water vapor in air are a major concern of meteorology and climatology. They will be discussed in Chapter 3 along with another unique feature of water vapor, namely, its ability to change to a liquid or solid state, releasing latent heat. The three forms of water in the atmosphere affect the transmission of solar radiation by absorbing, reflecting, and scattering certain wave lengths of the solar spectrum. Water droplets and ice crystals account for several optical phenomena such as rainbows, halos, or coronas. They also contain dissolved substances which may be precipitated in rain or snow.

Up to an altitude of 80 to 100 km the gaseous composition of the atmosphere remains fairly constant. The proportion of ozone increases to a maximum at about 20 km (Fig. 1.2). Carbon dioxide and water vapor decrease with altitude. Gases in the outer reaches of the atmosphere consist mainly of the two lightest elements, hydrogen and helium, but these are rare near the earth's surface.

In addition to the water droplets and ice crystals which occur as fog, haze, or clouds there are large quantities of solid particles collectively known as dust. The

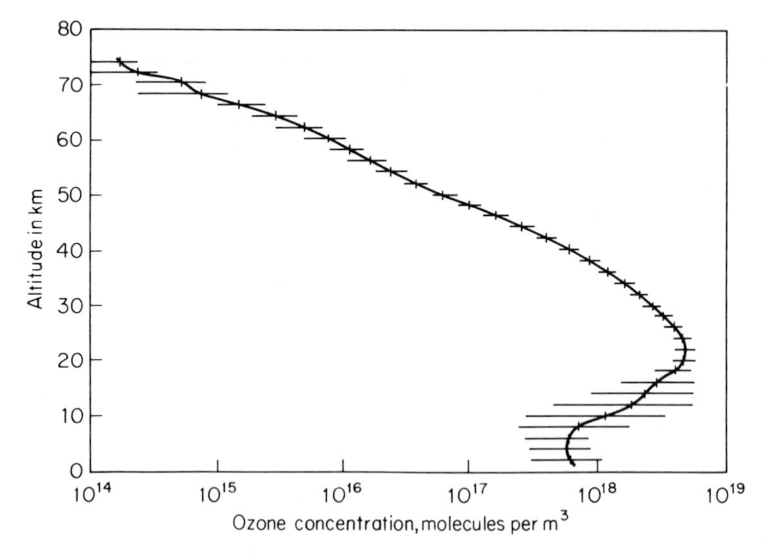

Figure 1.2 Variation of ozone concentration with height in the mid-latitudes. (*U.S. Standard Atmosphere 1976.*)

amount of atmospheric dust varies greatly over the earth, but even over the oceans hundreds of particles per milliliter have been counted. Most of the particles are invisible to the naked eye and many can be identified only with the aid of electron microscopes or other special analytical techniques (Fig. 1.3). Dust absorbs part of the incoming solar radiation as well as outgoing long-wave radiation and also is an agent of reflection and scattering. It is one of the factors in the intensity and duration of dawn and twilight. The blue color of the sky and the red of sunsets are due to selective scattering of the visible solar spectrum by gas molecules and dust. Condensation of water vapor begins on fine dust, salt crystals, and smoke nuclei. One might suppose that solids in the atmosphere would eventually be washed to earth by rains, but they are constantly being replenished. The sources are many and varied: dry soil (especially in deserts), salts from ocean spray, bacteria, pollens, seeds and

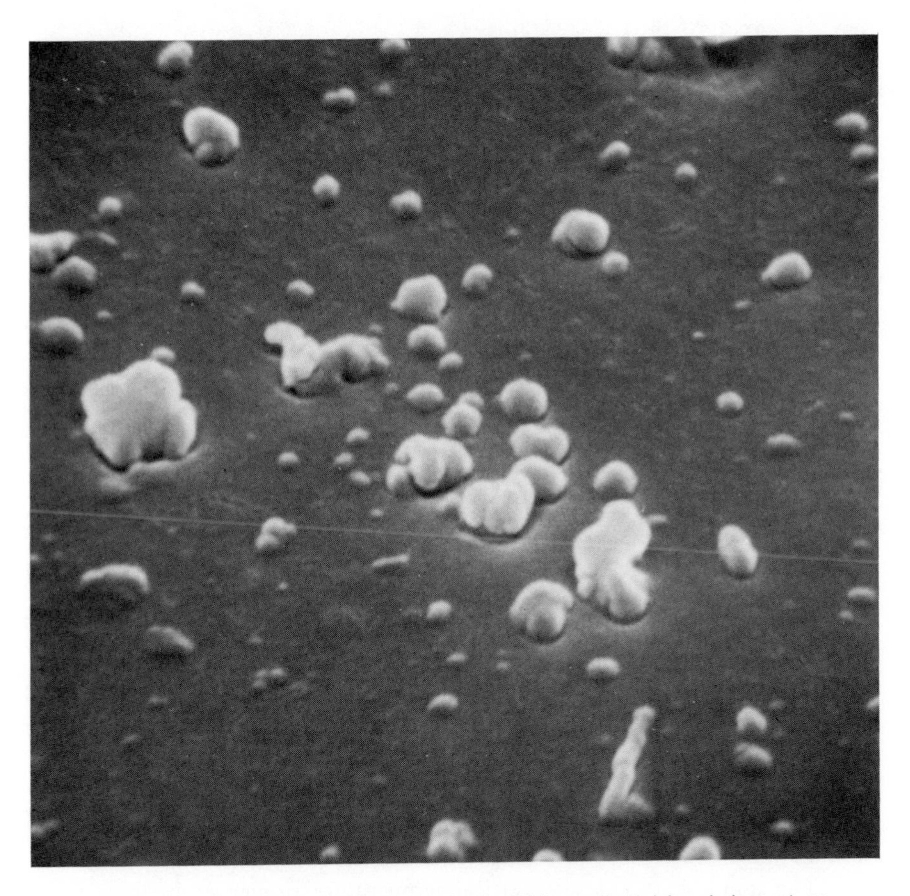

Figure 1.3 Highly magnified mineral dust particles collected by airplane at an altitude of 19.8 km over Alaska in July 1980. (Courtesy NASA and David C. Woods.)

spores, volcanic ash, meteoric dust, and industrial pollution. The number of particles is great in industrial centers and relatively greater in dry regions than in humid areas. Both density and size generally decrease with altitude. Atmospheric circulation carries the smallest particles to heights of several kilometers. Volcanic ash and nuclear detritus may be ejected to even higher altitudes, where they persist for long periods in upper-level flow patterns (Fig. 1.4). Meteoric dust is introduced at the outer limits of the atmosphere.

Figure 1.4 Sunlight reflected by Mount St. Helens volcanic ash illuminates the twilight sky above the Rocky Mountain Front Range in Colorado, 20 May 1980. (Bernard G. Mendonca, Geophysical Monitoring for Climatic Change, ERL/NOAA, Boulder, Colorado.)

VERTICAL THERMAL STRUCTURE OF THE ATMOSPHERE

The vertical extent of the atmosphere is difficult to ascertain, for it has no sharp boundary with extraterrestrial space. Atmospheric phenomena associated with the earth's magnetic and gravitational fields extend outward for several thousand kilometers to a vague zone of nebulous gases and radiation particles that become rarer and rarer until at last terrestrial characteristics of the atmosphere cease.

The atmosphere exhibits vertical temperature properties in four major layers, or ''shells'': the *troposphere, stratosphere, mesosphere,* and *thermosphere* (Fig. 1.5). The troposphere is the lower portion of the atmosphere, extending up to about 8 km

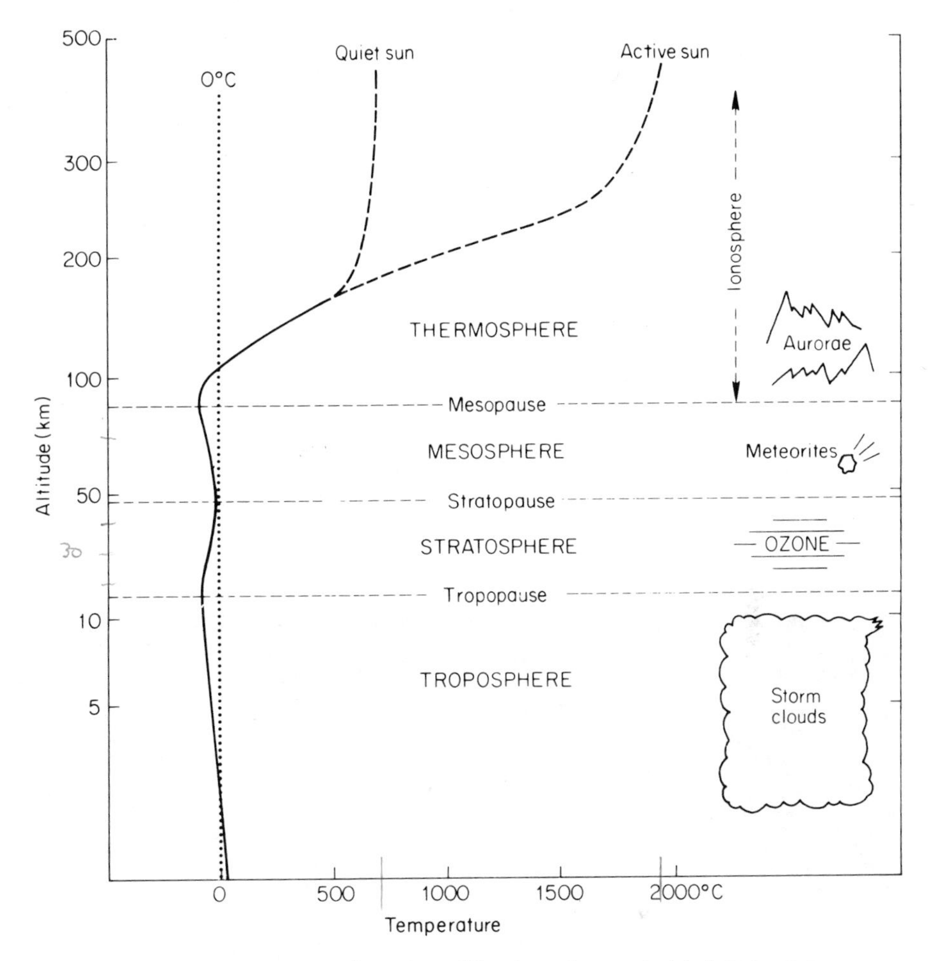

Figure 1.5 Vertical thermal structure of the atmosphere and related phenomena.

at the poles and 16 km at the equator. It contains three-fourths of the atmospheric mass and is the realm of clouds, storms, and convective motion. Thermal convection, being better developed in the low latitudes, is responsible for the greater vertical extent of the troposphere near the equator. Convective activity also explains the somewhat greater height of the troposphere in summer than in winter at a given latitude. The outstanding characteristic of the troposphere is the fairly uniform decrease in temperature with increase in altitude to minimums of −50° or −60°C. The zone marking the end of this temperature decrease is the *tropopause.* Above the tropopause lies the stratosphere, where temperature is nearly constant upward to about 20 km and then increases, owing to absorption of ultraviolet radiation by ozone. Sulfate particles are concentrated at the inner margin of the stratosphere. Near its outer limit, the *stratopause,* concentrations of ozone decrease. Beyond the stratopause, which has a mean altitude of about 50 km, is the mesosphere. Temperatures change slowly with altitude through the lower mesosphere but then decrease to a minimum near the *mesopause,* at 80 km. Most meteorites burn and disintegrate as they experience increasing friction in this layer. In the thermosphere the temperature again rises to values which vary with solar activity and at times approach 2,000°C at about 500 km. Such temperatures are not strictly comparable with those registered by thermometers at the earth's surface, however. Although the gas molecules exhibit high kinetic energy, and therefore have high temperatures, they are too sparse to transfer significant quantities of energy to an ordinary thermometer. The exposed hand of an astronaut would not feel hot in the thermosphere.

Additional vertical subdivisions of the atmosphere can be identified on the basis of chemical composition (for example, the ozonosphere) or physical properties other than temperature. Coinciding with the lower portion of the thermosphere is the *ionosphere,* an atmospheric layer at 100 to 400 km delimited on the basis of ionized particles and their effects on the propagation of radio waves. Little was known of the ionosphere until it was found that radio waves are reflected by ionized layers at great heights. The aurora borealis (Fig. 1.6) and its Southern Hemisphere

Figure 1.6 Aurora borealis and noctilucent clouds at Grande Prairie, Alberta, on the night of 18 July 1965. (Courtesy Benson Fogle.)

counterpart, aurora australis, apparently result from excitation of the ionosphere by streams of high-energy particles from the sun. Their maximum occurrence is near the magnetic poles, toward which the particles are deflected by the earth's magnetic field. Auroras have been observed to increase with an increase in sunspot activity. There is no proven relationship between auroral displays and weather in the troposphere.

In the chapters that follow we shall be concerned primarily with the lower part of the troposphere, for it is the sphere of our immediate climatic environment; but the unity of the atmosphere must be kept in mind. Variations of composition, density, temperature, or electrical fields high in the atmosphere may logically be expected to produce changes in lower layers. Much of what is not now fully understood about weather and climate may be explained ultimately with reference to the upper atmosphere.

QUESTIONS AND PROBLEMS FOR CHAPTER 1

1. How do the content, scope, and objectives of climatology differ from those of meteorology?
2. What is meant by the "climate system"?
3. Explain why today's atmosphere differs from that a billion years ago and a century ago.
4. Why do the proportions of major gases in air remain fairly constant through the troposphere?
5. Describe the climatic effects of carbon dioxide and ozone in the atmosphere.
6. How does atmospheric dust affect weather? Climate? Optical phenomena?
7. What causes fluctuations in the height of the tropopause?
8. Why do temperatures increase with altitude in the upper stratosphere?
9. Explain the absence of aurora in the atmosphere above the equator.

2

Energy and Temperature

The sun is the primary source of energy for the processes of change at the earth's surface and in the atmosphere. The amount of energy the earth receives from other celestial bodies is negligible by comparison. Radiant energy from the sun that strikes the earth is called *insolation*—a contraction of "incoming solar radiation." It is transmitted in various wave lengths of the solar spectrum, mainly in the ultraviolet, visible, and infrared bands. When components of the solar spectrum reach the earth, they are partially absorbed and converted to thermal energy. Only a small part of insolation is absorbed directly in the atmosphere, where it is transformed into heat and chemical energy. Radiation, convection, and conduction bring about heating and cooling of air by transferring energy between the earth's surface and the air and between different levels in the atmosphere. In addition, evaporation and condensation of water effect exchanges of latent heat between various surfaces and the overlying air. Eventually the energy which has provided "fuel" for the processes of weather and climate and created a variety of temperature conditions returns to outer space as long-wave terrestrial radiation, maintaining the earth's radiation balance.

THE SOLAR SOURCE

Solar radiation provides nearly all of the energy that reaches the earth, and it is the fundamental energy source to be considered in a study of the climate system. Radiation emanates from the visible surface of the sun, or *photosphere,* in wave lengths determined by the emitting temperature of about 6,000° Kelvin. The range of wave lengths, known as the *solar spectrum,* is part of the electromagnetic spectrum of radiant energy that also includes short-wave X-rays and gamma rays and the longer radio waves (Fig. 2.1). The solar spectrum comprises mainly ultraviolet, visible, and infrared wave lengths, which together transmit more than 95 percent of

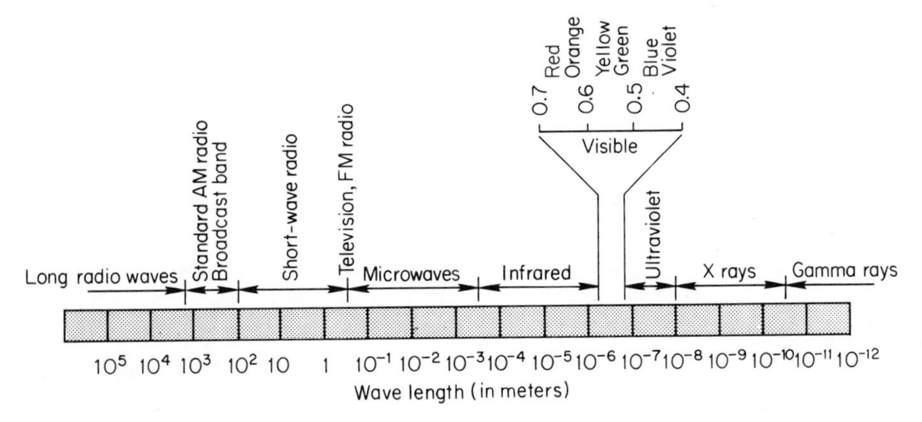

Figure 2.1 The electromagnetic spectrum.

the energy received by the earth (Fig. 2.2). The ultraviolet waves heat the upper atmosphere and produce photochemical effects. The longer waves at the other end of the spectrum are partially absorbed in the troposphere. Between these bands of in-

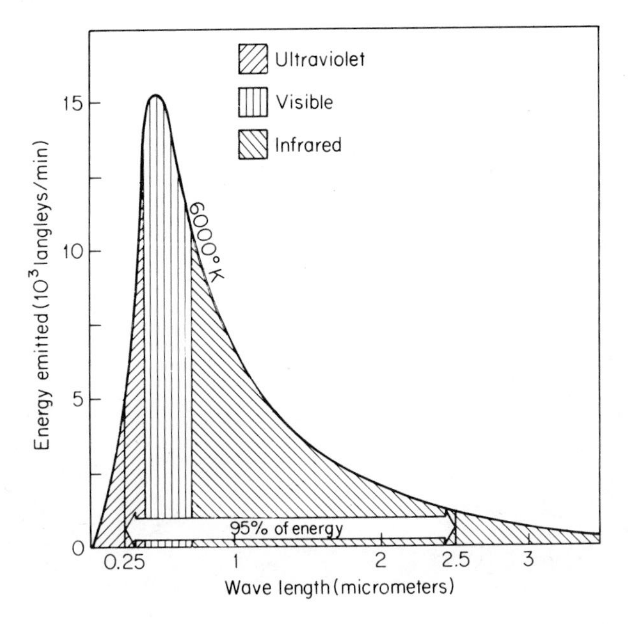

Figure 2.2 Distribution of energy emitted at different wave lengths of the solar spectrum.

visible radiation is visible sunlight composed of the colors violet, blue, green, orange, and red—the colors of the rainbow.

The earth intercepts only one two-billionth of the energy emitted by the sun, but this amount maintains an environment for terrestrial life. The average amount reaching the outer limits of the atmosphere is known as the *solar constant.* Although the actual amount varies slightly, measurements and calculations based on indirect effects agree remarkably on a mean value of 1.940 langleys per minute (one langley equals one gram-calorie per square centimeter) or 1,353 watts per square meter on a plane perpendicular to the solar beam. Variation from the average has not been found to affect daily weather greatly, but it is a minor factor in seasonal changes and may be one of the causes of climatic fluctuations. Variations in specific wave bands

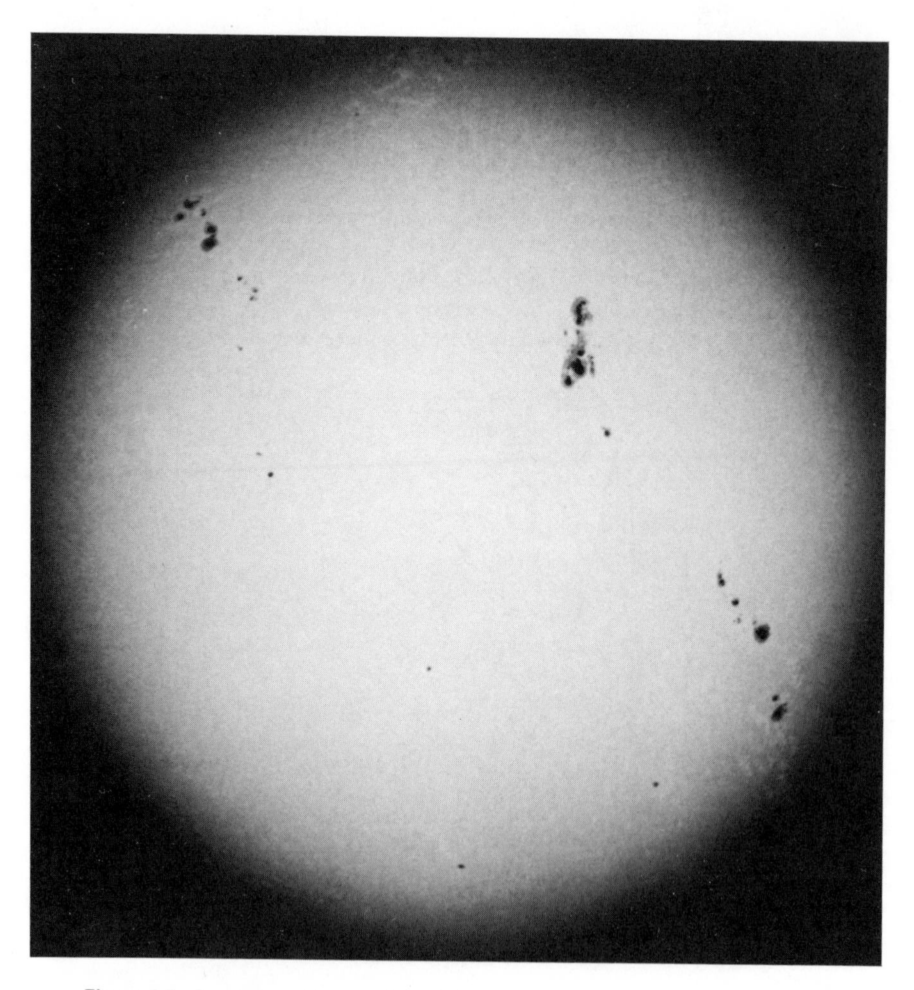

Figure 2.3 Sunspots on the solar disk. (© Association of Universities for Research in Astronomy, Inc., Sacramento Peak Observatory.)

of the solar spectrum, especially the ultraviolet, probably are even more important influences on atmospheric conditions.

In addition to radiation waves the sun emits energy as a "solar wind" consisting of ionized particles and magnetic fields. The effects of radiation are better understood than those of the solar wind, which nevertheless may influence weather and climate indirectly by disturbing the earth's magnetic field. The solar wind becomes strongest at times of increased sunspot and flare activity, when it is known to disrupt radio communication (Fig. 2.3). Research on climatic change suggests correlations between these solar disturbances and certain climatic events such as drought cycles, but a direct causal relation has not been firmly established.

THE RADIATION AND HEAT BUDGETS

Radiation is the means by which solar energy reaches the earth and the earth loses energy to outer space (Fig. 2.4). The global radiation budget has three major components: solar radiation incoming at the outer limits of the atmosphere (Q_s), the planetary albedo (a), and outgoing long-wave radiation from the earth to space (I). The basic form of the budget equation for the earth and its atmosphere is:

$$R = Q_s(1 - a) - I$$

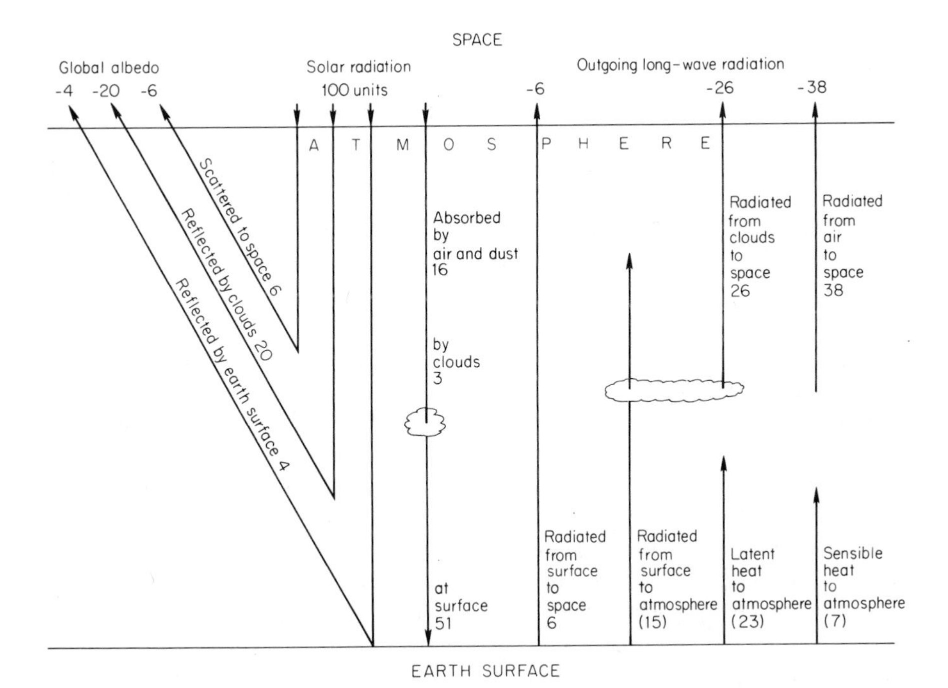

Figure 2.4 Energy balance of the earth and its atmosphere.

where R is the radiation balance (surplus or deficit) and $(1 - a)$ is the percentage of total insolation which is absorbed by the earth and atmosphere. Note that whereas insolation is intercepted on an area equivalent to the cross section of the earth and atmosphere terrestrial radiation is emitted from the entire sphere, an area four times larger (Fig. 2.5). Of the total income (Q_s) about 26 percent is reflected by

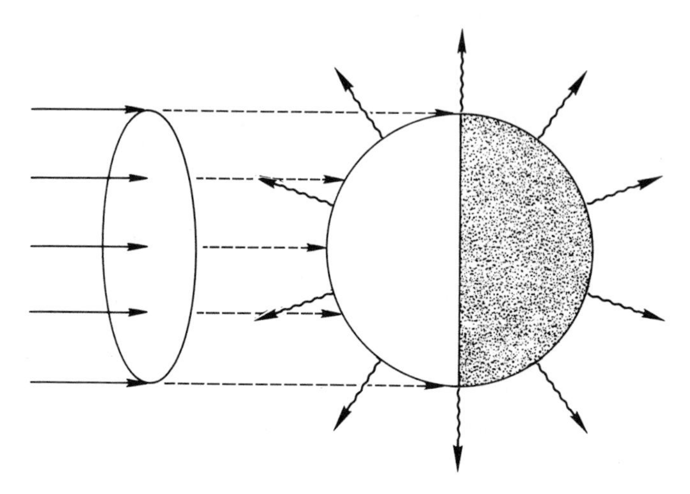

Figure 2.5 Contrast between global areas receiving and emitting radiation. The direct solar beam is intercepted by a circular cross section; the entire spherical surface emits long-wave radiation.

clouds or scattered back to space by clouds, dust, and gas molecules without heating the air; 4 percent is reflected to space from the earth's surface. The ratio of the amount of radiation reflected to the amount received on a surface is termed the *albedo.* The earth's mean albedo (known as the planetary albedo) is about 30 percent, although its value is greater in polar regions than at the equator, varying with the angle of incidence as well as the characteristics of the reflecting surfaces. The albedo of clouds varies widely with their thickness and composition. On the average clouds cover about half of the globe, giving them a major influence on the radiation budget. Table 2.1 lists the albedos of several kinds of surfaces; the values represent reflection from specific kinds of surfaces, not percentages of the global total. Much of the reflected insolation is absorbed in the atmosphere before being lost to space by radiation.

About 19 percent of insolation is absorbed in the atmosphere by gases, clouds, and suspended solids. Oxygen and ozone at high levels absorb most of the ultraviolet radiation to provide the main source of energy for circulation above 30 km. Water vapor, clouds, and dust are the principal absorbers of incoming long-wave radiation in the troposphere. The earth's surface absorbs 51 percent of insolation, either directly or after diffuse scattering downward by clouds and the atmosphere. Thus, approximately 70 percent of total insolation is effective in heating the earth and its atmosphere.

TABLE 2.1

Estimated Mean Albedos of Various Surfaces

Surface	*Albedo (%)*	*Surface*	*Albedo (%)*
Tropical forest	21	Dense cloud	70-80
Deciduous forest	18	Thin cloud	25-50
Coniferous forest	13	Oceans (60-70° lat.)	7-23
Savanna	15	Inland waters	2-78
Desert	28	Snow	40-90
Grain crops	10-25	Wet sand	30-35
Green grass	8-27	Bare rock	12-18

The earth also is a radiating body. In contrast to the sun, which radiates at a temperature of 6,000°K, the earth has an average temperature of only 288°K and emits terrestrial radiation at much greater wave lengths (Fig. 2.6). Although the constituents of the atmosphere collectively absorb only about a fifth of the incoming short-wave radiation, they can capture a large part of the outgoing long-wave radiation. This ability to admit most of the insolation yet retard losses by reradiation from the earth's surface commonly is known as the *greenhouse effect;* a more appropriate term is *atmosphere effect.* Besides the long-wave radiation from the earth's surface there is also radiation from cloud layers and from gases (especially water vapor and carbon dioxide) and dust to space. The total amount of energy reaching the earth over a considerable period of time is equalled by total outward losses. If this were not so the earth would soon become either very hot or very cold. Satellite observa-

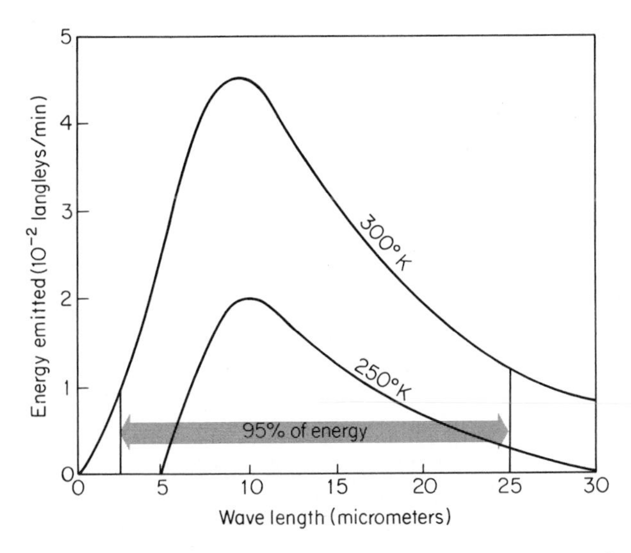

Figure 2.6 Energy emitted at different wave lengths of terrestrial radiation. Compare with Figure 2.2, noting different energy and wave length scales.

tions have confirmed a fairly stable global energy budget over relatively long periods. The potential effect of a radiative imbalance on climate is treated in Chapter 10. The radiation budget of a particular place is seldom in balance. Actually, there is an annual deficit at high latitudes and a surplus at low latitudes owing to differences in the angle of incidence and surface albedos. At any latitude variations in slope, exposure, and the character of surface cover (for example, water, land, or vegetation) produce regional differences in the radiation budget. In the middle and low latitudes a true radiative balance may exist briefly near the beginning and end of each daylight period, when deficits and surpluses replace one another. Figure 2.7 illustrates this in simplified form.

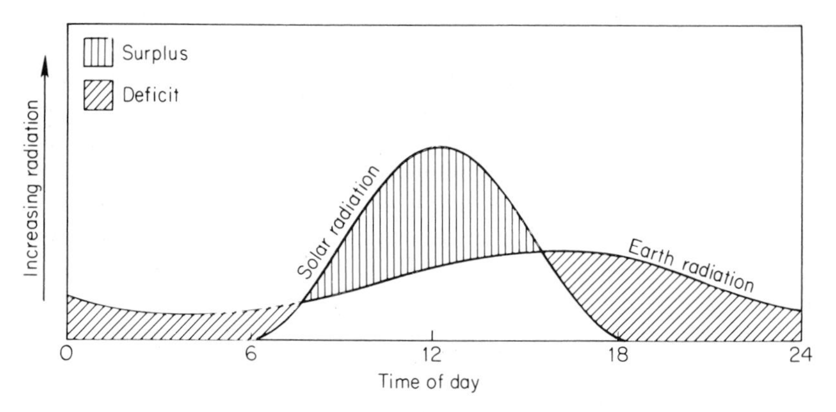

Figure 2.7 Diurnal regimes of incoming and outgoing radiation at the earth's surface during a day near the time of an equinox.

Energy transfer processes, including radiation, maintain a balanced heat budget at the earth's surface. They are represented in the equation:

$$R_s = H + LE + A$$

where R_s is the surface radiation balance (the difference between absorbed short-wave radiation and the net upward long-wave radiation from the surface), H the net transfer of sensible heat between surface and atmosphere by conduction and turbulent exchange, L the latent heat of vaporization, E the rate of evaporation, and A the flux of heat between the surface and lower layers of soil or water. Energy which raises the temperature of a substance is known as sensible heat; that which is transformed in the alteration of the physical state of matter (notably water) is latent heat.

Conduction transfers heat between adjacent molecules. Heat passes from warmer to colder substances as long as a temperature difference exists. Thus, when the surface absorbs radiation and warms above the air temperature, conduction transfers part of the heat to the lower layer of air. Figure 2.8 indicates the delays in surface air temperature responses which are common under clear skies. When the surface is cooler than the overlying air, the heat transfer is reversed and the air is

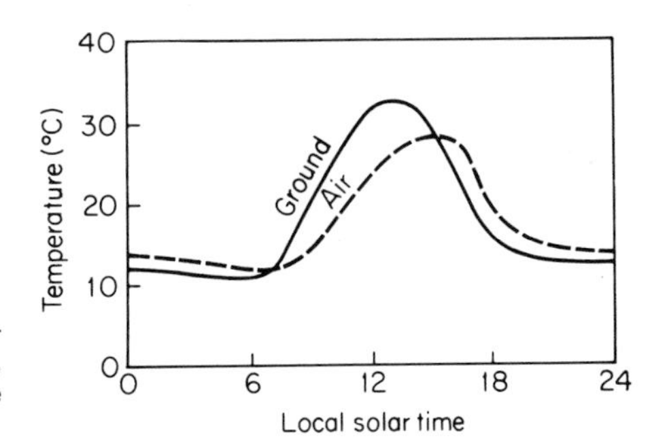

Figure 2.8 Diurnal temperature variations of bare ground and overlying air, generalized for a clear summer day in the mid-latitudes.

cooled. The latter phenomenon is common over land at night and during winter in middle and high latitudes.

Air itself is a poor conductor of heat. If conduction alone provided transfer of heat upward from the earth's surface, the air would be very hot along the ground on a summer day and quite cool a few meters above. Within the air heat conduction is insignificant compared with *convection,* which accomplishes transfer of heat through movement of the air itself. Air heated by contact with the warm earth surface expands and becomes less dense. The lighter air is displaced upward by cooler, heavier air and a convectional circulation is established with both horizontal and vertical components. The movement of air in the layer near the surface is erratic in speed and direction, consisting of eddies and bubbles. This transfer of heat and kinetic energy by turbulent motion depends on the difference in temperature between the air and the surface and is essential to development of more extensive circulation. Although much of atmospheric convection can be traced to heating at the earth's surface, it is important to remember that it also can be caused by cooling at higher levels. An example occurs when cold air temporarily pushes above warmer air, producing a sudden overturning as denser, cooler air from above replaces the lighter, warmer air below. Winter thunderstorms sometimes result from this kind of convection.

Liquids, being mobile like gases, also develop circulation systems. Surface winds and density variations due to differences in temperature and salinity generate large-scale ocean currents. Transport by winds and ocean currents prevents the progressive accumulation of heat in equatorial regions or in areas having a high absorptive capacity.

Part of the solar radiation absorbed at the earth's surface is transferred upward by long-wave radiation and as sensible heat by conduction and convection, but much of it reaches the atmosphere as latent heat when water is evaporated from land and water surfaces, especially from oceans. Moist air masses may move great distances before condensation of their water vapor releases huge quantities of latent heat. Thus, we note that the processes of heat and moisture exchange are closely interconnected. Their variations through the seasons are important features of the

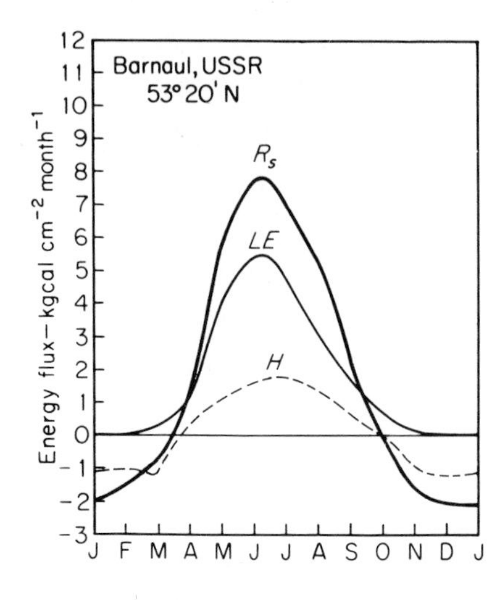

Figure 2.9 Annual march of the surface energy balance at Barnaul, U.S.S.R. (Data after Budyko.)

climate system. Figure 2.9 shows the annual march of the major heat balance components at Barnaul, U.S.S.R., an interior continental station in the mid-latitudes.

Together, the energy transfer processes effect complex exchanges between the earth's surface and the atmosphere. Local and regional energy budgets are kept in balance by transfer of radiative, sensible, and latent heat. The energy budget concept applies equally to the entire globe, geographic regions, buildings, plants,

TABLE 2.2

Annual Heat Budgets of the Continents and Oceans * (Kilolangleys Per Year)

Continent	*Net radiation*	*Latent heat flux*	*Sensible heat flux*	*Subsurface transport*
Africa	68	26	42	0
Antarctica	−11	0	−11	0
Asia	47	22	25	0
Australia	70	22	48	0
Europe	39	24	15	0
North America	40	23	17	0
South America	70	45	25	0
All land	49	25	24	0
Ocean				
Arctic	−4	5	−5	−4
Atlantic	82	72	8	2
Indian	85	77	7	1
Pacific	86	78	8	0

*After M. I. Budyko, *Climate and Life,* ed. David H. Miller (New York: Academic Press, Inc., 1974), p. 220; and William D. Sellers, *Physical Climatology* (Chicago: University of Chicago Press, 1975), p. 105.

animals, and even the human body. Annual heat budgets for the continents and oceans are summarized in Table 2.2.

VARIABILITY OF INSOLATION

The amount of insolation received on any date at a place on earth is governed by:

1. Solar radiation reaching the outer limits of the atmosphere, which depends on:
   a. Energy output of the sun
   b. Distance from the earth to the sun
   c. Interstellar dust in the solar system
2. Transparency of the atmosphere
3. Duration of the daily sunlight period
4. Angle at which the sun's noon rays strike the earth

To the extent that there may be changes in the quantity of energy emitted by the sun, the amount intercepted by our planet is altered correspondingly. An additional effect results from the ellipticity of the earth's orbit around the sun. The distance between the earth and sun varies from 152 million km at *aphelion* (July 4) to 147 million km at *perihelion* (January 3) (Fig. 2.10). Neither of these is a great de-

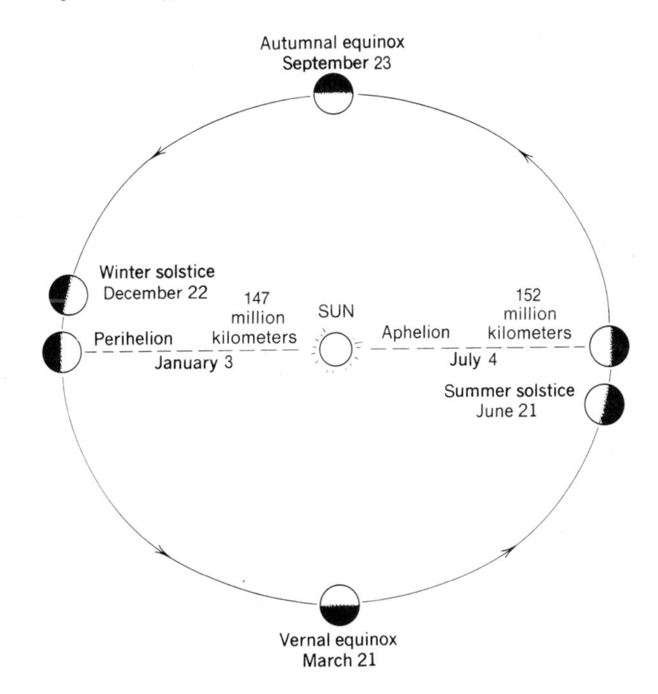

Figure 2.10 The earth's orbit around the sun.

parture from the average distance of 149.5 million km, and although the amount of radiation reaching the outer atmosphere varies by ±3.5 percent from the mean, other factors influencing the heat budget and temperature largely override this difference. Similarly, the effect of extraterrestrial dust is believed to be negligible.

Transparency of the atmosphere has a more important bearing on the amount of insolation which reaches the earth's surface. The effect of dust, clouds, water vapor, and certain gases in reflection, scattering, and absorption was noted previously. It follows that areas of dense cloudiness or polluted air transmit direct solar radiation less effectively to lower atmospheric layers or the earth's surface.

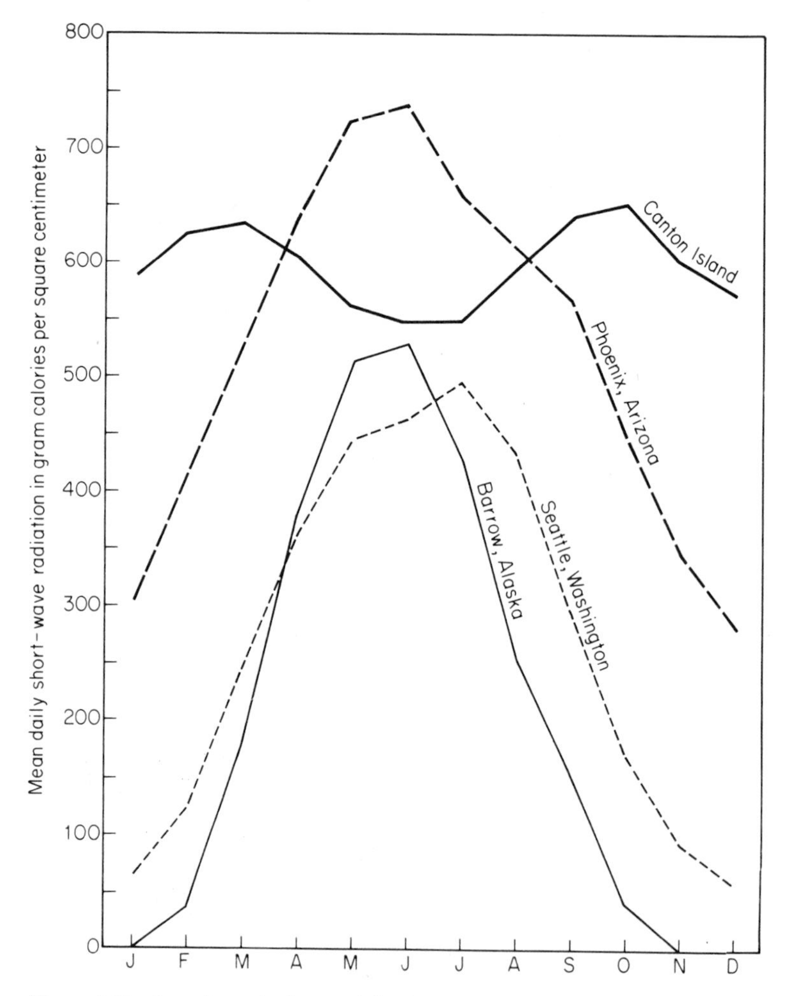

Figure 2.11 Annual march of mean daily short-wave solar radiation at selected stations.

Transparency is also a function of latitude, for at middle and high latitudes the solar beam must penetrate the reflecting–absorbing atmosphere at a lower angle than in tropical latitudes. This effect varies with the seasons, being greatest in winter when noon sun is lowest on the horizon (Fig. 2.11).

The duration of daylight (the *photoperiod*) also varies with latitude and the seasons, and the longer the photoperiod the greater is the total possible insolation (Table 2.3). At the equator day and night are always equal. In the polar regions the daily photoperiod reaches a maximum of 24 hours in summer and a minimum of zero hours in winter. At its summer solstice, under clear skies, a polar area may receive more radiation per 24-hour day than lower latitudes, although the net radiation used for heating is reduced because of the high albedo of ice and snow surfaces.

TABLE 2.3

Longest Possible Duration of Insolation

Latitude	0°	17°	41°	49°	63°	66½°	67°21′	90°
Daylight	12 hr	13 hr	15 hr	16 hr	20 hr	24 hr	1 mo	6 mo

The effect of the varying angle of the solar beam can be seen in the daily march of the sun across the sky. At or near solar noon the intensity of insolation at the earth's surface is greatest, but in the morning and evening hours, when the sun is at a low angle, intensity is reduced (Fig. 2.12). The same principle applies to latitude and the seasons. In winter and at high latitudes the sun's noon angle is low; in summer and at low latitudes it is more nearly vertical. The oblique rays of the low-angle sun are spread over a greater surface than are perpendicular rays and therefore produce less heating per unit area.

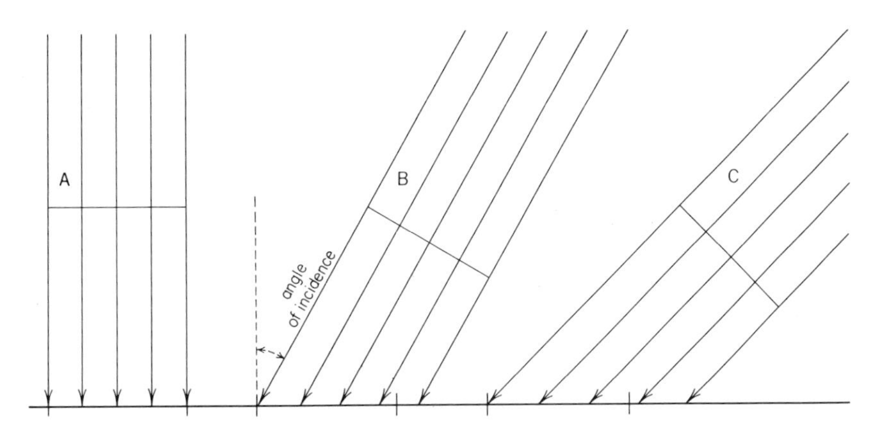

Figure 2.12 Relation of sun angle to effectiveness of insolation. *A*, sun directly overhead, highly effective; *B* and *C*, greater angles of incidence, less effective insolation per unit area.

The angle at which solar radiation strikes the earth's surface also depends on terrain features. In the Northern Hemisphere southern slopes receive a more direct solar beam, whereas northern slopes may be entirely in the shade. The possible hours of direct sunshine during winter in a deep valley may be reduced to zero by surrounding hills.

From the foregoing it is evident that the world distribution of possible insolation at the surface is closely related to latitude. At the equator the annual amount is about four times that at either of the poles (Fig. 2.13). As the direct solar beam shifts

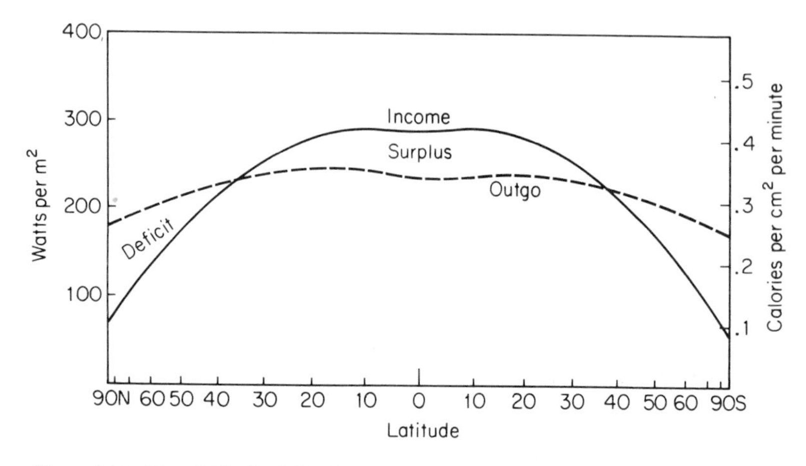

Figure 2.13 Mean latitudinal distribution of the earth's radiation budget. (After J. London and T. Sasamori.)

seasonally from one hemisphere to the other, the zone of maximum possible daily insolation moves with it. In tropical latitudes the amount of possible insolation is constantly great, and there is little variation with the seasons. But in its annual journey the sun passes over all places between the Tropic of Cancer and the Tropic of Capricorn twice, causing two maximums. In latitudes between 23½° and 66½°, maximum and minimum periods of insolation occur shortly after the summer and winter solstices, respectively. Beyond the Arctic and Antarctic Circles the maximum coincides with the summer solstice, but there is a period during which insolation is lacking. The length of the period increases toward the poles, where it is of six months' duration. Figure 2.14 shows daily amounts of solar radiation at different latitudes, assuming no atmospheric effects.

Observations of actual insolation at the earth's surface show a distribution that departs slightly from a simple latitudinal pattern. Maximum annual values are at about 20° latitude, where the drier air permits a greater proportion of the radiant energy to penetrate to surface levels. Cloudy regions receive less insolation at the surface than do areas with predominantly clear weather. In general, high plateaus

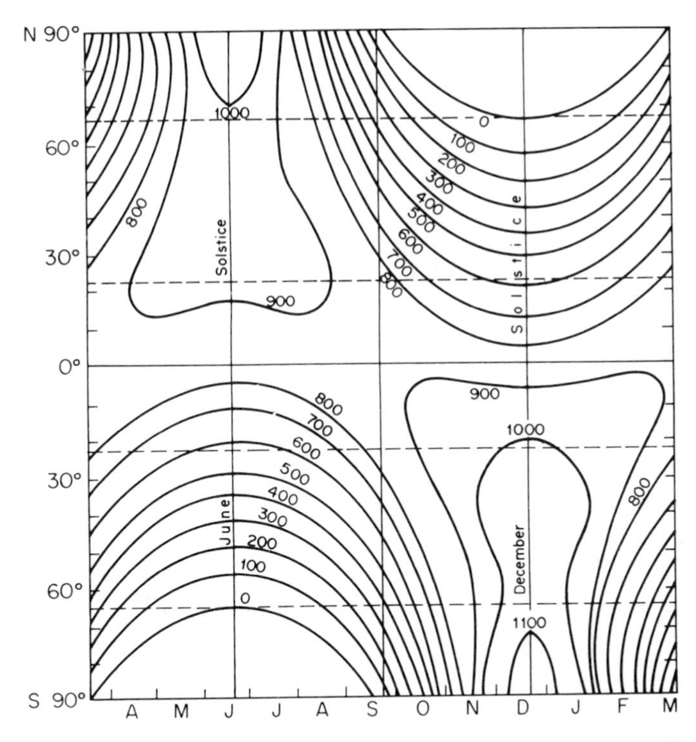

Figure 2.14 Total daily solar radiation in langleys during the year at different latitudes, assuming no atmospheric effects. (After Lamb, based on data calculated by Milankovitch.)

and mountains are favored by more effective insolation because of the relatively clearer and less dense air at high altitudes (Fig. 2.15).

MEASUREMENT OF SUNSHINE AND INSOLATION

Scientific understanding of phenomena and processes in the climate system begins with observation, which is the basis for description, analysis, and explanation of variations in time and space. Increasing attention to solar power as an energy resource in the late twentieth century has demonstrated the value of the climatic record for practical applications as well.

Several types of instruments measure the duration of sunshine (Fig. 2.16). The Campbell–Stokes recorder focuses the solar beam through a glass globe which, acting as a burning glass, chars a record of sunshine on a graduated strip of cardboard. The Foster switch employs a photoelectric principle to actuate an electric cir-

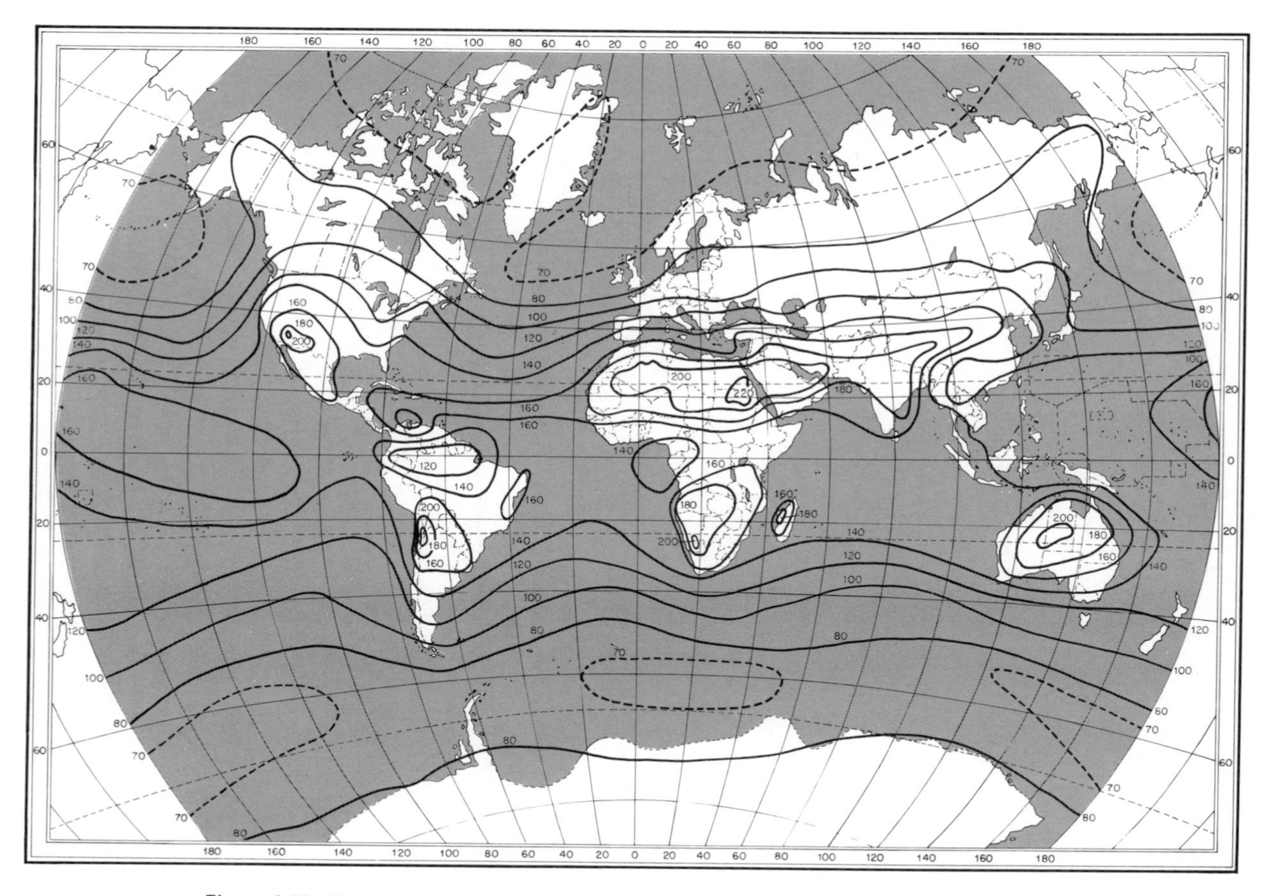

Figure 2.15 Mean annual total direct and diffuse radiation at the earth's surface. Values are in kilolangleys per year on a horizontal surface. (After Budyko and Landsberg; Van der Grinten projection courtesy A. J. Nystrom & Co., Chicago.)

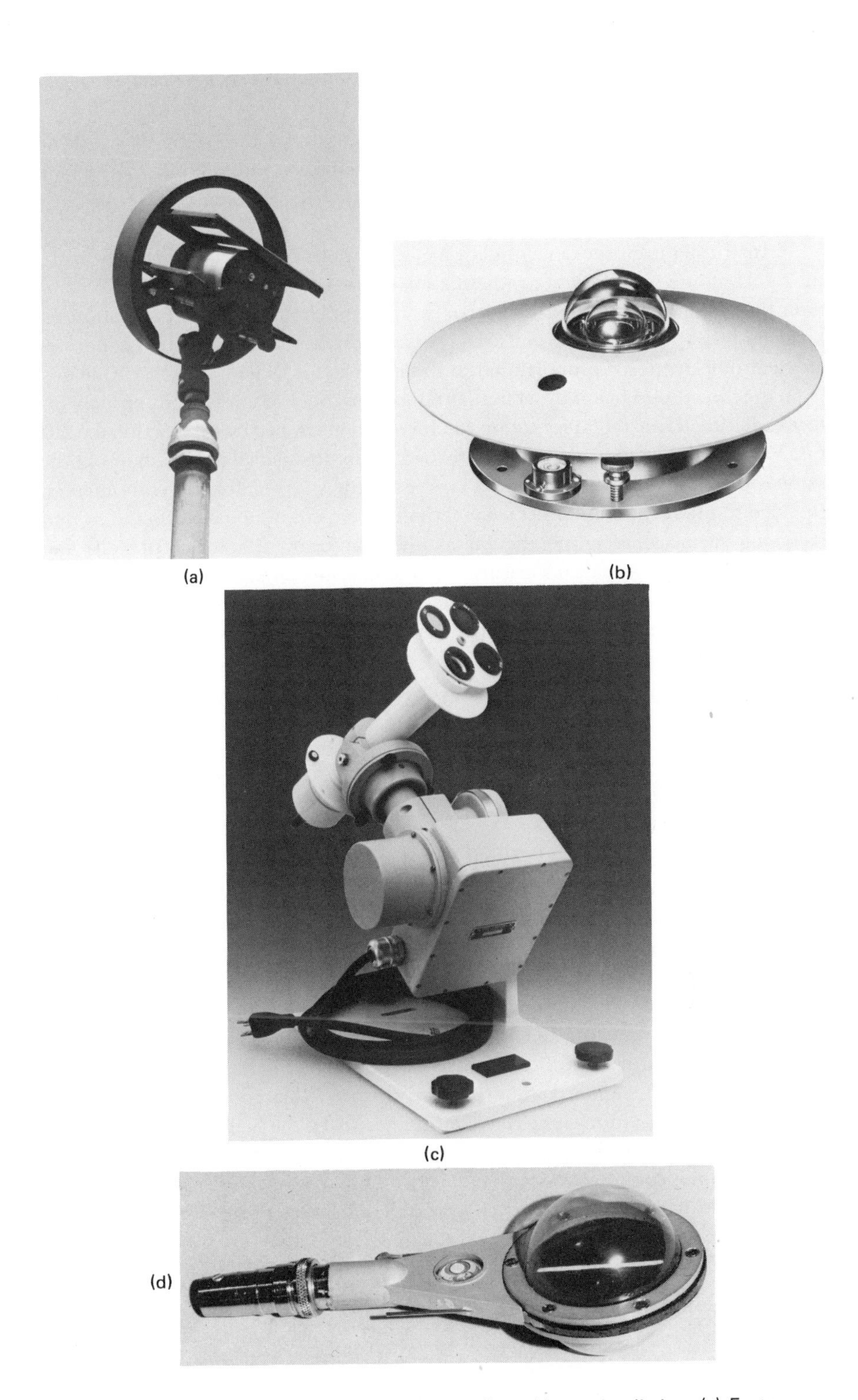

Figure 2.16 Instruments for measurement of sunshine and radiation. (a) Foster switch sunshine duration sensor; (b) total radiation pyranometer; (c) normal incidence pyrheliometer; (d) net radiometer. [(a) author; (b) and (c) courtesy Eppley Laboratory, Inc.; (d) courtesy Weather Measure Corporation.]

cuit when sunlight reaches a prescribed intensity. Another class of sunshine recorder depends on the fact that a black surface absorbs radiation at a faster rate than a white or translucent one. The resulting heat differential is translated mechanically or electrically to a recorder.

Instruments for the determination of insolation, called *pyranometers,* measure the total amount of short-wave radiation absorbed on a horizontal surface. A common type operates on a thermoelectric principle. The incidence of radiation on black and white surfaces of equal area produces differential heating that generates an electromotive force proportional to the intensity of radiation. When shielded from the direct solar beam by a circular band, a pyranometer responds only to diffuse radiation. Direct solar radiation can then be determined by subtracting the diffuse value from the total radiation absorbed by an unshielded pyranometer at the same observing site. *Pyrheliometers* are essentially specialized pyranometers, shielded to intercept only direct solar radiation and continuously oriented so that receiving surfaces are perpendicular to the solar beam. Photoelectric cells and photospectroscopic apparatus are also used to measure sunlight and to investigate selected components of the solar spectrum such as ultraviolet radiation. *Net radiometers* measure the difference between radiant transfer upward from the earth's surface and that directed downward.

AIR TEMPERATURE AND ITS MEASUREMENT

Temperature indicates the relative degree of molecular activity, or heat, of a substance. It is an index of sensible heat, not a direct measure of the quantity of energy. If heat flows from one body to another, the former has the higher temperature. To indicate the temperature of a body, an arbitrary scale of reference is employed. The Celsius scale, named for the Swedish astronomer Anders Celsius, is accepted internationally by scientists for reporting air temperature. Zero on this scale is the "triple point" temperature, at which the gaseous, liquid, and solid states of water are at equilibrium under standard atmospheric pressure. The boiling point of water under standard conditions is at 100°C. Another scale, the Kelvin, or Absolute, is based on *absolute zero,* the temperature at which molecular activity theoretically would cease. It is commonly used to indicate temperatures in the upper atmosphere and in studies involving energy exchange processes. The value of each degree on the Kelvin scale equals that of a Celsius degree, but Kelvin (or absolute) zero is at −273.16°C. For ordinary purposes we can convert Celsius to Kelvin simply by adding 273. The historical temperature records of several English-speaking countries include values on the Fahrenheit scale, which has its "triple point" of water at 32°

and boiling point at 212°. Fahrenheit temperatures may be converted to their Celsius equivalents by the formula: $C = \frac{5}{9}(F - 32)$.

The most common type of instrument for measuring air temperature is the mercury-in-glass or alcohol-in-glass thermometer. The accuracy of temperature observations depends on the care with which the thermometer is constructed and calibrated as well as on its exposure to the air. For official measurements of "surface" air temperature, thermometers are mounted in a louvered instrument shelter.

Another type of thermometer indicates temperature differences as they affect the shape of a bimetallic element whose curvature is translated into temperature values on a calibrated dial. A simpler version of this principle of expansion employs a metallic coil which produces varying tension on an indicating mechanism in response to temperature changes.

Thermocouples and *thermistors* indicate temperature electrically. Their accuracy and rapidity of response make them suitable for microclimatic observations in air, soil, plant tissues, clothing, and other specialized exposures. The thermocouple consists of a pair of junctions of two unlike metals. When one junction is kept at a constant temperature and the other is exposed to a different temperature, the electromotive force generated in the circuit can be measured by a potentiometer calibrated in degrees. The thermistor, a semiconducting ceramic element, offers less resistance to the flow of current as its temperature increases. Temperature can thus be indicated as a function of current.

For the purpose of climatic records, registering thermometers measure the maximum and minimum temperatures. The simplest *maximum thermometer* is a mercury thermometer with a constriction in the bore near the bulb (Fig. 2.17). The constriction allows the expanding mercury to pass as the temperature rises, but when cooling occurs the column of mercury breaks at the constriction leaving a part in the bore to register the highest temperature attained. The maximum thermometer is mounted horizontally and is reset by whirling so that centrifugal force pulls the detached thread of mercury down past the constriction. The fever thermometer used by physicians is a specially calibrated maximum thermometer.

The *minimum thermometer* has a large bore and its fluid is colorless alcohol. A tiny, dark index in the shape of a long dumbbell is placed in the bore below the top of the alcohol column. The minimum thermometer is mounted horizontally and as the alcohol contracts with the decreasing temperature the meniscus (concave surface) of the alcohol pulls the index down. When the meniscus moves up the bore, however, it leaves the index behind to register the lowest temperature. Resetting of the minimum thermometer is accomplished by inverting the stem until the index slides down to the meniscus.

Where a permanent, continuous record of temperature is desired, a *thermograph* is used (Fig. 2.18). The thermograph consists of an element responsive to temperature changes, a system of levers to translate these changes to a pen arm, and

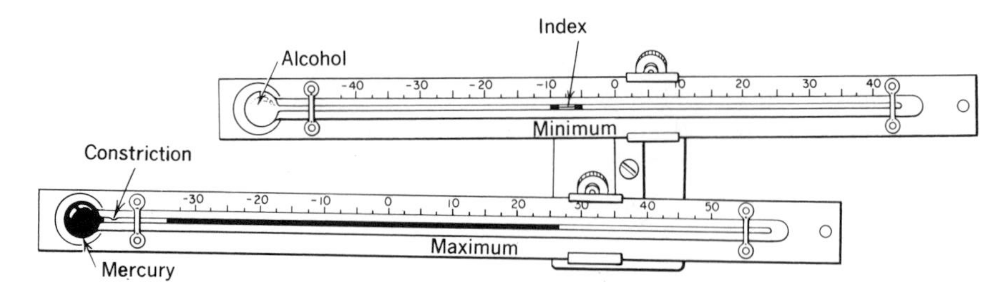

Figure 2.17 The maximum and minimum thermometers.

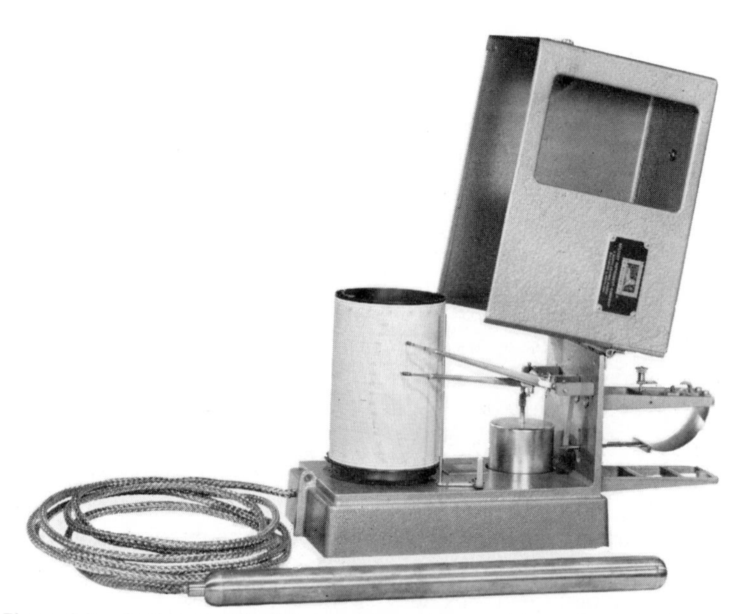

Figure 2.18 Double-recording thermograph. The Bourdon tube on the right activates a pen arm to record air temperature. The element in the foreground translates temperature changes to the second pen arm via a capillary tube from another location in air, water, or soil. (Courtesy Belfort Instrument Company.)

a cylindrical clock drum around which a calibrated chart is mounted. A common temperature element employed is the *Bourdon tube,* a flat, curved tube of phosphor bronze filled with alcohol. The tube changes its curvature in response to temperature changes. Bimetallic elements are also used in thermographs. The readings obtained from a thermograph are not as accurate as those from a mercury thermometer, but frequent checking of the thermograph traces against an accurate thermometer makes a corrected thermograph chart valuable for climatological work.

TEMPERATURE RECORDS

For climatological purposes several kinds of temperature values are desired. Probably the most used basic temperature value is the *daily mean,* from which monthly and annual average values can be derived. In practice the daily mean is found by adding the 24-hour maximum to the 24-hour minimum and dividing by 2, because comparatively few stations in the world take hourly or continuous temperature observations, which would afford a far better basis for a statistical mean. A thermograph trace provides a record of both hourly values and the daily march of temperature. The difference between the highest and lowest temperatures of the day is the *diurnal* (daily) *range.* Ordinarily, observations of the maximum and minimum thermometers are the bases for the daily mean temperature and the diurnal range. The *mean monthly* temperature is found by adding the daily means and dividing by the number of days in the month. Mean monthly values for the year indicate the *annual march* of temperature through the seasons (see Fig. 2.19). The *annual range* is the difference between the mean temperatures of the warmest and coldest months. For most stations in the Northern Hemisphere the warmest month is July and the coldest January. When corresponding temperature values for a number of years are averaged, a generalized value useful in climatic description is obtained. However, such averages tend to obscure the year-to-year variability and possible climatic fluctuations.

Especially in the middle and high latitudes the length of the *frost-free season* is commonly a part of the temperature record. The frost-free season is defined as the number of days during which temperatures are continuously above 0°C. The mean frost-free season is the difference in days between the mean date of the last frost in spring and the mean date of the first frost in autumn. It is often regarded as an important consideration in agriculture, but there is actually no simple, direct relation between plant growth and freedom from frost.

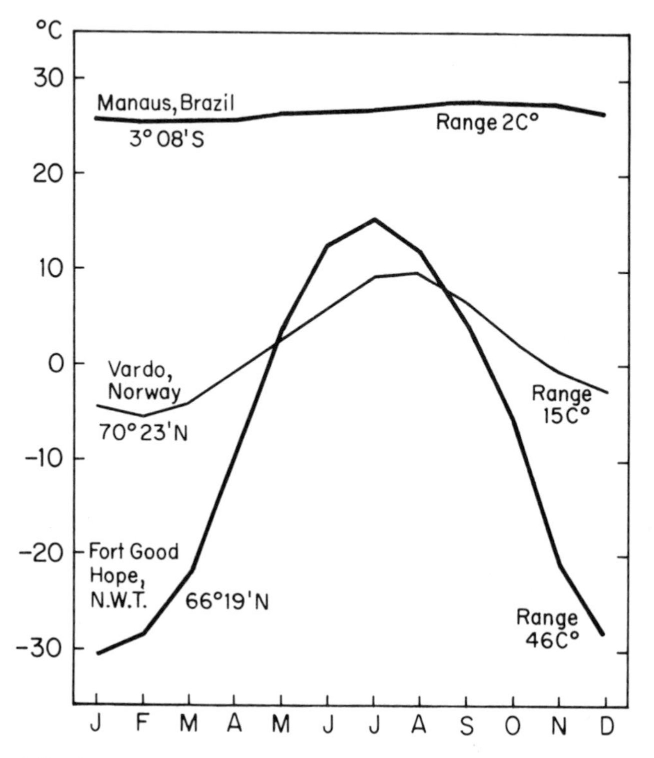

Figure 2.19 Annual variations of mean monthly temperature at selected stations.

Measurements of temperature at different depths in the soil or water bodies provide valuable data for studies of heat exchange and applications in a wide variety of practical problems.

HORIZONTAL TEMPERATURE DISTRIBUTION

The average global surface temperature is about 13°C, but local averages vary widely. An ever-expanding network of temperature-recording stations and improvements in remote sensing by satellite make it possible to describe areal differences with increasing accuracy. On maps, the horizontal distribution of temperature is commonly shown by *isotherms,* lines connecting points that have equal temperatures (see Figs. 2.20 and 2.21). On weather maps of small areas the

TABLE 2.4

Records of Extreme Temperatures

Record	°*C*	*Location*	*Date*
Highest official air temperature	58	Azizia, Libya	Sept. 13, 1922
Highest U.S. temperature	57	Greenland Ranch Death Valley, California	July 10, 1913
Highest mean annual temperature	31	Lugh, Somalia	13-year mean
Lowest official temperature in Northern Hemisphere	−68	Verkhoyansk, Siberia	Feb. 5 and 7, 1892
Lowest official temperature in Western Hemisphere	−66	On Greenland ice cap at 2,990 m	Dec. 6, 1949
Lowest temperature on North American continent	−63	Snag, Yukon	Feb. 3, 1947
Lowest U.S. temperature	−62.1	Prospect Creek Camp, Alaska	Jan. 23, 1971
Lowest 48-state temperature	−57	Rogers Pass, Mont.	Jan. 20, 1954
Lowest record by U.S. observers	−80.6	Amundsen-Scott Station (90° S)	July 22, 1965
Lowest world surface air temperature	−88.3	Vostok Soviet Station (78° 27′ S 106° 52′E at 3,420 m)	Aug. 24, 1960

actual observed temperatures are used as a basis for drawing isotherms, but on continental or world maps mean temperatures are frequently reduced to sea-level equivalents by adding about 6C° for each 1000 m of elevation. This adjustment essentially eliminates the effect of altitude on temperature and thus facilitates the mapping of horizontal temperature differences. Vertical distribution of temperature will be discussed in the next section.

The world pattern of temperature is determined by a number of factors. It has already been noted that the effectiveness of insolation in heating the earth's surface is largely determined by latitude. The general decrease in temperatures from the equator toward the poles is one of the most fundamental and best known facts of climatology (Table 2.5). If the effect of latitude were the only controlling factor affecting net radiation we would expect a world temperature map to have isotherms lying parallel to each other in the same fashion as parallels of latitude. Of course, such is not the actual case.

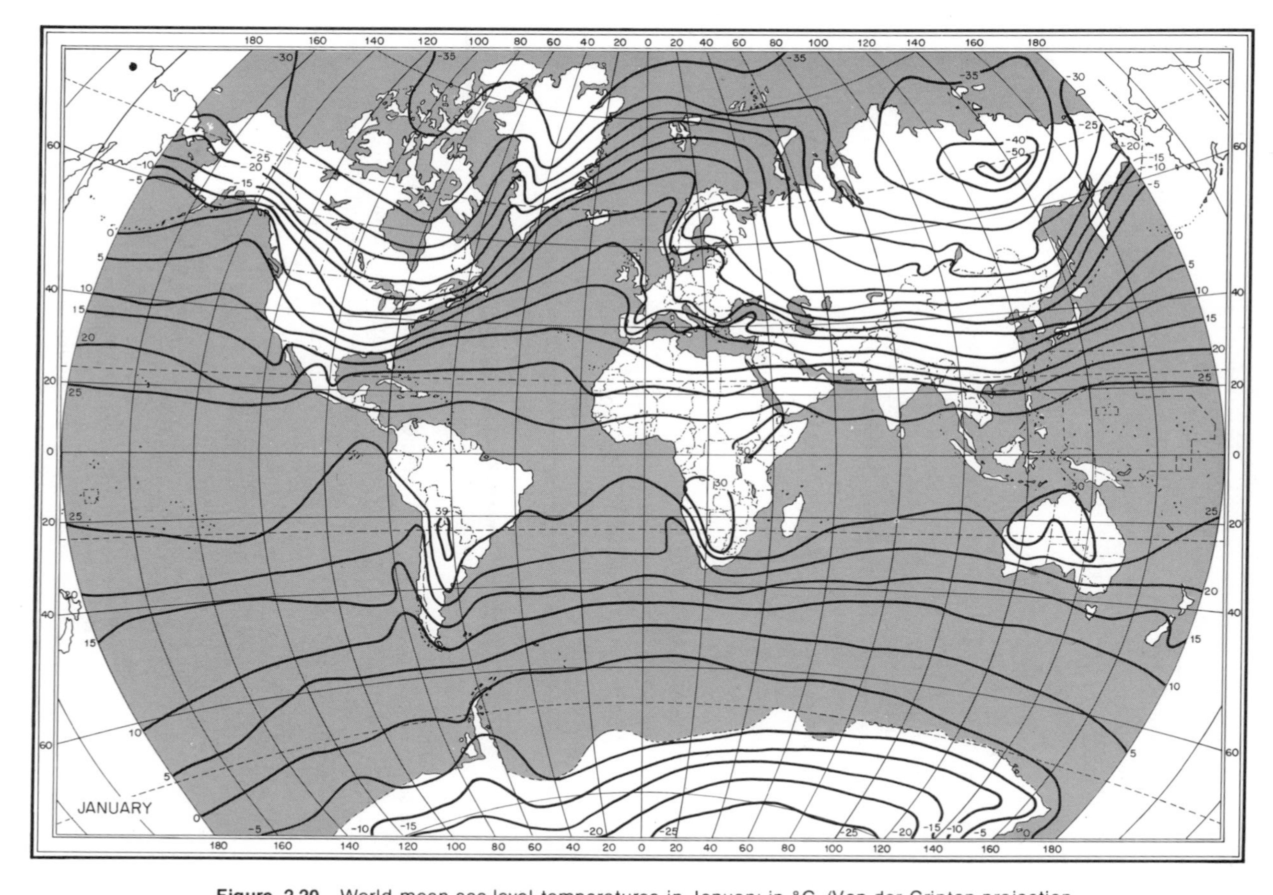

Figure 2.20 World mean sea-level temperatures in January in °C. (Van der Grinten projection courtesy A. J. Nystrom & Co., Chicago.)

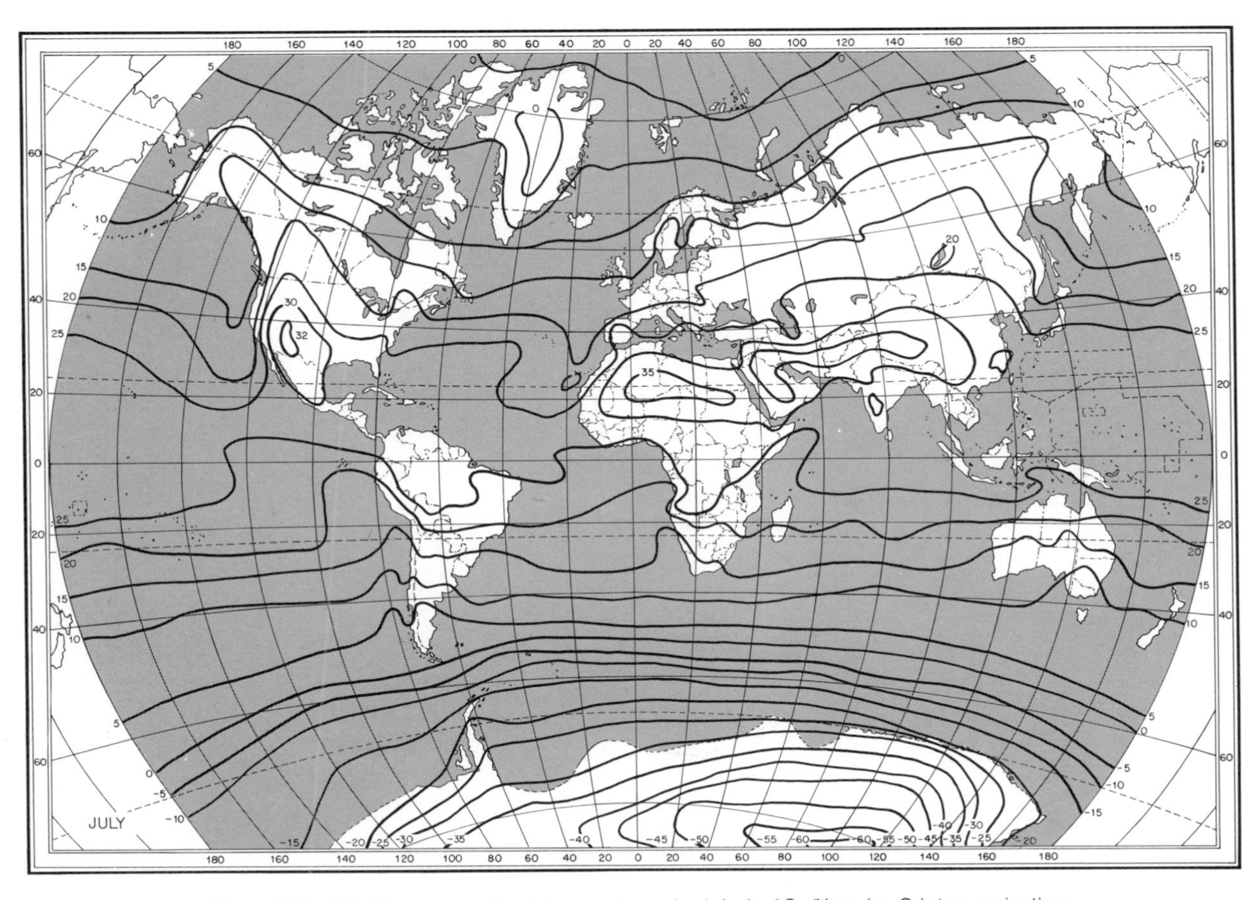

Figure 2.21 World mean sea-level temperatures in July in °C. (Van der Grinten projection courtesy A. J. Nystrom & Co., Chicago.)

TABLE 2.5

Mean Surface Temperatures (°C) at Selected Latitudes *

Latitude	*January*	*July*
35°	10	36
40°	5	24
45°	−2	21
50°	−7	18
55°	−11	16
60°	−16	14
65°	−23	12
70°	−26	7
70°	−29	3
80°	−32	2
85°	−38	0
90°	−41	−1

*After W. Meinardus, adapted from A. A. Borisov, *Climates of the U.S.S.R.* (Chicago: Aldine Publishing Company. 1965), p. 38.

The irregular distribution of land and water on the earth's surface tends to break up the orderly latitudinal arrangement. Land areas warm and cool more rapidly than do bodies of water, with the result that annual temperature ranges are greater over land. There are three primary reasons for the contrasts in land and water temperatures. (1) Water is mobile and experiences both vertical and horizontal movements which distribute energy absorbed at the surface throughout its mass, whereas insolation is absorbed by land only at the surface and is transmitted downward slowly by conduction. (2) Water is translucent and is penetrated by radiant energy to a much greater depth than is opaque land. Thus, a given quantity of insolation must be distributed through a greater mass of water than of land, even though their surface areas are the same. (3) The *specific heat* of water is higher than that of land. That is, a given mass of water requires more heat to raise its temperature 1° than does an equal mass of dry land. Consequently, the same amount of insolation will produce a higher temperature on a land surface than on a water surface. Conversely, in cooling, water loses a greater amount of heat than does land to produce the same drop in temperature.

The general effect of the contrast in heating of land and water areas is to produce cooler winters and warmer summers in the center of continents than along coasts and over oceans. Coastal or marine climates tend to be moderate, experiencing no great extremes in either daily or annual temperature changes.

Transport of ocean water in the form of currents and drifts carries heat from one part of the earth to another. Thus, an ocean current traveling poleward warms overlying air, producing air temperatures higher than would normally be expected for the latitude. A current moving toward the equator will produce lower air temperatures. An outstanding example of the effect of an ocean current on temperature is that of the North Atlantic Drift off northwest Europe. The January isotherms for −5° and 0°C intercept the U.S. east coast in the vicinity of latitude 40°N, but the North Atlantic Drift carries warm water into the northeast Atlantic so that these two isotherms cross the coasts of Great Britain and the Scandinavian Peninsula at a much higher latitude. Examples of the effect of a cold ocean current can be seen in the temperature distribution along the west coasts of South America and Africa. The Peru (South America) and Benguela (Africa) Currents flow northward along the west coasts. The cooling effect is enhanced by upwelling produced mainly by offshore winds that skim away the warmer surface water and cause temperatures to be lower than in corresponding latitudes on the east sides of the continents.

The horizontal distribution of sea-surface temperatures results from the effects of latitude and the seasons as well as of ocean currents. Extremes range from below 0°C in high latitudes to about 30°C in certain tropical gulfs. Annual variations in temperature of the sea surface are greatest along the east coasts in the middle latitudes of the Northern Hemisphere, the extremes occurring in February and August. There is thus a lag of about two months in heating and cooling of the ocean surfaces behind the periods of high and low sun.

It is obvious that neither oceans nor ocean currents can have their maximum effect on temperature over land unless the prevailing winds blow onshore. This is well illustrated by the winter flow of air across the United States. The west coast experiences the relatively high temperatures characteristic of air blown from the Pacific Ocean. The eastern part of the nation receives cooler air from the continent in the general west-to-east movement of storms and air masses and the immediate effect of the Atlantic Ocean upon east coast temperatures is somewhat restricted.

Still another influence on horizontal temperature distribution is the mountain barrier effect. Mountain ranges tend to guide the movement of cold air masses. In the United States the Rocky Mountain barrier assists the prevailing westerlies in diverting most of the cold outbursts from Canada to the east. Similary, the Himalayas in Asia and the Alps in Europe protect the regions to the south from polar air.

On a local scale, topographic relief exerts an influence upon temperatures. In the Northern Hemisphere north-facing slopes generally receive less insolation than south-facing slopes, and temperatures are normally lower. Drainage of cold air into valleys at night also affects local temperature distribution.

VERTICAL DISTRIBUTION OF TEMPERATURE

The permanent snow caps on high mountains, even in the tropics, indicate the decrease of temperature with altitude. Observations of the temperature in the upper air by means of instruments attached to kites, balloons, airplanes, and rockets have shown that there is a fairly regular decrease in temperature with an increase in altitude. The average rate of temperature decrease upward in the troposphere is about 6C° per km, extending to the tropopause. This vertical gradient of temperature is commonly referred to as the standard atmosphere or *normal lapse rate,* but it varies with height, season, latitude, and other factors. The normal lapse rate represents the average of many observations of vertical temperature distribution and should not be confused with the *actual lapse rate,* which indicates temperature values above a given location at a given time. Indeed, the actual lapse rate of temperature does not always show a decrease with altitude. Where observations indicate no change with altitude, the lapse rate is termed *isothermal*. Such a condition never occurs over a very great vertical range nor for a long period of time in the troposphere (see Fig. 2.22).

Certain processes in the lower troposphere may create an actual increase in temperature with an increase in altitude, that is, a *temperature inversion*. Temperature inversions may be produced in five ways. (1) When the earth's surface loses more heat by radiation than it gains by any of the energy transfer processes, as on a clear night or at high latitudes in winter, it cools and consequently lowers the temperature of the adjacent layer of air. Inversions due to radiational cooling (*radiation inversions*) develop best in calm air under clear skies over flat terrain and often are accompanied by ground fog or surface concentration of pollutants. They are not likely to form over water, which is slow to cool. (2) Because of its greater density,

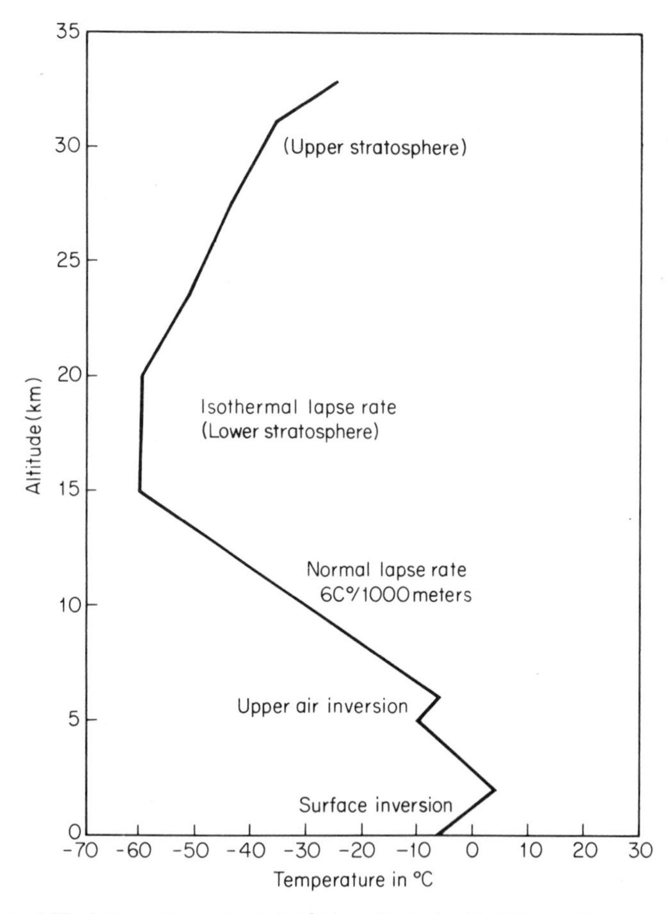

Figure 2.22 Lapse rate components of a hypothetical vertical temperature sounding.

cold air from hilltops and slopes tends to flow downslope to collect in valley bottoms, creating an inverted lapse rate up the slopes as well as in the free air over the valley floor. *Air drainage inversions* are frequently associated with spring frosts in middle latitudes, and for this reason fruit growers prefer gentle slopes to valley bottoms for orchard sites. (3) When two air masses with different temperature characteristics come together, the colder air, being more dense, tends to push underneath the warmer air and replace it. The boundary zones along which two air masses meet are called *fronts,* and the inverted lapse rate which results is a *frontal inversion.* Frontal inversions are not confined to the lower layers of the troposphere as are the types mentioned previously; they may form at upper levels wherever cold air underruns warm air or warm air advances above cold air. (4) *Advection* of warm air over a cold surface creates an inversion in the lower layers of the air mass as the warm air is cooled by conduction. This process is common when warm air passes over a cold water surface, cold land, or snowfields, where fog may result. (5) Another type of inversion, the *subsidence inversion,* forms in an air mass when a large body of air subsides and spreads out above a lower layer. In the process the air heats dynamically more in the upper portion than at its base. Inversions of this type may develop at considerable altitudes. They are especially common in the region of the trade winds and in large, slow-moving high-pressure areas.

QUESTIONS AND PROBLEMS FOR CHAPTER 2

1. Explain the processes that heat and cool the atmosphere.
2. Describe the effects that are produced by different wave lengths of the solar spectrum.
3. Assuming the mean value of the solar constant, calculate the total amount of solar radiation intercepted by the earth during an hour.
4. The radiation budget at a particular place is seldom in balance. Why?
5. Summarize the processes that transfer energy within the climate system.
6. Outline the factors that determine the amount of insolation received on a particular area of the earth's surface on a given date.
7. Explain the operating principles of the following: pyranometer, pyrheliometer, maximum and minimum thermometers, thermistor.
8. How is the mean monthly temperature for a climatic station calculated?
9. Compare and explain the effects of land and water surfaces on the global distribution of temperature.
10. Outline the causes of temperature inversions and explain the associated processes.

3

Atmospheric Moisture

Moisture in the atmosphere plays such a significant role in weather and climate that it commonly is treated separately from the other constituents of air. In one or more of its forms atmospheric moisture is a factor in humidity, cloudiness, precipitation, and visibility. Water vapor and clouds affect the transmission of radiant energy both to and from the earth's surface. Through the process of evaporation water vapor becomes a medium for conveying latent heat into the air, thus giving it a function in exchanges of heat. Atmospheric water is gained by evaporation but lost by precipitation after complex intervening processes of horizontal and vertical transport and changes in physical state. Only a minute fraction of the earth's water is stored as clouds and vapor in the atmosphere at any one time. The net amount at the end of any given period for a particular column of air is the algebraic sum of the amount stored from a previous period, the gain by evaporation, the loss by precipitation, and the gain or loss by horizontal transport. This relation expresses the *water budget of the atmosphere.*

WATER VAPOR

Humidity is a general term connoting the amount of water vapor in the air. As a mixture of gases, air itself has no inherent capacity to "hold" moisture; space rather than air contains the water vapor molecules. It is nevertheless convenient to speak of air space, or simply air, as containing a certain amount or proportion of water vapor. *Specific humidity* is the ratio of the mass of water vapor actually in the air to a unit mass of air, including the water vapor. Thus, a kilogram (1,000 g) of air of which 12 g are water vapor has a specific humidity of 12 g per kg. A closely related value is *mixing ratio,* which is the mass of water vapor per unit mass of dry air. The mixing ratio is essentially a recipe for a mixture of water vapor and dry air. A value of 12 g per kg would entail a total of 1,012 g for the mixture. For most conditions, specific humidity and mixing ratio differ insignificantly. Another humidity expression based on separate considerations of air and water vapor is *vapor pressure.* This is the partial pressure exerted by water vapor in the air, and it is expressed in the same units used for barometric pressure, for example, millimeters of mercury or millibars.

When air space contains the maximum amount of water vapor possible at a given temperature and pressure it is considered to be *saturated,* and its actual vapor pressure nearly equals its *saturation vapor pressure.* The air is then at its *dew point* temperature, or *frost point* if below freezing. Both conditions depend on the amount of impurities in the air. Saturation vapor pressure varies with temperature and at temperatures below 0°C it is slightly lower over ice than over liquid water (Fig. 3.1).

The best known and most used popular reference to water vapor is *relative humidity,* the ratio of the amount of water vapor actually in a volume occupied by air to the amount the space could contain at saturation. Thus, if the space containing a kilogram of air at constant pressure has a capacity for 12 g of water vapor at a certain temperature but contains only 9 g at that temperature, it has a relative humidity of 75 percent. When the temperature of air is increased, the saturation vapor pressure increases in response to heightened molecular activity. If no further moisture is added, the result will be a decrease in relative humidity. Conversely, when the temperature of a volume of air is decreased, its capacity for moisture decreases and its relative humidity increases. These inverse relations are highly significant in heating or cooling of buildings, human comfort, and other aspects of applied climatology.

Relative humidity usually reaches its diurnal maximum in the early morning hours when temperatures are low and then decreases to a minimum in the early afternoon. It tends to be greater in winter over land, except where there are strong summer incursions of moist air as in regions affected by monsoon winds. Over the oceans relative humidity reaches a slight maximum in summer. Mid-latitude mountains tend to have summer maximums because of stronger convective flow of moist air. Figure 3.2 shows the approximate mean latitudinal distribution of mean relative humidity. Although the actual pattern is modified by other factors in the climate system, maximums generally prevail near the equator and minimums in the belts of high pressure at subtropical latitudes. Poleward, toward the westerly wind belts, there is again an increase owing to decreasing temperatures.

The zonal pattern of average specific humidity or vapor pressure contrasts somewhat with that of relative humidity (Fig. 3.2). Even the air of the tropical deserts has considerable water vapor, although its *relative humidity* is low. In the polar regions low mean temperatures reduce the air's capacity for moisture. Figure

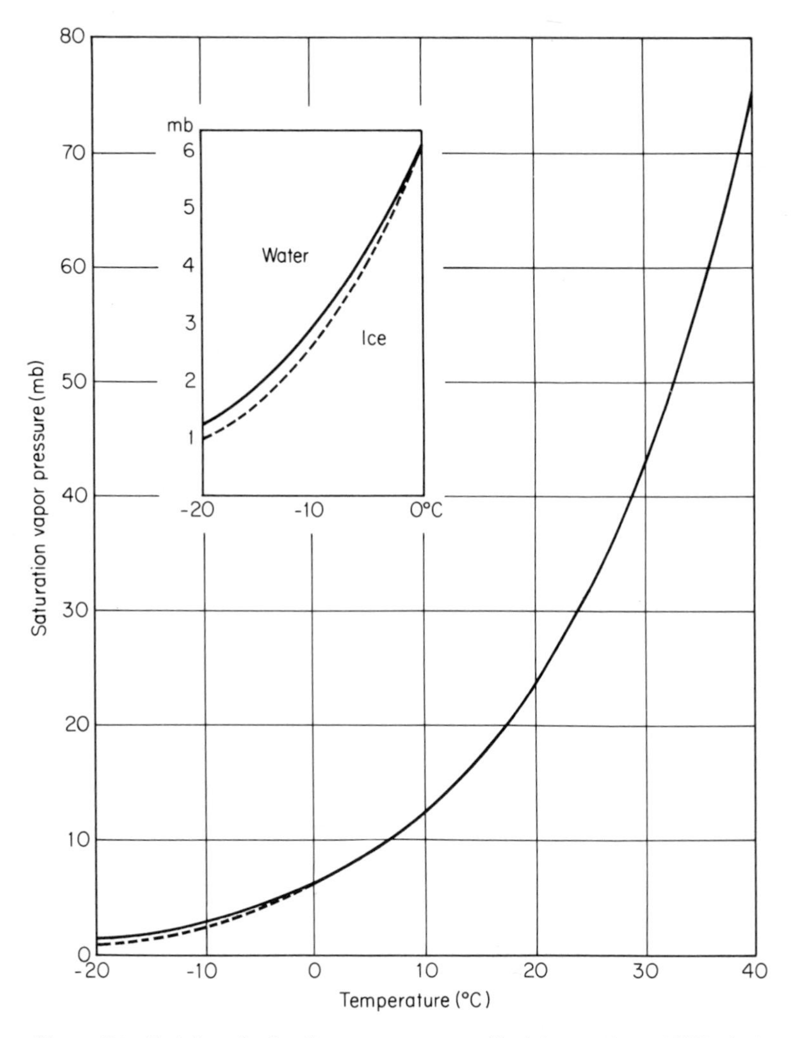

Figure 3.1 Variation of saturation vapor pressure with air temperature at 1000 mb atmospheric pressure. Expanded vertical scale of inset emphasizes the difference between saturation values over liquid water and ice at a given temperature.

3.3 shows the mean global distribution of water vapor in the atmosphere expressed as the liquid equivalent (precipitable water) for January and July.

HUMIDITY MEASUREMENTS

Direct measurement of the actual amount of water vapor in the air is not feasible for ordinary observations. Instead, most humidity values are determined indirectly from the *psychrometer*. In its simplest form this instrument consists of two ther-

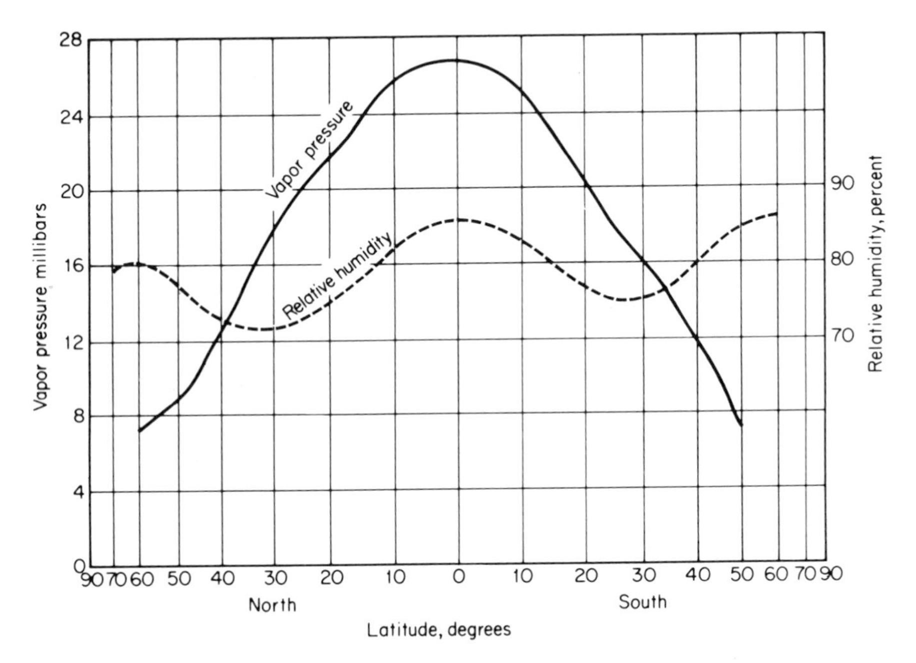

Figure 3.2 Mean latitudinal distribution of relative humidity and vapor pressure.

mometers on the same mounting (Fig. 3.4). The *wet-bulb thermometer* is mounted a little lower than the other and has its bulb covered with a cloth wick that can be wetted for the observation. The other is the *dry-bulb thermometer.* When the psychrometer is swung freely in the air or aerated by a fan, evaporative cooling lowers the wet-bulb reading. The difference between the readings is the *wet-bulb depression.* The dry-bulb temperature, wet-bulb depression, and atmospheric pressure may be used to determine any of the standard expressions of humidity from psychrometric tables which are derived from their mathematical relations (see Tables B.1 and B.2, Appendix). If there is no depression of the wet-bulb, the air is saturated and the relative humidity is 100 percent.

Relative humidity is one of the humidity values for which there is an instrument that makes a direct measurement. The *hair hygrometer* operates on the principle that human hair lengthens as relative humidity increases and contracts with decreasing relative humidity. The tension of several strands of hair is linked to an indicator. Unfortunately, there is such considerable lag in the response, especially at low temperatures, that the hair hygrometer is a much less accurate instrument than the psychrometer. For meteorological purposes, its principle is employed primarily in the *hygrograph,* a recording hygrometer which has a clockdrum and pen arrangement just as in the thermograph.

A device known as the *infrared hygrometer* employs a beam of light projected through the air to a photoelectric detector. Two separate wave lengths of light are used; one is absorbed by water vapor and the other passes through undiminished.

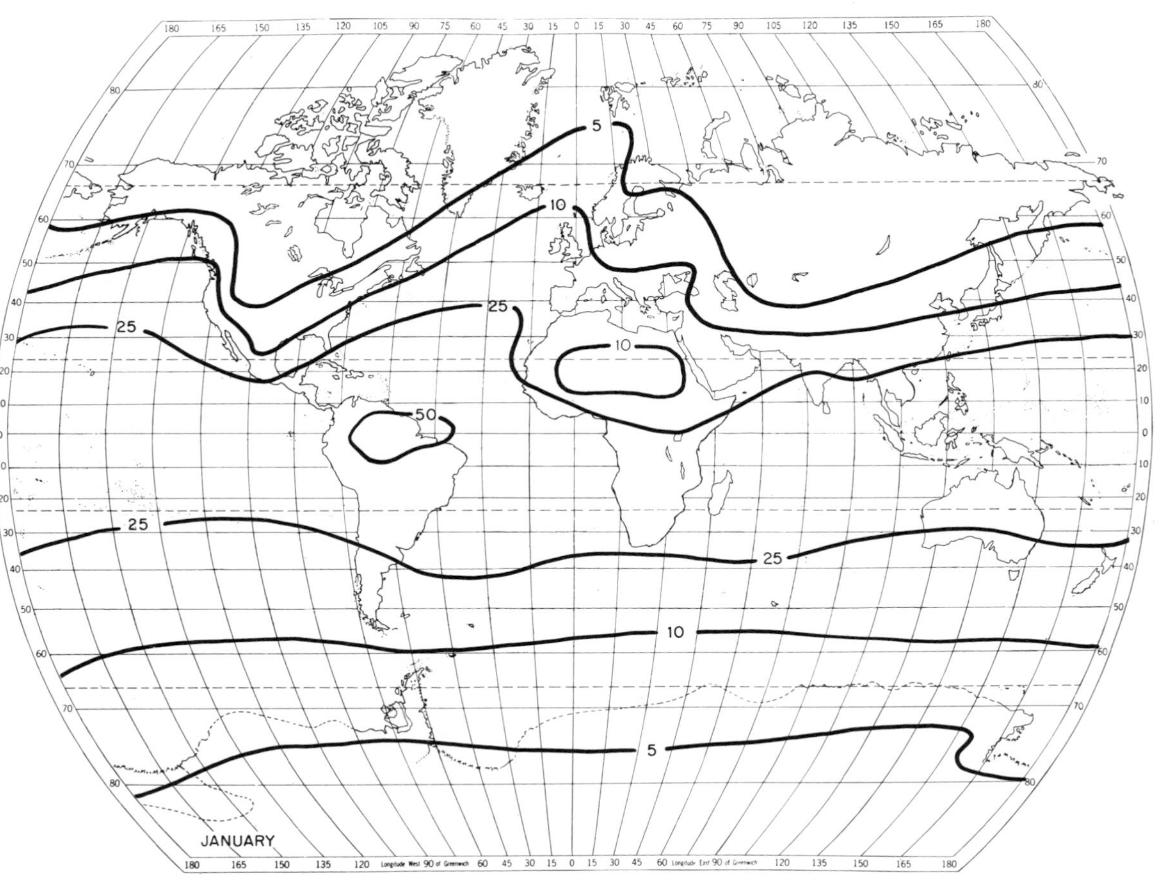

Figure 3.3(a) Mean distribution of precipitable water in January. Values are liquid equivalent of water vapor in mm. (After Tuller and Lamb; modified Van der Grinten projection.)

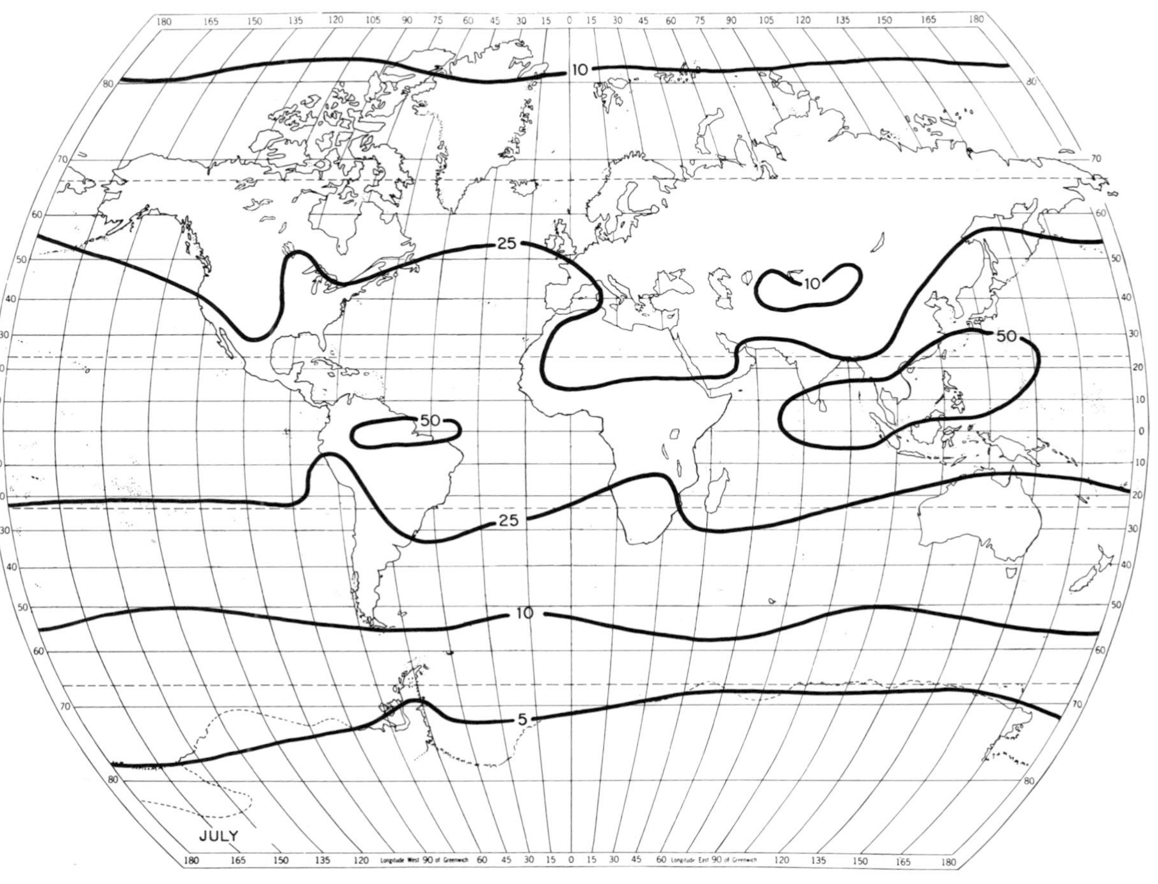

Figure 3.3(b) Mean distribution of precipitable water in July.

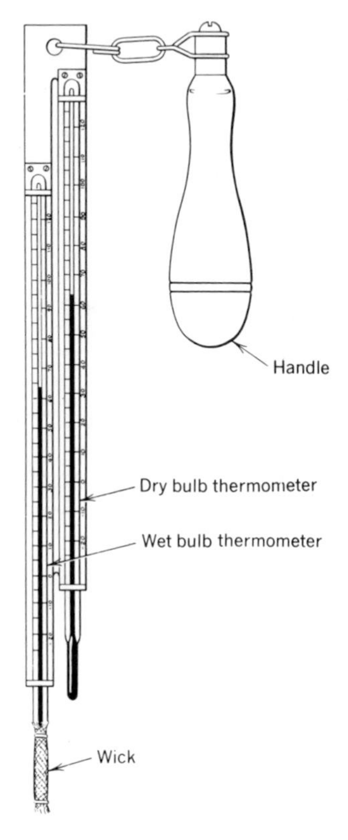

Figure 3.4 Sling psychrometer.

The ratio of the infrared light transmitted by the different wave lengths is a direct indication of the amount of water vapor in the light path.

A type of humidity gauge useful where electrical transmission of the variation is desirable depends upon the fact that the passage of electrical current across a chemically coated strip of plastic is proportional to the amount of moisture absorbed at its surface. This type of hygristor element is in common use in remote-indicating instruments such as the radiosondes used for upper air observations. Still other devices known as *dew point hygrometers* employ artificially cooled surfaces to determine the dew point temperature directly.

PHYSICAL CHANGES OF STATE OF WATER

The chemical compound H_2O occurs naturally in the atmosphere in all three physical stages—gas, liquid, and solid—and it may change from one state to any other, always requiring exchange of latent heat (Fig. 3.5).

Evaporation, the change of state from liquid to vapor, results when molecules escape from any water surface, whether it be the surface of water bodies, droplets in

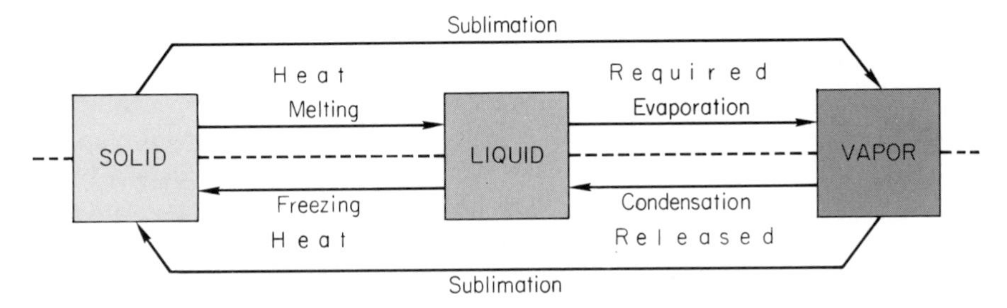

Figure 3.5 Physical changes of state of water.

clouds or fog, or thin films on solids such as soil particles. The process requires energy and is the means by which vast amounts of latent heat are transferred from the earth's surface to the atmosphere. Its rate is dependent upon three major factors: vapor pressure, temperature, and air movement. Evaporation increases as the saturation vapor pressure at the water surface becomes greater than the actual vapor pressure of the adjacent air. It therefore takes place more rapidly into dry air than into air with a high relative humidity. The rate of evaporation also increases with rising water temperature, other factors being equal. When the temperature of the water is higher than that of the air, evaporation always takes place. Finally, wind movement and turbulence replace air near the water surface with less moist air and increase evaporation. The heat used in carrying out evaporation is retained by the water vapor as the *latent heat of vaporization* and does not raise the temperature of the vapor. The latent heat of vaporization ranges in value from nearly 600 calories (2,510 joules) per gram of water at 0°C to about 540 cal (2,259 joules) at 100°C because as temperature increases the difference between the kinetic energy of the escaping molecules and the average kinetic energy of all the water molecules decreases. The process of evaporation therefore becomes less selective and requires less heat as a growing proportion of the liquid molecules gain the necessary energy to escape from the water surface.

A special case of evaporation is *transpiration,* which entails a loss of water from leaf and stem tissues of growing vegetation. The combined losses of moisture by evaporation and transpiration from a given area are termed *evapotranspiration.*

The complex measurements required to determine the actual amount of water evaporated from soil surfaces and transpired from vegetation prove difficult in practice. Direct measurement of the loss from an exposed pan or tank, adjusted for precipitation, affords an approximation but does not take into account true energy and moisture conditions of a vegetated surface. Other types of *evaporimeters* known as *atmometers* employ constantly wetted porous substances that are exposed to the free air in an attempt to simulate surface conditions. Indirect approaches are based on the surface water budget equation:

$$ET = P - (R + dS)$$

where ET is the loss by evapotranspiration, P the gain by precipitation, R the loss to runoff, and dS is the gain or loss of water stored in the soil or water bodies. Thus, if precipitation, runoff, and storage can be measured or estimated adequately, we can calculate evapotranspiration as a difference.

The importance of solar energy in the evaporation process is evident in the latitudinal distribution of evapotranspiration (Fig. 3.6). Mean annual rates are greatest in the vicinity of 20° north and south latitude, where net radiation is at a maximum under the generally clear skies and subsiding air associated with subtropical high-pressure areas. The greater evapotranspiration in the Southern Hemisphere results from a larger proportion of ocean surface south of the equator.

Condensation is the change of state from water vapor to liquid water. When moist air comes in contact with cool surfaces it may be cooled to the *dew point*, or *con-*

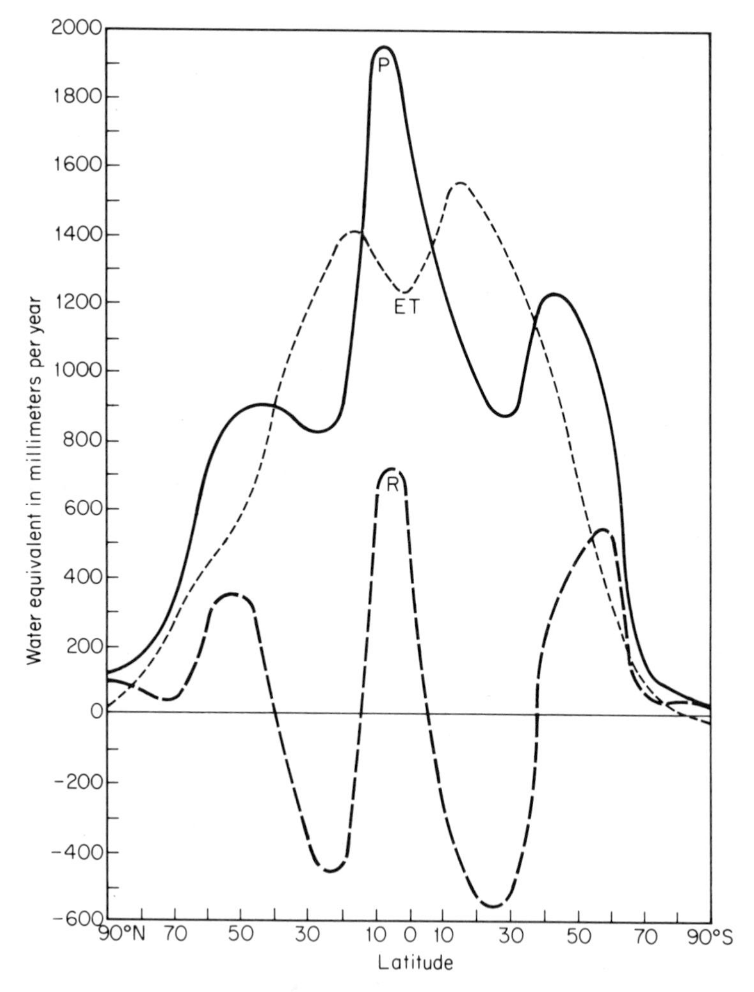

Figure 3.6 Mean annual latitudinal distribution of water budget elements. *ET*, evapotranspiration; *P*, precipitation; *R*, runoff. (After Sellers.)

densation temperature. A part of the vapor then condenses into a liquid form on the cool surface, producing *dew.* When this happens the latent heat of vaporization, in this process called the *latent heat of condensation,* is released. Liberation of the latent heat of condensation always tends to slow down the cooling process. It provides one of the main sources of heat energy for atmospheric processes. The dew point temperature is not a definite boundary at which all condensation takes place. Condensation also results from cooling in the free air, but it requires the presence of very small *condensation nuclei.* The most active nuclei are hygroscopic particles, which have an affinity for water, and upon which condensation may take place even at temperatures above the dew point, that is, when the relative humidity is less than 100 percent. If sufficient water droplets are formed in this way to become visible, *haze* will result, becoming fog or possibly clouds as the number, size, and density of droplets increase. Among the hygroscopic materials in the atmosphere are salts liberated from bursting bubbles in the foam of ocean spray and numerous chemical compounds in industrial smokes. Nonhygroscopic dust particles are less effective nuclei, although they may have greater significance in the formation of ice crystals at high levels.

As cooling and condensation proceed various degrees of *supersaturation* are required to maintain the growth of droplets. The saturation vapor pressure must be considerably exceeded in the air surrounding the smaller droplets and those with nonhygroscopic nuclei. Larger droplets and those having soluble nuclei gain increments of condensation more readily. Supersaturation of cloudy air with respect to liquid water rarely exceeds 1 percent. Since there is normally an abundance of condensation nuclei in the lower atmosphere, visible condensation products can be expected to form whenever the relative humidity reaches 100 percent.

Superficially, the processes of *melting* and *freezing* appear to be remarkably uncomplicated. Ice melts at a temperature of 0°C when warmed, but liquid water does not always freeze at that temperature, contrary to popular belief. Foreign material dissolved in the water, suspended in it, or covering it, the amount of water involved in the process, and electrical charges all may lower the freezing point. In free air, small water droplets are known to exist at temperatures far below the "normal" freezing point. This is especially true at upper atmospheric levels. The supercooled condition apparently can be discontinued by lowering the temperature below a critical point of about −38°C or by introducing artificial ice nuclei. In the rainmaking experiments of recent years dry ice has been used to produce low temperatures. Silver iodide crystals introduced into a supercooled cloud result in formation of ice particles when temperatures are below −4°C, being most effective at about −15°C. The smaller the droplets of water the lower the temperature at which they will freeze spontaneously. The processes both of freezing and melting involve the *latent heat of fusion* (80 cal per g at 0°C), which is absorbed when ice melts and released when water freezes.

At temperatures below freezing, water may bypass the liquid form in its change of state. When dry air with a temperature well below freezing comes in contact with ice, molecules of the ice (H_2O) pass directly into the vapor state by means of the process of *sublimation.* Sublimation may remove part of the snow from the ground in dry winter weather. The reverse process is also known as sublimation, in which both the heat of vaporization and the heat of fusion are released. If sublimation occurs on a cold surface the visible product is *frost*. In the free air, sublimation nuclei (also known as ice nuclei) must be present to produce ice fog or clouds, and the ice crystals may continue to grow as water vapor sublimates on their surfaces. Since air that is saturated with respect to liquid water is highly supersaturated with respect to ice, sublimation nuclei are active in dispersing clouds of supercooled water droplets. A leading theory is that they initiate the formation of small ice crystals from the vapor in the air adjacent to the supercooled droplets. A part of the supercooled water then evaporates into the drier air and the resulting vapor in turn sublimates on the ice crystals. Thus, the supercooled droplets do not freeze directly on the crystals, as might be supposed.

PROCESSES OF COOLING TO PRODUCE CONDENSATION AND SUBLIMATION

Some type of cooling process normally is required to initiate and sustain condensation or sublimation. The cooling processes are as follows:

1. Adiabatic processes
   a. A decrease in barometric pressure at the surface, in which case fog might form
   b. Rising air, which in turn results from
      (1) Convection
      (2) Convergence of wind currents or air masses (as along fronts)
      (3) Orographic lifting
2. Diabatic processes
   a. Loss of heat by radiation. Direct radiation from moist air may produce fog or clouds.
   b. Contact with a cold surface (conduction). Dew, frost, or fog are the normal products. The process may be associated with the movement of air across a cold surface (advection).
   c. Mixing with colder air. If the mixture has a temperature below its dew point clouds or fog will form, assuming sufficient effective nuclei are present.

Of all the types of cooling which air may experience, those due to lifting are by far the most important for condensation in free air. A rising parcel of air cools because part of its internal energy is used for expansion in response to decreasing environmental pressure. Conversely, subsiding air undergoes an increase in pressure and warms. These temperature changes, occurring without any heat being added to or subtracted from the air, are termed *adiabatic*. *Diabatic,* or nonadiabatic, processes involve the exchange of heat with an external source or sink. The rate at which unsaturated air cools adiabatically as it rises is 10C° per 1,000 m. If rising and cooling continue until condensation begins, however, the latent heat of condensation released into the air reduces the rate of cooling. The new rate of cooling is the *wet,* or *pseudoadiabatic, rate*; it is the *dry adiabatic rate* modified by the latent heat of condensation. The wet adiabatic value varies with temperature and pressure (and therefore with heights reached by ascending air), averaging about 6C° per 1,000 m.

Adiabatic cooling due to lifting is accomplished by any of the three ways outlined above or by any combination of them. Convection has been discussed previously as a method of heat transfer. As a process leading to adiabatic cooling it is common whenever the earth's surface is warmer than the air above. Convergence occurs when winds of different directions or speeds meet one another. In tropical areas convergence frequently involves air currents with similar temperature characteristics. In middle latitudes there is likely to be a considerable difference between the temperatures of converging winds, and the warmer air will ride over the cooler. This special case of convergence is known as *frontal lifting. Orographic lifting* is ascension of air caused or intensified by any topographic obstruction on the earth's surface. Ranges of mountains and hills are most effective in orographic lifting, forcing winds which blow against them to adopt vertical components. Even a slight rise in land elevation, as on a coastal plain, can induce some lifting. Assume that the wind blows in from the sea across a flat coastal plain. The frictional drag of the land is greater than that of the sea, and the lower layers of air will consequently be slowed. The air coming from seaward will be forced to rise over this barrier of congested air in order to proceed inland. Because of the tendency of air to pile up against the barrier, the effect of any surface that induces orographic lifting extends for some distance to the windward.

CLOUDS: THEIR FORMATION AND CLASSIFICATION

Clouds are visible aggregates of water droplets, ice particles, or a mixture of both along with varying amounts of dust particles. A typical cloud contains billions of droplets having diameters on the order of 0.01 to 0.02 mm; yet liquid or solid water accounts for less than 10 parts per million of the cloud volume. Most clouds result from cooling due to lifting of air. Those associated with strong rising air currents have vertical development and a puffy appearance, whereas those produced by gentler lifting or other methods of cooling tend to form in layers. Although their method of formation may affect their appearance, clouds are classified primarily on the basis of their height, shape, color, and transmission or reflection of light. The World Meteorological Organization has established a detailed cloud classification system with categories for genera, species, and varieties and with provisions for naming clouds that have undergone changes in form.

It will be sufficient here to divide all clouds of the troposphere into four families: high, middle, low, and clouds with vertical development. Average heights of cloud bases vary with latitude, being lower in polar regions and in winter. High clouds are the *cirrus* types, always composed of ice crystals, whose average heights range from 5 km above the earth's surface to the tropopause. Cirrus is nearly transparent, white, and fibrous or silky (Fig. 3.7). *Cirrocumulus* clouds are a cir-

Figure 3.7 Cirrus (*Cirrus uncinus*). (Photograph by author.)

riform layer or patch of small white flakes or tiny globules arranged in distinct groups or lines. They may have the appearance of ripples similar to sand on a beach. *Cirrostratus* is a thin white veil of cirrus. It is nearly transparent so that the sun, moon, and brighter stars show through distinctly. Cirrus clouds frequently can be detected by the *halos* which they create around the sun or moon and can thus be distinguished from haze or light fog. Halos result from the refraction of light by ice crystals suspended in the air.

The middle clouds predominate at heights ranging from 2 to 8 km above ground level. They are composed of water droplets, ice crystals, or mixtures of both. The principal types are *altocumulus* and *altostratus*. Altocumulus contains layers or patches of globular clouds, usually arranged in fairly regular patterns of lines, groups, or waves (Fig. 3.8). Vertical air currents in the layers occupied by altocumulus may cause the clouds to build upward. Isolated altocumulus may form in the wave of an air stream as it rises over a mountain or a convective column, provided the rise is sufficient to cool the air below its dew point. The banner clouds above peaks (or above local ''chimneys'' of convection) are of this type. Because of

Figure 3.8 Altocumulus (*Altocumulus stratiformis*). (Photograph by author.)

their lens shape they are named *altocumulus lenticularis*. Altostratus is a fibrous veil, gray or blue-gray, and thicker than the higher cirrostratus, although it may merge gradually into the latter. It ordinarily does not exhibit halo phenomena, but another optical effect, the *corona,* sometimes appears as a smaller circle of light around the sun or moon in both altostratus and altocumulus. The corona results from the diffraction of light by water droplets, and its diameter is inversely proportional to the size of the cloud droplets. The red of the faint color band is on the outside, that is, away from the sun or moon; in the halo the red is on the inside. It is common for altostratus to change into altocumulus and vice versa. Precipitation may fall from altostratus, or altocumulus, but it does not necessarily reach the ground. Sometimes this *virga* is visible hanging from the bottom of the cloud in dark, sinuous streaks. When the cloud layer lowers to become somewhat thicker and darker and falling rain or snow obscures its base it is called *nimbostratus.*

The base level of low clouds varies from very near the ground to about 2,000 m. The basic type of this family is the *stratus,* a low, uniform layer resembling fog but not resting on the ground. Stratus is frequently formed by the lifting of a fog bank or by the dissipation of its lower layer (Fig. 3.9). If broken into fragments by

Figure 3.9 "Table Cloth" on Table Mountain, Cape Town, South Africa. The stratus layer forms as southeasterly winds lift moist air over the mountain. The cloud evaporates in the descending flow on the lee (near) side. (Photograph by author.)

the wind it is called *fractostratus.* Heat loss by radiation from the top of a stratus layer can cause condensation in the cooled layer above the cloud so that it builds upward. This should not be confused with vertical development due to rising air. When associated with nimbostratus, stratus sometimes builds downward as well. Rain falling from the upper cloud layer may evaporate on falling through dry air and then recondense in lower saturated layers to form stratus. Precipitation from stratus, if it occurs, is usually light.

Stratocumulus clouds form a low, gray layer composed of globular masses or rolls which are usually arranged in groups, lines, or waves. If the aggregates of thick stratocumulus fuse together completely so that the structure is no longer evident, the clouds are classified as stratus. Note that altocumulus clouds have the same general characteristics. Altocumulus viewed from a mountain or an airplane may be difficult to distinguish from stratocumulus for this reason.

Clouds with vertical development fall into two principal categories: *cumulus* and *cumulonimbus.* Cumulus clouds are dense, dome-shaped, and have flat bases. They may grow to become cumulonimbus, the extent of vertical development depending upon the force of vertical currents below the clouds as well as upon the amount of latent heat of condensation liberated in the clouds as they form. Cumulus with little vertical development and a slightly flattened appearance are commonly associated with fair weather (Fig. 3.10). Viewed from satellites the

Figure 3.10 Fair weather cumulus (*Cumulus humilis*) on the Salisbury Plain, England. (Photograph by author.)

distinctive patterns of cumuliform clouds are valuable aids in the identification of widespread convective activity (Fig. 3.11).

Sometimes large groups of cumulus develop a nearly complete covering of the sky and become stratocumulus. Conversely, stratocumulus occasionally break apart into separate clouds and become cumulus.

Cumulonimbus clouds exhibit great vertical development, towering at times to 18 km or more, where they spread out to leeward and form an anvil of cirrus (Fig. 3.12). These are the thunderhead clouds that produce heavy showers of rain, snow, or hail, often accompanied by lightning and thunder. To an observer directly beneath, a cumulonimbus cloud may cover the whole sky and have the appearance of nimbostratus. In this case its true nature is revealed by the much heavier precipitation that falls from the cumulonimbus or by the preceding evolution of the cloud cover. Turbulence along the front of an advancing cumulonimbus often creates a roll of dark fractocumulus or fractostratus.

CLOUD OBSERVATIONS

Because of their obvious relation to the energy and water budgets and to motion in the air, cloud observations are among the most valuable types of information that can be used in climatological analyses. *Cloud types* are determined by visual observa-

Figure 3.11 Satellite view of eastern Pacific, February 17, 1979. All cloud families are represented. (National Environmental Satellite Service.)

tion, the accuracy of which depends on the observer's experience and knowledge of cloud-forming processes. *Sky cover,* or cloud cover, is obtained visually by viewing the clouds against the dome of the sky and estimating the fraction of the sky covered. In the United States five terms are in common use to express sky cover.

Clear, no clouds or less than $\frac{1}{10}$ of sky obscured by clouds
Scattered, $\frac{1}{10}$ to less than $\frac{6}{10}$ cover
Broken, $\frac{6}{10}$ to $\frac{9}{10}$ cover
Overcast, more than $\frac{9}{10}$ cover
Obscured, sky and clouds obscured by fog, smoke, etc.

Weather records include the coverage of separate layers of clouds at different heights. For example, an overcast of cirostratus may be observed with 2/10 of low

Figure 3.12 A line of cumulonimbus along an advancing cold front. (National Weather Service.)

clouds near the horizon. Of course it is not always possible to see upper cloud decks, especially if they lie above an overcast. Aircraft, satellites, and radiosonde observations greatly assist the ground observer in such cases. The percentage of time the sky is covered by clouds is useful for indirect determination of solar radiation. Figure 3.13 shows the mean annual cloudiness at different latitudes without regard to cloud types.

Cloud height is the distance between the ground and the base of the cloud. The height to the lowest cloud layer which creates a broken, overcast, or obscured sky cover is termed the *ceiling,* a value especially useful in airport operations. Cloud heights may be estimated visually, but other methods provide greater accuracy. In the daytime, gas-filled *ceiling balloons* with a known rate of ascent are timed until they enter the cloud layer to determine height. Under a dark cloud layer or at night, the *ceiling light* projects a vertical beam onto the base of the cloud. The angle of elevation of the light spot as measured at a known distance from the projector permits trigonometric determination of cloud height. The *ceilometer* operates on the same principle as the ceiling light but has a photoelectric element that reacts selectively to the light spot to indicate height.

Cloud direction and *speed* can be estimated visually, although more precise measurements are obtained by means of sighting devices, radar, or photography. Observations of cloud movements, shape, and orientation aid the estimation of winds at upper levels.

Increasingly, the composition and motions within clouds are being observed

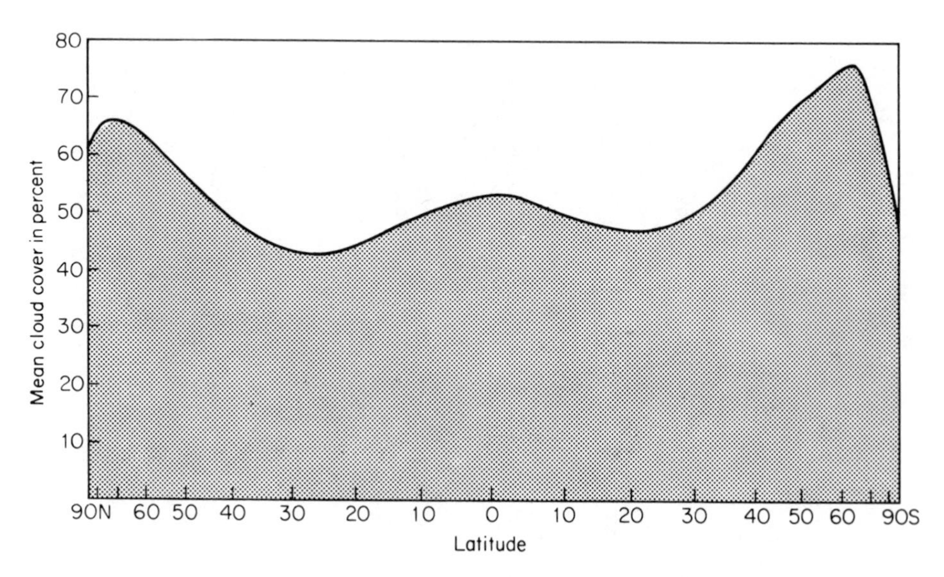

Figure 3.13 Mean annual percentage cloud cover at different latitudes. (Data from Sellers.)

by means of radar techniques that can determine the size of particles and distinguish liquids from solids. Doppler radars measure droplet speeds and thereby assist understanding of cloud development. Lidar, an application of the laser, employs an optical principle to detect clouds and measure their height and rates of development. Observations from satellites provide extensive data on the global distribution of cloud cover, making it possible to relate clouds to the general atmospheric circulation and to storm systems.

FOG

Fog results when atmospheric water vapor condenses or sublimates (or when supercooled droplets freeze) to the extent that the new forms, water droplets or ice crystals, become visible and have their base in contact with the ground. Saturation of the air and sufficient nuclei are ordinarily prerequisite to formation of fog, but hygroscopic nuclei may "force" the process at relative humidities below 100 percent. Very light fogs formed in this way are sometimes referred to as *damp haze.* Dense fogs restrict visibility and constitute a hazard. The collection of fog particles on vegetation and other obstacles may produce *fog drip,* which contributes to the receipt of moisture at the ground surface.

Observations of fog for meteorological records take into account the composition of the fog (water droplets or ice crystals) and its effect on visibility. *Visibility* in a given direction is the greatest distance at which common objects like buildings or trees are visible to the unaided eye. *Prevailing visibility* is the greatest visibility value that prevails over at least one-half of the horizon. For determination of visibility at airport runways a device known as the *transmissometer* measures the transmission of light over a fixed path. Visibility can be affected by dust, smoke, and other suspended particles in addition to or apart from fog.

The principal processes that cause saturation are cooling of the air and evaporation of water into it. Accordingly, fogs are classified into two main categories:

1. Fogs resulting from evaporation
   a. Steam fog
   b. Frontal fog
2. Fogs resulting from cooling
   a. Ground, or Radiation fog
   b. Advection fog
   c. Upslope fog
   d. Mixing fog
   e. Barometric fog

Steam fog forms when intense evaporation takes place into relatively cold air. As water vapor is added to the cool air saturation occurs and condensation produces fog. Steam fog is most commonly observed over bodies of water in middle and high latitudes. On the oceans it is also known as "sea smoke," and in the arctic builds to heights of 1,000 or 1,500 m. Steam fog sometimes forms on a small scale over warm, wet land immediately after a rain. In the tropics after a thundershower the air is temporarily cooled so that evaporation from soil and vegetation produces saturation and subsequent condensation.

Along the sloping boundary between two air masses, evaporation from warm rain falling through the drier air below may be followed by saturation and condensation in cooler layers to form *frontal fog,* also known as *rain fog.* Turbulence in the lower air reduces the extent and persistence of such a fog. If the condensation layer is above ground level, the result is a stratus cloud.

Ground fog, or *radiation fog,* develops when fairly calm moist air is in contact with ground that has been cooled by nighttime radiation. If the air is nearly motionless, only dew or frost is likely to form. Turbulence increases the depth of the fog, but if violent enough it will mix the excess moisture into upper, warmer layers and prevent fog formation, or dissipate that already in existence. Valleys are particularly susceptible to ground fogs as well as to frosts because of frequent temperature inversions. This type of fog does not occur at sea, where there is a negligible daily variation of water temperature. In winter, mainly in polar regions, and usually at temperatures below −30°C, ground fogs may be composed of ice crystals. These *ice fogs* are sometimes called *diamond dust* because the ice crystals glitter in the light.

Advection fog forms as moist air is transported over a cold surface. It is especially common at sea where moist air masses move over the colder water of high latitudes or across cold ocean currents. In summer, along arid west coasts, the cold water is provided by upwelling offshore, making fogs of this type frequent and persistent. Onshore winds tend to carry both temperature and humidity characteristics, extending the fog inland (Fig. 3.14). Advection fog also may form where moist air flows across cold lake surfaces such as those of the Great Lakes. During mid-latitude winters the introduction of moist air of tropical or subtropical origin over cold land or snow surfaces likewise can result in advection fog.

The gradual orographic ascension of moist air up a sloping plain or hilly region can cool the air adiabatically to form *upslope fog,* provided the air is already near saturation. If the ascent is too rapid or there is convective turbulence in the air, condensation will likely take place above ground level to form clouds. Upslope fog is fairly common on the high plains east of the Rocky Mountains. In southeastern Wyoming it is locally known as "Cheyenne fog."

When warm moist air meets cool moist air, the mixture in the boundary zone may have a temperature low enough to produce saturation and condensation. If this takes place at the earth's surface, a *mixing fog* is the result. The most common occurrence is at fronts between air masses of maritime origin. Convection or convergence will cause it to dissipate or lift, and it is usually difficult to separate mixing as a cause from the other processes which generate fog at a front.

Figure 3.14 Advection fog formed over cold water offshore drifts through Golden Gate into San Francisco Bay. (NOAA)

Barometric fog is extremely rare, although the cooling process associated with it possibly intensifies other types of fog. If the general pressure distribution over an area undergoes such a change that a layer of moist air at the ground level experiences a lowering of barometric pressure, the resultant adiabatic cooling could lead to condensation. These conditions are most probable in a valley or basin filled with stagnant air which is not immediately replaced as less dense air moves overhead.

The above outline of fog types should be regarded more as a classification of fog causes than as distinct kinds of fog. Although most fogs have a principal cause, their depth and density usually are influenced by one or more additional processes which lead to saturation. Figure 3.15 shows world areas having frequent fogs.

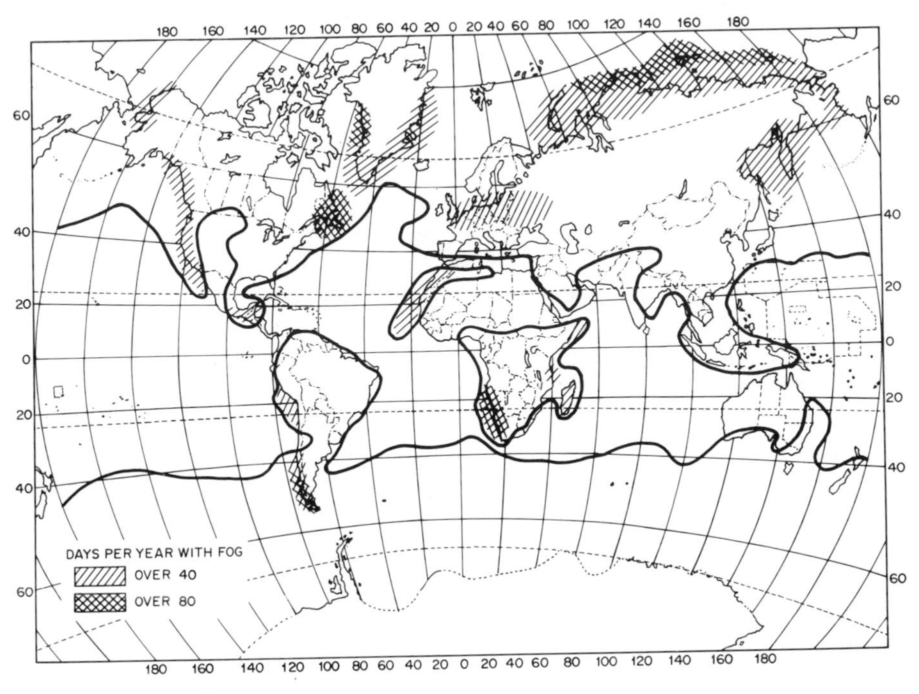

Figure 3.15 World areas having a high incidence of fog. The isolines represent a mean incidence of five days per year. (After Blüthgen; Van der Grinten projection courtesy A. J. Nystrom & Co., Chicago.)

PRECIPITATION: CAUSES, FORMS, PROCESSES, AND TYPES

Precipitation is defined as water in liquid or solid forms falling to the earth. It is always preceded by condensation or sublimation or a combination of the two, and is primarily associated with rising air. Although they contribute to the transfer of moisture from the atmosphere to the earth's surface, fog drip, dew, and frost are not considered to be precipitation. The common precipitation forms are *rain, drizzle, snow, hail,* and their modifications. Of these drizzle and light snow are the only ones likely to fall from clouds having little or no vertical development.

It is obvious that not all condensation, even in rising air, is followed immediately by precipitation. A complete explanation of the mechanics by which some clouds produce rain or snow and others do not is still lacking. A cloud is physically an *aerosol,* that is, a suspension of minute water droplets or ice crystals and other particles in air. In order to fall from the cloud the water forms must grow to sizes that can no longer be buoyed up by the air. This process is initiated by condensation on nuclei and continues as water molecules pass from slightly supersaturated air to the droplet surfaces. Coalescence may be achieved through collision of falling drops, turbulence, or electrical attraction between droplets. Most growth probably occurs as water vapor lost by small droplets migrates to larger or cooler droplets or from liquid droplets to ice crystals. The latter process is favored by the lower saturation vapor pressure over ice than over supercooled water (see Fig. 3.1). This ice crystal theory appears to explain heavy precipitation in many rainstorms of the middle and high latitudes. In the upper levels of vertically developed clouds supercooled water droplets often do not have an active direct role in precipitation. If ice crystals form in the supercooled cloud, water vapor crystallizes onto them, and the water droplets evaporate into the resulting drier air. Ultimately, all liquid water disappears as the ice crystals grow large enough to begin falling. If they fall through cold air, they may reach the ground as snowflakes; if through warmer air, they melt to fall as raindrops. Unfortunately, this theory does not explain tropical precipitation from clouds having temperatures above freezing throughout. Nor can it explain the drizzle that falls from relatively thin warm stratus. One or a combination of the other processes of coalescence apparently initiates these kinds of precipitation.

Precipitation is classified in two ways: according to the physical state of the falling water or on the basis of the processes which lead to its formation. The most common precipitation, rain, falls from clouds formed in rising air when the temperature, at least at lower levels, is above 0°C. Raindrops that reach the ground range from less than 1 mm to 5 mm in diameter; larger drops break apart as they encounter the resistance of the air. They may begin as snow crystals which melt while descending through warmer air. If the crystals grow and reach the ground, they produce snow. Large snowflakes are composed of several hexagonal crystals adhering to one another. Rain that freezes as it passes from warm air through a cold layer near the surface becomes *ice pellets,* or sleet. Another form of ice pellets consists of compacted snow covered with ice and falls from turbulent air in showers. During winter, rain may fall through subfreezing air, supercool, and freeze as it strikes cold surfaces to form *freezing rain.* Also known as glaze, ice storm, or "silver thaw," it ordinarily presages a warmer mass of air that follows a warm front.

Drizzle consists of numerous uniformly minute droplets of water (less than 0.5 mm in diameter) that seem to float in response to the slightest movement of air. It may fall continuously from low stratus-type clouds, never from convective clouds, and often accompanies fog and poor visibility. These characteristics distinguish it from light rain, which falls from thick stratiform clouds (generally nimbostratus) that have progressively lowered. Drizzle is sometimes called "Scotch mist," or simply mist. If supercooled it will produce *freezing drizzle* upon striking cold surfaces.

Strong, rising convective currents, as in a cumulonimbus cloud, carry the raindrops formed by intense condensation and coalescence of water droplets to higher levels where they freeze to become *hail*. These frozen drops fall again after reaching a level of decreasing convection and take on a coat of ice as they collide with supercooled droplets. Repeated ascent and descent results in concentric layers of alternately clear and crystalline ice. Another theory of hail formation holds that a frozen nucleus falls and alternately encounters supercooled water droplets and snow crystals which adhere to the hailstone, accounting for the concentric layers. The size of the stone depends on the amount of ice and snow it collects in one continuous descent. Hailstones weighing more than a kilogram have been recorded, although they were probably two or more separate hailstones frozen together. An exceptionally large hailstone that fell at Coffeyville, Kansas, on September 3, 1970, weighed 766 g. Hail rarely falls at high latitudes; in the tropics it is most common in highlands but occasionally occurs even at sea level.

Under less violent convection *snow pellets* may fall. They are small, opaque balls of compacted snow crystals, white in appearance, and do not have an ice coat. Snow pellets are more common in convective storms of winter and spring, whereas true hail comes with violent summer thundershowers.

Another ice form that may, in one sense, be regarded as precipitation is *rime*. Rime forms on cold surfaces that are exposed to moving supercooled fog or clouds, in contrast to frost, which is a sublimation product. It has a crystalline form and builds to the windward in feather-like shapes. A special case of rime is the rime icing of aircraft in supercooled clouds. It is most prevalent in temperatures below −9°C.

Having previously dealt with the methods of cooling which lead to condensation and sublimation, we can now outline the principal genetic types of precipitation according to their formation processes:

1. *Convectional precipitation,* resulting from convective overturning of moist air. Heavy, showery precipitation is most likely to occur. Rain or snow showers, hail, and snow pellets are the forms associated with convective precipitation.

2. *Orographic precipitation,* formed where air rises and cools because of a topographic barrier. It is doubtful that much of the world's precipitation is formed by orographic lifting alone; on the other hand, it is an important factor in triggering the precipitation process and intensifying rainfall on windward slopes, and it therefore affects areal distribution. The greatest annual totals of rainfall in the world occur where mountain barriers lie across the paths of moisture-bearing winds. A famous example is Cherrapunji on the southern margin of the Khasi Hills in Assam, India. This station averages 1,144 cm of rainfall annually. In 1873 the total was only 719 cm; in 1861, when the annual total was a fantastic 2,299 cm, 930 cm fell in the month of July alone. A station atop Mt. Waialeale on the island of Kauai in the Hawaiian group has an annual average of more than 1,200 cm. These

and other record rainfalls in mountainous areas should not be ascribed entirely to a simple orographic effect. Besides forcing moist air aloft, orographic barriers hinder the passage of low pressure areas and fronts, promote convection due to differential heating along the slopes, and directly chill moist winds which come in contact with cold summits and snowfields. Even this combination of effects cannot induce precipitation unless wind and moisture conditions are favorable.

3. *Convergence precipitation,* produced where air currents converge and rise. In tropical regions where opposing air currents have comparable temperatures, the lifting is more or less vertical and is usually accompanied by convection. In middle latitudes, frontal convergence is characterized by the more gradual sloping ascent of warm air over cooler. However, convectional activity frequently occurs along fronts where the temperatures of the air masses concerned are quite different. Mixing of air along the front also probably contributes to condensation and therefore to the frontal precipitation.

Most precipitation results ultimately from a combination of cooling processes which produce condensation or sublimation. An additional process is necessary to cause raindrops or ice crystals to grow large enough to fall from the clouds.

OBSERVATIONS OF PRECIPITATION

A great deal of important information about precipitation is obtained by visual observation of its form, type, and duration. It is obviously impractical to measure all the precipitation which falls over large areas. Instead precipitation is sampled at a number of places. In the United States alone there are several thousands of sampling stations. At a given place, precipitation is measured by the *rain gauge.* The standard rain gauge used in the United States is a cylindrical container, usually fitted with a covering funnel which directs the water into a measuring tube and which also reduces loss by evaporation. A graduated stick is used to determine depth, or the water may be poured into a graduated beaker. Amounts less than 0.25 mm are recorded as a trace, or simply "T." Values to the nearest whole millimeter are adequate for most climatological records. When snow, hail, or other ice forms are anticipated, the receiver funnel and measuring tube are removed and only the overflow cylinder is used. To determine the water equivalent of snow and ice a known quantity of warm water is added to the gauge. After melting is complete, the amount of the warm water is subtracted from the measured total. (Because snow density varies greatly use of a standard ratio between melt water and snow depth is not acceptable for precipitation records.) Rain and snow tend to be blown in eddies about the gauge, making collection of a true sample questionable, especially in gusty winds. Slatted wind shields or other apparatus around the gauge orifice reduce this effect but do not ensure accuracy of the sample (Fig. 3.16).

Modifications of the standard rain gauge are the *tipping-bucket gauge* and the

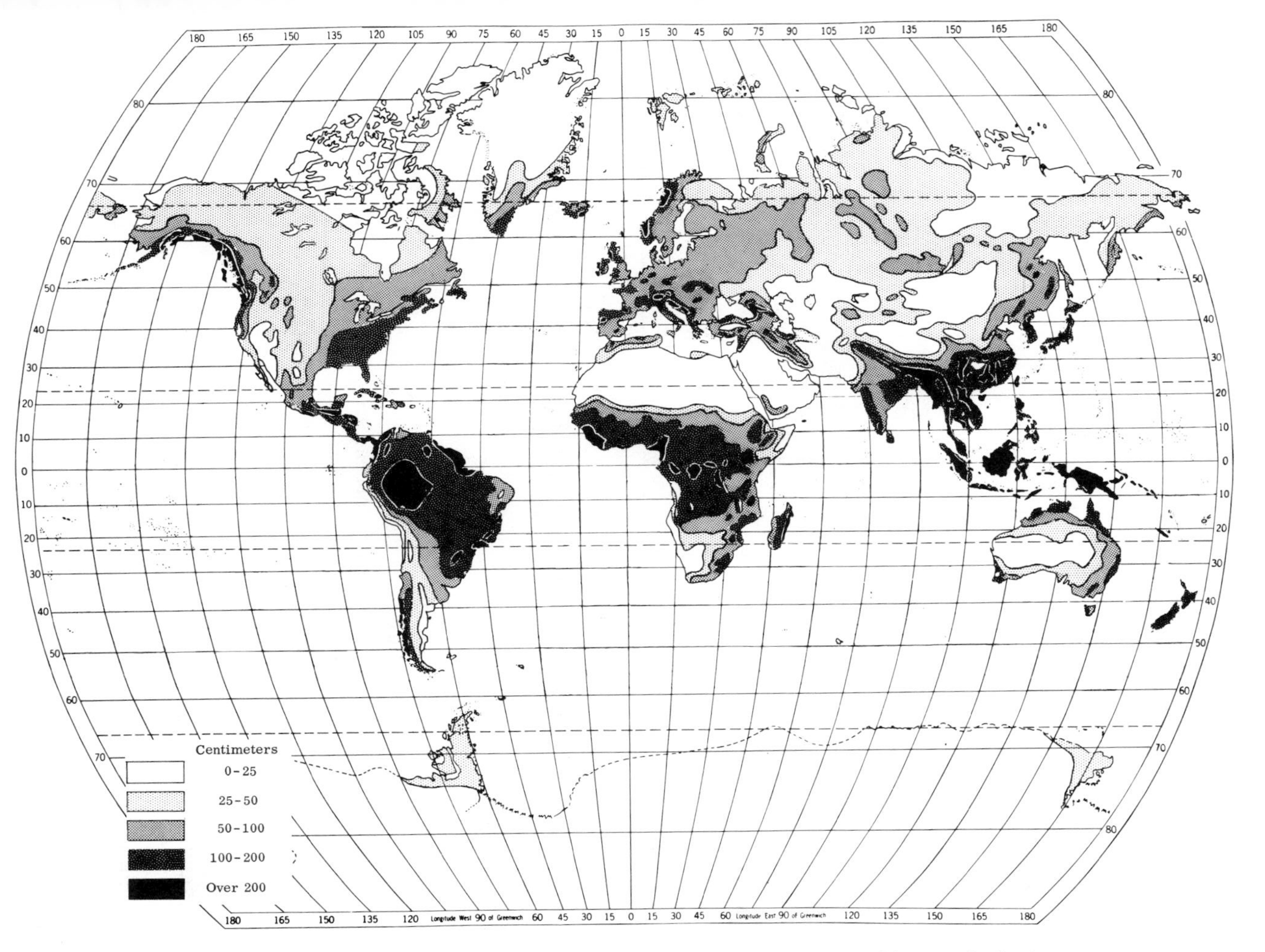

Figure 3.17 Mean annual precipitation on the continents. (Modified Van der Grinten projection.)

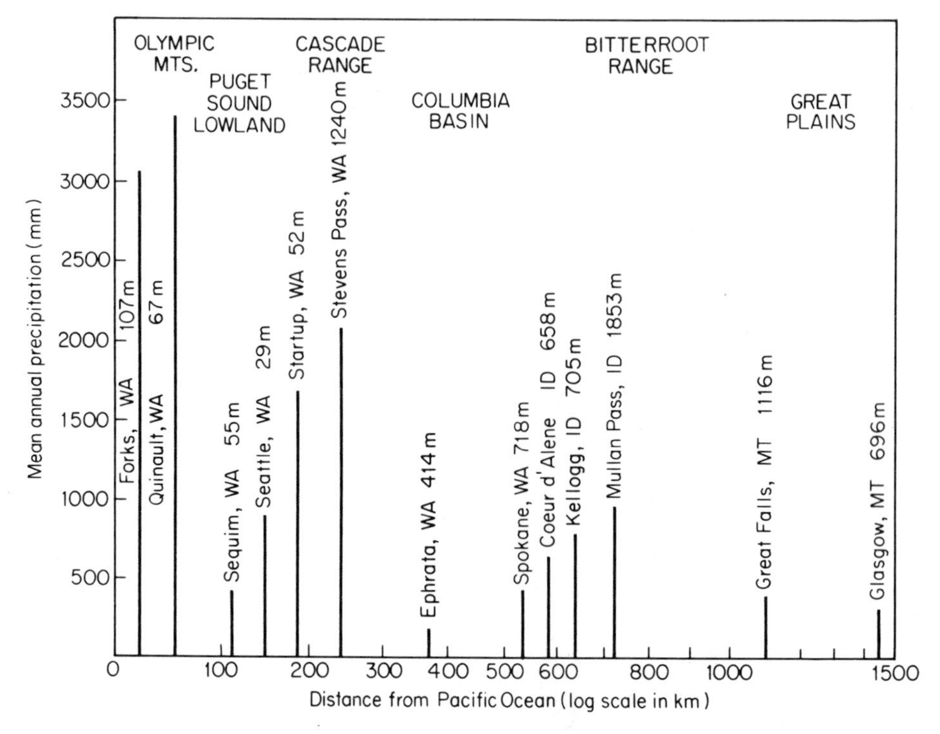

Figure 3.18 Cross section of mean annual precipitation and major relief features along the 48°N parallel in western North America. (Data from National Climatic Center, NOAA.)

tion may decrease again on high peaks under the influence of upper-air inversions or circulation systems that inhibit widespread lifting, but it is likely to increase upward in the path of vigorous storm systems.

Since the precipitation of a particular area results from several interacting processes, regional distribution is likewise determined by complex causes. "Normal precipitation" is a hypothetical concept with respect to both horizontal distribution and the time factor.

SEASONAL VARIATION OF PRECIPITATION

Quite as important as the precipitation totals accumulated in a particular year are the characteristics of seasonal variation from month to month through the year, the *dependability* from year to year, the *frequency,* and the *intensity*. The conditions for precipitation in a region do not exist in the same combination throughout the year. Latitudinal migration of the general pattern of wind and pressure systems is discussed in the section on general circulation in Chapter 4. Precipitation areas associated with the belts of convergence tend to shift poleward in the summer and equatorward in winter. In those equatorial regions which are constantly under an influence of converging winds monthly rainfall is fairly well distributed through the year, but a few degrees north or south areas dominated alternately by convergence and subsidence have wet summers and dry winters. Certain mid-latitude west coasts, for example, in California, have dry summers and wet winters as they experience the effects first of the subtropic highs, and then of the westerlies with their cyclonic storms. The monsoon circulation brings even more striking seasonal contrasts, producing wet summers as winds blow on shore and dry winters when the circulation is reversed. Monsoonal variation in rainfall is especially well developed in India and southeast Asia. The effect of mountain barriers on seasonal variation can only be to increase the amount in the rainy period when other factors are favorable. Because of its significance to human activities, especially agriculture, the annual *regime of precipitation* (monthly distribution through the year) is often more useful than the annual mean. Precipitation graphs for Allahabad, India, and Washington, D.C., in Fig. 3.19, clearly illustrate the difference in regimes although these stations have approximately the same annual average.

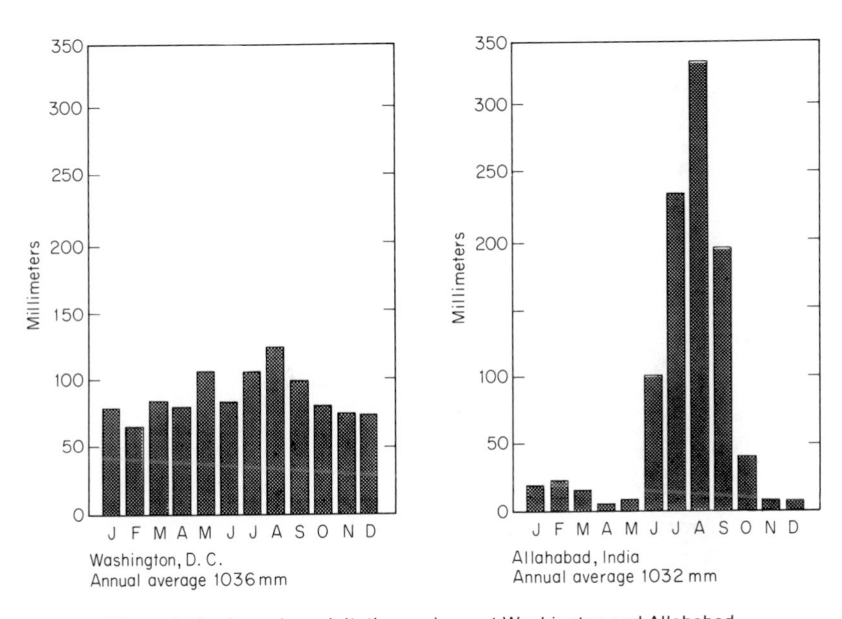

Figure 3.19 Annual precipitation regimes at Washington and Allahabad.

Dependability of rainfall refers to the possible deviation from the average, that is, the variation from the normal. The actual annual total for a station may be far above or below the statistical annual average. There may also be a considerable deviation from the normal regime. Generally speaking, dependability of precipitation is relatively high in humid climates and decreases toward the regions with lower annual averages. (Compare Figs. 3.17 and 3.20). The variability of rainfall (or snow) is a matter of great concern to farmers in subhumid and semiarid lands,

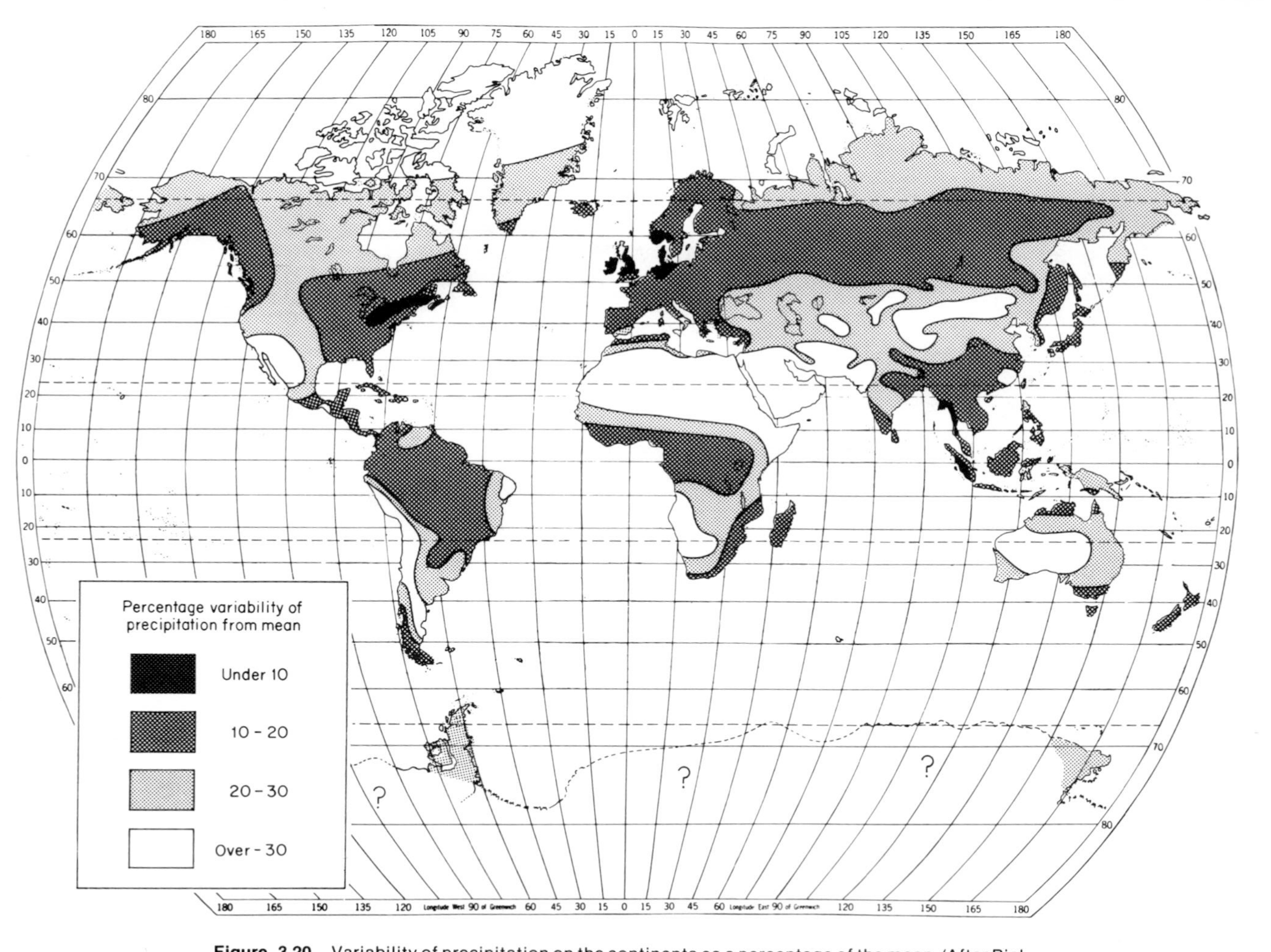

Figure 3.20 Variability of precipitation on the continents as a percentage of the mean. (After Biel and others.)

where negative departure from the average annual rainfall—or even from the monthly means in certain critical months—can cause crop failure.

Frequency of rainfall is usually expressed as the number of precipitation days per year. Whether a station's total falls in a few heavy thundershowers or as light rain over a great number of days is of vital importance in many economic activities. Comparison of the amount of precipitation with the time during which it fell yields precipitation *intensity*. This ratio may be expressed for an individual storm or for longer periods. Table 3.1 shows some observed rainfalls of exceptional intensity.

TABLE 3.1
World Maximum Observed Point Rainfalls*

Duration	*Millimeters*	*Location*	*Date*
1 min	31.2	Unionville, Maryland	July 4, 1956
5 min	63.0	Pòrto Bello, Panama	Nov. 29, 1911
8 min	126.0	Füssen, Germany	May 25, 1920
15 min	198.1	Plumb Point, Jamaica	May 12, 1916
20 min	205.7	Curtea de Arges, Romania	July 7, 1889
42 min	304.8	Holt, Missouri	June 22, 1947
9 hr	1,086.9	Belouve, Réunion	Feb. 28, 1964
12 hr	1,340.1	Belouve, Réunion	Feb. 28–29, 1964
24 hr	1,870.0	Cilaos, Réunion	Mar. 15–16, 1952
2 days	2,499.9	Cilaos, Réunion	Mar. 15–17, 1952
7 days	4,110.0	Cilaos, Réunion	Mar. 12–19, 1952
15 days	4,797.6	Cherrapunji, India	June 24–July 8, 1931
31 days	9,300.0	Cherrapunji, India	July 1861
2 mo	12,766.8	Cherrapunji, India	June–July 1861
6 mo	22,454.4	Cherrapunji, India	April–Sept. 1861
1 yr	26,461.2	Cherrapunji, India	Aug. 1860–July 1861
2 yr	40,768.3	Cherrapunji, India	1860–1861

**Daily Weather Map,* Dec. 13, 1962; J. L. H. Paulhus, "Indian Ocean and Taiwan Rainfalls Set New Records," *Monthly Weather Review,* 93, 5 (May 1965), 335.

QUESTIONS AND PROBLEMS FOR CHAPTER 3

1. Why is mean relative humidity generally high in polar areas, whereas vapor pressure in the same latitudes is low? (See Fig. 3.2.)
2. Outline the atmospheric processes that may cause cooling and subsequent condensation or sublimation of water vapor.
3. Why do the dry and wet rates of adiabatic cooling differ? What is the significance of the difference in cloud formation?
4. What criteria are used in cloud classification? Of what value is a classification of clouds?
5. How can ground fog be distinguished from advection fog? From steam fog?

6. Explain the high incidence of fog off the north coast of Siberia and along the southwestern coast of Africa (see Fig. 3.15).
7. An observation indicates a dry-bulb temperature of 20°C, wet-bulb 14°C, and barometric pressure 1,000 mb. What is the vapor pressure of the air? Relative humidity? (See Table B.1 and B.2, Appendix.)
8. Explain the ice crystal theory of precipitation.
9. State and explain the surface water budget equation.
10. Why is hail rare in equatorial regions?
11. What factors produce the large differences in annual precipitation throughout the world?
12. Explain the operating principles of the ceilometer, psychrometer, and evaporimeter.

4

Motion in the Climate System

The atmosphere and oceans are held to the earth by gravity and rotate as part of the planet, but they have complex internal motions of their own relative to the earth. Variable forces keep them in a state of constant agitation. One of the basic atmospheric forces, air pressure, is the main link between solar energy and motion. Ordinarily, we cannot perceive small changes in pressure, but together with other forces they activate the winds that transfer energy and moisture from one area to another, often in association with storms. At sea, differences in salinity and temperature combine with tidal changes and surface winds to initiate motion. Both atmospheric and oceanic motions respond to the rotation of the earth, the effect varying with the speed of movement and the latitude. Mountain barriers and surface friction influence the direction and speed of air, and air streams themselves interact to complicate further the patterns of flow. Similarly, movement in the oceans is subject to the shape of the ocean floor and shorelines and to conflicts among currents.

ATMOSPHERIC PRESSURE MEASUREMENTS

Atmospheric pressure observations are essential for the analysis of world patterns of winds, storms, and related climatic phenomena, although pressure itself usually is not regarded as a major descriptor of climate. The most accurate instrument for air pressure measurements is the *mercurial barometer*. Its operating principle is the balancing of the force exerted by air pressure against a column of mercury in a sealed glass tube (Figs. 4.1 and 4.2). Under standard conditions at sea level the at-

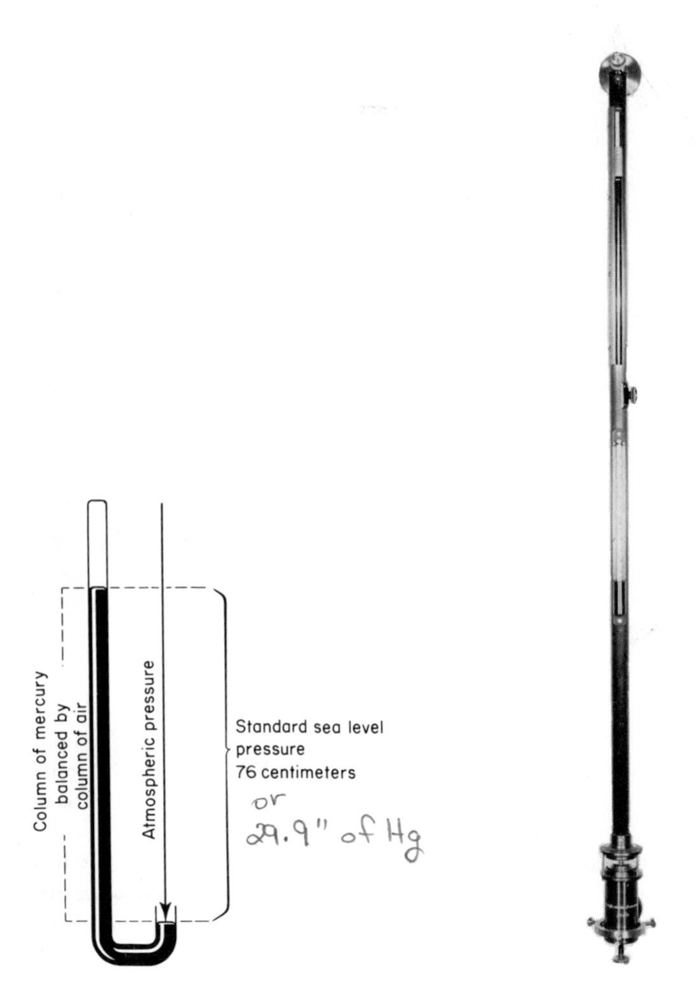

Figure 4.1 Principle of the mercurial barometer.

Figure 4.2 Fortin-type mercurial barometer. (NOAA)

mosphere balances a column of mercury 76 cm high. Determination of actual pressure requires corrections for temperature responses of the mercury column and brass scale, errors in instrument construction, and the variation of gravitational force with latitude and altitude. The corrected reading is the *station pressure.* Obviously, the station pressure on a high mountain will always be much lower than that in a nearby valley, and the two will be of questionable value in determining the lateral distribution of pressure. For direct comparison of pressures at different altitudes station pressure is converted to *sea-level pressure.* It is assumed that the atmospheric column above the barometer extends downward to sea level, and a figure to represent the additional mass of air is added to the station pressure (or subtracted in those rare instances where the station is below sea level).

The height of the column in the mercurial barometer indicates pressure in length units, which are unsatisfactory for expression of a force. Measurements that employ the concept of pressure as a force are necessary for meteorological calculations. The pressure unit most widely used in the United States is the *millibar* (mb), which represents the force exerted by 1,000 dynes on a square centimeter. (A *dyne* is the force which, acting for 1 sec on a 1-g mass at rest, imparts to it a velocity of 1 cm per sec.) The international standard (SI) pressure unit is the *pascal,* a force of one newton per square meter. In practice atmospheric pressure is expressed in kilopascals. (One kPa equals 1,000 Pa.) The value of mean sea-level pressure is 1,013.25 mb or 101.325 kPa.

The *aneroid* barometer indicates pressure using the principle of the *sylphon cell,* a partially evacuated metal wafer (Fig. 4.3) that expands or collapses in response to

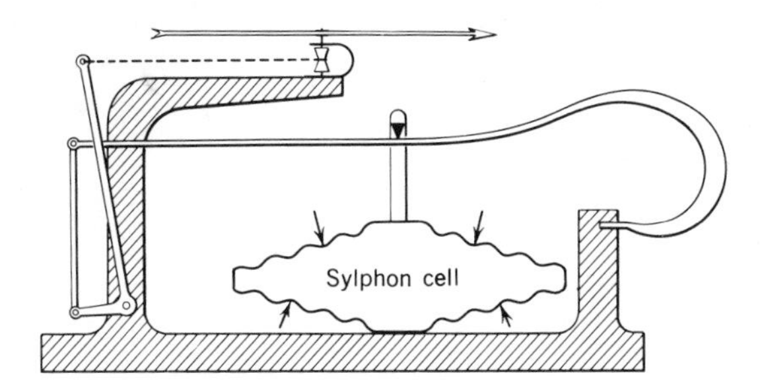

Figure 4.3 Principle of the sylphon cell in the aneroid barometer.

changing pressure. The fluctuation is linked mechanically to an indicator on a calibrated dial. The common aneroid barometer found in many households usually has such terms as *rain, windy,* or *fair* printed along the circumference. Presumably, the aneroid can forecast weather; but pressure is only one of the weather elements, and it alone cannot be relied upon to indicate future weather with any certainty. The principle of the altitude barometer, or *altimeter,* is the same as that of the aneroid, but calibration is in units of height rather than pressure. The *barograph* employs sylphon cells to activate a pen arm that makes a continuous chart record. Specially made *microbarographs* use several sylphon cells and a carefully constructed system of mechanical linkages to produce a more accurate record (Fig. 4.4).

PRESSURE-HEIGHT RELATIONS

Perhaps the most consistent property of the atmosphere is that pressure always decreases with height because of the progressive reduction of mass above a point. However, because the mass per unit volume (density) of air depends on its temperature, composition, and pressure and all of these factors vary, there is no simple relation between altitude and pressure. In other words, although atmospheric pressure always decreases with altitude, it does not always decrease at

Figure 4.4 Microbarograph. Part of the assembly has been removed to show the sylphon cells. The pen arm traces a continuous record of pressure on a chart drum which sits on the clock at left. (Courtesy Belfort Instrument Company.)

the same rate. Table 4.1 gives the average pressures and temperatures at selected levels under conditions that have been widely adopted as the *standard atmosphere* for the purposes of calibrating altimeters, for aerospace engineering, and for other scientific calculations. The standard atmosphere assumes a constant vertical structure of the atmosphere and is therefore an approximation of actual conditions.

TABLE 4.1

Standard Pressure-Height-Temperature Relations*

Pressure (mb)	*Pressure (mm Hg)*	*Altitude (m)*	*Temperature (°C)*
1,013.25	760.0	Sea level	15.0
898.76	674.1	1,000	8.5
795.01	596.3	2,000	2.0
701.01	525.8	3,000	–4.5
616.60	462.5	4,000	–11.0
540.48	405.4	5,000	–17.5
264.00	198.8	10,000	–49.9
11.97	9.0	30,000	–46.6
0.22	0.16	60,000	–26.1
0.0003	0.0002	100,000	–78.1

*Adapted from *U.S. Standard Atmosphere, 1976* (Washington, D.C.: NOAA, 1976), pp. 53-68.

Seasonal variations at upper levels are greatest in high-latitude regions. Other changes with time also result from solar activity.

HORIZONTAL PRESSURE DISTRIBUTION

Atmospheric pressure varies from time to time at a given place; it varies from place to place over short distances and on a worldwide scale; and it decreases with increasing altitude. The mass of a column of air above a point determines the atmospheric pressure at that point, but pressure also represents a force resulting from the kinetic energy of the gaseous molecules, and it is exerted in all directions. Horizontal pressure differences result primarily from differential heating that produces density contrasts and from redistribution of mass by the atmospheric circulation. An increased proportion of water vapor in the air can alter pressure slightly because water vapor is less dense than the mixture of other constituents of air. Differences from place to place in the force of gravity also affect atmospheric pressure slightly. Solar radiation and the gravitational pull of the sun and moon induce atmospheric tides and regular variations of pressure with time (Fig. 4.5), but tidal fluctuations frequently are obscured by the passage of storm and pressure systems.

Comparison of barometric pressure values that have been converted to a standard level reveals small horizontal differences that are highly significant in the analysis of weather patterns. On a worldwide scale mean sea-level pressure may vary from less than 990 mb to more than 1,030 mb. Agata, Siberia, recorded the world's greatest sea-level pressure on December 31, 1968, when the barometer reached 1,083.8 mb. A record low sea-level pressure of 870 mb was calculated from

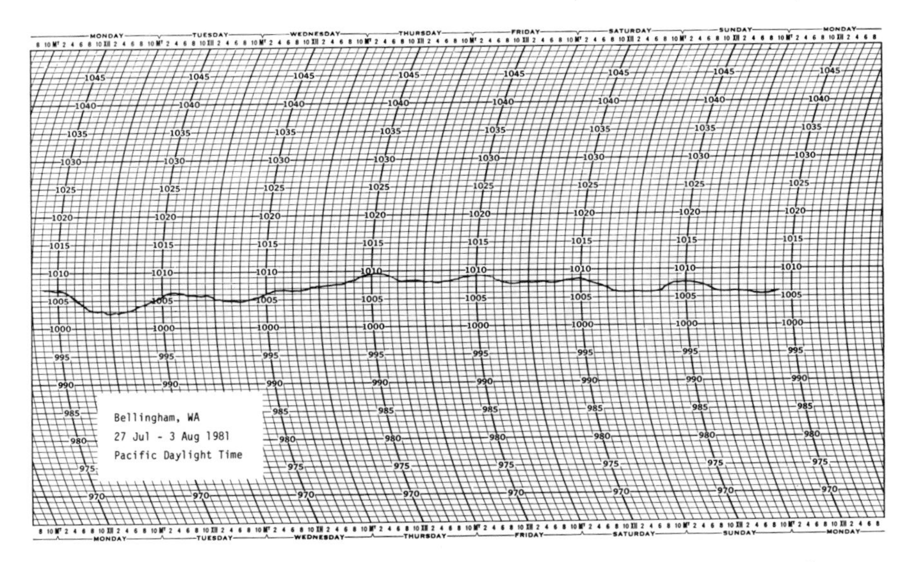

Figure 4.5 Microbarograph record of diurnal atmospheric tide maximums near solar noon during a week of settled weather at Bellingham, Washington.

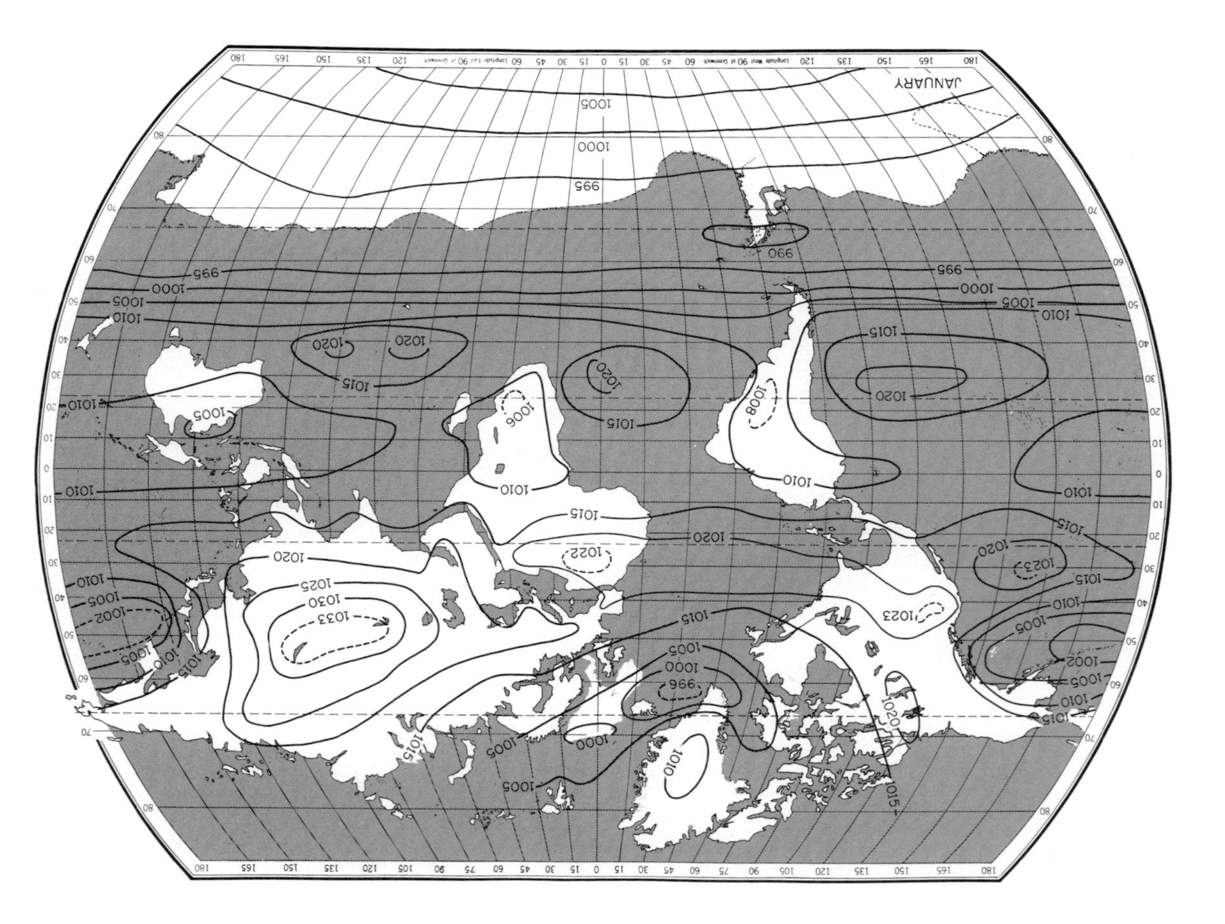

Figure 4.6 World sea-level pressure (mbs) in January. (Modified Van der Grinten projection.)

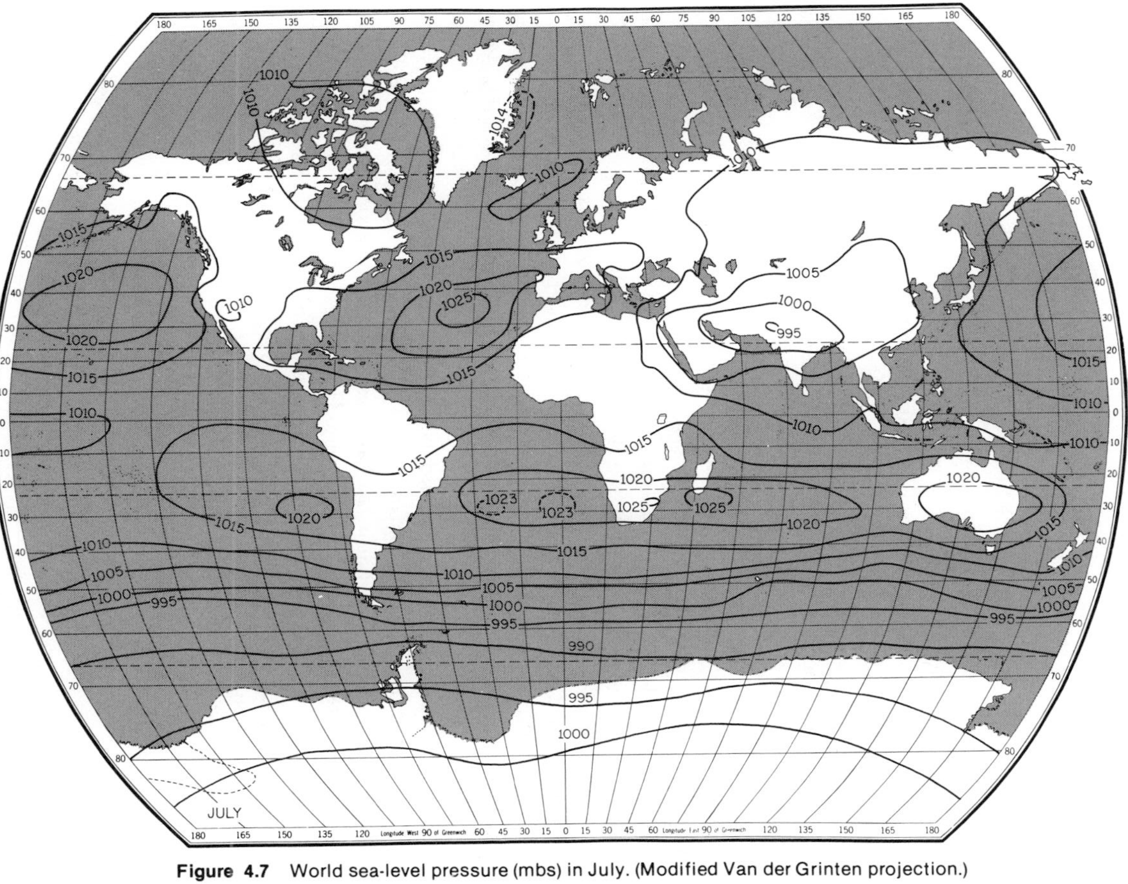

Figure 4.7 World sea-level pressure (mbs) in July. (Modified Van der Grinten projection.)

an aerial sounding in the eye of Typhoon *Tip* west of the Mariana Islands on October 12, 1979.

To map the horizontal pattern of pressure we use *isobars*. An isobar is a line connecting points that have equal values of pressure. Maps of surface conditions ordinarily use sea-level pressure data, but they may also be drawn to show pressure distribution at a constant elevation in the upper atmosphere. Isobars are analogous to the contour lines on a relief map. In fact, another method of mapping pressure is the constant-pressure chart, on which contours indicate the heights of a given pressure value (see Fig. 5.2). Constant-pressure surfaces are convenient levels for meteorological calculations and for airplanes which use aneroid altimeters.

The underlying causes of most pressure differences at the bottom of the atmosphere are the same factors which affect the horizontal distribution of temperature, latitude and land–water relationships being the most important. It should be remembered that the terms *high* and *low* as applied to temperature and pressure are relative. Low pressures in the hot equatorial zone are not necessarily as low in actual value as the low pressures associated with middle-latitude cyclonic storms. The effect of latitude, through temperature, upon pressure is to produce a more or less symmetrical pattern of pressure zones on the earth. Along the equator lies a belt of low pressure known as the *equatorial low* or *doldrums*. In the cold polar latitudes are the vaguely persistent high-pressure areas, the *polar highs*. Centered at about 60° to 70° north and south latitude are the *subpolar low-pressure belts*, and at 25° to 35° north and south latitude are the *subtropic highs*. These intermediate pressure zones result ultimately from temperature differences, but it is not possible to regard temperature alone as the direct cause of pressure distribution, since wind plays an important part in redistributing density characteristics from one latitude to another. Furthermore, the global circulation includes vertical motions that may cause dynamic pressure differences. For example, the piling up of air by winds blowing along convergent paths, as happens at upper levels over the subtropic highs, increases surface pressure. Nor should these pressure "belts" be regarded as permanent. They are greatly affected by differences in net radiation resulting from seasonal migration of the sun and from variations in heating of land and water surfaces. All are subject to incursions of air masses with contrasting temperature and density properties from other latitudes.

The effect of the irregular distribution of land and water upon pressure is best seen in seasonal contrasts, especially in the middle latitudes. In winter the continents are relatively cool and tend to develop high-pressure centers; in summer they are warmer than the oceans and tend to be dominated by low pressure. Conversely, the oceans are associated with low pressure in winter and high pressure in summer. By reference to Figs. 4.6 and 4.7, which show the average sea-level pressures for January and July, it can be seen that the North Atlantic, for example, is under the dominance of high pressure in summer and low in winter. In contrast, the Southern Hemisphere, which has a more homogeneous surface (mostly water) does not show such great seasonal differences. Note that the isobars in the latitudes 40° to 70° south are nearly parallel and aligned along parallels of latitude. The tendency to greater persistence of the subtropic high and the more orderly arrangement of isobars in the Southern Hemisphere are both results of the smaller effect of land–water differences.

PRESSURE AND CIRCULATION

The rate of change in atmospheric pressure between two points at the same elevation is the pressure gradient or isobaric slope. It is proportional to the difference in pressure and is the force that initiates horizontal air movement from high to low pressure (see Fig. 4.8). The pressure gradient is said to be steep when the rate of change with distance is great, and the steeper the gradient the more rapid will be the flow of air.

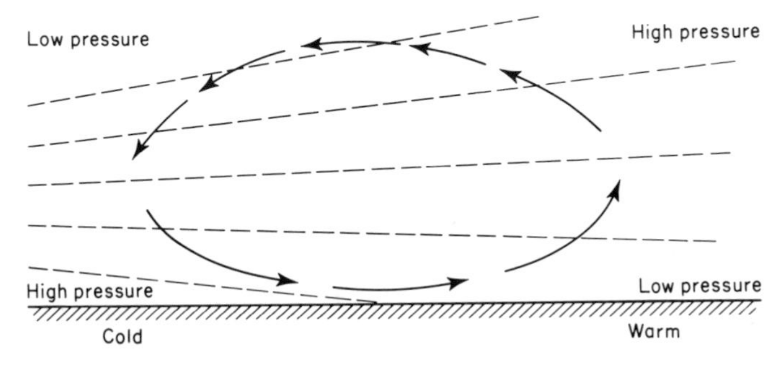

Figure 4.8 Schematic relation between pressure and wind. Dashed lines represent planes of equal pressure.

Horizontal flow in the air is called simply wind, whereas the term *advection* is applied to the horizontal transfer of atmospheric properties such as heat, moisture, or momentum. Small-scale vertical motion occurs in turbulent eddies and local convection currents; convergent ascent, orographic lifting, and subsidence are move evident in large-scale circulation. Any of these displacements may trigger additional vertical motion by altering the internal energy and buoyancy of air parcels. Their effects on weather conditions are treated in the section "Stability and Instability," Chapter 5. In combination the horizontal and vertical movements form complex motion systems, ranging from tiny thermal or frictional whirls through regional and global patterns. The smaller systems are embedded in the larger and act interdependently with them, transferring mass and energy from one system to another.

Atmospheric circulation systems are analogous to the movement of air in a room by hot radiators (thermal convection) or by forced air ventilation (mechanical convection). As long as sufficient differences in air density prevail at different parts of the system air will continue to flow in response to the pressure gradient. Converging and rising air streams are associated with low pressure at the base of the system; over high-pressure areas, the air will subside and diverge (Fig. 4.9). Strong gra-

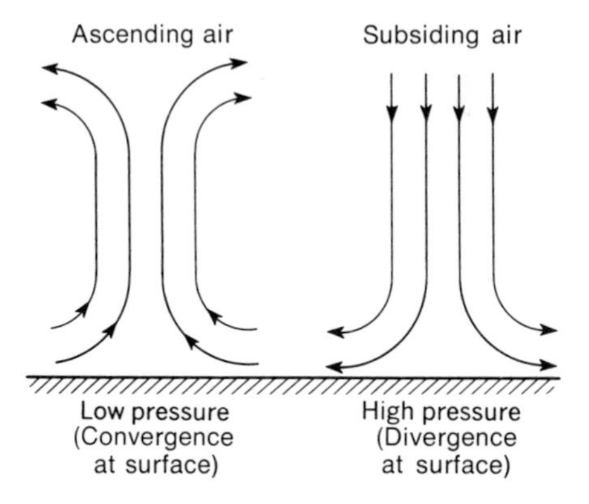

Figure 4.9 Schematic relation between vertical air currents and pressure.

dients between the two are indicated by closely spaced isobars on a pressure map, and the resulting winds will be of correspondingly high velocity.

FACTORS AFFECTING WIND DIRECTION AND SPEED

Because wind results basically from the pressure gradient, the initial determinant of its direction is the force exerted by the pressure gradient. However, as soon as air begins to move along the earth's surface its direction is altered by certain other factors acting together. Most of the winds of the earth follow a generally curved path rather than a straight one because of these factors.

One of the most potent influences on wind direction is the deflection caused by the earth's rotation on its axis. Demonstrated by Gaspard de Coriolis in 1844 and known as the *Coriolis force,* it is, strictly speaking, not a force but an effect resulting from the rotational movement of the earth and the movement of air relative to the earth (Fig. 4.10). But because we live on the earth and are a part of its rotation, the apparent effect is that of a force which turns winds from the paths initiated by the pressure gradient. The Coriolis effect causes all winds (indeed all moving objects) in the Northern Hemisphere to move toward their right and those of the Southern Hemisphere to move to their left with respect to the rotating earth. At the equator the effect has a value of zero and it increases regularly toward the poles. It acts at an angle of 90° to the horizontal direction of the wind and is directly proportional to horizontal wind speed. (Vertical motion also is subject to a Coriolis effect which has a horizontal component, but vertical velocities often are so small that the effect is negligible.) When the pressure gradient has initiated motion the resulting wind is deflected to the right (left in the Southern Hemisphere) until it may blow parallel to the isobars, that is, at right angles to the pressure gradient. If the isobars are assumed to be straight and parallel, the resultant is the *geostrophic wind.* If the air moves along curved isobars, a net centripetal acceleration tends to pull it toward the center of curvature, producing the *gradient wind* and a rotating mo-

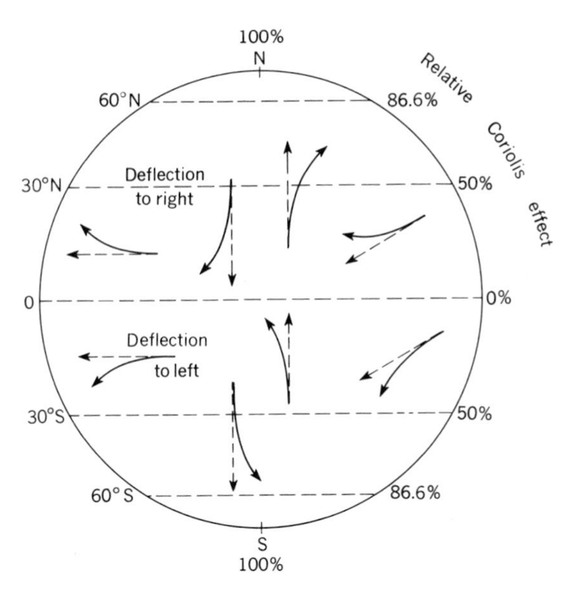

Figure 4.10 Deflection of winds by the Coriolis effect.

tion relative to the earth's surface (Fig. 4.11). The geostrophic and gradient wind concepts are useful for calculating and forecasting flow patterns. A counterclockwise flow is termed *cyclonic* in the Northern Hemisphere but *anticyclonic* in the Southern Hemisphere; clockwise rotation is anticyclonic in the Northern and cyclonic in the Southern Hemisphere.

Recognizing the relationship of wind direction to pressure distribution, the nineteenth-century Dutch meteorologist Buys Ballot formulated the rule: If you stand with your back to the wind in the Northern Hemisphere, pressure is lower on your left than on your right. In the Southern Hemisphere, again with your back to the wind, lower pressure will be on your right and higher pressure on your left.

The rotation or spin of any fluid about an axis, that is, its *vorticity,* depends on the orientation of the axis and the speed of rotation. As air converges toward a vertical cyclonic axis it conserves angular momentum that it has developed along a larger circumference and flows at an increasing speed. This principle explains the greater wind speeds usually encountered near the center of a low-pressure (cyclonic) system. (It also is illustrated by the more rapid spin of a figure skater when she draws her arms and legs toward the axis of rotation.)

Along and near the earth's surface wind does not move freely in a horizontal plane. Irregularities in surface relief and local differences in thermal convection cause moving air to take on correspondingly irregular motion so that it undergoes abrupt changes in speed and direction. This fluctuating wind action, known as *turbulence,* is associated with lulls, gusts, and eddies and increases with increasing wind speeds. The effects of surface turbulence are not very great above 500 m, and at

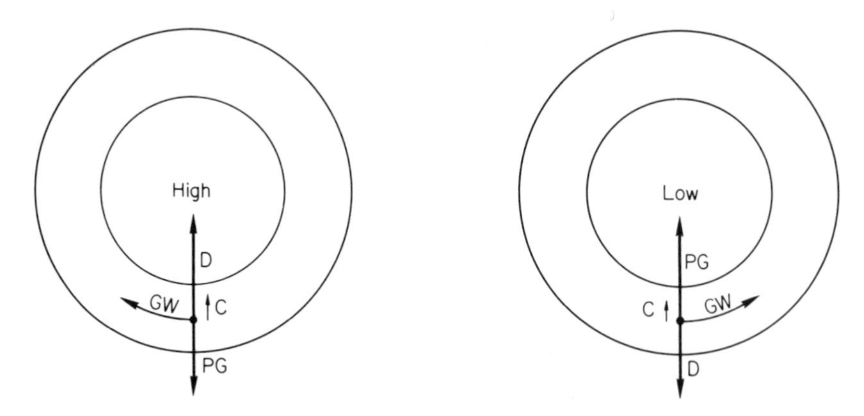

Figure 4.11 Forces affecting the gradient wind around high- and low- pressure centers in the Northern Hemisphere. *PG*, pressure gradient; *D*, deflective (Coriolis) effect; *C*, centripetal acceleration; *GW*, gradient wind resulting from balanced forces.

about 1,000 m actual wind direction and speed approximate the theoretical gradient wind, which may be calculated from the pressure gradient and other forces. The effect of surface friction is to reduce wind speed, in turn reducing the Coriolis effect so that the wind moves across the isobars at an angle toward lower pressure.

Other factors being equal, the difference in wind speed and direction between the surface and upper levels is greatest over rough land surfaces. Over water the surface wind more nearly equals the gradient wind. On the average, low-speed winds cross the isobars over land at an angle of about 45° with a speed about 40 percent of the gradient wind speed; over oceans the angle is about 30° and the surface speed about 65 percent of the gradient wind speed.

WINDS ALOFT

In free air the effect of increasing altitude is to reduce the kind of turbulence induced by surface friction. Ordinarily, wind direction changes with altitude so as to become more nearly parallel to the isobars and wind speed increases as it more nearly approximates the theoretical gradient wind. Strong convection currents and strong *wind shear* along the boundaries between air streams having converging directions or different speeds may cause turbulence at high levels, however. The swift passage of wind over the top of relatively calm air in a temperature inversion is likely to produce a turbulent layer.

Considerable variation of wind direction with altitude may result from convergence of two air masses, particularly if their temperatures are different. Along the boundary where cold air advances against warmer air the lighter, warmer air will rise and tend to spread out above the heavier, cold air so that winds in the upper air may be blowing directly opposite the cold surface winds. This phenomenon is commonly associated with the cyclonic storms of middle latitudes, but it is by no means confined to those areas.

Because local winds are all comparatively shallow and have pressure gradients only vaguely related to the general circulation, vertical soundings through and above them often show great variation of both direction and speed. If the conditions responsible for the local wind create a pressure gradient in the same direction as the major circulation, the two simply reinforce one another.

Although the considerations of upper winds so far have been chiefly with respect to free air, surface winds on hills and mountains are governed by most of the same principles. Rising above the maximum friction layer, mountain peaks are buffeted by the higher-speed winds of upper levels, as the condition of trees near the timberline and the experiences of mountain climbers attest. The summits of mountains tend to increase further the speeds at upper levels. Air blowing across the top of a mountain is constricted between the summit below and the layers of air above; being thus funneled, it increases in speed until it passes the summit. A dramatic example of wind force near a mountaintop occurred at Mt. Washington (1,908 m) in northern New Hampshire on April 12, 1934. Wind speeds of more than 130 knots (67 m per sec) were recorded for most of the afternoon and at one time reached 200 knots (103 m per sec).

DIURNAL VARIATION OF WIND SPEED

If we consider the air movement in an extensive wind system and disregard wind shifts or locally induced winds, we find a fairly regular daily variation of wind speed at the surface. The maximum speed usually occurs in the early afternoon and the minimum in the early morning hours just before sunrise. Probably everyone has experienced the phenomenon of winds dying down in the evening only to rise again with renewed force in the morning. In order to find the reason for this, it is necessary to recall that wind speed increases with increasing altitude. During the daytime, convection caused by heating of the surface air layers brings about an exchange of momentum between lower and higher levels, and a more nearly uniform vertical distribution of speed exists. At night the air near the ground is cooled and, being heavy, tends to remain at the low levels, where because of the greater frictional drag it resists being carried along by the fast-moving winds above. On high hills and mountains, especially if they are isolated, the effect may be the reverse. During the day the slowing effect of frictional drag is transferred to the upper levels by turbulence to produce a midday minimum of wind speed. At night the air rides above the cool layer lying in depressions and reaches its maximum speed for the 24-hour period. This is illustrated by the graphs in Fig. 4.12, which compare winds speeds at three levels on the slopes of Mt. Fuji. Note that in July, when daytime heating is at a maximum, the diurnal variation of wind speed is greater at the low and high levels than at the intermediate height.

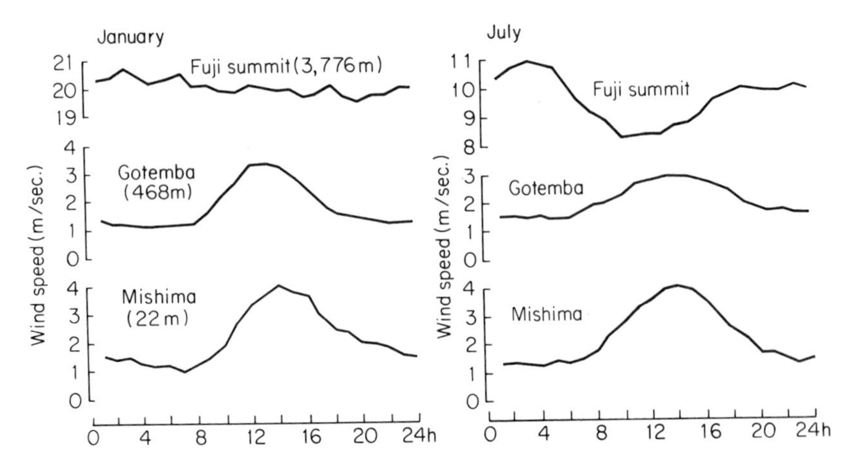

Figure 4.12 Diurnal variation of mean wind speeds at three levels on Mt. Fuji, Japan, in January and July. (After Fujimara; adapted from Masatoshi M. Yoshino, *Climate in a Small Area,* 1975.)

WIND OBSERVATIONS

Wind direction may be ascertained quite simply by watching the movement of clouds, vegetation, smoke, waves on water surfaces, and so forth. This type of visual observation yields important information on winds, and, with some experience, an observer can also estimate the speed of surface winds. In 1805 Admiral Francis Beaufort introduced a wind force scale which was based upon the response of certain objects to the wind. In applying the Beaufort Scale the extent to which smoke is carried horizontally or to which trees bend before the wind is used as an index of speed. At sea the condition of waves, swell, and spray in addition to response of sails and masts is the basis for windspeed estimates.

Most weather stations are equipped with a wind observation unit which consists of a wind vane and an anemometer mounted at the top of a steel support column. The wind vane points into the wind, but it should be noted that winds are named for the direction from which they come. A wind blowing from north to south is a north wind. Meteorologists commonly use degrees of azimuth measured from north at 0° or 360° through east (90°), south (180°), and west (270°). Ordinarily the wind vane is electrically connected to show wind direction on an indicator inside the weather station, since the instrument support may have to be mounted at some distance from the roof immediately above the station. Adaptation of the movement resulting from changing wind direction to a pen arm makes possible a continuing record on the *wind register.*

A unit of wind speed in common use is the *knot,* that is, 1 nautical mi per hr. Smaller units such as meters per second are also used in research or microclimatic observations, and they increasingly appear in climatic records. One knot equals 0.5148 m per sec or 1.854 km per hr. Practical applications of wind observations in

or 6076.115 ft = 1 nautical mi.

(a)

(b)

(c)

Figure 4.13 Instruments for wind measurements. (a) Anemometer bivane. The vane fins orient the propeller to measure vertical as well as horizontal air movement. (b) Cup anemometer and wind vane. Infrared lamps prevent freezing of the anemometer. (c) *UVW* anemometer. Propellers mounted at right angles respond to air movement in three dimensions.
(Photographs (a) and (c) courtesy R. M. Young Company; (b) courtesy Bureau of Reclamation, U.S. Department of the Interior.)

structural engineering and conversion of wind energy require wind power data, which can be calculated as a function of air density and velocity. Because wind power is proportional to the cube of the wind speed, small increments of speed yield significant increases in power.

Anemometers are instruments for measurement of wind speed. A common type is the Robinson rotating-cup anemometer, which registers both instantaneous speed and the accumulated flow past the instrument (Fig. 4.13B). A ring of cups (usually 32) in the *bridled anemometer* is held, that is, bridled, by a spring that translates wind force to a wind speed indicator. It is commonly used on ships, where allowance has to be made for direction and speed of the ship. The *pressure-tube anemometer* operates on the same principle as the *pitot tube* used on aircraft to determine air speed. Mounted in a wind vane, it responds to the force of wind entering the tube and registers wind speed as a function of pressure. The dynamic principle employed in the pressure-tube anemometer causes it to lag in indicating gusts.

Air movement can also be measured electrically. The rate of cooling of an exposed hot wire, a thermistor, or a thermocouple is a function of air movement and therefore an index of wind speed. Vertical as well as horizontal speed and direction of air flow is important in microclimatological studies. Bivanes and various assemblies that are sensitive to forces from all directions provide the desired information.

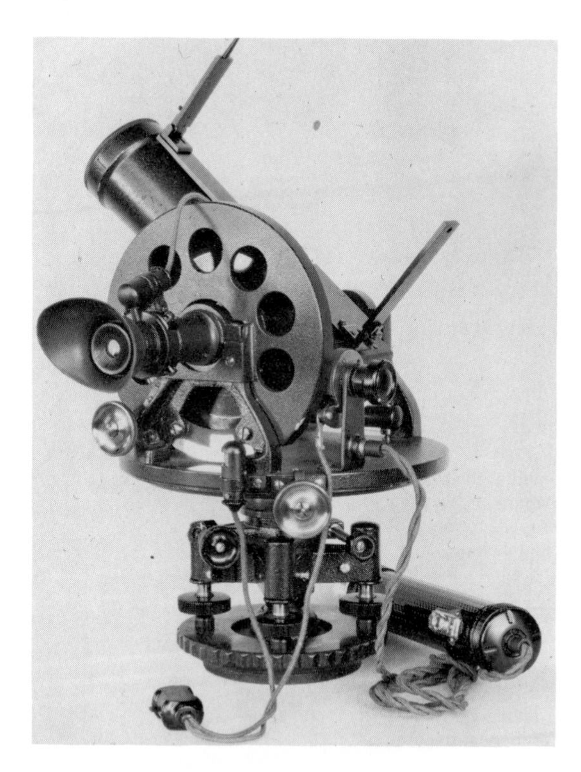

Figure 4.14 Pilot balloon theodolite. (Courtesy Teledyne Gurley.)

For purposes of analyzing storms and air masses as well as for planning air travel, observations of upper winds are as important as those of winds at the surface. Clouds are an obvious indicator of direction of winds aloft, and if their height is known their drift can be used to estimate their velocity.

The most common method of obtaining upper wind data is by means of the *pilot balloon,* or "pibal." A balloon filled with hydrogen or helium is released from the earth's surface and as it floats aloft it rides the winds to reveal their direction and speed. The *theodolite,* a right-angled telescopic transit which is mounted so as to make possible reading of both azimuthal (horizontal) and vertical angles, is used to track the balloon (Fig. 4.14). The latter's progress is plotted minute by minute on a special plotting board, and the direction and speed for successive altitudes can be computed. When the sky is obscured by a low cloud cover, the pilot balloon is useless for upper-air wind observations. Radar has provided the answer to this problem. As a larger balloon with greater lift carries a metal target upward a radar transceiver or radio theodolite tracks the ascent. This technique is known as *rawin* observation.

MAPPING WIND DATA

Maps of winds ordinarily indicate direction by means of arrows pointing in the direction of flow, that is, flying with the wind. Straight arrows represent observed directions; curved *stream lines* show general air trajectories or theoretical flow patterns. Although wind speed may be entered in numerals, it is more often designated by short barbs or feathers at the tail of the arrow, the head of the arrow being the station circle (Fig. 4.15). *Isotachs* connect points that have equal wind speed and show the distribution of wind force on maps.

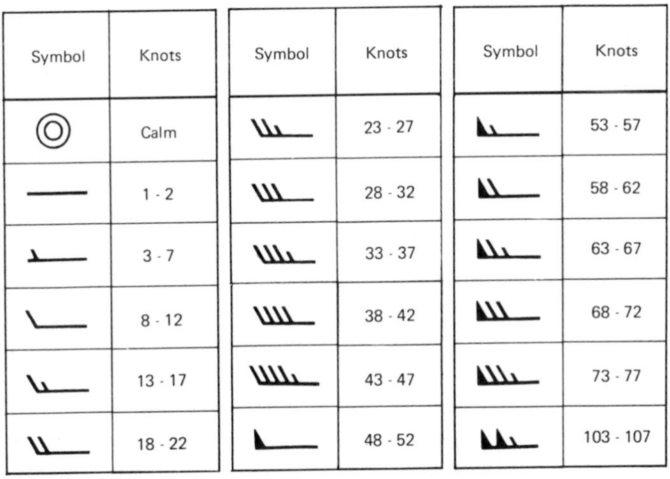

Symbol	Knots	Symbol	Knots	Symbol	Knots
◎	Calm		23 - 27		53 - 57
	1 - 2		28 - 32		58 - 62
	3 - 7		33 - 37		63 - 67
	8 - 12		38 - 42		68 - 72
	13 - 17		43 - 47		73 - 77
	18 - 22		48 - 52		103 - 107

Figure 4.15 Wind symbols used on weather maps.

Wind roses portray climatic aspects of winds graphically on maps, charts, and diagrams. The relative lengths of lines radiating from the station circle indicate percentage frequency of principal directions (Fig. 4.16). The same data may be plotted as a polar graph in the *wind frequency polygon,* preferably with radii lengths proportional to the square roots of percentage frequency. Barbs, varying line widths, or bars on the basic wind rose permit display of wind speeds from different directions.

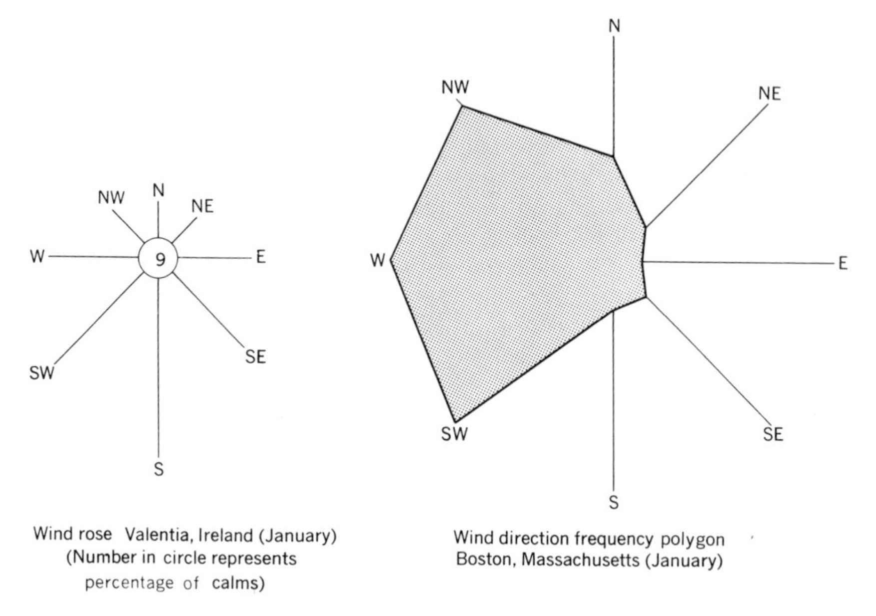

Figure 4.16 Methods of graphing mean wind directions.

LOCAL WINDS

In order to develop an understanding of the factors that influence wind systems it is helpful to begin with relatively simple small-scale motion, such as winds generated by local temperature differences. Local winds usually affect small areas and are confined to the lower levels of the troposphere. Their air movements may be generated either by heating or cooling of a particular area. Sometimes they represent virtually complete convectional circulation; in other cases they merge into the stronger wind systems of the larger circulation patterns.

The *sea breeze* develops along seacoasts or large inland water bodies in summer when the land heats much faster than the water on a clear day and a pressure gradient is directed from high over the water to the low over the land (Fig. 4.17). The circulation which follows brings cool air onto the land, but the system is not entirely a self-contained unit because the air which returns to the sea at upper levels spreads out and does not necessarily all return at the surface. The sea breeze rarely has a

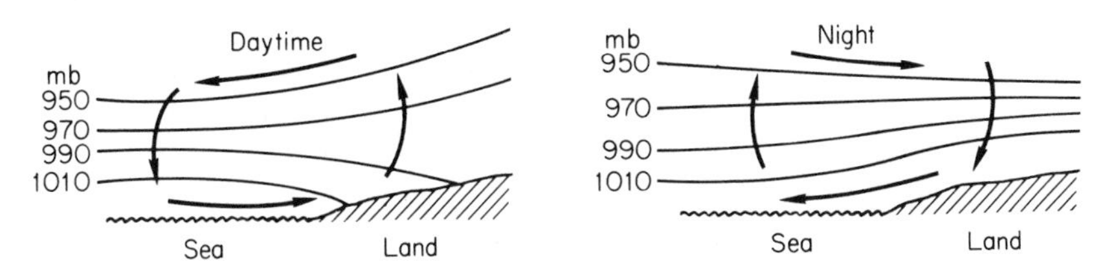

Figure 4.17 Principle of the land and sea breezes.

depth of more than 1,000 m and at its maximum strength does not extend inland more than 50 km. (The distance is much less around lakes, where the wind is more properly called a lake breeze.) It begins offshore in the late morning hours and gradually extends inland to decrease afternoon temperatures. Toward evening it subsides. Because of the sea breeze, areas on the immediate coast in middle latitudes and the tropics may have lower temperatures than a few miles inland.

At night a reversal of the sea breeze may occur but with somewhat weaker characteristics. The *land breeze* becomes established during the night as the land cools to temperatures lower than the adjacent water, setting up a pressure gradient from land to sea (Fig. 4.17). Because temperature differences between land and water are rarely as great at night as in the daytime, the land breeze is less extensive both vertically and horizontally than the sea breeze. It usually attains its maximum intensity in the early morning hours and dies out soon after sunup.

Another combination of local winds that undergoes a daily reversal consists of the *mountain* and *valley breezes.* On mountain sides under a clear night sky the higher land radiates heat and is cooled, in turn cooling the air in contact with it. The cool, denser air then flows down the mountain slopes into the valleys and lowlands. Since it blows from the mountain, this air flow is termed a *mountain breeze.* By morning it may produce temperature inversions in depressions so that the valley bottoms are colder than the hillsides. But, where the mountain breeze is funneled into a narrow valley or a gorge, it may gain considerable velocity and generate enough turbulent mixing to break up the inversion. On warm sunny days, the heating of mountain slopes may generate an upslope flow of air called a *valley breeze.* As the warm air moves up the mountain, it is replaced by cooler air from above the valley, and surface temperatures are moderated slightly.

A type of wind known as a *gravity* or *katabatic wind* occurs in several parts of the world. Gravity winds result from the drainage of cool air off high plateaus or ice fields. The *mistral* flows onto the Mediterranean coast of France in winter from the higher lands and snow-capped mountains to the north. It is channeled somewhat by the Rhone Valley, and its coolness, dryness, and velocity sometimes detract from the otherwise attractive climate of the Riviera in winter. Along the northern coast of the Adriatic Sea, a cold, northeast wind known as the *bora* flows down from the plateau region in Yugoslavia onto the narrow coastal plain.

In numerous instances the winds associated with particular storm or pressure conditions occur with some regularity and are characteristic of the climate of a given locality. The Guinea Coast of Africa experiences the dry, dusty *harmattan* in winter when air from the relatively cool Sahara replaces humid tropical air. The harmattan has a marked cooling effect and is sometimes known as the *doctor.* In spring, as the subtropic high moves northward, the *sirocco* blows from the south across the Mediterranean lands. In Egypt, where it is known as the *khamsin,* and in other parts of North Africa, it is hot and dry. Along European coasts of the Mediterranean the sirocco has more moisture in its lower layers, but its high temperatures often have a withering effect on crops. *Leveche* is the name given to this wind in Spain. In the eastern Mediterranean northerly winds that persist at times during summer are known by the classical Greek name *etesians.*

Winds that bring abrupt changes in temperature or moisture conditions are likely to take on local names. The *norther* marks the onset of cold, stormy weather in winter over Texas, the Gulf of Mexico, and the western Caribbean. The *pampero* is a squally wind from the northwest that blows over the Argentine pampas in winter. The *chinook* of North America, the *foehn* of the European Alps, and the New Zealand *nor'wester* are warm, dry, gusty winds induced by mountain ranges. They are discussed in more detail in connection with storm effects in Chapter 5.

GENERAL ATMOSPHERIC CIRCULATION

The net effect of the differential heating of the earth by insolation is to produce density differences that set the atmosphere in three-dimensional motion. Much of the energy for maintaining the global circulation comes from the tropical oceans, where evaporation transfers large amounts of latent heat to the air. Although the influence of latitude upon heating might be expected to create a simple circulation between tropical and polar areas, the effect of the earth's rotation diverts the wind into gigantic whirling systems, or vortices, that are aligned more or less in latitudinal belts. The resulting zonal flow patterns have prevailing winds with strong easterly or westerly components and comprise the basic motion systems of the *general circulation.* If, for the time being, we neglect the influences resulting from differences in heat and moisture exchange over land and water surfaces we find a hypothetical arrangement of surface wind and pressure belts, as shown in Fig. 4.18.

In the vicinity of the equator, where pressures are low, winds converge and rise, and the surface winds are light and variable. This *intertropical convergence zone* (ITCZ), also known as the *doldrums,* fluctuates in position and intensity and is at times a weak, discontinuous belt. On either side of the ITCZ, blowing into it (converging), are the *trade winds.* They are named the *northeast* and *southeast trades* in the Northern and Southern Hemispheres, respectively. Note that although the pressure gradient is directed from the subtropic high toward the intertropical convergence the winds are deflected by the earth's rotation so that they approach the equator at acute angles rather than at the perpendicular. The sources of the trades are the subtropic highs, sometimes called the *horse latitudes,* where much of the upper air arriving from the equatorial zone converges and piles up, then subsides and

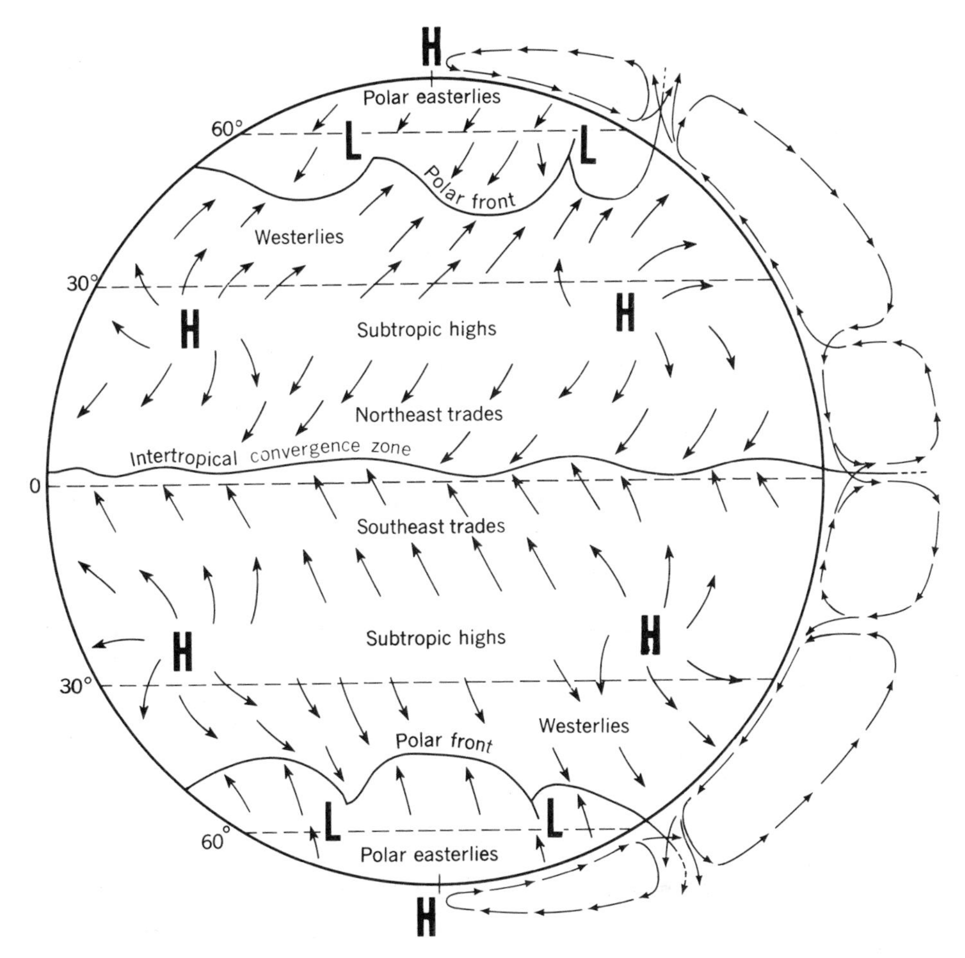

Figure 4.18 Schematic arrangement of winds and pressure in the general circulation. Effects of differential heating of land and water surfaces are neglected.

diverges near the surface. The subtropic highs are not actually continuous belts, but are broken up into "cells," having their best development over the oceans. Because much of the air movement is downward and pressure gradients are weak a tendency to calms or variable winds characterize these cells at the surface. A part of the diverging air becomes the trades; that which flows toward the pole forms *prevailing westerlies.* In the Northern Hemisphere the westerlies are from the southwest and in the Southern Hemisphere from the northwest. The Coriolis effect accounts for the westerly component in each case. The westerlies are zones of cyclonic storms, and although strong winds in these storms may blow from any direction of the compass, zonal westerlies predominate. The cyclonic storms themselves move in a general west to east direction. Land masses disrupt the

cyclonic meaning frontal

westerlies considerably in the Northern Hemisphere, but in the Southern Hemisphere, where there is a virtually unbroken belt of water between 40° and 60° south, the westerlies are strong and persistent. They are often called the "roaring forties" in this zone, a carryover from sailing days.

The westerlies and the *polar easterlies* meet and converge at the *subpolar lows,* or *polar fronts.* Here there is frequently a great contrast between the temperatures of the winds from subtropical and polar source regions, giving rise to the cyclonic vortices or "lows" that are carried along in the westerlies. The polar easterlies carry air outward from the *polar highs,* which are regions of subsidence of air from higher levels.

UPPER-LEVEL WAVES AND JET STREAMS

Completion of the circulation pattern of the major wind systems requires an exchange of air at the upper levels to compensate for any net transfer horizontally at the bottom of the troposphere. Strong air streams in the upper troposphere, lower stratosphere, and possibly even in the mesosphere help to achieve a balance in the global circulation. At the higher tropospheric levels the mean flow is more nearly along parallels of latitude than is the case at the surface. Nevertheless, there is a slight poleward flow in latitudes 0° to 30° and 60° to 90° and an equatorward flow in the middle latitudes. The mean vertical motion that completes the three-dimensional circulation is upward in the intertropical convergence zone and along the subpolar low; it is downward in the subtropical high-pressure cells and near the poles. These vertical transfers are of small magnitude compared to horizontal motion, but they are quite essential for maintenance of the general circulation and development of storms.

The winds at several kilometers above the surface follow giant, undulating paths around the earth in the latitudes of the westerlies (Fig. 4.19). These waves ap-

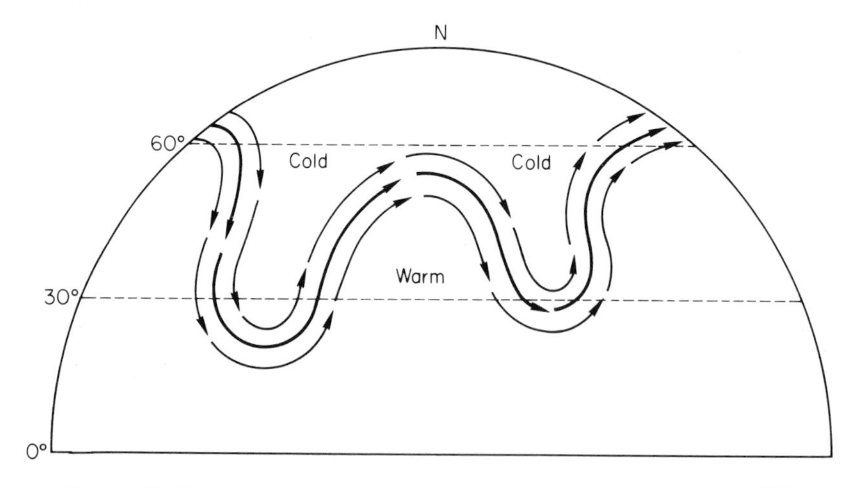

Figure 4.19 Wave trajectories (Rossby waves) in the upper westerlies of the Northern Hemisphere. Heavy arrows represent typical jet-stream paths.

parently result from the tendency of winds in large-scale motion systems to retain a constant spin (angular momentum) about the earth's axis of rotation. A stream of air moving toward the equator adopts a cyclonic curvature (counterclockwise in the Northern Hemisphere) relative to the surface at the lower latitude as the distance from the axis of rotation increases. Eventually, the curved path turns the wind back toward the pole. Passing its original latitude, the wind takes an opposite (that is, anticyclonic) curvature relative to the earth as it comes closer to the polar axis, where the rotational velocity is less. (These effects are easily demonstrated on a globe.) The resulting *Rossby waves* may have lengths of 3,000 to 6,000 km, permitting three to six circumpolar waves that correspond to surface paths of depressions and anticyclones in the westerlies (see Fig. 4.20). Their length, amplitude, and position are influenced by differential heating at the surface and by extensive mountain bar-

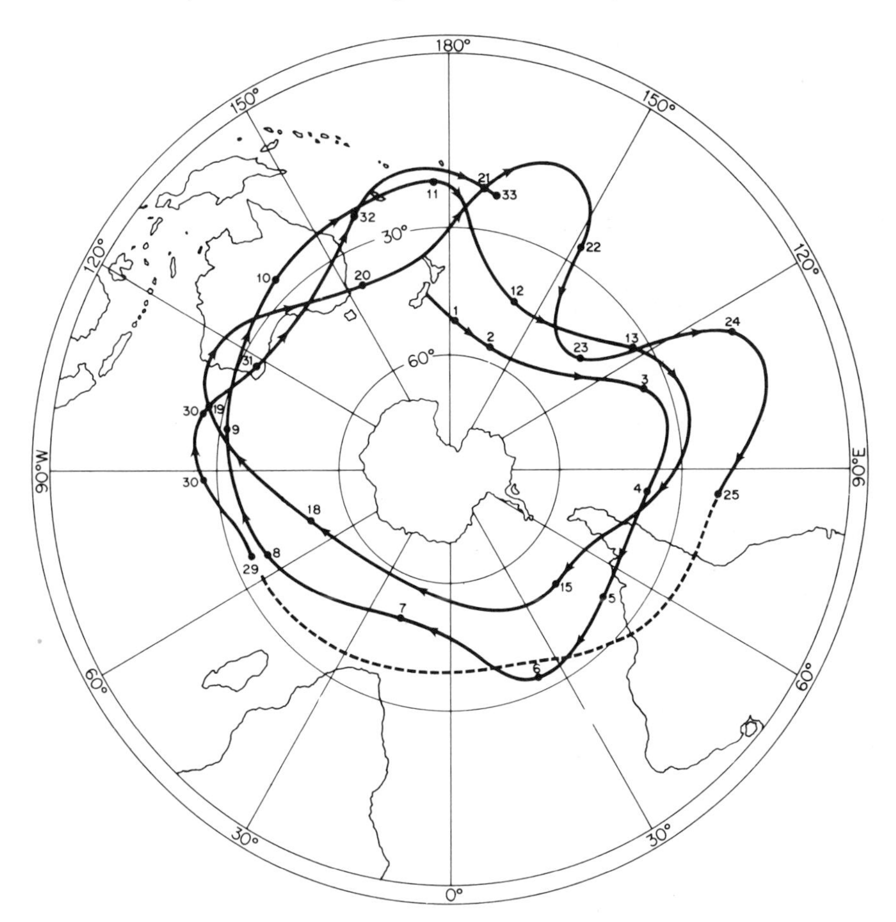

Figure 4.20 Path and daily positions of a constant-level balloon in the Southern Hemisphere upper westerlies. Launched at Christchurch, New Zealand, 30 March 1966, the *GHOST* balloon drifted at an altitude of about 12 km in upper-level waves for 33 days. Compare with Figure 4.19. (NCAR [National Center for Atmospheric Research], Boulder, Colorado.)

riers. Major waves tend to persist in the westerlies above the Rockies, the Andes, and the plateaus of Central Asia and South Africa. Long waves often remain stationary for considerable periods, although winds within the wave pattern move at great speeds.

Winds in upper-level tropospheric waves reach maximum speeds in the *jet streams,* narrow bands of high-velocity winds that follow the wave path near the tropopause at elevations of 8 to 15 km. The *polar front jet stream* achieves its maximum force and extent in winter, when there may be two or even three distinct currents having wind speeds of 100 knots or more at their cores. In the Northern Hemisphere the main jet stream system undulates from north to south as part of the upper-level westerly wave, often swerving far equatorward over the continents. Sometimes it divides into apparently discontinuous segments. Its position relative to the polar front suggests that it may be a guiding mechanism for cyclonic storms as well as a means of transporting great quantities of air across the continents (see Fig. 4.21). In summer the polar front jet has a mean position at higher latitudes, and its velocity and extent are reduced. A similar jet stream system in comparable latitudes of the Southern Hemisphere crosses South America, Australia, and New Zealand.

Above the subtropic highs a high-speed westerly flow, the *subtropical jet stream,* persists through most of the year. Its wind speeds commonly exceed 100 knots; maximums of more than 300 knots have been recorded in both hemispheres, for example, over Japan and the South Indian Ocean.

Near the equator upper-level wave motion is not so well developed as at higher latitudes owing to a much smaller Coriolis effect and a relatively steady flow of air that does not push vigorously across parallels of latitude. Nevertheless, in the

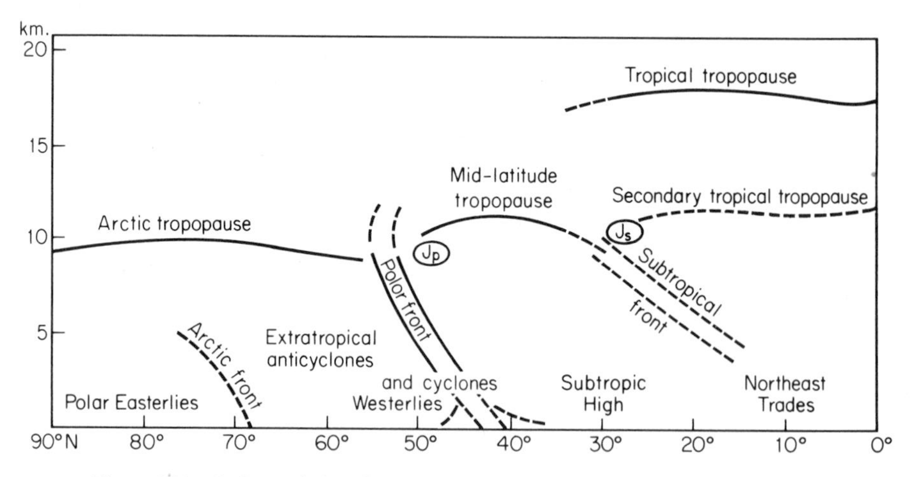

Figure 4.21 Surface wind and pressure systems in relation to the fronts and jet streams of the Northern Hemisphere. Dashed lines represent zones of major air mass exchange. (After Defant and Taba, Newton and Persson, and Palmén.)

tropical stratosphere an *easterly jet stream* develops south of Asia during the Northern Hemisphere summer. Although its direction is steady, it varies in speed, averaging about 70 knots. The tropical easterly jet is thought to be associated with thermal conditions arising from summer heating of the Asian continent. Also known as the Krakatoa easterlies, it probably carried volcanic dust westward after the eruption of Krakatoa in 1883.

Another stratospheric jet stream is the *polar night jet.* During winter the sun does not heat a cone of atmosphere at the pole, and the resulting strong temperature gradient in the stratosphere creates westerly winds at heights averaging 60 km. Rocket observations have recorded speeds exceeding 330 knots at a height of nearly 80 km in the polar night jet stream above the arctic.

SEASONAL CHANGES IN THE GENERAL CIRCULATION

The global patterns of temperature, pressure, and circulation migrate seasonally along the meridians. Figure 4.22 shows contrasting seasonal positions of the intertropical convergence. The shift is basically the result of seasonal changes in the heat

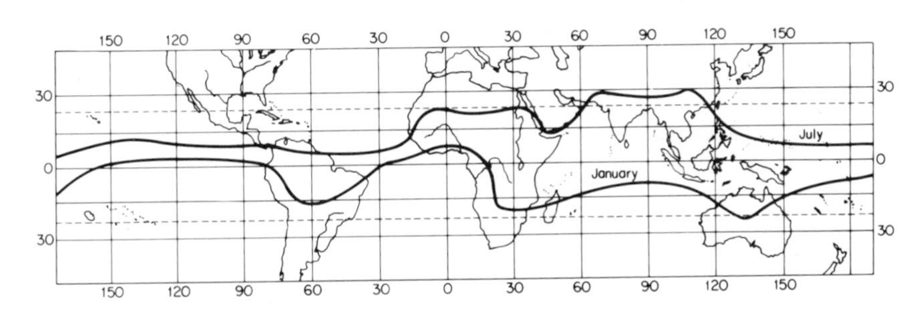

Figure 4.22 Mean positions of the intertropical convergence in January and July.

budget at different latitudes. Not only is there a continual change in the declination of the sun, but continents and oceans also assume different roles in response to insolation. The tendency is for motion systems to shift northward a few degrees in the Northern Hemisphere summer and southward in winter. It should be emphasized, therefore, that the distribution of wind and pressure systems in the general circulation, outlined earlier in this chapter, represents idealized pressure distribution and prevailing air movements that are modified by actual conditions. It has been previously pointed out that pressures tend to be higher in winter over land and lower over adjacent waters. This is most evident in the middle latitudes. The result may be creation of complete reversals of pressure gradients over the continents with corresponding seasonal shifts in wind direction. The mean flow patterns of surface winds in January and July are shown in Figs. 4.23 and 4.24.

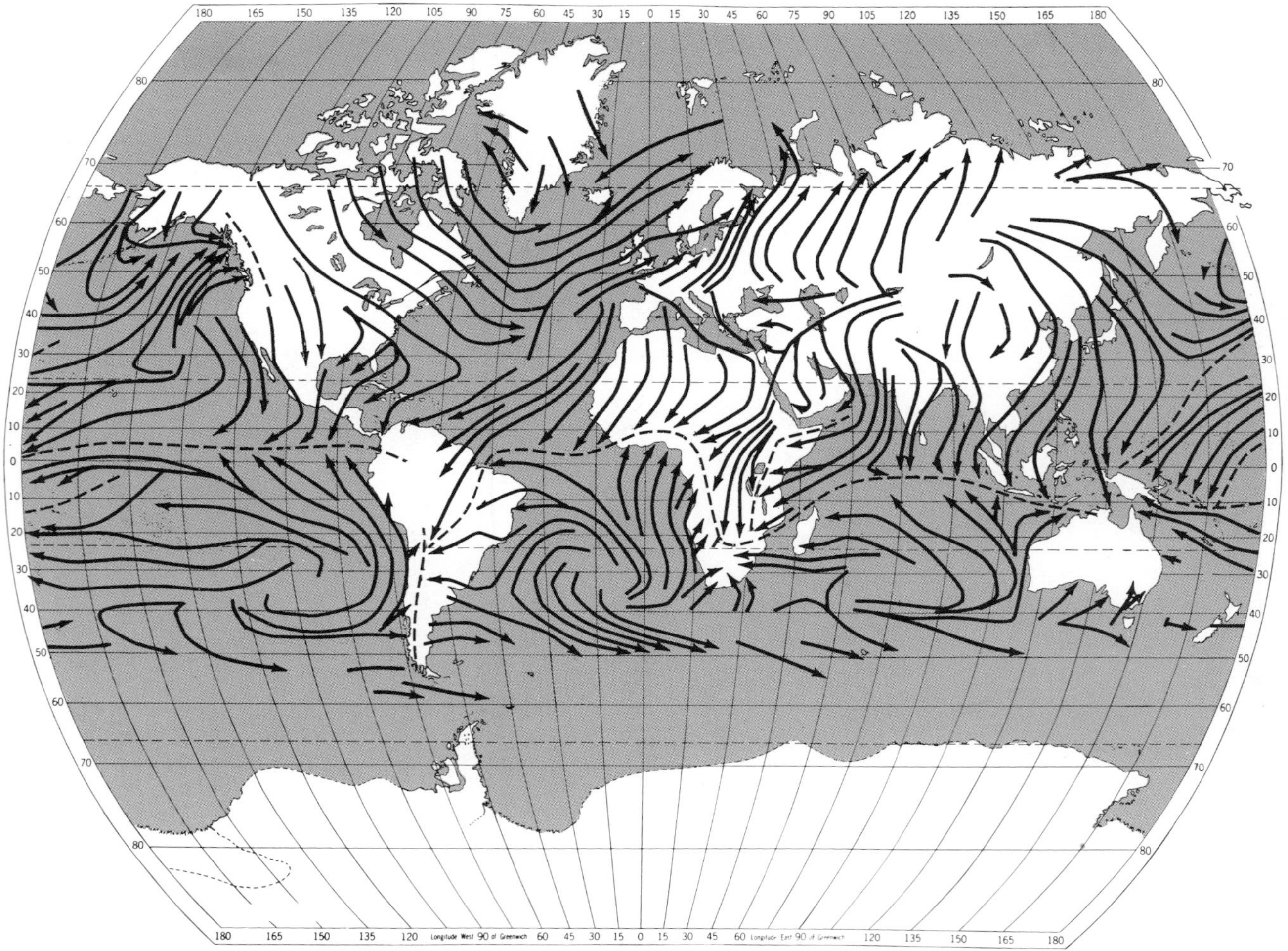

Figure 4.23 Mean flow patterns of surface winds in January. (Modified Van der Grinten projection.)

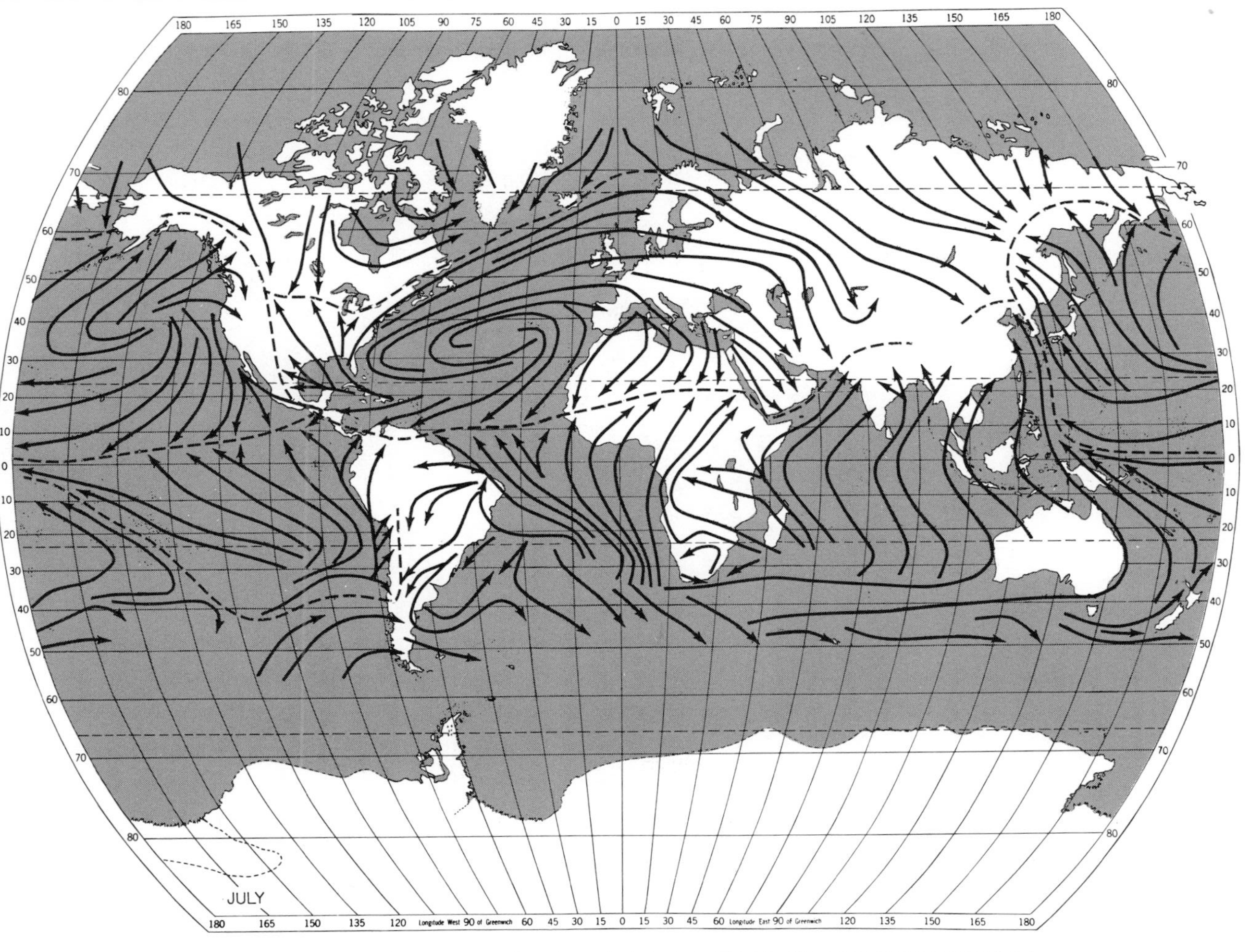

Figure 4.24 Mean flow patterns of surface winds in July. (Modified Van der Grinten projection.)

THE MONSOONS

In several parts of the world seasonally prevailing winds known as *monsoons* blow from approximately opposite directions in summer and winter. These wind systems are best developed in India and adjacent southeastern Asia, where there is a persistent flow from the Indian Ocean onto the land in the summer half of the year and where winds blow out from the continent in winter. This reversing circulation has been ascribed traditionally to the heating and cooling of the continent with the changes of season and resulting changes in the distribution of pressure. Increased understanding of upper-air flow in recent years has led to a modification of this simple explanation. In winter the high pressure over the Asian continent forces the westerly jet stream southward so that a portion of it lies south of the Himalayas. Subsiding air along the tropical margins of the jet moves toward the Indian Ocean, blocking incursions of maritime tropical air and producing the dry winter monsoon. In summer the jet stream shifts with the general circulation and lies north of the Himalayas, allowing a series of low-pressure disturbances to bring moist tropical air of the wet summer monsoon onto the continent. Northern Australia experiences a similar, though less well developed, monsoonal circulation which is related to seasonal migration of components of the general circulation.

Along the east coast of Asia and across Central Africa there are other monsoonal effects that merge into the shifting pattern of the general circulation (see Fig. 4.25). Outbursts of polar air in the westerlies result in a predominantly westerly flow off northeastern Asia in winter, but in summer maritime air moves inland

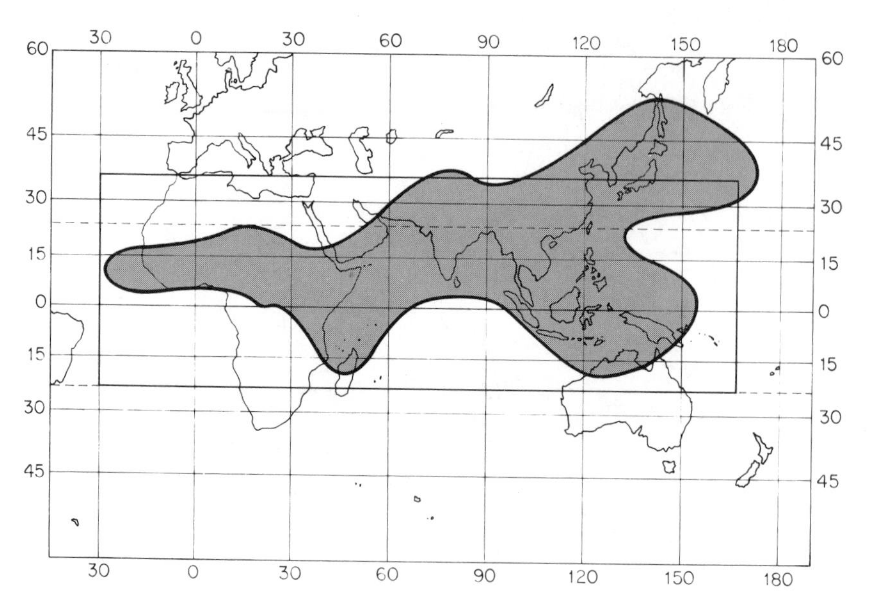

Figure 4.25 The monsoon region. (Shaded area after Khromov; rectangle after Ramage.)

from the subtropic highs that lie over the adjacent Pacific Ocean. In Africa the alternating flow arises in part from seasonal changes in heating and pressure over the deserts of both hemispheres. There is a marked summer maximum of rainfall along the Guinea Coast, where humid air moves onshore toward thermally induced low pressure over the continent. In East Africa, however, dry air dominates both phases of the monsoon, and most of the rainfall occurs during transitional periods when the ITCZ passes over the region.

Delays in the onset, or "burst," of the wet monsoon have disastrous consequences for agriculture and water supplies in the monsoon region, especially in southern Asia. For social and economic, as well as purely scientific, reasons meteorologists continue to seek the physical causes of monsoons and attempt their prediction. One of the probable causes of a delayed summer monsoon over northern India is heavy winter snow on the Tibetan Plateau, where the resulting increase in surface albedo retards spring warming and normal development of the required circulation pattern.

OCEANIC CIRCULATION

Although the atmosphere is the chief agent for transfer of energy from the tropics to higher latitudes, the oceans assist this function on a global scale (Fig. 4.26) and thereby influence world climates. In addition, energy is exchanged between ocean and atmosphere by transfer of sensible and latent heat and mechanical action along the ocean–air boundary. Oceanic circulation is forced mainly by density differences due to variations in water temperature and salinity and by winds. The direction of ocean currents is modified by the Coriolis effect and the configuration of ocean basins and shorelines as well as by the initial driving forces. Large-scale, slow movements of ocean water usually are known as *drifts* rather than currents. They are best developed in the belt of prevailing westerlies in the Southern Hemisphere (Fig. 4.27). The map of world ocean currents reveals the close correlation between general atmospheric circulation and oceanic circulation in the middle and low latitudes. In the Northern Hemisphere the circulation is clockwise; in the Southern Hemisphere it is counterclockwise. Ocean currents that are strongly affected by surface winds respond to seasonal changes in the general circulation. For example, the Somali Current, off the Horn of Africa, is a strong flow from the southwest during the summer monsoon, but in autumn it reverses as the northeast trades return to the area.

The net effect of most surface oceanic circulation is to carry cold water toward the equator along the eastern margins of the oceans and warm water toward the poles along the western margins. At high latitudes the pattern of flow is disrupted by the arrangement of land masses. Cold water subsides in some of these areas and moves toward the equator at great depths.

Along some coasts prevailing winds skim away the surface water to produce *upwelling.* When the relatively warmer surface water is removed by wind action the

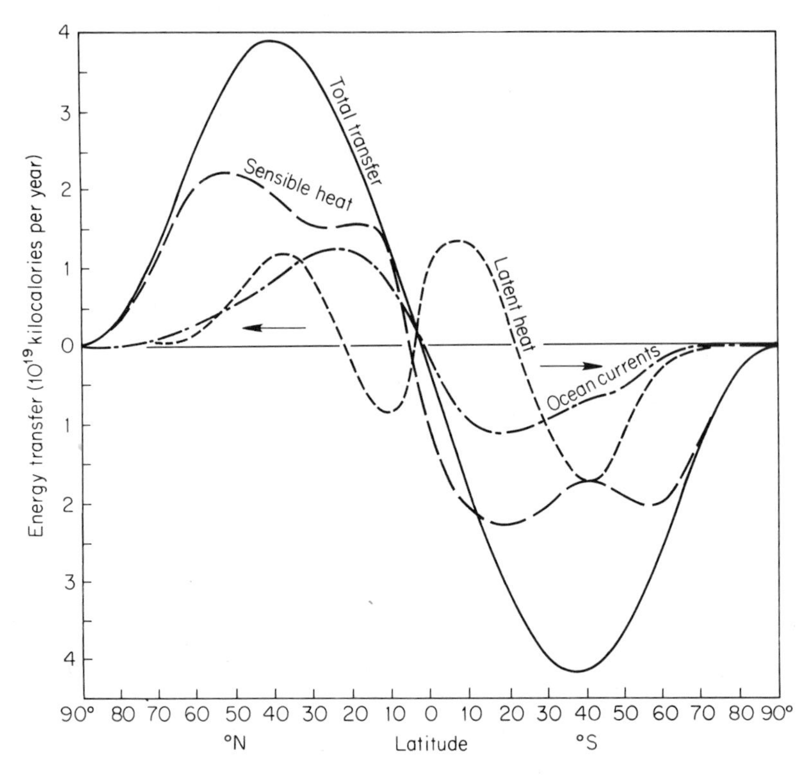

Figure 4.26 Components of the mean annual poleward energy transfer. (After Sellers.)

cooler water from below replaces it. Upwelling is an especially important influence on temperatures along the coasts of southern California, Peru and northern Chile, northwest Africa, and southwest Africa. In each of these areas the Coriolis effect turns the ocean current away from the shore. Since the surface water has a greater speed than that at greater depths it is deflected at a greater rate, enhancing upwelling of the colder subsurface water. Advection fog frequently forms as moist air moves across the cold surface. Where the prevailing winds change direction during the year, as in the monsoon region, upwelling may be a seasonal phenomenon. This is the case off the Somali Coast of Africa, where the sea temperatures are markedly lower during the summer (southwesterly) monsoon than in winter.

Tidal action in the oceans (and inland water bodies) is fundamentally a gravitational effect, but tide levels are modified by atmospheric processes. The passage of low-pressure centers causes a rise in water level; higher pressure causes a fall. Superimposed on normal tides these effects produce a *surge* that can be disastrous in shallow coastal waters. Strong onshore winds also cause surges and coastal flooding, especially when they accompany high tides. Steep pressure gradients and high-velocity surface winds associated with local storms can generate surges in lakes and inland seas, where the normal tidal range is small.

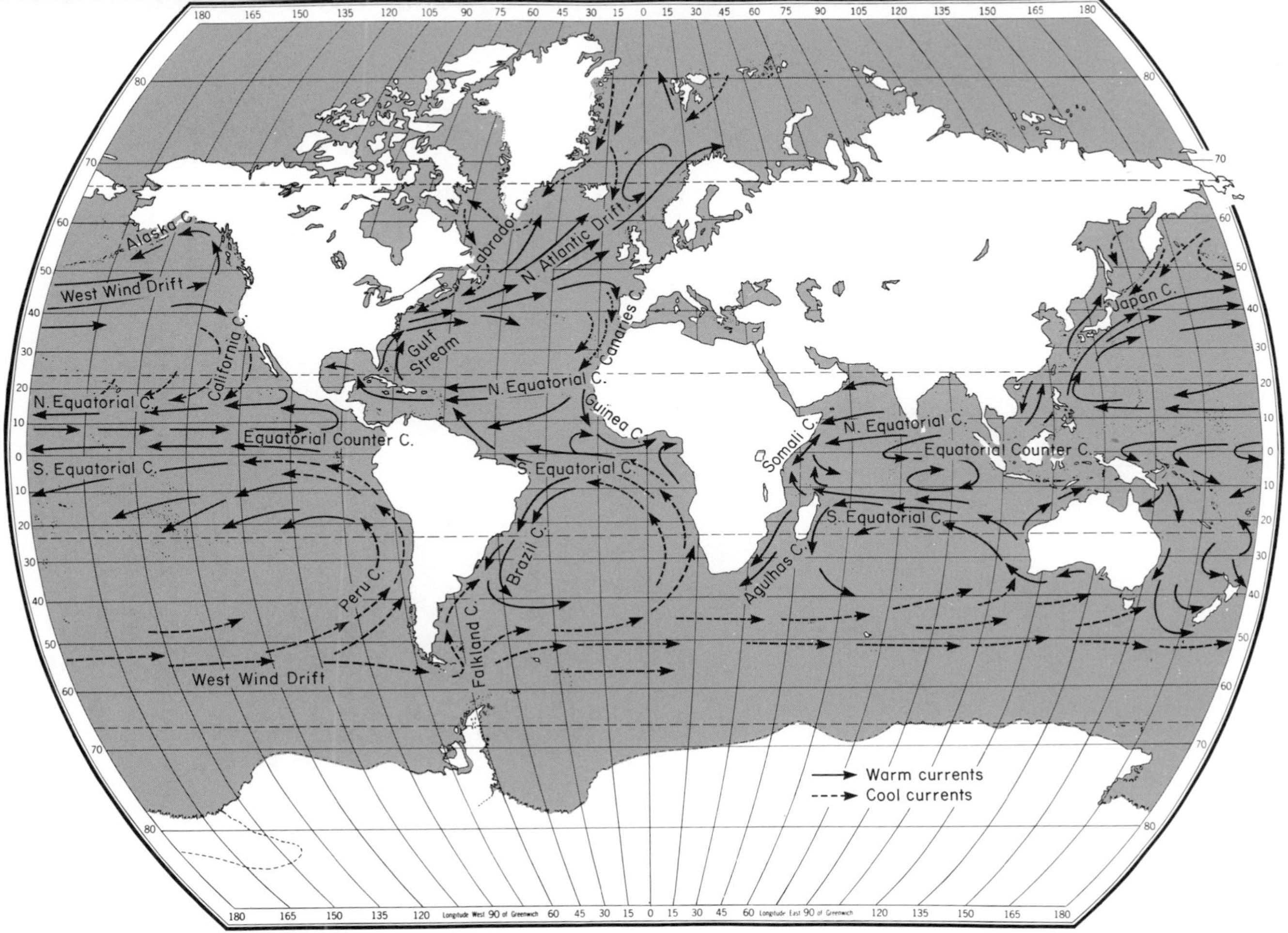

Figure 4.27 Major surface currents of the oceans. (Modified Van der Grinten projection.)

QUESTIONS AND PROBLEMS FOR CHAPTER 4

1. What causes horizontal variations of atmospheric pressure?
2. Diagram the forces that produce (a) geostrophic, (b) gradient, and (c) surface winds.
3. Why do wind speeds tend to be greater by day in valleys but greater at night on mountain tops?
4. Draw isobars and wind arrows to represent the pressure distribution and flow pattern around a "low" in the Southern Hemisphere.
5. Assuming that you have no instruments, how would you estimate local wind speed and direction at the surface? In the upper air?
6. Draw and label a cross section to illustrate the essential features of mountain and valley breezes.
7. In what respects does the schematic model of general atmospheric circulation (Fig. 4.18) depart from reality?
8. Why does the jet stream in the upper westerlies follow an undulating path?
9. Compare and contrast the Asian monsoon circulation with land and sea breezes.
10. Summarize the processes that initiate ocean currents and influence their speed and direction.
11. What causes upwelling of ocean water? How does upwelling influence climate?
12. What conditions are necessary for the generation of katabatic winds?

5

Weather Disturbances

Traveling along as part of the large-scale motion systems of the general circulation are the air masses, storms, and lesser systems that bring daily weather and, by their cumulative effect, produce different climates. Most weather changes in the middle and high latitudes accompany the advances and interaction of air masses and the processes within the air masses themselves; in the tropics weather often is associated with mesoscale waves, cyclones, or local squall lines. Mass transport of moisture from the oceans accounts for much of the water that later precipitates over the continents, and air in motion is largely responsible for the transfer of sensible and latent heat from one region to another. An understanding of the climate system depends on dynamic aspects of weather disturbances as well as the patterns of individual climatic elements.

PROPERTIES OF AIR MASSES

An *air mass* is an extensive portion of the atmosphere having characteristics of temperature and moisture which are relatively homogeneous horizontally. For a large body of air to acquire temperature and moisture properties that are approximately the same at a given level, that air must rest for a time over a *source region,* which must itself have fairly homogeneous surface conditions of temperature and moisture availability. A large land or water area which has evenly distributed insolation affords a good source region, but a second prerequisite for a distinctive air mass is large-scale transfer of heat and moisture between the surface and the overlying air. Air that resides over a homogeneous source region gradually becomes homogeneous itself and tends to retain its acquired characteristics when it moves away. The heat and moisture properties of the air mass begin to change, however, as it moves over other surface conditions. Zones of convergence and rising air are inimical to the production of air masses because the general movement of winds is toward these areas at surface levels, bringing a constant renewal of air with heterogeneous temperature and humidity properties. In contrast, the subsiding air in an anticyclone is more favorable to air mass formation.

AIR MASS IDENTIFICATION AND ANALYSIS

Once an air mass has left its source region, it undergoes changes which often make it difficult to identify. For example, if it is formed over a cold surface and subsequently passes over a warm ocean its temperature and moisture content increase. Local surface conditions created by ocean currents, land relief, minor water bodies, or nighttime radiation can produce quite different values of temperature and humidity at the bottom of an air mass. Consequently, it is necessary to analyze conditions in the upper air and to understand the processes that accompany the change in air mass properties.

Upper-air observations to provide information on the vertical distribution of meteorological elements are carried out by means of the *radiosonde* (Fig. 5.1). For the sake of convenience, radiosonde observations are called *raobs.* The radiosonde consists of a lightweight box fitted with a radio transmitter and sensing devices for pressure, temperature, and relative humidity. Although there are variations in the construction of radiosondes, they commonly employ a sylphon cell to indicate pressure, a thermistor for temperature, and a carbon-coated hygristor for relative humidity. The instrument package rides aloft under a gas-filled balloon and descends on a parachute after the balloon bursts at an altitude of about 25 to 30 km. At a ground station the signals are received, recorded, and translated into usable data. Analysis of raobs from a series of stations enables the forecaster to determine the properties of air masses that prevail over a region. At some stations the radiosonde is followed by radar to obtain wind direction and speed at various levels as in the case of rawin. The combined technique is known as *rawinsonde.*

Another type of airborne instrument, the *transosonde,* is a balloon-borne radiometeorological device whose altitude can be controlled by releasing ballast or gas from the balloon as it floats along at a constant pressure level for several days. Its advantage is that it can obtain data from positions over the oceans, where other sources of information are limited. Giant adaptations known as GHOST (Global Horizontal Sounding Technique) balloons have provided valuable data from upper levels, especially in the Southern Hemisphere westerlies (see Fig. 4.20).

Identification of air masses is based upon three kinds of information: (1) the history of change in the air since it left its source region; (2) horizontal characteristics at certain levels in the upper air; and (3) the vertical distribution of temperature, winds, and humidity. Because the number of raob reporting stations

Figure 5.1 Radiosonde in flight. (The Bendix Corporation, Environmental Science Division.)

is limited and because major changes in air masses may occur between the times of radiosonde observations, the historical approach alone is inadequate. Maps plotted with temperature, wind, and humidity values for one or more upper levels show a minimum of the influences due to local surface conditions and reveal moisture and energy components of the upper part of the air mass. They represent a horizontal "slice" through the atmosphere. Constant-level charts may be plotted either for a certain height or for a constant-pressure level, the latter being most often used by meteorological services (Fig. 5.2). Common pressure levels for constant-pressure analyses are 850, 700, 500, 300, and 100 mb.

Just as charts showing the horizontal distribution of weather elements in the upper atmosphere help to identify air masses and their boundaries, so vertical cross

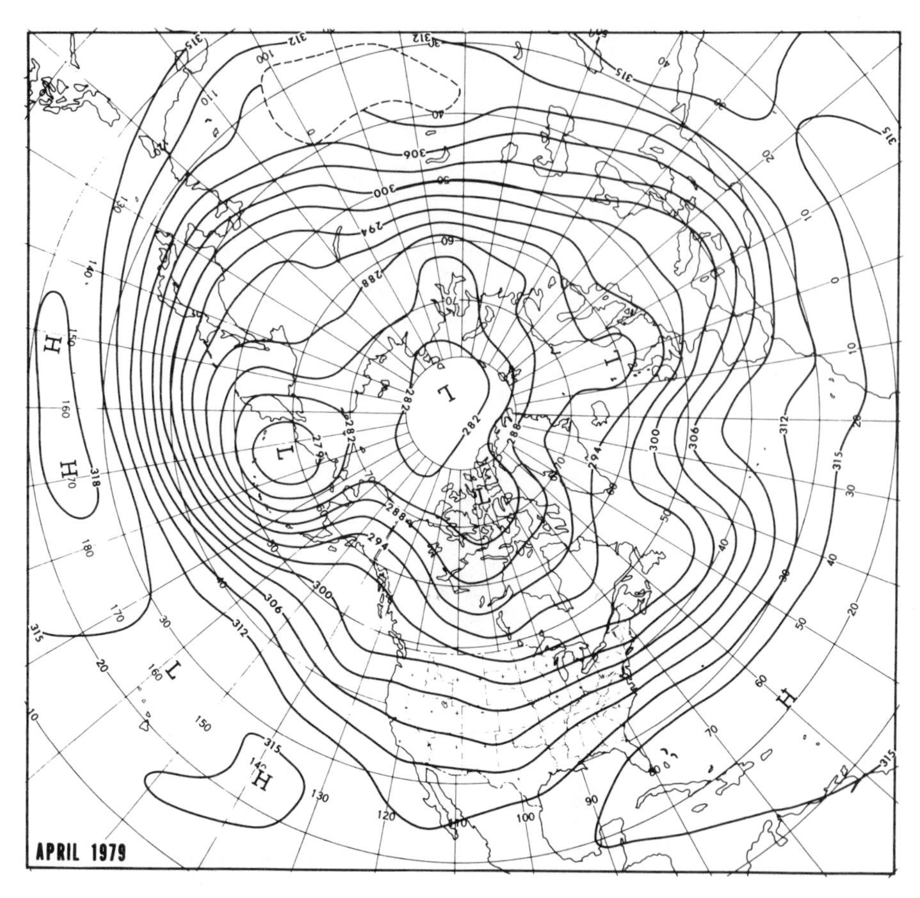

Figure 5.2 Constant pressure chart. Contours show mean height of the 700 mb level in dekameters. (From A. James Wagner, "Weather and Circulation of April 1979," *Monthly Weather Review,* Vol. 107, p. 948.)

sections based on raob and upper wind reports make it possible to analyze vertical differences in the air, especially those that affect vertical motion. Wind direction and velocity in particular are useful in judging air mass movements and in flight planning. Temperature and humidity values from the raob are analyzed as a basis for forecasting temperature changes, cloudiness, storms, and precipitation.

AIR MASS SOURCE REGIONS AND CLASSIFICATION

An air mass source region has already been defined as a large area with approximately homogeneous temperature and moisture properties, generally where there is subsidence and divergence of air. These conditions are best developed in the semipermanent high-pressure belts of the earth (Fig. 5.3). In the belt of low

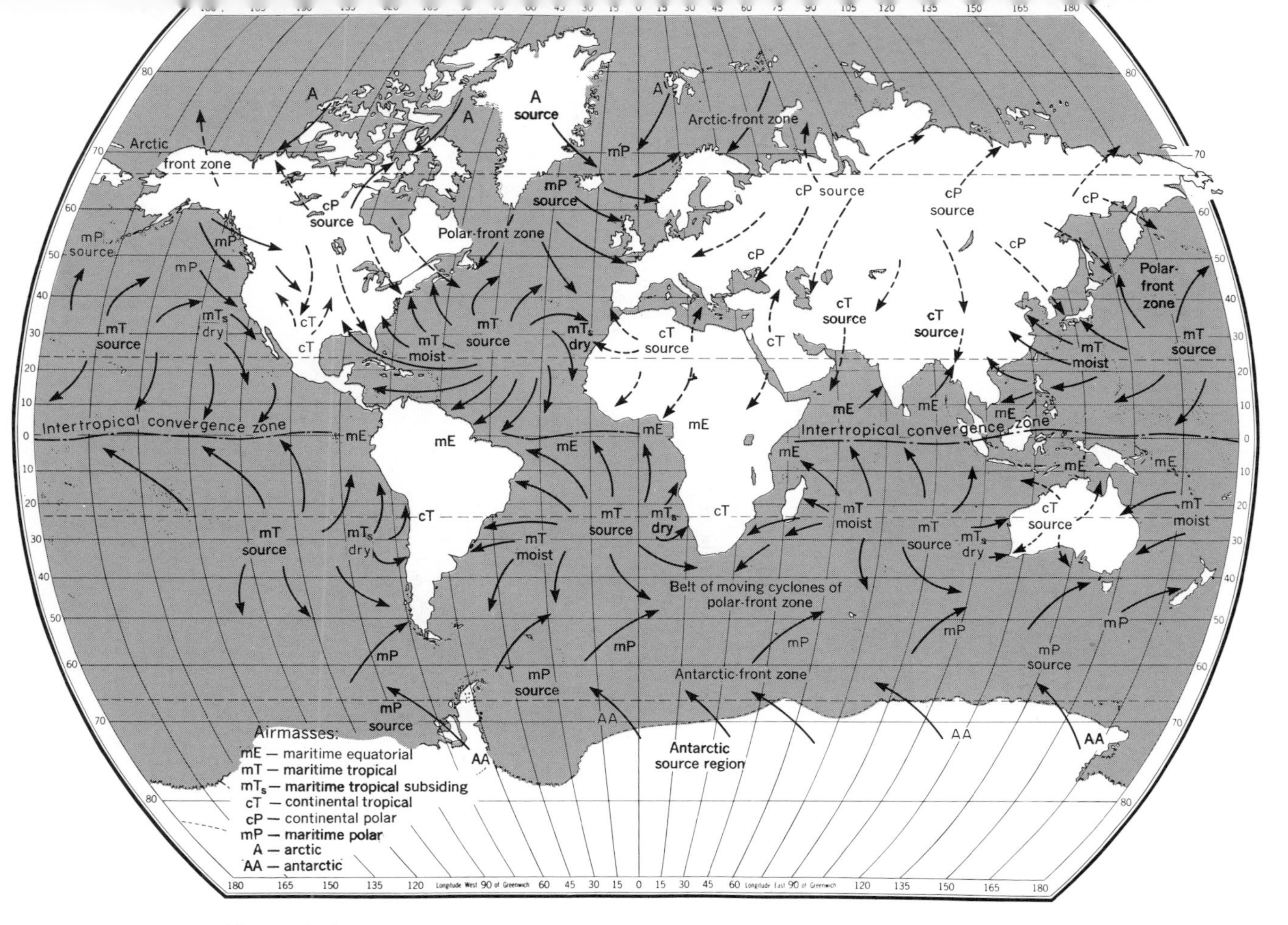

Figure 5.3 Principal air masses and source regions of the world. (Data after Strahler, courtesy John Wiley & Sons, Inc.)

pressure along the equator, however, the equatorial convergence may be weakly defined and stagnation of air will produce equatorial air masses. Where development of high pressure is seasonal, for example, over mid-latitude continental regions in winter, the source regions will likewise have seasonal maximum development.

Classification of air masses is based primarily upon their source regions and secondarily upon temperature and moisture properties. The two main categories are tropical or subtropical and polar or subpolar, because the great source regions are located at high and at low latitudes. Subdivision of these groups is made according to whether the source region is oceanic or continental, and further, according to what modifications the masses experience as they move from their source regions. Eventually, air masses become modified to such an extent that special designations are necessary. For climatological purposes this classification is of great importance because the extent to which air masses dominate different regions determines the climate of those regions.

In practice, letter symbols are used to designate air masses. Ordinarily *c* and *m* are used for *continental* and *maritime* and they are placed first in the designation. Following that the source region is indicated: *tropical* (*T*), *polar* (*P*), *equatorial* (*E*), *arctic* (*A*), and *antarctic* (*AA*). To indicate modifications of air masses due to transfer of heat between the bottom of the mass and the surface over which it passes, another symbol is appended: *k* (for the German *kalt*) for air colder than the underlying surface or *w* for air warmer than the surface. A generalized map of the principal world

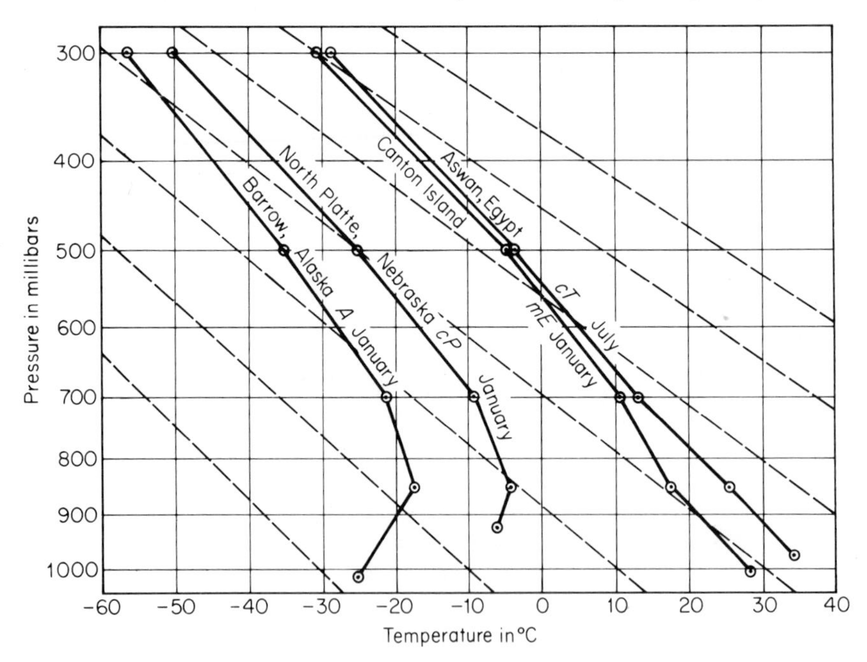

Figure 5.4 Mean vertical temperature soundings in contrasting air masses.

air masses is shown in Fig. 5.3. Note that arctic air masses have their sources north of polar masses. This incongruity is explained by the history of air mass study. The term "polar" had already come into wide use to designate air masses in subpolar regions before the distinctive character of "arctic" air had been discovered. Figure 5.4 shows examples of temperature lapse rates in different air masses.

STABILITY AND INSTABILITY

Vertical movements in an air mass are of great significance because nearly all precipitation is associated with adiabatic cooling and condensation in rising air. The external causes of lifting that leads to cloud formation have been discussed in Chapter 3. Let us now examine the conditions within large bodies of air that affect the buoyancy of smaller portions and their tendency to exhibit vertical motion. If a parcel of air tends to remain in its position, or to return to that position when displaced by an outside force, the larger air mass is said to be *stable*. If vertical displacement results in a tendency to further movement, the air mass is termed *unstable*. If conditions are such that a parcel neither resists nor aids vertical motion, the air mass is in a state of *neutral equilibrium*. In each of these cases it is assumed that the air of the parcel does not mix with its surroundings and that there are no compensating motions in the parent air mass.

An air column having colder, drier air in its surface layers than aloft is likely to be stable, and any process which cools the air at the bottom of an air mass tends to make it more stable. Radiational cooling at night, for example, produces a stable condition, that is, a temperature inversion. Advection of air over a cold surface will result in increasing stability, as when warm oceanic air passes over a cold ocean current or onto cold land. While these processes can produce condensation (fog or dew) in the lower layers of air, they do not induce vertical currents and therefore are not likely to produce precipitation. Subsidence of air also increases its stability. As the bottom of a layer of subsiding air is compressed and diverges, the top of the layer descends farther and is warmed more adiabatically than the base, thus producing a smaller difference in temperature between the top and base of the layer, that is, a decreased lapse rate. Upper-air inversions, which exhibit extremely stable lapse rates, are often caused by subsidence.

The vertical lapse rates of temperature and humidity determine the degree of stability of an air column (Fig. 5.5). If the actual temperature of the air increases with altitude, the colder, denser air below will tend to stay in place under the warmer, lighter air—*absolute stability*. If the actual temperature is exactly the same vertically through a layer of air, that layer is also stable. When an outside force causes a parcel of air to rise in a stable environment, the air cools at the dry adiabatic rate (10C° per 1,000 m) as long as condensation does not take place and mixing does not occur. At a higher altitude (lower pressure), it will have a temperature lower than the surrounding air and will descend if the original lifting force is withdrawn. This effect is best seen by the comparison of the lapse rate of stable air

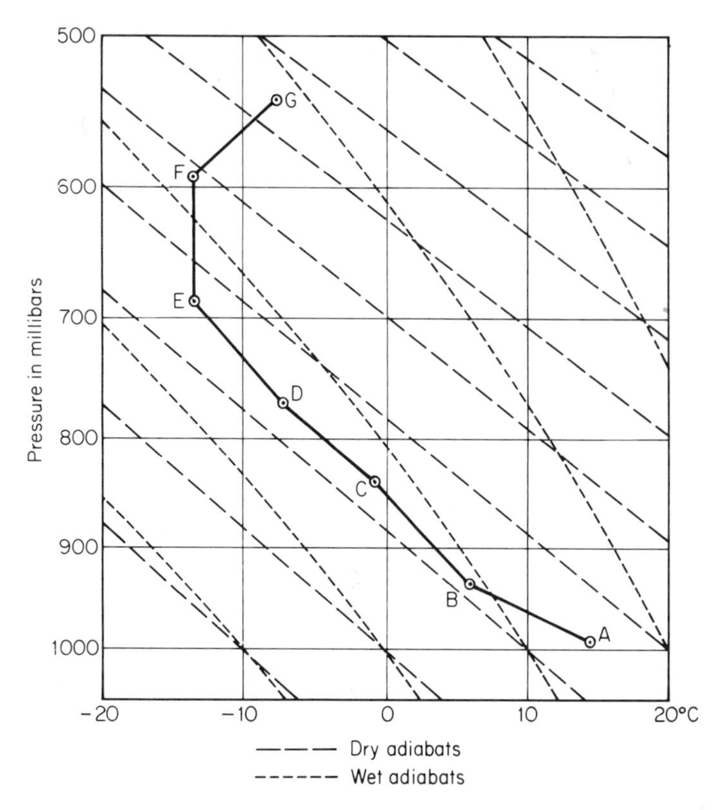

Figure 5.5 Hypothetical lapse rates of temperature showing different stability conditions. *AB*, superadiabatic, absolutely unstable; *BC*, conditionally unstable; *CD*, neutral equilibrium (unsaturated air); *DE*, neutral equilibrium (saturated air); *EF*, isothermal lapse rate, absolutely stable; *FG*, inversion, absolutely stable.

with the dry adiabatic rate (Fig. 5.5). As long as condensation does not take place, air is stable if its lapse rate is less than the dry adiabatic rate. Because of its tendency to stay at or to return to its position, stable air does not move readily over mountain barriers but flows along them and pours out through passes and gorges. The weather associated with stable air is generally fair, but visibility is often poor owing to ground fog or concentration of smoke and haze near the surface. In winter stable air masses may be extremely cold in middle and high latitudes.

Absolute instability in air occurs when its lapse rate is greater than the dry adiabatic rate of cooling. If air in an unstable layer is forced aloft, it will be warmer than the surrounding air and will continue to rise. The initial force may result from heating at the earth's surface, from orographic lifting, or from convergence. As long as the actual lapse rate of temperature is greater than the dry adiabatic rate, the air will continue to rise and cool adiabatically. When its saturation temperature is reached, condensation produces clouds, and precipitation may result. Warm, humid air often is unstable, especially when it lies over a warm surface. Any column of air may become unstable if its lower layers are warmed or their moisture content is increased. Cooling (or introduction of dry air) at upper levels also can induce instability. Thunderstorms, tornadoes, and hurricanes are associated with intensely unstable conditions that can be developed by a combination of these processes.

It frequently happens that air is stable while it rises and cools at the dry rate (that is, without condensation) but becomes unstable when condensation begins. Consider a parcel of mildly stable air being forced aloft. It cools at the dry adiabatic rate until it reaches saturation (see Fig. 5.5). The release of the latent heat of condensation gives added heat to produce further lifting and the air then cools at the slower wet rate (6C° per 1,000 m). Strong convective activity and heavy precipitation is the probable result. Humid air which is initially stable but which becomes unstable when condensation takes place within it is termed conditionally unstable. *Conditional instability* can be identified by the fact that the actual lapse rate in the air is less than the dry rate and greater than the wet rate. That is, it is stable with respect to the dry rate but unstable with respect to the wet rate. Instability is thus "conditional" upon the presence of considerable water vapor in the air.

A state of *neutral equilibrium* exists when the actual lapse rate in dry air equals the dry adiabatic rate. Air rising and cooling at the dry rate has the same temperature as the surrounding air at every level. In saturated air, neutral equilibrium is achieved when the lapse rate equals the wet adiabatic rate. Again vertical displacement results in the same temperature as that of the surrounding air. In both cases the conditions within the air produce a tendency neither to sink nor to rise.

In view of their importance in connection with vertical motion in air masses, stability and instability are significant properties for air mass analysis and forecasting. By studying the temperature and humidity values for different levels, the meteorologist can ascertain the degree of stability and the associated potential weather of an air mass. The lapse rate as plotted from raob data is a profile of the upper air; the vertical temperature and humidity gradients in various layers of air are fundamental clues to stability conditions.

THE EXTRATROPICAL CYCLONE

The general atmospheric circulation favors divergence and air mass development in subtropical and polar regions, but it is characterized by convergence of dissimilar air masses in the mid-latitude zone of cyclonic storms. During the first quarter of the twentieth century the Norwegian meteorologist Vilhelm Bjerknes and his son Jakob developed a *polar front theory* of mid-latitude cyclone formation that has proved useful for both weather forecasting and dynamic explanations of climate.

When air masses having different temperature and moisture properties meet, they do not mix readily but maintain a boundary surface of discontinuity for some time as the warmer, less dense air is forced aloft over the colder mass. The sloping boundary surface between contrasting air masses is called a *front*. It is indicated on a weather map by a line which represents the intersection of the frontal boundary with the earth's surface, but it is three-dimensional. It extends vertically as well as horizontally and has a typical thickness of 50 to 100 km. The frontal zone always slopes upward over the colder, denser air, and it exhibits abrupt temperature discontinuities throughout its vertical extent (Fig. 5.6). Discovery of these and other vertical properties of air led to the *baroclinic wave theory* of cyclones following World War II. A modification of the polar front theory, it has provided more satisfactory explanations of wave formation and is more amenable to mathematical treatment in numerical forecasting. As indicated in Fig. 5.6, the constant-density surfaces, represented by isotherms, are not parallel to the constant-pressure surfaces (isobars). That is, temperature differs along isobaric surfaces. This condition, known as a baroclinic atmosphere, or baroclinity, tends to cause an increase in both the pressure gradient from cold toward warm air and the resulting wind speeds as altitude increases. Where rising air is replaced by a converging flow, cyclonic vorticity develops in response to the earth's rotation and a cyclonic storm may form. Retaining their angular momentum, winds accelerate as they approach the axis of spin.

Between the prevailing westerlies and the polar easterlies lies a more or less

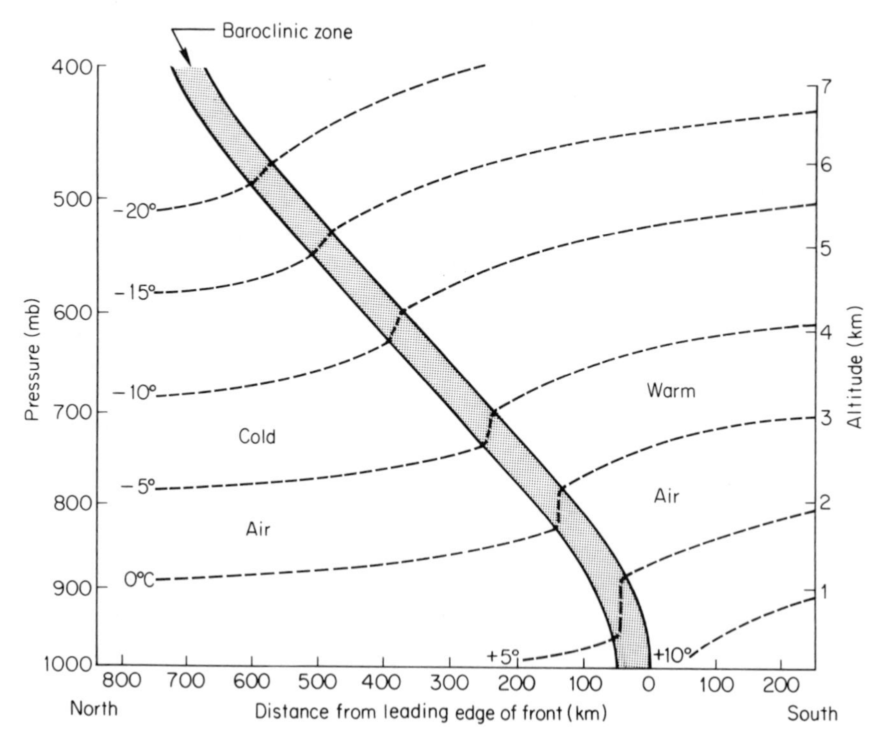

Figure 5.6 Vertical cross section through a baroclinic (frontal) zone in the Northern Hemisphere.

permanent, undulating baroclinic zone known as the polar front, along which most extratropical (mid-latitude) cyclones originate. In much the same way that whirlpools are generated between adjacent currents of water moving at different velocities, giant waves or whirls form along the polar front (Fig. 5.7). Other waves

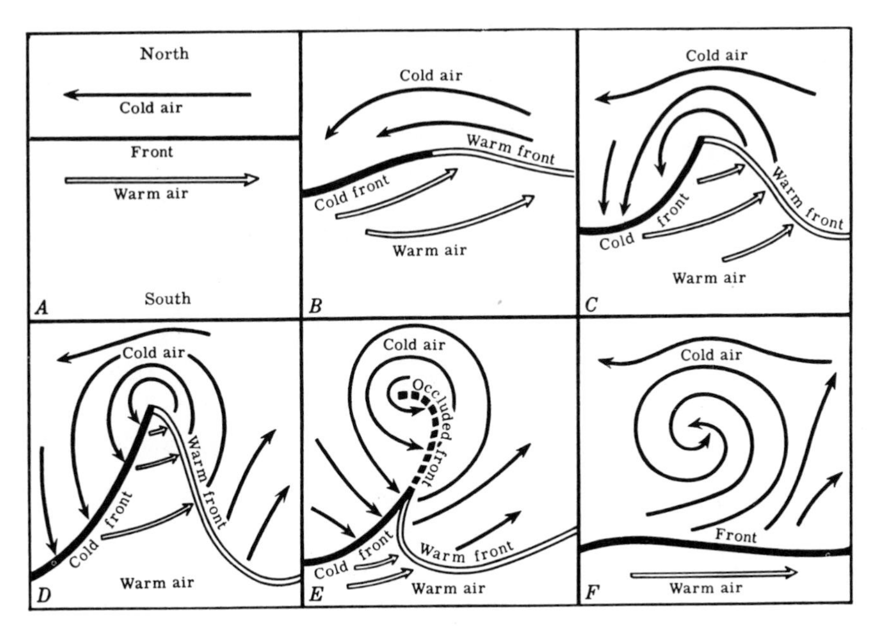

Figure 5.7 Plan view of stages in the development and occlusion of an extratropical cyclone along the polar front in the Northern Hemisphere. *A*, initial stage; *B*, beginning of cyclonic circulation; *C*, warm sector well defined between fronts; *D*, cold front overtaking warm front; *E*, occlusion; *F*, dissipation. (NOAA)

at higher latitudes may develop in the arctic front between arctic and polar air. A cyclone family sometimes results from a succession of waves along one of these quasi-stationary fronts (Fig. 5.8).

The cold air mass in the typical Northern Hemisphere cyclone pushes southward under the warmer air, which advances northward. As the cyclonic wave increases in amplitude the pressure gradient is directed toward the low center and the pattern of air flow takes on a counterclockwise, or cyclonic, circulation with surface winds crossing isobars at angles of about 20° to 40°. Low pressure is accompanied by convergence and rising air at the center and along the front; for this reason cyclones often are called "lows" or "depressions." Note that the Coriolis effect deflects winds to the right of the pressure gradient as they approach the low center (Fig. 5.9). (In the Southern Hemisphere cyclonic circulation is clockwise and toward the low-pressure center.) The "tongue" of warm air advancing from equatorward is known as the *warm sector.* Where the advancing warm air mass replaces colder air at a given level, the boundary is a *warm front.* Windward of the warm sector the leading edge of the cold air is the *cold front.* If a front neither ad-

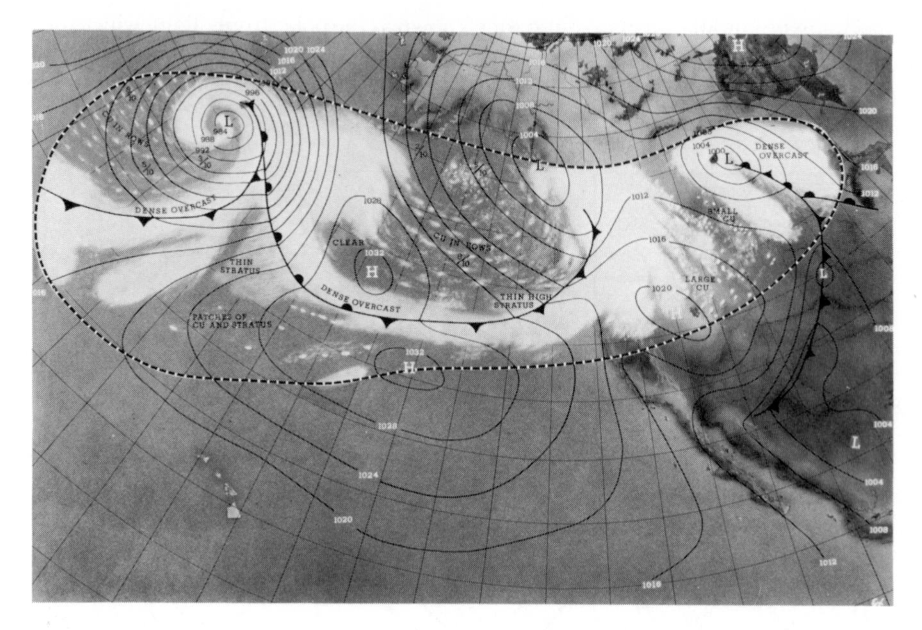

Figure 5.8 Satellite view and map analysis of a cyclone family over the North Pacific Ocean and western North America. (Vincent J. Oliver, NOAA, National Weather Service)

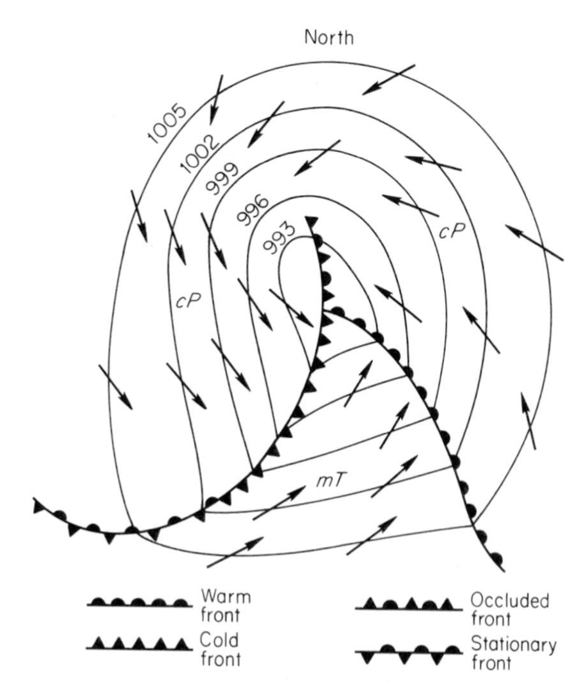

Figure 5.9 Model of a mature extratropical cyclone in the Northern Hemisphere.

vances nor retreats across the earth's surface, it is a *stationary front,* and it may become a warm or a cold front, depending on the direction of subsequent movement.

Extratropical cyclones vary in diameter from 200 to as much as 3,000 km. Most are in the range of 800 to 1,500 km. Their form may be roughly circular or elongated and oval, or they may be broad, shallow, weak depressions. Their general direction of movement is from west to east in the mid-latitude westerlies, but specific paths are often curved and sometimes erratic. The surface tracks of most cyclones closely follow the wind pattern at the 500-mb level, and they apparently are controlled by the jet streams. Over the North American continent they tend to curve toward the southeast into the Mississippi Valley and then toward the northeast (Fig. 5.10). The speed of an extratropical cyclone also varies; the average is about 30 to 50 km per hr or from 800 to 1,100 km per day, about half that of the winds at the 500-mb pressure surface, and normally greater in winter than in summer. As long as the discontinuity in temperature and moisture is maintained along its fronts the cyclone is likely to persist, but a cyclonic pattern of pressure and winds is also essential. Some cyclones pass from central North America out across the Atlantic and enter western Europe; others dissolve because of modifications of their energy and moisture properties.

FRONTS

The warm front in a Northern Hemisphere extratropical cyclone normally extends east and southeast from the low center. A gradual clockwise (veering) wind shift occurs as the front passes. Maximum wind speeds, perhaps with gusts, are experienced slightly in advance of the front. Barometric pressure decreases gradually until the front passes and then tends to level off. The slope of the front is gradual; 300 km ahead of its intersection with the ground it may be only 1,500 or 2,000 m above the surface. The great areal expanse of rising air produces a vast cloud system with the highest clouds lying far in advance of the front. Since the vertical movement along most of the warm-front boundary is not violent the types of clouds are predominantly stratiform. High wispy cirrus, often in the form of "mares' tails," appear first. As the front approaches the clouds lower and thicken progressively to cirrostratus, altostratus, and nimbostratus (see Fig. 5.11).

Precipitation from the warm front is light to moderately heavy and continuous, extending over a wide zone. Fogginess and poor visibility are common in the precipitation area. Typical warm front precipitation occasionally is interspersed with heavy showers where convection takes place in unstable warm air as it rises rapidly over the cold air just ahead of the front. In winter, ice pellets or freezing rain may occur when rain falls from the warm air through the cold mass below. The succession of weather can include snow, ice pellets, and then rain as the front approaches, clouds lower, and temperature rises. Sometimes the temperature does not rise above freezing during the entire passage of a warm front. Nevertheless, an appreciable rise in surface temperature normally does accompany the passage of

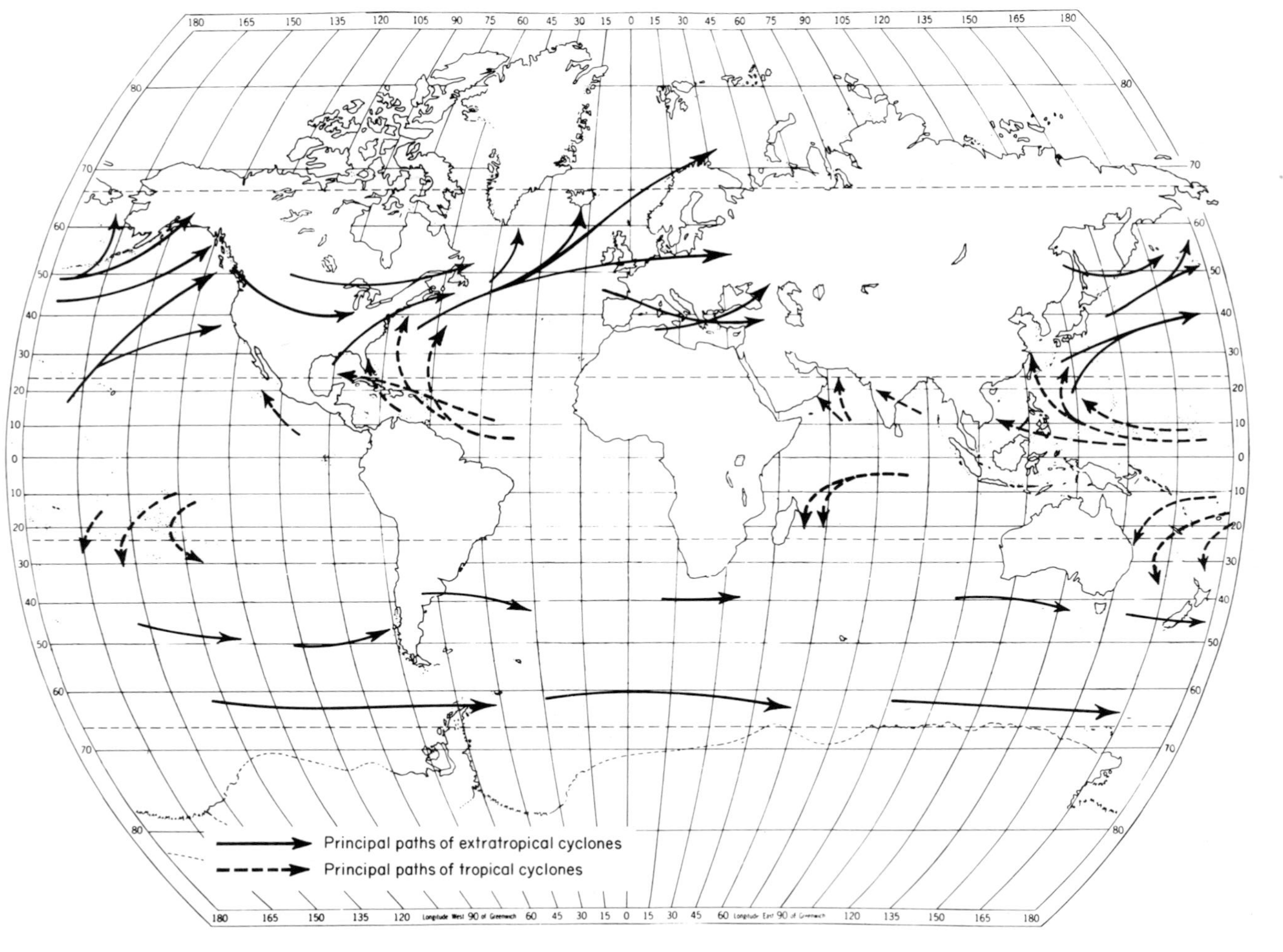

Figure 5.10 Common paths of cyclones. (After Petterssen.)

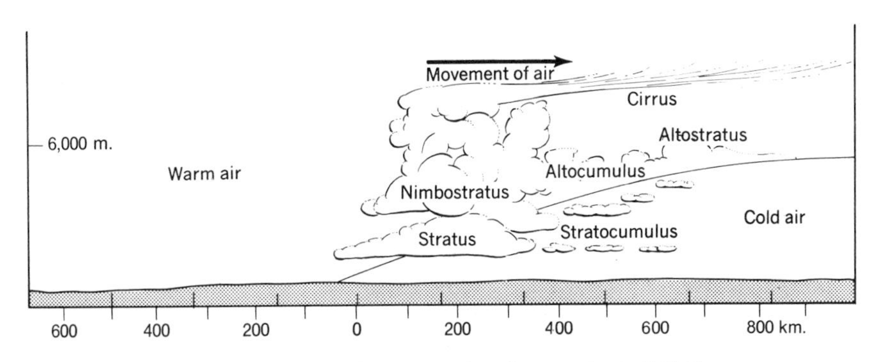

Figure 5.11 Vertical cross section of a warm front. (NOAA)

the front and the relatively warmer, moister air of the warm sector then begins to dominate the weather. The air which occupies the warm sector is likely to be conditionally unstable and therefore produce convective showers, especially in summer over land that is being heated.

The cold front is the leading edge of an intrusion of cold air into territory previously occupied by warmer air. It pushes under the warmer air after the fashion of a wedge, forcing the warm air to rise. Frictional drag along the ground retards the advance of the cold air and it develops a relatively steep forward surface in contrast to the gentler slope of the warm front (see Fig. 5.12). Hence it is accompanied by clouds with vertical development and heavy, showery precipitation. At times, cold air in the upper levels overruns the warm air to create extreme instability and overturning. Convective clouds and showers then precede the surface advance of the front.

The most significant identifying feature associated with the cold front, in addition to the temperature change, is the veering wind shift from a southerly or southwesterly direction to the northwest or north at its passage. Falling pressure heralds the approach of the cold front, followed by a sharp drop just as the front passes. This is the basis for the popular association of a falling barometer with a

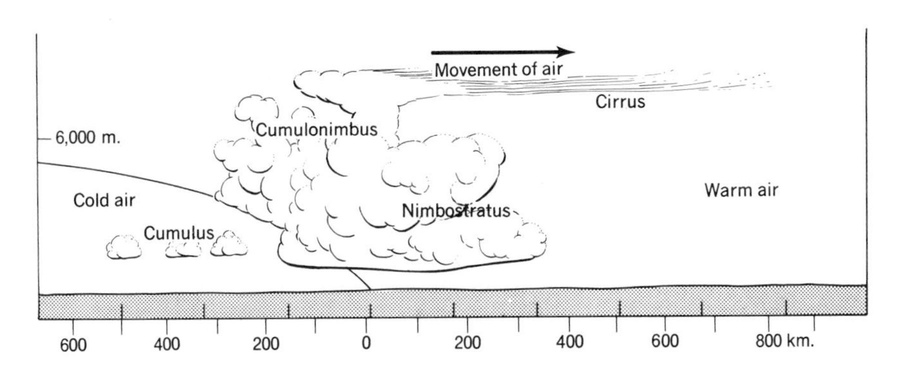

Figure 5.12 Vertical cross section of a cold front. (NOAA)

coming storm. As the front moves on, the barometer rises again. A drop in temperature ordinarily can be expected with the passage of a cold front, although surface temperatures are not so reliable as those in the upper air for detecting the transition from warm to cold air. Because of lower temperatures and its probable polar origin, the air behind the front has a lower moisture content. Therefore, humidity values also aid location of the front. The zone of precipitation usually is much wider than the discontinuities of wind, temperature, and humidity would seem to indicate, yet narrower than at the warm front. A few hours after the front has passed clearing skies normally can be expected. The following cold air mass moves over warmer ground which is probably also moist, so that sufficient instability may develop in its lower layers to produce scattered showers. If the cold air mass is unstable and moist, and especially if it is moving across mountainous terrain, intermittent cloudiness and showers may continue for some time. When the precipitation is snow the combination of low temperatures, strong winds, and blowing snow along and following the cold front is called a *blizzard* in North America. A similar storm in the Soviet Union is known as the *buran*.

In the cyclonic storm, the cold front usually advances faster than the warm front, eventually overtaking it. When the air mass behind the cold front comes into contact with the air in advance of the warm front, the air of the warm sector, being less dense, is pushed aloft. This process is called *occlusion,* and the front remaining at the surface is an *occluded front.* There are two ways in which an occlusion may form, depending upon the relative temperatures of the air masses within the cyclone. Over continents and east coasts of continents the air behind the cold front is normally colder than that ahead of the warm front because it has had a shorter path across relatively warm surfaces. As the occlusion develops, the cold front moves under the cool air to form a *cold-front occlusion* (Fig. 5.13). Prefrontal weather ahead of the cold-front occlusion is similar to that of the warm front. The lifting of conditionally unstable air in the warm sector may be the "trigger action" to produce thunderstorms. At the front there is typical cold-front activity, and after the occlusion has passed a marked improvement in conditions occurs. In their later stages,

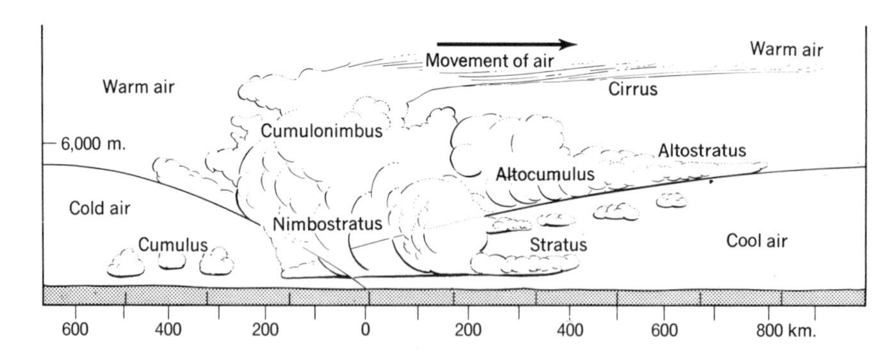

Figure 5.13 Vertical cross section of a cold-front occlusion. (NOAA)

cold-front occlusions often become well-defined cold fronts when they move over a cold continent. This results from the introduction of increasingly colder air behind the front and increasingly warmer air in the "cool" area ahead of the front.

Where the air behind the cold front is warmer than that ahead of the warm front, the former (cool) overruns the latter (cold) when occlusion occurs, producing a *warm-front occlusion* (Fig. 5.14). In winter along west coasts where cool air flowing onshore is usually warmer than the cold air over the land, warm-front occlusions are especially prevalent. In its initial stage, this type of occlusion shows most of the characteristics of the warm front. Then follows moderate frontal activity in connection with the *upper cold front.* As the occlusion process continues, the whole system tends to dissipate because there is a lack of sharp, persistent differences in air mass properties.

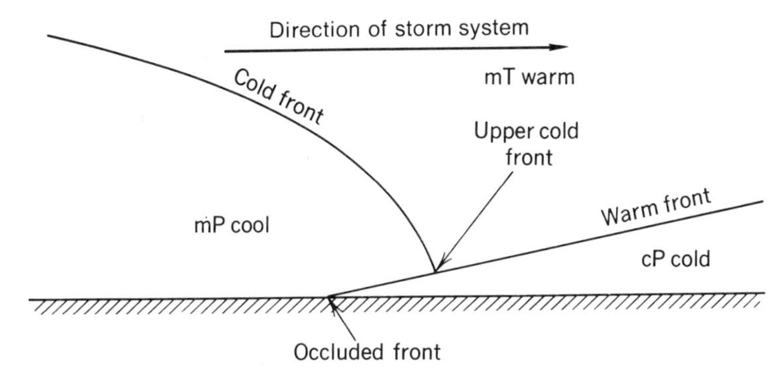

Figure 5.14 Vertical cross section of a warm-front occlusion.

MOUNTAIN BARRIERS AND SURFACE FRONTS

Mountain barriers lying across the path of a moving cyclone tend to induce occlusion by delaying the movement of the cold air in advance of the warm front and by promoting more rapid lifting of the air in the warm sector. The tendency to occlude depends on the relative stability of the air masses involved. Stable air is dammed more effectively than unstable air, which may become even more unstable when forced up mountain slopes. This phenomenon is common on the west coast of North America, where numerous cyclones occlude as they move against the mountain ranges. When, as is frequently the case, a warm-front occlusion is formed, the moist unstable air off the ocean moves over the mountain barrier and may continue for some distance to the leeward as an upper cold front, riding above the cold continental air mass which preceded the original warm front (see Fig. 5.15).

In addition to retarding the advance of frontal systems, mountains produce greater precipitation on their windward slopes than on the leeward as a result of the orographic effect.

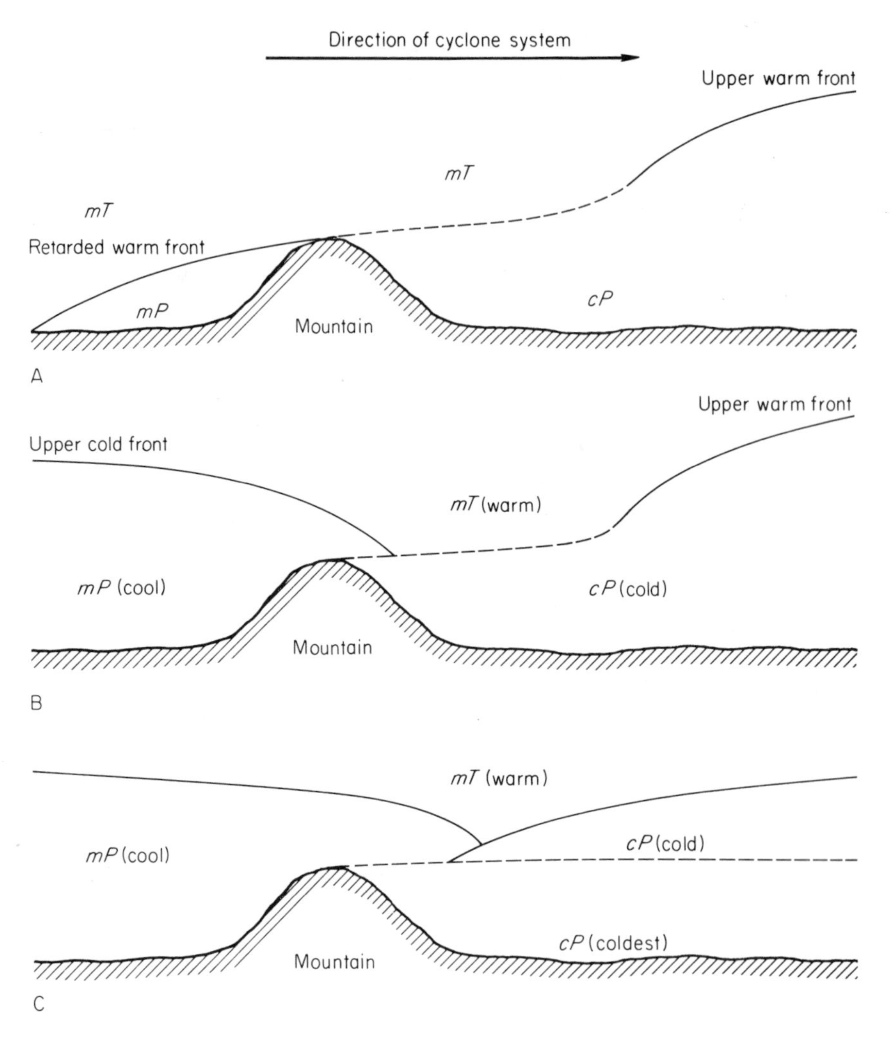

Figure 5.15 Stages in occlusion and formation of upper fronts induced by a mountain barrier. (After Willett.)

ANTICYCLONES

The term *anticyclone* implies characteristics opposite those of the cyclone. Barometric pressure is highest at its center and decreases outward. Consequently, the anticyclonic wind system blows out from the center, and because of the Coriolis effect it has a clockwise circulation in the Northern Hemisphere (counterclockwise in the Southern Hemisphere).

Further, the anticyclone is composed of subsiding air which renders it stable in contrast to cyclones or other low-pressure systems. It rarely has fronts or definite wind-shift lines, but a gradual change in wind direction takes place as it passes. Except where winds converge along converging isobars of a local pressure pattern or where convection is induced by thermal instability there is little cloudiness and therefore lack of precipitation in the typical anticyclone. Maritime stratus below the subsidence inversion of subtropic highs along continental west coasts sometimes produces light drizzle. Wind shear and convergence may generate frontal activity in the trough of low pressure between two anticyclones, however (Fig. 5.16). The resulting front often has a north–south alignment and is known as a *meridional front.* Meridional cold fronts are especially common in the mid-latitudes of the Southern Hemisphere, where they sometimes develop a wave form and become extratropical cyclones.

During middle- and high-latitude winters anticyclones are the sources of cold stable air masses that invade cyclonic systems behind cold fronts. These "cold" anticyclones, or cold waves, are confined to the lower troposphere. Diameters range from a few hundred to a few thousand kilometers. Like the cyclone, the anticyclone usually has a pressure pattern represented by circular or oval isobars, but it may assume various shapes. Ordinarily, it travels at a rate appreciably slower than the typical cylcone, but its direction is even more erratic. The paths of highs are roughly similar to those of lows across North America except that they typically do not turn northeastward over the Mississippi Valley but rather proceed more directly eastward. Where strong mid-latitude anticyclones remain nearly stationary for several days, they control the movement of adjacent depressions. Persistent "blocking highs" can exert a profound influence on regional weather, for example, during late winter and spring in the zone of westerlies over the northeastern quadrants of the Atlantic and Pacific Oceans.

In summer anticyclones over the continents may move slowly and be warmed by subsidence and surface heating, producing "heat waves." Under such conditions the high temperatures, light winds, and pollution attending stable air can create a hazard to health and economic activities. Visibility becomes progressively poor, and low relative humidity accentuates fire danger.

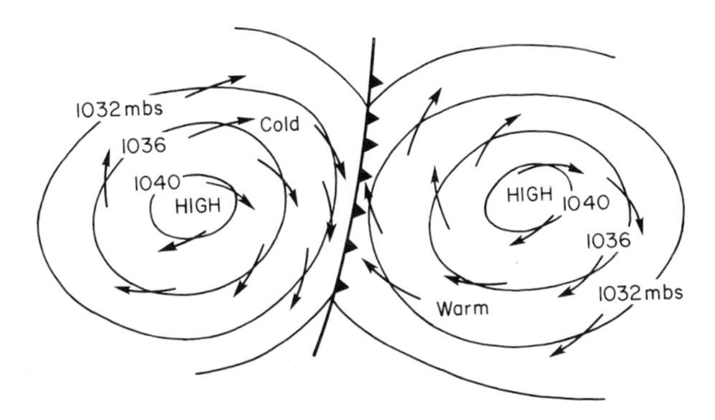

Figure 5.16 Meridional cold front in a trough of low pressure between adjacent cyclones, Northern Hemisphere.

The anticyclones of the subtropics are associated with the subtropic highs in the general circulation and are warmer than those of the higher latitudes. Their pressure results from the piling up of a great mass of air; subsidence accounts in part for their warmth. Although they tend to be more nearly stationary, portions sometimes break away to move along the margins of the westerlies. They are frequently reinforced by polar anticyclones that merge with them. Because of the difference in temperature between the two, a front may be formed in the trough of lower pressure that separates the high cells before they completely merge. Along their trade-wind margins waves and cyclonic vorticity may develop in the pressure pattern, leading to cloudiness and precipitation.

TROPICAL WEATHER

Although they have other features similar to those of mid-latitude storms, tropical disturbances generally do not exhibit sharp discontinuities of temperature. Many have weak pressure gradients and lack well-defined wind systems. Extensive, shallow lows occasionally bring long periods of overcast weather with continuous rain. In the intertropical convergence there may be convective activity and thunderstorms, the smallest and most frequent type of tropical disturbance. Convergence tends to increase when the equatorial trough of low pressure moves poleward in summer, producing bands of cumulonimbus clouds and high overcasts of cirrus.

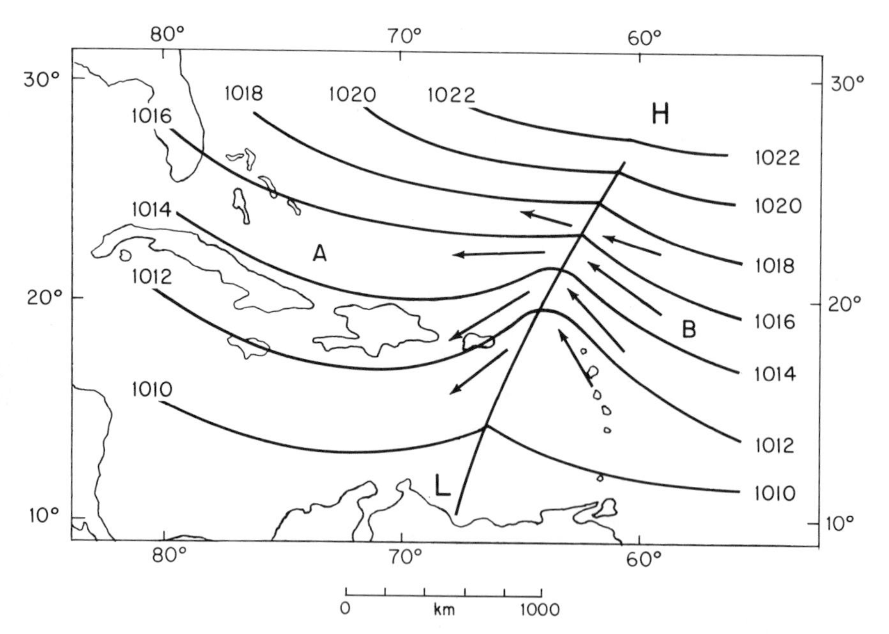

Figure 5.17 Pressure distribution in an easterly wave in the Caribbean. Winds blowing from the direction of *B* converge toward the wave; they diverge in the region of *A*.

A common feature of tropical weather is the *easterly wave,* which normally forms in the convergent flow of trade winds and moves slowly from east to west (see Fig. 5.17). Fair weather and divergent winds precede the wave, followed by convergent air flow and an extensive belt of towering clouds that may appear similar to a mid-latitude cold front. Squally weather and precipitation frequently accompany such a disturbance. Some easterly waves move poleward and curve toward the east to become extratropical cyclones. Others may develop vortices, become tropical cyclones, and even grow to hurricane intensity.

The violent and destructive forms of tropical cyclones are much better known than the weaker variety although the former are, fortunately, much less common. They originate over the tropical oceans only. In the Caribbean and off the Pacific Coast of Mexico they are known as *hurricanes* (Fig. 5.18); in the seas off China, the Philippines, Japan, and the other islands of the western Pacific they are called *typhoons;* in the Indian Ocean they are simply called *cyclones,* a term which should not be confused with cyclones in general. In the Southern Hemisphere they occur east of the African coast and along the northwest and northeast coasts of Australia. Off northwest Australia the associated strong winds are locally known as *willy-willies.*

Figure 5.18 Satellite photograph of *Hurricane Frederic* in the Gulf of Mexico. Note the central eye. (NOAA, NESS)

Elsewhere the term *tropical cyclone* is generally applied. To avoid possible confusion with weak tropical cyclones and the extratropical cyclones, the term *hurricane* will be used in the following paragraphs unless otherwise specified. Hurricanes are apparently absent from the South Atlantic, presumably because the equatorial belt of convergence seldom moves far enough south of the equator in that region. In contrast, of 761 tropical cyclones recorded over the North Atlantic from 1886 to 1977, nearly 60 percent reached hurricane intensity.

A hurricane is a giant heat engine that derives its energy mainly through the transfer of sensible and latent heat from sea to air. It acquires spin (vorticity) from the Coriolis effect and consequently is most likely to form in the ITCZ when the latter is located at least 5° to 10° from the geographic equator, that is, in late summer or early autumn. At the equator the deflective force is inadequate to generate the violent vortex. Most hurricanes evolve from pre-existing disturbances such as incursions of cold air in the upper troposphere, easterly waves, or weak tropical depressions that become charged with warm, moist air and migrate poleward. Off the east coasts of Asia and North America they often move northwestward and then turn away toward the northeast, although the actual paths vary widely and may be erratic (see Fig. 5.10). The diameter of the typical hurricane is 150 to 1,000 km, and it increases as the storm moves away from the low latitudes. Some hurricanes, however, have diameters as small as 30 km. The size of the disturbance has no predictable relation to intensity, and its rate of travel appears to be unrelated to either its size or the wind speeds within the system. An advance of 15 to 30 km in an hour is typical, but hurricanes are erratic in their general rate of progress as well as in direction.

Several features distinguish tropical hurricanes from the cyclones of the midlatitudes. The pressure distribution, which is represented by isobars, is more nearly concentric and circular and the isobars are closely spaced with very low values at the center, indicating the steep pressure gradient which produces the high-velocity winds. There are no fronts or wind-shift lines, but at the center of the whirl there may be a calm "eye" 10 to 50 km in diameter in which air is descending and which, therefore, is comparatively clear and warm. The lack of introduction of contrasting air masses results in fairly even distribution of temperature in all directions from the center. Rainfall is also relatively evenly distributed, especially if the storm is stationary; in a moving hurricane, rainfall is slightly greater in the forward half of the storm. In either case rain is torrential. Because of the tropical nature of the air and the high freezing level in the latitudes of hurricanes, hail does not occur.

The outstanding feature of the tropical hurricane is its wind force. At the outer margins of the system winds are moderate, but velocities increase rapidly toward the center, owing to conservation of angular momentum. They reach a maximum along the outer edge of the central eye. Technically, wind speed must reach 65 knots (33 m per sec or 120 km per hr) to be classified as of hurricane force; speeds of 130 knots have been recorded, with gusts estimated as high as 220 knots.

The destructiveness of hurricanes is due to extreme wind force, intense rainfall, and flooding. At sea, hurricanes produce a distinctive heavy swell that affects ocean shipping. Swells in advance of an approaching hurricane come at longer intervals than ordinary storm swells and are one of the most valuable warning signs to mariners because they appear before the winds of the hurricane are encountered. Large waves accompanying the hurricane swell can cause coastal damage even while the storm center is well out to sea. As the storm approaches a coast, the piling up of water by strong winds may produce a disastrous *storm surge.* A normal high tide augmented by a storm surge in the Bay of Bengal caused approximately 300,000 deaths among delta residents of Bangladesh on November 13, 1970.

When hurricanes pass inland, frictional effects and the loss of sustaining energy from warm water surfaces cause them to lose much of their force. Although of interest to the meteorologist, this phenomenon may be a small consolation to people in the affected region. As it moves to higher latitudes, where prolonged rain may produce flooding long after winds have moderated, a hurricane often draws in air from poleward sources and transforms into an extratropical cyclone.

As a matter of convenience for tracking the paths of hurricanes and typhoons the weather services name the storms in alphabetical sequence, using a different name series for each of the major regions of occurrence.

THUNDERSTORMS

Several thousand thunderstorms occur every day, mainly in the tropics. The number is smaller over oceans than over land, owing to lack of intense convection above water surfaces. In polar regions they are virtually unknown. They are always associated with unstable air and strong vertical motion that produce clouds of the cumulonimbus type. A great deal of the energy for their development comes from the release of the latent heat of condensation in rising humid air. It is important to note in this connection that an unstable lapse rate and overturning of air may result from either warming of surface layers or introduction of cold air aloft.

The common processes that initiate thunderstorm development are (1) heating and convection in moist air over warm land surfaces; (2) passage of cold, moist air over warm water; (3) forced ascent of conditionally unstable air along zones of convergence or at mountain barriers; (4) radiational cooling at upper levels; or (5) advection of cold air aloft. Moisture and rising air are the essentials for any thunderstorm. Most thunderstorms consist of several convective cells that grow and dissipate erratically. When the force of rising air is great enough the storm can reach to heights from 4 to more than 20 km (see Figs. 5.19 and 5.20). Cooling proceeds at the dry adiabatic rate up to the base of the storm cloud and continues to be rapid as dry air is entrained in the violent updrafts. Inertia in the convective "chimney" may carry air beyond the level where equilibrium is established between the convective mass and the surrounding air aloft. The height of a thunderstorm is related to latitude and season of the year; the greatest heights are developed in summer and in the tropics. Typical diameters of thunderstorms vary from 3 to 40 km. A well-developed cumulonimbus thunderhead has an anvil-

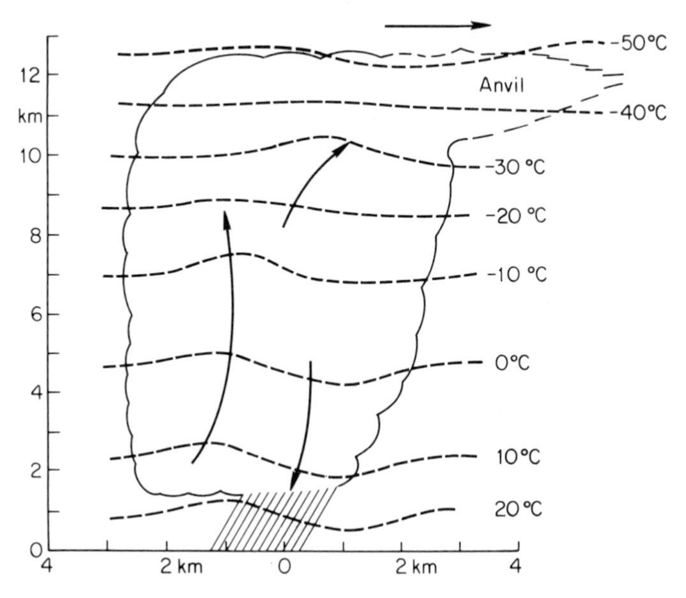

Figure 5.19 Mature stage of a thunderstorm. (After Byers.)

shaped crest that points in the direction of the storm's movement, and there are often shelves of clouds in advance of the main thunderhead. Along its base the storm is dark and ominous, and it may be preceded by roll clouds of a *squall line,* created by air currents moving in opposing directions. Precipitation from the mature storm is intense and composed of large raindrops, literally a cloudburst. If the updrafts have sufficient force and penetrate well above the freezing level, hail may fall from the cloud, usually from the leading edge. Under favorable temperature conditions, precipitation may be in the form of snow or snow pellets.

Thunderstorms caused by surface heating over land are most common in summer and in the afternoon or early evening. Over the oceans the temperature difference between the water and cooler air above is greatest at night, and thunderstorm activity resulting from this set of conditions is accordingly greater at night. Along and above mountains, maximum incidence of thunderstorms is usually in the afternoon or early evening, when the combined effects of daytime heating and orographic lifting are at a maximum.

Along zones of convergence thunderstorms develop when air is forced up rapidly. In the mid-latitudes they may be associated with the cold front, warm front, or upper fronts and are often accentuated by surface heating, orographic lifting, or overrunning by cold air at upper levels. Along a cold front they are generally closer to the ground and more violent than the warm or upper-front types. In contrast with localized convective thunderstorms, which tend to be spotted about in a random pattern, the frontal types concentrate in a zone 20 to 80 km in width and perhaps hundreds of kilometers in length. Because the upper portions of cumulo-

Figure 5.20 Aerial view of a mature thunderstorm. The top of the anvil is about 15 km above the ground. Note the convective humps on the upper surface of the cumulonimbus and the shelf clouds extending toward the lower right. (NOAA, National Severe Storms Laboratory.)

nimbus clouds are frequently obscured by lower clouds, frontal thunderstorms are sometimes difficult to identify from the ground. Pilots flying at high altitudes have some advantage in this respect. A line of air mass thunderstorms along the crest of a mountain range should not be confused with frontal thunderstorms, although it does indicate convergence of air above ridges and peaks.

For the purposes of meteorological records, thunder must be heard or overhead lightning or hail observed at a station before a thunderstorm is reported. Although thunder and lightning accompany the typical mature thunderstorm and often are its most dramatic manifestations, their roles in the development of precipitation and other storm characteristics are not clearly understood. Lightning discharges may take place from cloud to cloud horizontally, between different levels in a cloud, or from the cloud base to the ground. Thunder is the explosive sound created as the air expands suddenly in response to the great heat of the lightning discharge and then rapidly cools and contracts.

TORNADOES AND WATERSPOUTS

Tornadoes are the most violent storms of the lower troposphere. A tornado is a tight vortex, gyrating around a center of extremely low pressure, usually associated with a severe thunderstorm. Accurate wind observations are impossible within the maelstrom, but velocities of 200 to 300 knots are probably common. Estimates of wind speed are based on surveys of damage or analyses of radar and motion picture records. The direction of rotation usually is cyclonic, but in rare cases it may be anticyclonic. A tornado can be distinguished by its writhing, funnel-shaped cloud, which extends downward from the base of a cumulonimbus or a turbulent cloud layer (Fig. 5.21). Multiple funnels occasionally descend from the same cloud, and what appears to be a single funnel may contain two or more vortices. If it reaches the ground a tornado can cause incredible destruction along its path. Buildings seem to explode as a result of wind force and the sudden decrease of outside pressure. A tor-

Figure 5.21 Tornado near Enid, Oklahoma, 5 June 1966. The upper funnel consists of condensation particles; the lower column is dust and debris raised by the whirling vortex. (Photograph by Leo Ainsworth, courtesy NOAA, National Severe Storms Laboratory.)

nado that traveled through Missouri, Illinois, and Indiana on March 18, 1925, caused 689 deaths and more than 2,000 injuries. Most tornadoes are only a few hundred meters in diameter at the ground, but swaths 3 to 5 km wide have been recorded. In mid-latitudes the funnels most commonly form in the warm sector of cyclonic systems just ahead of the cold front where dry air is overrunning aloft. Tornadoes also occur in hurricanes; an astounding total of 115 were observed in the company of Hurricane *Beulah* over southeastern Texas in September 1967. Thunderstorm activity, rain or hail, and lightning accompany most tornadoes. Low-level convergence and abnormally strong convective turbulence in moist, unstable air generate the vortex. Paths of tornadoes usually are determined by the prevailing flow in the lower and middle troposphere. This rule has exceptions, however, for they have been observed to make U-turns and even complete circles. Their rate of travel at the ground may reach 50 knots or more or they may remain stationary for short periods. Some contact the ground and then lift, only to strike again at a distance of several kilometers; others travel only a few meters before rising and dissipating.

In North America the greatest frequency of tornadoes is in the Mississippi Valley and Great Plains, but they also occur in the Gulf States, western Lake States, and the Canadian Prairie Provinces. One was sighted at Yellowknife, N.W.T., Canada (lat. 62°28' N), on June 19, 1962. Every state in the United States has experienced tornadoes. A total of 1,109 were reported in the United States in 1973, but only 421 in 1953. They are most likely to appear in spring and early summer, when contrasts of temperature and humidity are at a maximum in the air masses along cold fronts. The diurnal maximum incidence is in the afternoon, especially between 3 and 7 P.M. Although tornadoes are apparently less common elsewhere in the world, they have been reported at widely scattered locations in the mid-latitudes and tropics, notably in Europe, Japan, and Australia. Many undoubtedly have escaped official climatic records.

At sea tornadoes are known as *waterspouts,* having much the same characteristics except that they are usually smaller in diameter. Waterspout frequency is greatest where cold continental air pushes over warm water, as off the east coasts of China, Japan, and the United States. When in contact with the surface a waterspout picks up some spray, but its funnel is composed primarily of condensed water vapor in the low-pressure vortex.

OTHER AIR MASS AND STORM EFFECTS

Some weather phenomena do not exhibit the dramatic features ordinarily associated with storms and are not accompanied directly by precipitation. Yet they incorporate distinctive "weather," and a few are referred to as storms, at least locally.

Perhaps the best known of these phenomena is the *foehn effect,* which produces a warm, dry, and often gusty wind on the lee side of mountain ranges. The name *foehn* originated in Austria and Germany, where the foehn wind is frequently experienced in valleys of the Alps. In the western United States and Canada the same type of mountain-induced effect is called the *chinook.* The explanation of the relative warmth of the typical chinook rests upon two principles. Ordinarily, a chinook wind is accompanied by cyclonic activity which produces clouds and precipitation on the windward side of the mountain range (for example, the Rockies). The latent heat released to the air by the condensation process warms the air which passes across the range, and, because the air has lost some of its moisture, it will also be drier (Fig. 5.22). However, the latent heat of condensation alone does not account for the temperatures which occur on the leeward. The mountain barrier creates a frictional drag which tends to pull the air from higher levels down on the leeward. Air forced down in this way is heated adiabatically, and therefore its relative humidity is lowered. This action is by no means regular, especially in its earlier stages, but comes in surges which are experienced at the ground as gustiness. If the wind is to affect an extensive area to the leeward of the mountains, the general pressure gradient must be such that the cold air will gradually move out ahead of the chinook. When fully developed the chinook can remove snow cover in a short time, and it is in winter that it is most often recognized. It is well to keep in mind that the temperature of the foehn or chinook is *relatively* warm, that is, it replaces colder air at the surface. Its actual temperature may occasionally be below freezing, but because it is dry it can remove snow or ice by sublimation; however, if it replaces air colder than itself, it is properly called a chinook.

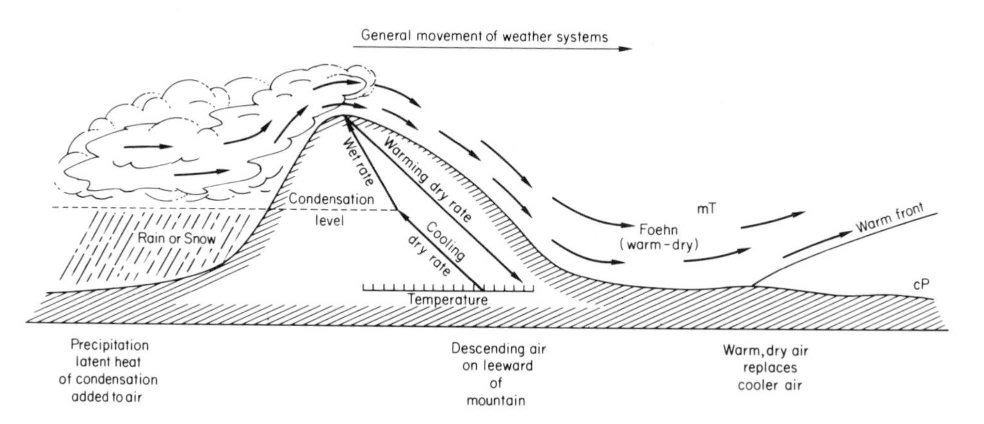

Figure 5.22 Typical conditions which produce a foehn wind.

The danger of lending too much importance to precipitation and to the release of the latent heat of condensation on windward slopes as the cause of the foehn wind can best be illustrated by examining the *stable-air foehn effect.* Consider a mass of cold stable air moving against a mountain range, as might happen in winter when a polar air mass out of Canada travels along the eastern margin of the Rockies (see Fig. 5.23). It is a characteristic of stable air that in its upper levels the air is

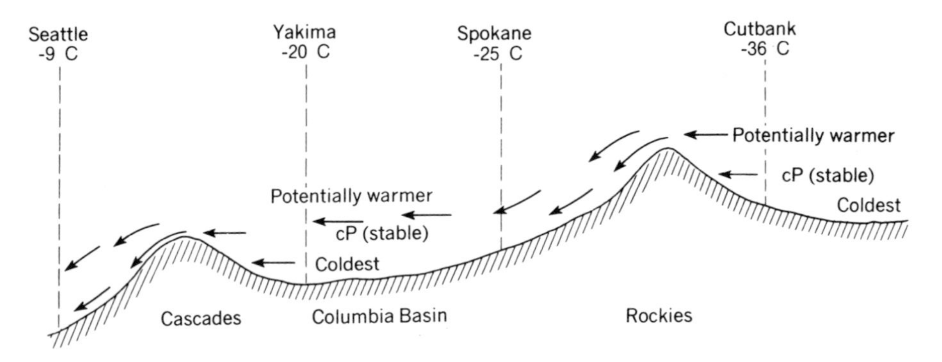

Figure 5.23 Stable-air foehn induced by mountain barriers.

potentially warmer, that is, if the upper air is brought down to lower altitudes it will be warmed adiabatically to temperatures greater than the surrounding air in the mass. When stable air moves against a mountain barrier, it is this potentially warmer air which spills over. It warms as it descends, and, in the lowlands, has somewhat warmer temperatures than the surface air in the parent air mass. West of the Rockies it will normally replace warmer air and will therefore be experienced as a cold wave in contrast to the chinook. Nevertheless, the temperatures on the west side of the barrier are not so low as those on the east. The stable-air foehn effect may be reinduced subsequently by the Cascade Range so that that portion of the cold air mass which reaches the Pacific Coast is further warmed. In making regional comparisons of temperature under such conditions it is, of course, necessary to take into account the different altitudes involved and the residual heat of the surface west of the mountains. Both tend to accentuate the temperature differences, although they can hardly be considered a part of the stable-air foehn effect.

In summer the stable-air foehn brings hot, dry air to the Pacific slope when an anticyclone lies inland; this creates an extreme forest fire danger. It functions in the opposite direction when the relatively stable air blowing out of the Pacific high cell moves against the coastal ranges and the Cascades, and the potentially warmer air of the higher levels crosses the barriers first. Upon descending on the lee side, it is warmed adiabatically and thus contributes to the higher temperatures of the interior.

The foehn wind unquestionably occurs in many mountainous areas of the world, but is best known and most easily recognized where high mountain chains lie approximately at right angles to prevailing winds. Similarly, the stable-air foehn effect requires the proper juxtaposition of a mountain barrier and a stable air mass.

A large group of so-called storms are the *dust storms* of dry climates and drought-stricken areas. They are often associated with true storm conditions, for example, a mid-latitude cyclone, but all too rarely are they accompanied by subsequent precipitation. Dust storms result from the action of strong winds, picking up loose earth material and carrying it to great heights and perhaps for great distances as well. During the drought years of the early 1930s, dust storms ravaged a large area of the Great Plains of the United States as well as central Australia. In desert regions, where there is little protective cover, dust or sand storms are generated by sustained high-velocity winds.

The much smaller *dust-devil,* or *whirlwind,* is an eddy that forms over hot, dry land as a result of intense local daytime convection and surface friction. It is a fair-weather phenomenon identified by the whirling column of dust, usually not more than a few meters in diameter and a few tens of meters high. Ordinarily the dust-devil is not destructive, but it moves loose trash and is unpleasant for anyone caught in its path. Its whirling eddy of air operates on a small scale and may turn in either direction in response to the conflicting air currents that set it in motion. Dust-devils probably form in a manner similar to fair-weather waterspouts, but, lacking the added energy due to release of latent heat, they are generally somewhat smaller.

OBSERVATIONS OF CIRCULATION SYSTEMS AND STORMS

A great deal of the information of value in forecasting weather and for climatic records comes from general storm observations. Details of temperature, pressure, humidity, wind, cloudiness, and so on are essentially the symptoms of the storms; it is their peculiar combination that constitutes a specific disturbance. It is one thing to measure directly the various weather elements but quite another to undertake objective observation of an entire storm or circulation system. Knowledge of areal extent, height, speed of movement, and intensity is desirable, however. Of these, intensity is the most difficult to express in quantitative terms. In connection with storm research, indices combining two or more measured quantities, such as wind force and rate of precipitation, have been established.

Radar (radio detection and ranging) techniques make it possible to track clouds and precipitation at distances of 300 km or more, and even to discern the size of water droplets or ice particles in the air (Fig. 5.24). As a result, storms can be followed continuously and their evolution traced over areas where satisfactory ground observations are lacking. *Doppler radar,* which has the additional capability of detecting motion in the atmosphere, is used to observe thunderstorm and tornado development or even turbulence above airport runways.

Lidar, an application of the laser, employs an optical principle to determine cloud characteristics. It also can detect solid particles and track their movement in otherwise clear air. The acoustic echo sounder depends on the backscattering of sound waves from air parcels having different density characteristics. Known as *sodar,* it is useful for remote sensing of disturbances in low tropospheric layers.

An obvious but expensive—and often dangerous—method of storm and air mass observation is aerial reconnaissance. Flights into the upper air often reveal a general storm or an air mass property which is not evident from isolated radiosonde and winds-aloft reports. Furthermore, the flight path can be altered to correspond to the location of a storm, and successive flights can follow the storm—an impossible maneuver for either ground stations or ground-based radiosondes. Instrumen-

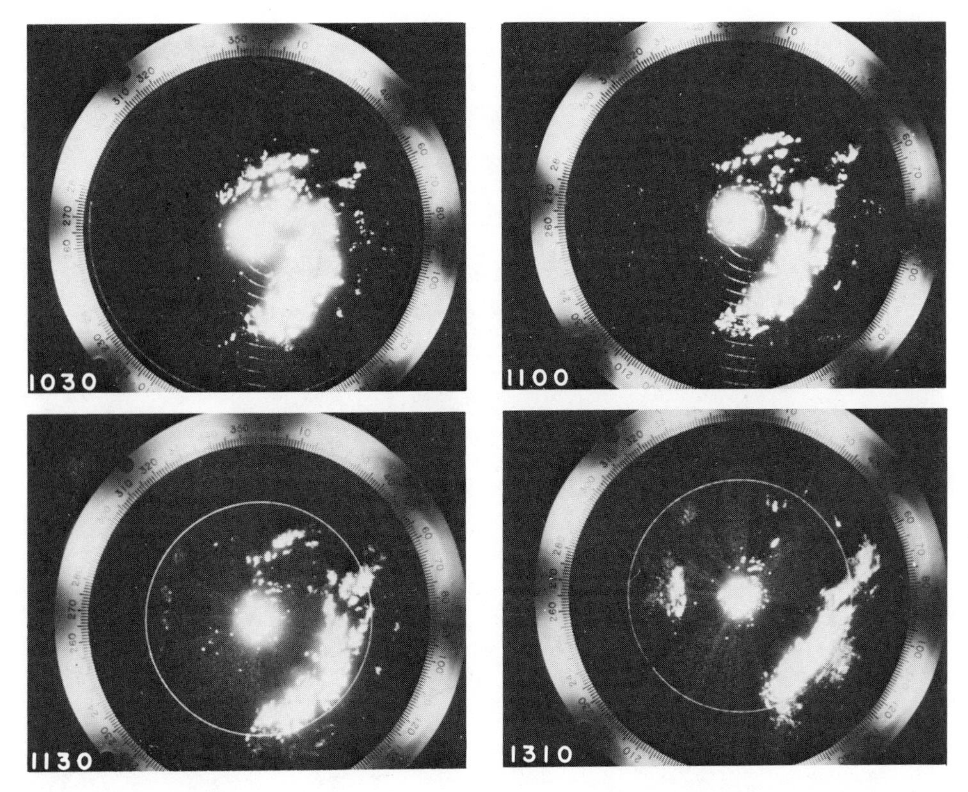

Figure 5.24 Radar images of a moving line of thunderstorms. Range circles in the lower frames represent a sweep radius of 220 km. Note the new storm forming to the west at 1130 and 1310. (NOAA)

tal and visual observations along weather reconnaissance routes have been especially valuable in weather research and forecasting. In both the Atlantic and Pacific areas, hurricanes (or typhoons) have been studied by means of airplane flights into the storms. Strong aircraft and careful navigation are necessary during reconnaissance of a hurricane, but the information gained aids in forecasting the future path of the storm as well as in understanding the general principles which govern hurricanes. Weather planes are elaborately equipped with meteorological instruments, including radarscopes. Some planes periodically release *dropsondes* to obtain data on conditions in the air below the flight level.

Rockets and satellites, by providing information from ever-greater heights, have opened a new range of techniques in storm observation. Since the world's first meteorological satellite (TIROS-I) was launched in 1960 a growing number of satellites incorporating television and radar have scanned the atmosphere regularly from above to furnish data on the extent and movement of storm systems as well as certain associated phenomena. Polar-orbiting and geostationary satellites provide photographs of all parts of the earth and its cloud cover (see Fig. 5.25). Interpreta-

Figure 5.25 Satellite photograph centered on the Amazon Basin, recorded on 13 October 1970. Cloud patterns indicate major cyclonic disturbances in the North and South Atlantic. (NOAA, National Environmental Satellite Service.)

tion of satellite photos by *nephanalysis* (cloud analysis) reveals the development of storms and features of the atmospheric circulation. Weather satellites also facilitate the measurement of radiation, atmospheric temperature and moisture, sea-surface temperature, albedo of the earth and clouds, and the extent of ice and snow cover, all of which have wide use in climatology. Rockets equipped with meteorological sensors extend the range of upper-air observation well beyond that of conventional balloon ascents and thereby provide data at levels intermediate between the latter and satellites.

Developments in remote sensing, communications, and computer data processing have improved the study of atmospheric processes at various scales of time and space. Since the organization of a World Weather Watch under the auspices of the World Meteorological Organization in the 1960s many of the world's nations have cooperated to extend and intensify atmospheric surveillance. Assembly of data at World Meteorological Centers in Melbourne, Moscow, and Washington facilitates compilation of global weather maps. A closely related effort, the Global Atmospheric Research Program (GARP), has led to special studies such as GARP Atlantic Tropical Experiment (GATE), Monsoon Experiment (MONEX), the First GARP Global Experiment (FGGE), and the Joint Air–Sea Interaction Experiment (JASIN). These projects have been concerned with refinement of observing techniques as well as the collection of raw data for forecasting and research.

REGIONAL WEATHER PATTERNS

The world's weather at any instant is a kaleidoscope of many motion systems, each incorporating a unique combination of atmospheric conditions. As time passes each system changes in response to internal dynamic processes, exchanges of energy and moisture with the earth's surface, and interaction with other systems. Although these modifications often appear to be random in nature, they are phenomena that tend to follow regional patterns with some regularity. Thus, it is common to identify tropical, monsoonal, arctic, and other weather types. Over a period of years the recurrence of characteristic weather sequences creates distinctive climates. The chapters of Part II treat regional climates of the world.

Regional weather patterns are of immediate significance in forecasting, which is concerned with predicting their changing character. The basic map for display and analysis of weather data is the *synoptic chart,* which shows the surface distribution of weather conditions at a given time and is therefore a valuable addition to the climatic record (Fig. 5.26). The scale and area covered by the map depend on its intended use, ranging from a local forecast region to an entire hemisphere. Synoptic weather maps are compiled from reports of observations at many stations throughout the world by agreement through the World Meteorological Organization. Prompt data collection requires efficient electronic communication; computers and automatic plotting devices speed analysis. Entire charts of both surface and upper-air conditions are transmitted from main analysis centers to facsimile recorders or computer-graphic consoles for local use. A computer-based system known as Automation of Field Operations Services (AFOS) links regional forecast centers in the United States, making it possible to receive a variety of graphic and numerical data on request.

Knowledge of regularities, probabilities, and anomalies in weather constitutes the climatological aspect of forecasting. To the extent that weather sequences repeat themselves, analogous patterns in the past can assist prediction. This principle has enabled countless "old-timers" to achieve remarkably accurate short-term predictions of local weather. The more formal *analog method* of

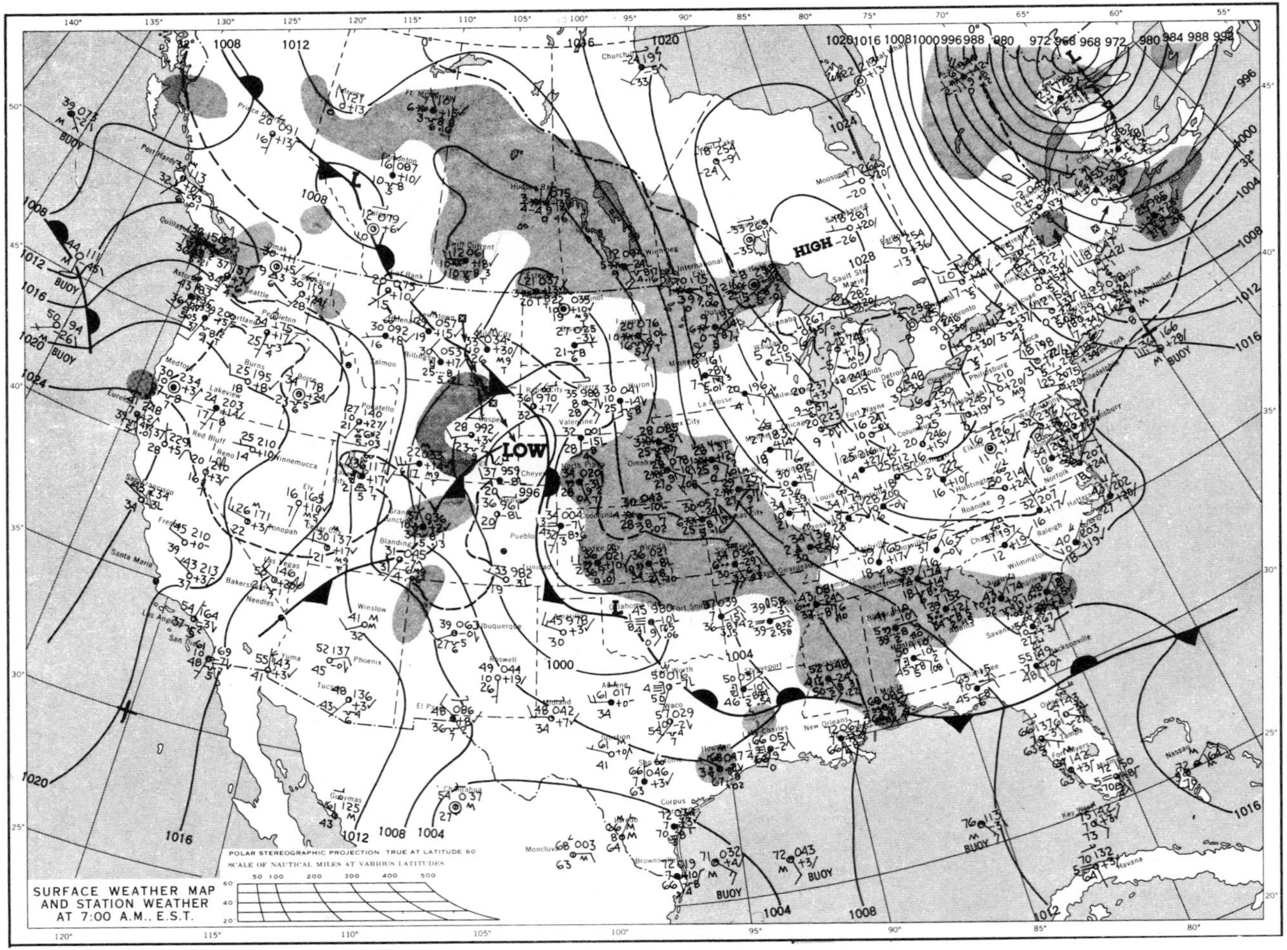

SURFACE WEATHER MAP AND STATION WEATHER AT 7:00 A.M., E.S.T.

Figure 5.26 Daily Weather Map, 11 March 1980. (NOAA, Environmental Data and Information Service.)

forecasting depends on reference to historical weather maps and records (Fig. 5.27). Classification of weather types aids the forecaster's selection of similar weather patterns from the past and examination of the succession of weather as a guide to determining future trends.

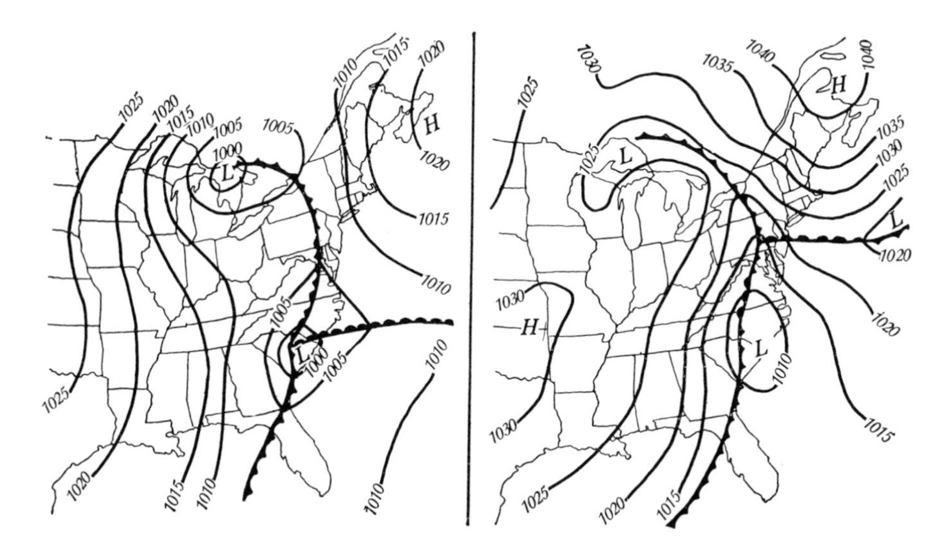

Figure 5.27 Synoptic analogs, Eastern United States. The simplified weather map on the left shows the pressure pattern and fronts on 8 November 1913; that on the right is for 25 November 1950. An intense storm and heavy snow accompanied both of these cyclonic systems. (After Mook and Smith, from *Water, 1955 Yearbook of Agriculture.*)

A related application of climatology in forecasting is based on mean atmospheric flow patterns, or circulation types, and their associated weather. A synthesis of synoptic maps, upper-air charts, and vertical cross sections over an extended period makes it possible to identify regional or local weather conditions that usually accompany a particular motion system. This approach, known as *synoptic climatology,* enables the forecaster to characterize weather events in terms of their causes rather than merely the statistical averages of individual climatic elements. Although quantified expressions of entire motion systems thus far have been crude, they offer a promising bridge between the daily weather forecast and long range predictions. Synoptic climatology also can enhance dynamic explanation and the classification of different climates.

QUESTIONS AND PROBLEMS FOR CHAPTER 5

1. Describe the characteristics of a good air mass source region.
2. How are air masses identified and distinguished from one another?
3. Prepare a graph to illustrate examples of radiosonde observations that indicate

absolute stability, conditional instability, neutral equilibrium, and absolute instability in the atmosphere.

4. Why does smoke normally rise in the atmosphere? Under what conditions is it likely to concentrate at low levels?
5. Draw a plan view of a meridional front in the Southern Hemisphere. Indicate air masses, fronts, winds, and cloud types.
6. What conditions might lead to precipitation in an anticyclone?
7. Why are tornadoes often associated with thunderstorms? Why has tornado damage increased markedly in the twentieth century?
8. How does a hurricane differ from an extratropical cyclone? How can a hurricane change to become an extratropical cyclone?
9. Explain the major differences between the foehn wind and the stable-air foehn effect.
10. What important information needed for weather forecasting is lacking on the synoptic chart? How is this information obtained?
11. How does an understanding of the general atmospheric circulation aid daily weather forecasting?
12. Summarize the contributions of the following to daily weather forecasting and to the climatic record: rawinsonde, radar, lidar, sodar, satellites, nephanalysis, synoptic analogs.

Part II

PATTERNS OF WORLD CLIMATE

6

Climatic Classification

Different combinations of processes in the earth's climate system produce many variations in climate from place to place and from time to time. An area of the earth's surface over which the combined effects result in an approximately homogeneous set of climatic conditions, that is, a climatic type, is termed a *climatic region*. To facilitate description and mapping of climatic regions it is necessary to identify and classify the respective types. Regional climatology is concerned with this task, employing analytical and descriptive techniques in its search for order in the world climatic pattern. Changes of climate through time suggest a historical methodology in which time scale divisions, for example, eras or periods, are analogous to regions in the spatial scale and graphs may replace maps. Integration of climatic phenomena with respect to both their spatial and temporal dimensions is necessary for a complete understanding of the climate system.

APPROACHES TO CLIMATIC CLASSIFICATION

As a fundamental tool of science, classification has three interrelated objectives: to bring order to large quantities of information, to speed retrieval of information, and to facilitate communication. Classification of climate shares these objectives. It is concerned with organization of climatic data in such a way that both descriptive and analytical generalizations can be made, and it attempts to store information in an orderly manner for easy reference and communication, often in the form of maps. The value of a systematic arrangement of climates is determined largely by its intended use; a system that suits one purpose is not necessarily useful for another. For example, a classification based on critical temperature and moisture limits for growth of a certain plant or animal organism might serve the needs of a biological study, but it is not likely to be satisfactory for weather forecasting, which relies more on such factors as the general circulation, storm types, and weather probabilities. Thus, in the design of a climatic classification we should begin by defining the purpose. Three broad approaches are equally feasible: (1) empirical, (2) genetic, and (3) applied. Together they constitute a classification of classifications, but the features of all three may be incorporated in a single system.

Empirical classifications are based on the observable features of climate, which may be treated singly or in combination to establish criteria for climatic types. Temperature criteria, for example, might yield "hot," "warm," "cool," and "cold" climates, each of which can be defined in terms of strict mathematical limits. Adding precipitation and other elements to the criteria, the number of possible combinations rapidly multiplies, and soon the system becomes unwieldy. It is, therefore, necessary to select the criteria that are most significant in light of the intended purpose. Heat and moisture factors have dominated empirical classification, but all elements are inherently significant for one purpose or another.

Genetic classification attempts to organize climates according to their causes. Ideally, the criteria employed in the differentiation of climatic types should reflect their origins if climatology is to be explanatory as well as descriptive. In practice, however, explanations are often theoretical, incomplete, and difficult to quantify. Genetic classification also is subject to theoretical biases; a system based on causes tends to perpetuate faulty or over-generalized theories. The ancient Greeks recognized a relationship between latitude and temperature and devised a system of *klimata,* or zones (torrid, temperate, and frigid), that have persisted in writings to the present day in spite of evidence that net radiation does not vary solely with latitude and that other factors affect the world patterns of temperature. Besides latitude, features of the general circulation, including winds, air masses, and storm types; terrain features such as elevation, slope, and mountain barriers; and the distribution of land and water are other bases for genetic classification. A common genetic approach attempts to distinguish the relative *continentality* or *maritimity* (sometimes termed *oceanity*) of a climate. In practice, indices to express the influences of land or water surfaces have been determined from various empirical data, mainly temperature, precipitation, wind, and air mass frequency. The most widely accepted criterion is the mean annual range of temperature, which tends to be greater over the continents than over oceans. Since the annual temperature range is also a function of latitude, compensating adjustments are needed. The map, Fig. 6.1, shows North American regions of relative continentality derived mainly from temperature ranges and latitude. Outside the middle latitudes it is difficult to isolate influences of land or sea in terms of temperature or other climatic variables. In the tropics annual temperature ranges usually are small even in continental interiors; at high latitudes complications are introduced by the polar night and ice cover. The concept of relative continentality–maritimity has its greatest taxonomic utility between about 30° and 60° latitude. Several formulas have been

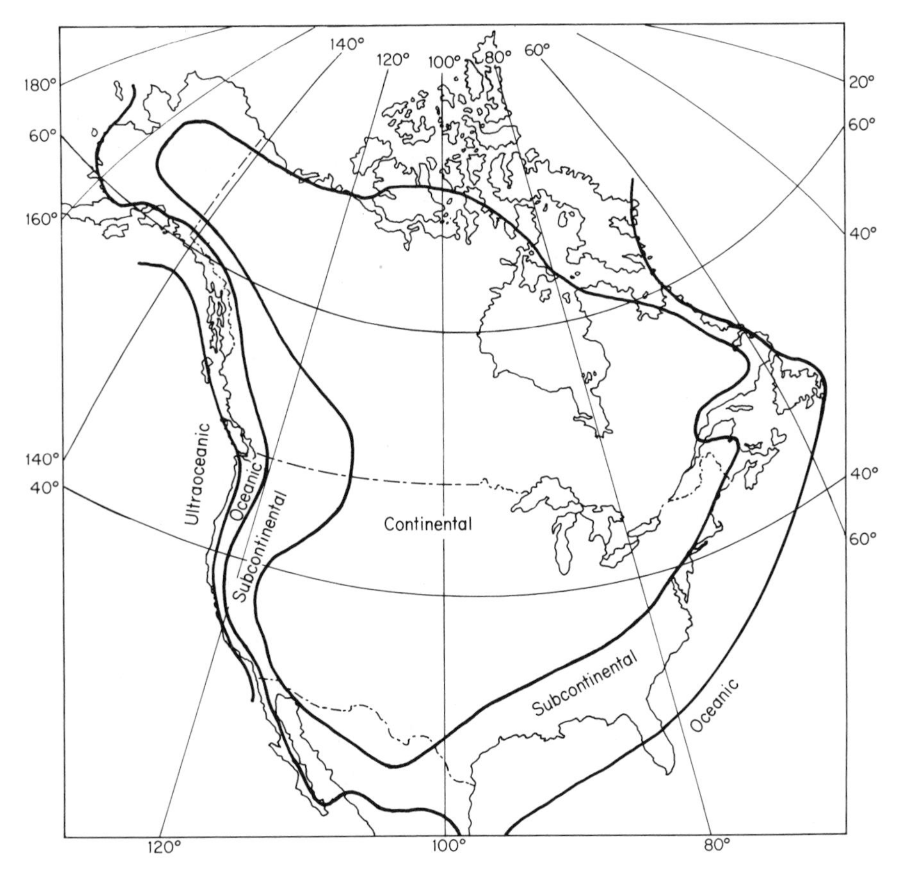

Figure 6.1 Continentality in North America. (Reproduced by permission from the ANNALS of the Association of American Geographers, Volume 64, 1974, p. 274, Donald R. Currey.)

devised to correct for latitude as well as eliminate negative index values and allow for asymmetric annual marches of temperature. Early indices expressed continentality as a percentage, implying the existence of a wholly continental or wholly maritime climate, which is a genetic impossibility in the dynamic atmosphere.

Applied (also known as technical or functional) classifications of climate assist in the solution of specialized problems that involve one or more climatic factors. They define class limits in terms of the effects of climate on other phenomena. Outstanding among modern attempts at climatic classification are those that seek a systematic relationship between climatic factors and the world pattern of vegetation. Natural vegetation integrates certain effects of climate better than any instrument that has so far been designed, and it is thus an index of climatic conditions. By referring to the major plant associations, biologists have tried to determine the climatic factors that correspond with areal differences in vegetation. Numerous correlations between vegetation and heat or moisture factors have been discovered, permitting the use of temperature or moisture indices as criteria for climatic types. The resulting types and their regional boundaries approach reality in terms of the associated vegetation, while retaining a climatic basis. Commonly, classifications of this kind employ vegetation terms. Rain forest, desert, steppe, and tundra are names that have a climatic connotation. In each case there are climatic limits beyond which the characteristic plant association (or a specific indicator species) does not occur naturally. Whereas fluctuations in climate and in nonclimatic influences create problems in delineation of static areal boundaries, they are of great significance in interpreting vegetation changes through time.

An improvement on the use of simple rainfall or temperature limits in climatic classification recognizes the relation between heat and moisture factors. Under high temperatures plants require more precipitation to meet the needs of evapotranspiration. Rains totaling 250 mm annually may support little plant life in a hot, tropical desert, but that amount may be sufficient for coniferous forests in the cool higher latitudes. Thus, natural vegetation is an expression of the adequacy of moisture under a given set of temperature conditions. Precipitation-evaporation ratios and aridity or humidity indices have been introduced to indicate this more complex relationship and establish criteria for climatic types. Variability and seasonal distribution of precipitation and temperature are additional factors which influence plant growth and must be taken into account in any classification that derives from climate–vegetation relations.

Following a period of experimentation with the definition and mapping of precipitation effectiveness and thermal efficiency in the 1930s, the American climatologist C. Warren Thornthwaite (1899–1963) developed a classification based on the concept of *potential evapotranspiration,* that is, the amount of moisture that would be evaporated from the soil and transpired from vegetation if it were available. Thornthwaite regarded potential evapotranspiration (*PE*) as a climatic factor equal in importance to precipitation, reasoning that if more water were available in a hot desert there would be more vegetation and more water would be transferred by evaporation and transpiration. The water need (*PE*) generated by available energy is greater in summer than in winter and greater in hot climates than in cold. By comparing the amount of water available from precipitation with the water need, it is possible to assess moisture conditions to determine seasonal surpluses or deficits and whether a climate is truly wet or dry. Inasmuch as *PE* represents a transfer of both heat and moisture to the atmosphere and is primarily a function of energy received from the sun, it is an index of *thermal efficiency* as well as water loss, combining the heat and moisture factors in climate. Thus, a climate may be regarded as an expression of the energy and water budgets at the earth's surface. (For a discussion of water budgeting, see Chapter 11.)

Actual measurements of potential evapotranspiration are inadequate in number and duration for a worldwide classification of climates. For calculating values Thornthwaite devised a complex empirical formula in which temperature and daylight hours (a function of latitude) are the variables. (Other methods of determining potential evapotranspiration are treated in Chapter 11.) In the classification as slightly revised in 1955 potential evapotranspiration (PE) and precipitation (P) are the bases of four climatic criteria: *moisture adequacy, thermal efficiency, seasonal distribution of moisture adequacy,* and *summer concentration of thermal efficiency.* Each is represented by an index value, and boundaries are set quantitatively. When adjustments are made for the storage of water in the soil the difference between mean monthly amounts of P and PE is the monthly surplus (S) or deficit (D). Moisture adequacy may then be expressed by the monthly *moisture index* (I_m) in the formula:

$$I_m = 100 \frac{S - D}{PE}$$

If soil moisture is assumed to be constant the equation is simply:

$$I_m = 100 \frac{P}{PE} - 1$$

The sum of the 12 monthly values of I_m gives the annual moisture index. Table 6.1 shows the annual index limits for nine climatic moisture types.

TABLE 6.1

Climatic Moisture Types *

	Type	*Moisture Index*		
A	Perhumid	100 and above		
B_4	Humid	80	to	100
B_3	Humid	60	to	80
B_2	Humid	40	to	60
B_1	Humid	20	to	40
C_2	Moist subhumid	0	to	20
C_1	Dry subhumid	−33.3	to	0
D	Semiarid	−66.7	to	−33.3
E	Arid	−100	to	−66.7

*Data in Tables 6.1, 6.2, and 6.3 are adapted from Douglas B. Carter and John R. Mather, "Climatic Classification for Environmental Biology," *Publications in Climatology,* 29, 4 (1966).

Thermal efficiency is simply the potential evapotranspiration in centimeters. Again, the annual index is the total of monthly values. Note that it is an index of energy in terms of water depth, but that it is derived from thermal data. Table 6.2 gives the nine thermal efficiency types.

Indices of aridity and humidity are used to determine the seasonal regime of moisture adequacy. In moist climates the *aridity index* is the annual water deficit taken as a percentage of annual PE. In dry climates water surplus as a percentage of

TABLE 6.2

Thermal Efficiency and Its Summer Concentration

Thermal Efficiency			*Summer Concentration*	
TYPE		INDEX (cm)	TYPE	CONCENTRATION (%)
A′	Megathermal	114 and above	a′	below 48.0
B'_4	Mesothermal	99.7 to 114.0	b′	48.0 to 51.9
B'_3	Mesothermal	85.5 to 99.7	b'_3	51.9 to 56.3
B'_2	Mesothermal	71.2 to 85.5	b'_2	56.3 to 61.6
B'_1	Mesothermal	57.0 to 71.2	b'_1	61.6 to 68.0
C'_2	Microthermal	42.7 to 57.0	c'_2	68.0 to 76.3
C'_1	Microthermal	28.5 to 42.7	c'_1	76.3 to 88.0
D′	Tundra	14.2 to 28.5	d′	above 88.0
E′	Frost	below 14.2		

annual *PE* gives the *humidity index.* Inspection of monthly data reveals the season of surplus or deficit (see Table 6.3).

TABLE 6.3

Seasonal Moisture Adequacy

Moist Climates (A, B, C_2)		*Aridity Index*
r	little or no water deficit	0 to 10
s	moderate summer deficit	10 to 20
w	moderate winter deficit	10 to 20
s_2	large summer deficit	above 20
w_2	large winter deficit	above 20
Dry Climates (C_1, D, E)		*Humidity Index*
d	little or no water surplus	0 to 16.7
s	moderate winter surplus	16.7 to 33.3
w	moderate summer surplus	16.7 to 33.3
s_2	large winter surplus	above 33.3
w_2	large summer surplus	above 33.3

The percentage of the mean annual *PE* that accumulates in the three summer months is the summer concentration of thermal efficiency (see Table 6.2). By combining the four elements of the classification it is possible to represent the climate of a place by four letter symbols (see examples in Table 6.4). The complexity of the system has made it difficult to display the great number of different climates cartographically, although continental maps of the separate elements have been prepared. The water budget concept has proved to be a more useful tool in the management of water resources, making it a truly applied approach to classification.

TABLE 6.4

Thornthwaite Classification of Selected Stations

Annual Water Budget Criteria	*Alice Springs, Australia* $(EA'da')$	*San Francisco, California* $(C_1 B'_1 da')$	*Moscow U.S.S.R.* $(C_2 C'_2 rb'_1)$
Precipitation (cm)	24.5	55.1	63.1
Thermal efficiency (*PE* in cm)	116.2	70.2	55.4
Summer concentration (% of *PE*)	43.8	33.3	64.4
Surplus (cm)	0	3.6	10.4
Deficit (cm)	91.7	18.7	4.5
Humidity index (%)	0	5.1	18.8
Aridity index (%)	78.9	26.6	8.1
Moisture index	-78.9	-21.5	10.7

Human health and comfort suggest another possible approach to defining climatic types, with potential applications in clothing design, housing, physiology, and medicine. In much the same way that heat and moisture data are used to determine critical boundaries for natural vegetation or crops, optimum and limiting values of climatic elements afford a basis for classification in terms of human response. Everyone is aware that the reaction of the body to a given air temperature is conditioned by wind, humidity, and sunshine. An individual's state of health, emotional outlook, type of clothing, degree of acclimation, and a host of other factors also influence personal reaction to climate. One's own reaction to a climate is perhaps a satisfactory basis for a "personal" classification, but as a scientific approach to the problem it is attended by great complexity. The human body appears to be a far less dependable instrument for integrating climatic elements than is a plant. Nevertheless, fruitful studies in the field of applied climatology have related climate to health, clothing, diet, or human perception of the environment. Maps which show the regional distribution of such relationships are in effect very specialized climatic classifications. A bioclimatic classification introduced by the American climatologist Werner H. Terjung in 1966 defined "physiologic climates" in terms of human responses to temperature, relative humidity, wind chill, and solar radiation (Fig. 6.2). Still other applied classifications organize information on various economic and social adjustments to climate or the impact of human activities on the environment. They are especially useful in studies of climatic change.

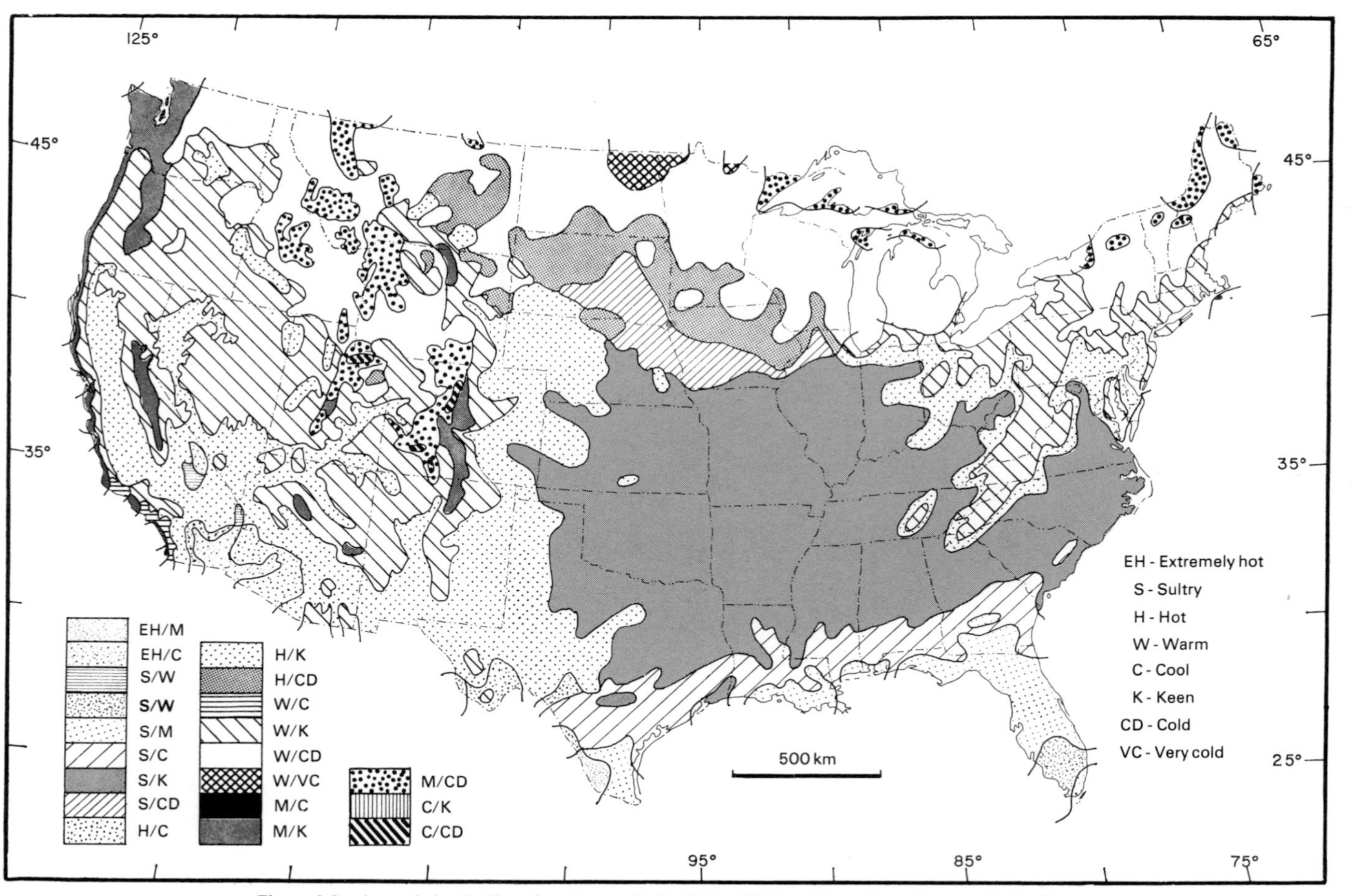

Figure 6.2 Annual physioclimatic extremes in the United States. (Reproduced by permission from the ANNALS of the Association of American Geographers, Volume 56, 1966, p. 176, Werner H.

It should be evident that there are many possible classifications of climate, for classification is a product of human ingenuity rather than a natural phenomenon. A complete classification should provide a system of pyramiding categories, ranging from the innumerable *microclimates* of exceedingly small areas (and often restricted to a shallow layer at the earth's surface), through *mesoclimates,* to highly generalized *macroclimates* on a world scale. But the description of world climates is not easily accomplished as the summation of a great number of microclimates, nor are microclimates easily fitted into the pattern of major world climatic regions. The higher categories of any classification system are necessarily generalizations; the lowest category must include individuals. Great difficulties attend the delimitation of an ''individual climate,'' for climates vary as a continuum over the entire earth. The concepts of *topoclimate,* representing the climatic response to local topographic conditions, and *ecoclimate,* the climatic environment of a living organism, are useful approximations. In any event, an individual climate is the synthesis of all the climatic elements in a unique combination that results from interacting physical processes. Since the exchanges of energy and mass between the air and the earth's surface are basic climatic processes, both surface and atmospheric conditions are appropriate criteria for classification of climate. Large-scale generalizations which take into account horizontal as well as vertical transport processes can be derived from a synthesis of atmospheric flow patterns. Sequences of synoptic maps, upper-air charts, and vertical cross sections afford a basis for classifying mean regional circulation types and their associated weather (see examples in Figs. 4.21, 4.23, 5.2, and 5.10). This approach, known as synoptic climatology, has great utility in long-range weather forecasting and aids the study of both genesis and effects of climate.

In order to achieve objectivity in defining the categories of a system, it is useful to have quantitative measurements of the climatic elements. In the past the lack of adequate records with respect both to periods covered and to worldwide distribution has presented a serious obstacle. The more than 100,000 surface weather-observing stations of all types in the world are by no means evenly distributed, and many of them record only one or two climatic elements during short or irregular periods. Although 3,500 ships take meteorological observations that are of great value in weather forecasting, the transient nature of these stations limits the use of their records for climatological analyses. The small number of stationary weather ships provide only a token record of climate for huge expanses of water. Eventually, these problems should be overcome by satellite technology, which already provides observations of surface temperatures and offers the promise of a thorough charting of world climates. But it still will be necessary to infer climatic data from the evidence of geology, glaciology, biology, archaeology, or early historical accounts in order to reconstruct climates of the distant past.

Some knowledge of the common classifications of climate is needed for an understanding of the objectives and methods of organizing climatic information on a regional basis. The following section outlines the main features of a classification scheme which has been used widely in climatology and physical geography and which combines features of the empirical and applied approaches.

KOEPPEN'S CLASSIFICATION

The most widely used system of climatic classification in its various modified forms is that of Wladimir Koeppen (1846–1940), a Russian-born meteorologist and climatologist of German descent who devoted most of his life to a scientific career in Germany. Koeppen aimed for an applied scheme that would relate climate to vegetation but provide an objective, numerical definition of climate types in terms of climatic elements. He devised his first classification (1900) largely on the basis of vegetation zones and later (1918) revised it with greater attention to temperature, rainfall, and their seasonal characteristics.

The Koeppen system includes five major categories which are designated by capital letters as follows:

A Tropical forest climates; hot all seasons
B Dry climates
C Warm temperate rainy climates; mild winters
D Cold forest climates; severe winters
E Polar climates

In order to represent the main climatic types, additional symbols are added. Except in the dry climates the second letter refers to rainfall regime, the third to temperature characteristics, and the fourth to special features of the climate. Table 6.5 shows the main climatic types of a modified Koeppen system. Tables 6.6 and 6.7 explain the symbols and boundary criteria used in designating the subdivisions. The distribution of major climatic types on the continents is shown in Fig. 6.3.

In actual application of the system to climatic statistics a great number of subdivisions are possible using the symbols noted in the tables. The numerical limits for certain of the subdivisions are different for each of the higher categories, making the detailed use of the system rather complicated. Several climatologists and geographers have made modifications of the Koeppen classification. The German climatologist, Rudolf Geiger, collaborated with Koeppen on revisions of the world climatic map. He and others continued to modify it after Koeppen's death. It is inevitable that boundary revisions will need to be made as new climatic data become available. One of the best known modifications of the Koeppen system in the United States is that by Glenn T. Trewartha, who has simplified the world climatic map and redefined several climatic types.

CLIMATIC REGIONS OF THE WORLD

Considering the world distribution of climatic types as primarily the result of heat and moisture regimes, we can recognize broad categories based on the exchange of heat and moisture between the surface and overlying air masses, which tend to

TABLE 6.5

Main Climatic Types of the Koeppen Classification *

Af	Tropical rain forest. Hot; rainy all seasons
Am	Tropical monsoon. Hot; seasonally excessive rainfall
Aw	Tropical savanna. Hot; seasonally dry (usually winter)
BSh	Tropical steppe. Semiarid; hot
BSk	Mid-latitude steppe. Semiarid; cool or cold
BWh	Tropical desert. Arid; hot
BWk	Mid-latitude desert. Arid; cool or cold
Cfa	Humid subtropical. Mild winter; moist all seasons; long hot summer
Cfb	Marine. Mild winter; moist all seasons; warm summer
Cfc	Marine. Mild winter; moist all seasons; short cool summer
Csa	Interior Mediterranean. Mild winter; dry summer; hot summer
Csb	Coastal Mediterranean. Mild winter; dry summer; short warm summer
Cwa	Subtropical monsoon. Mild winter; dry winter; hot summer
Cwb	Tropical upland. Mild winter; dry winter; short warm summer
Dfa	Humid continental. Severe winter; moist all seasons; long, hot summer
Dfb	Humid continental. Severe winter; moist all seasons; short warm summer
Dfc	Subarctic. Severe winter; moist all seasons; short cool summer
Dfd	Subarctic. Extremely cold winter; moist all seasons; short summer
Dwa	Humid continental. Severe winter; dry winter; long hot summer
Dwb	Humid continental. Severe winter; dry winter; warm summer
Dwc	Subarctic. Severe winter; dry winter; short cool summer
Dwd	Subarctic. Extremely cold winter; dry winter; short cool summer
ET	Tundra. Very short summer
EF	Perpetual ice and snow
H	Undifferentiated highland climates

*The data included in Tables 6.5, 6.6, and 6.7 are based on W. Koeppen, *Grundriss der Klimakunde* (Berlin: Walter de Gruyter Company, 1931); W. Koeppen, "Das geographische System der Klimate," Vol I, Part C, of W. Koeppen and R. Geiger, *Handbuch der Klimatologie* (Berlin: Gebruder Borntraeger, 1936); and modifications by R. Geiger, R. J. Russell, Glenn T. Trewartha, and others.

dominate the climates of different regions. The major groups thus defined are: (I) Climates dominated by equatorial and tropical air masses; (II) climates dominated by tropical and polar air masses; and (III) climates dominated by polar and arctic-type air masses. Other groups in this category are (IV) highland climates, which have distinctive features arising from the influence of altitude, and (V) climates of the oceans. Criteria for subdivision of these five groups into climatic types are the regional distribution of climatic elements—especially temperature and precipitation—and their seasonal variations. The subheadings of Chapters 7, 8, and 9 comprise an outline of the principal types. The schematic relation of

TABLE 6.6

Criteria for Classification of Major Climatic Types in Modified Koeppen System * (Based on annual and monthly means of precipitation in cm and temperature in °C)

Letter symbol 1st	2nd	3rd	Explanation
A			Average temperature of coolest month 18°C or higher
	f		Precipitation in driest month at least 6 cm
	m		Precipitation in driest month less than 6 cm but equal to or greater than $10 - r/25$
	w		Precipitation in driest month less than $10 - r/25$
B			70% or more of annual precipitation falls in warmer six months (April through September in the Northern Hemisphere) and r less than $2t + 28$
			70% or more of annual precipitation falls in cooler six months (October through March in Northern Hemisphere) and r less than $2t$
			Neither half of year with more than 70% of annual precipitation and r less than $2t + 14$
	W		r less than ½ upper limit of applicable requirement for B
	S		r less than upper limit for B but more than ½ that amount
		h	t greater than 18°C
		k	t less than 18°C
C			Average temperature of warmest month greater than 10°C and of coldest month between 18° and 0°C
	s		Precipitation in driest month of summer half of year less than 4 cm and less than ⅓ the amount in wettest winter month
	w		Precipitation in driest month of winter half of year less than 1/10 of amount in wettest summer month
	f		Precipitation not meeting conditions of either **s** or **w**
		a	Average temperature of warmest month 22°C or above
		b	Average temperature of each of four warmest months 10°C or above; temperature of warmest month below 22°C
		c	Average temperature of from one to three months 10°C or above; temperature of warmest month below 22°C
D			Average temperature of warmest month greater than 10° and of coldest month 0°C or below
	s		Same as under **C**
	w		Same as under **C**
	f		Same as under **C**
		a	Same as under **C**
		b	Same as under **C**
		c	Same as under **C**
		d	Average temperature of coldest month below −38°C (**d** is then used instead of **a**, **b**, or **c**)
E			Average temperature of warmest month below 10°C
	T		Average temperature of warmest month between 10° and 0°C
	F		Average temperature of warmest month 0° or below
H			Temperature requirements same as **E**, but due to altitude (generally above 1,500 m)

*In formulas t is the average annual temperature in °C; r is average annual precipitation in centimeters.

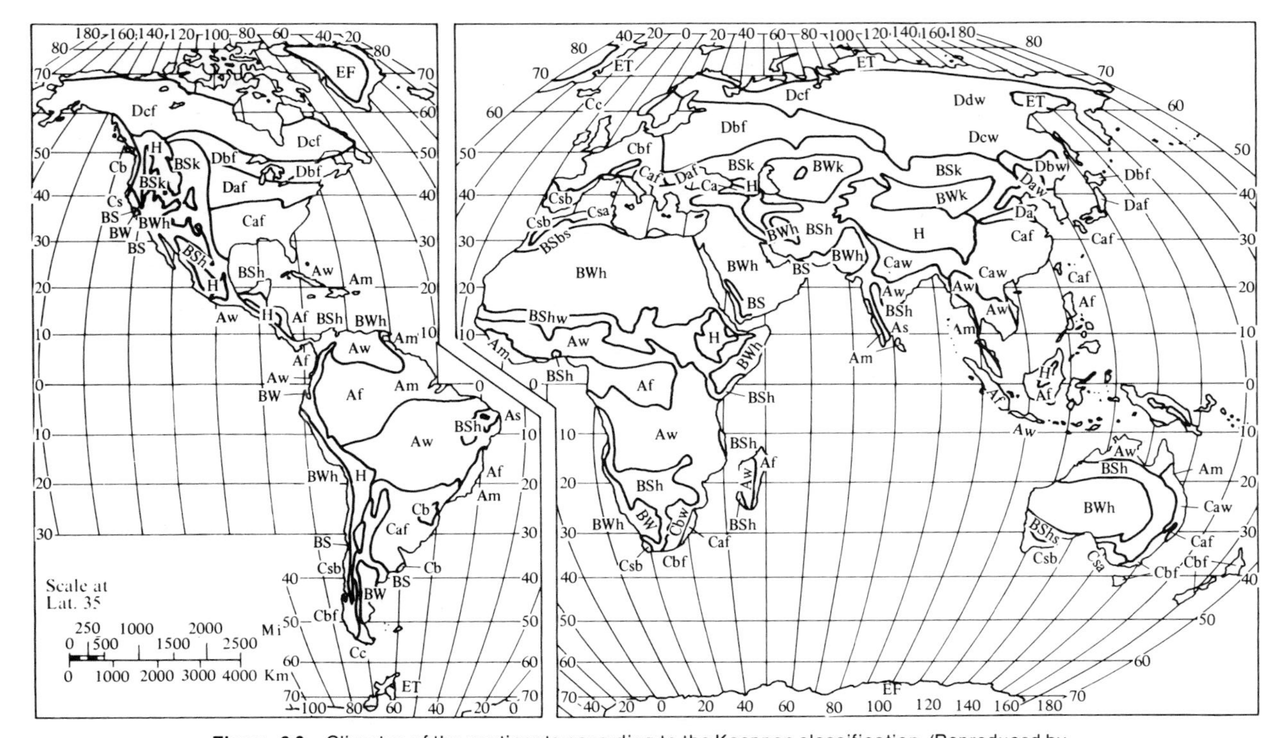

Figure 6.3 Climates of the continents according to the Koeppen classification. (Reproduced by permission from *Elements of Meteorology* by Albert Miller and Jack C. Thompson. Columbus, Ohio: Charles E. Merrill Publishing Company. Copyright 1970, 1975 by Bell & Howell Company.)

TABLE 6.7

Supplementary Subdivision Symbols and Criteria in Koeppen Classification

Symbol	Criteria
g	Ganges type of temperature regime; maximum before the summer rainy season
i	Annual temperature range less than 5°C
k′	Same as **k** but average temperature of warmest month less than 18°C
l	Mild; average temperature in all months between 10° and 22°C
n	Frequent fog
n′	Infrequent fog; high humidity, low rainfall, warmest month temperature below 23°C
p	Same as **n′**, except warmest month between 23° and 28°C
p′	Same as **n′**, except warmest month above 28°C
u	Coolest month after summer solstice
v	Warmest month in autumn
w′	Rainy season in autumn
w″	Two distinct rainfall maximums separated by two dry seasons
x	Maximum rainfall in spring or early summer, dry in late summer
x′	Same as **x** but with infrequent heavy rains in all seasons

climatic types over land is diagrammed in Fig. 6.4, and the front endpaper shows their geographic pattern on the continents.

This system for organizing climatic types is designed to facilitate the explanatory description of world climates. It follows the approach of Koeppen, relying heavily on temperature and precipitation, their seasonal regimes, and (on land) the response of natural vegetation as criteria for subdivision. It is in no sense a rational classification arrived at by mathematical computation. Nor is it readily susceptible to refinement to identify the many climatic subtypes at the meso- and microclimatic scales. Rather, it is intended to provide a framework for understanding the major climates of the earth and for examining climatic relations in the biosphere. At the same time it introduces a minimum of departure from the terminology one is likely to encounter in the literature of geography, climatology, or other sciences that treat world patterns of spatial distribution. Classifications of this kind have been used widely by geographers and climatologists; although they may not always be in agreement as to specific boundaries and subdivisions, the fundamental pattern of climates recurs for the reason that a fairly well-ordered system of climates does prevail on the earth in response to the effects of latitude, land–water distribution, and the global circulation of the atmosphere and oceans.

CLIMATIC TIME SCALES

Whereas regional climatology treats the climate system in its spatial dimensions, the history of climate is concerned with changes through time. Time, like space, is a continuum, and its divisions are products of human imagination. Any of the approaches to classification (empirical, genetic, or applied) might be employed logically to identify patterns of climatic change, but far less is known about the

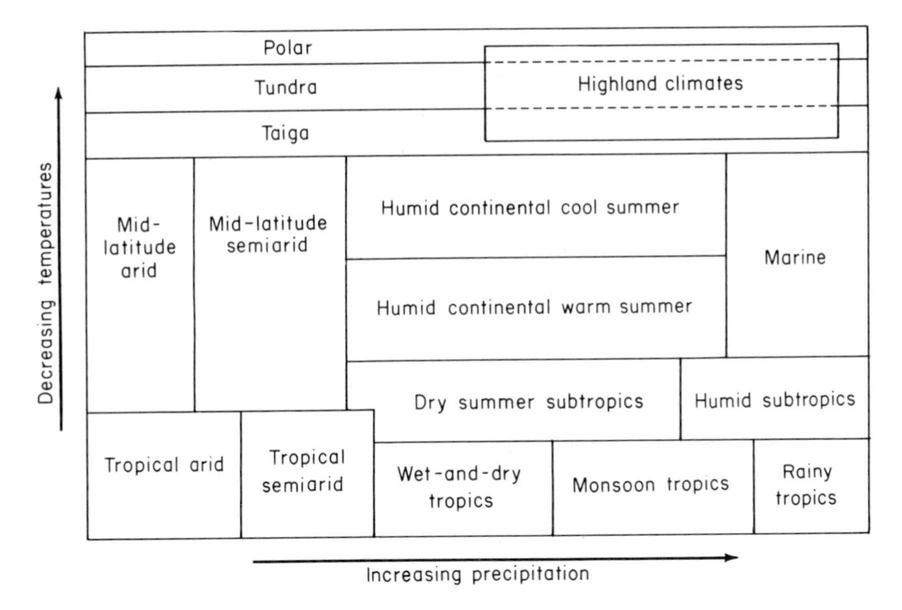

Figure 6.4 Schematic relation of major climatic types over land to temperature and precipitation.

nature and causes of climates in the distant past than those of the present. Consequently, time scales of climate commonly are based on the kinds of available evidence and theoretical explanations of change (Fig. 6.5).

Climate as a summation of all atmospheric events over an infinite time span does not change. Climatic change has no meaning unless we establish a time scale for comparing specific periods and analyzing trends. For the purpose of examining past climatic changes, periods of time and the associated evidence fall into four broad, overlapping categories:

1. Periods on the order of millions of years—*paleoclimates.* Since the climatic evidence is almost entirely inferred from rocks, fossils, and related deposits, the geologic time table is the conventional scale for classifying time in this category.

2. Periods in the past several thousand years, whose climates may be interpreted in terms of natural and cultural phenomena during the present geologic epoch, including recent ice ages, vegetation changes, and archaeological evidence.

3. Historic time, extending from the early traces of human settlements to the present and incorporating early written accounts of weather, climate, and their effects, such as famines, floods, or other anomalies.

4. The instrumental period, dating mainly from the nineteenth century. Within this period subdivisions often are marked by observed trends in temperature, precipitation, or their effects on the natural environment and human activities. Technological improvements in instrumentation, communication, and statistical analysis have permitted increasingly precise identification of spatial and temporal variations in the climate system. Averages for an arbitrary number of

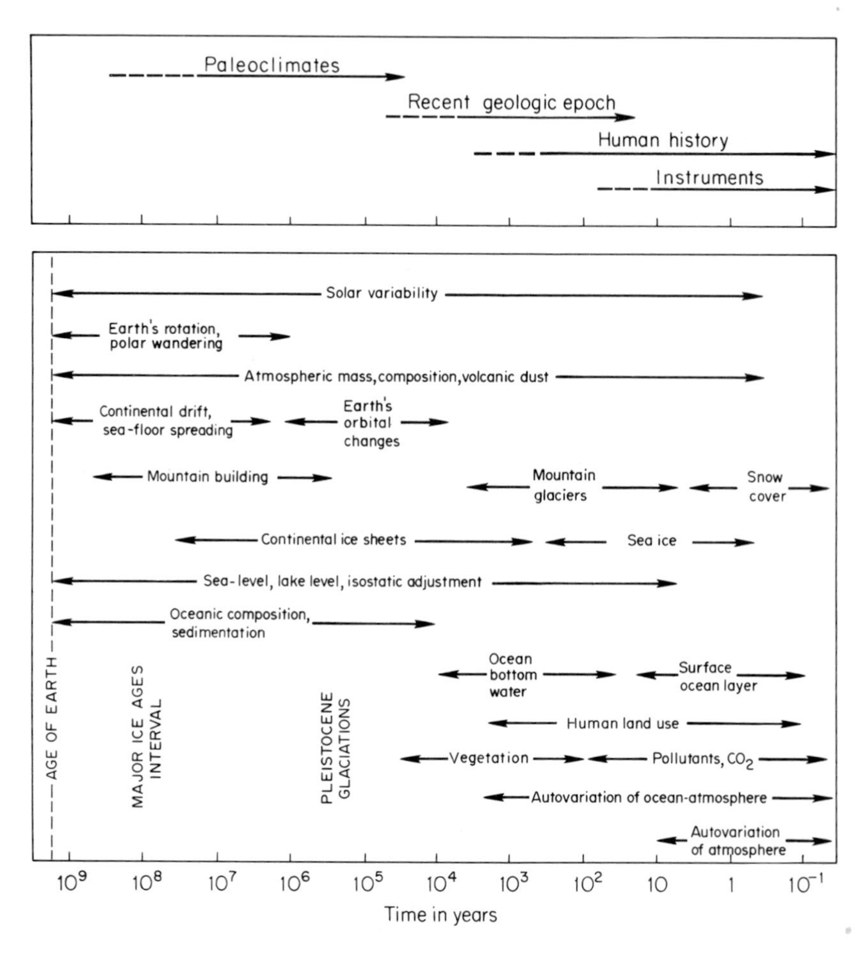

Figure 6.5 Relative time scales of evidence and possible causes of climatic change. (Adapted from *Understanding Climatic Change,* Washington: National Academy of Sciences, 1975.)

years (commonly 10 to 30 years) facilitate studies of trends. The monthly means of temperature and precipitation for three successive decades are taken as *climatic normals* for comparisons in the United States, although where the record is short or long-term analyses are desirable all years of the instrumental record may be required.

Much empirical research on climatic change has tended to emphasize one climatic element, usually temperature or precipitation, whose values can be graphed as a function of time. This approach is analogous to spatial classification on the basis of a single element and has similar shortcomings as an attempt to characterize climate. Criteria such as the energy and water budgets or the general circulation achieve better integration of climatic conditions through time as well as areally, and they are also preferable in applied studies of natural and cultural phenomena. Unfortunately, most climatic records are limited to temperature and precipitation data.

To the extent that a regional classification scheme expresses the essential features of climate, it also is a potential aid in identification of climatic change. Both statistical and cartographic techniques can be employed to differentiate prevailing climatic types during selected periods for which comparable data are available. Sequences of regional climate maps are especially useful for demonstrating that climatic change has spatial as well as temporal dimensions. Maps of successive vegetation patterns afford an indirect approach to this kind of analysis.

QUESTIONS AND PROBLEMS FOR CHAPTER 6

1. What is the purpose of a climatic classification? Explain why there is no single, universally accepted classification of climate.
2. Outline the various criteria that might be considered in making each of the major kinds of climatic classifications.
3. Contrast the taxonomic criteria of the Koeppen system with those of Terjung's physiologic climates.
4. Using climatic data for a station in your vicinity, determine the Koeppen classification of the climate. Why might the result be different from that derived by inspection of a world map of Koeppen climates?
5. What special problems are involved in a genetic approach to classification?
6. Critically examine the use of the term "humid" in description of a climate.
7. Evaluate the concept of continentality–maritimity as a basis for classification of world climates.
8. How might a regional classification of climates be used to study climatic change through time?

7

Climates Dominated by Equatorial and Tropical Air Masses

Climates under the influence of equatorial and tropical air have small annual ranges of temperature, and except in highlands temperatures are high throughout the year. These climates lie at low latitudes in the zones of the intertropical convergence, the trade winds, and the subtropic highs, where temperature contrasts between horizontal air streams are not great. A year-round net radiation surplus accounts for the major climatic features, yet there are marked regional variations in monthly and annual water budgets, ranging from some of the world's wettest areas to the driest deserts. Air masses within the tropics usually are distinguished more by their moisture properties than by temperature, and weather disturbances are associated with convection, convergence, or wave development rather than well-defined fronts.

THE RAINY TROPICS

The rainy tropical type of climate prevails in lowlands on and near the equator and along tropical coasts that are exposed to trade winds but backed by interior highlands (see front endpaper). The combination of constantly high temperatures with abundant rainfall well distributed through the year makes this a climate literally without seasons. Monthly temperatures average 25° to 28°C and, because there is little variation in the net gain of radiation throughout the year, the annual range of temperature is small. The annual march of the surface energy balance at Manaus, Brazil, is typical (Fig. 7.1).

The diurnal range of temperature in the rainy tropics is normally much

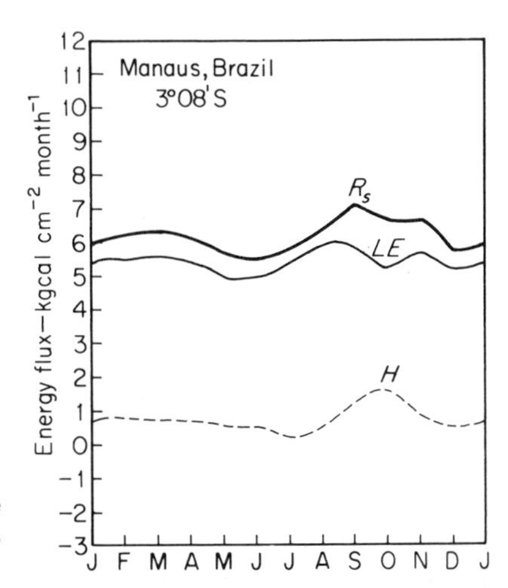

Figure 7.1 Annual march of surface energy balance components at Manaus, Brazil. (After Budyko.)

greater than the annual range between the maximum and minimum monthly means. Annual ranges are commonly less than 2 or 3C°, whereas diurnal ranges may be 8 or 10C°. Nighttime cooling of moist air under clear skies produces saturation and copious dew or perhaps morning fog. These condensation products evaporate into the warmed air shortly after sunup. Average daytime maximums are usually below 32°C. Singapore has mean daily maximums ranging between 30 and 32°C. Extreme maximums in the rainy tropics are lower than those of many mid-latitude stations, rarely exceeding 38°C. Another feature of temperatures is the small interdiurnal variation, which nevertheless may be perceived as significant by native residents. Whereas in higher latitudes the climate has a more or less regular seasonal rhythm, in the rainy tropics the dominant cycle is the day.

Annual precipitation in the rainy tropics exceeds 1,500 mm at most stations. No month is exceptionally dry, yet a graph of the precipitation regime is by no means as smooth as the annual march of temperature (see Fig. 7.2). Some stations have a definite maximum of precipitation in one month; a few have regimes with two maximums during the year, resulting from the seasonal migration of the ITCZ. Most rainfall is of the convectional type and comes with thunderstorms. The convergence of moist, tropical air and the intense daytime radiation create ideal conditions for thunderstorm formation. Usually, the thundershowers are concentrated in small areas, are of short duration, and often are of high intensity. In many areas their maximum occurrence is in the afternoon, and they are frequently preceded and followed by clear, sunny weather (see Fig. 7.3). Coastal locations and islands near the poleward margins of the rainy tropics may experience hurricanes, but these violent tropical cyclones do not occur along the equator nor in interior areas (Fig. 5.10). Hurricanes may account for a considerable proportion of the autumn rainfall of coastal areas visited by this type of storm.

Although thunderstorms are the dominant storm type, shallow cyclonic circulation sometimes develops in the rainy tropics. Two air streams of equatorial or

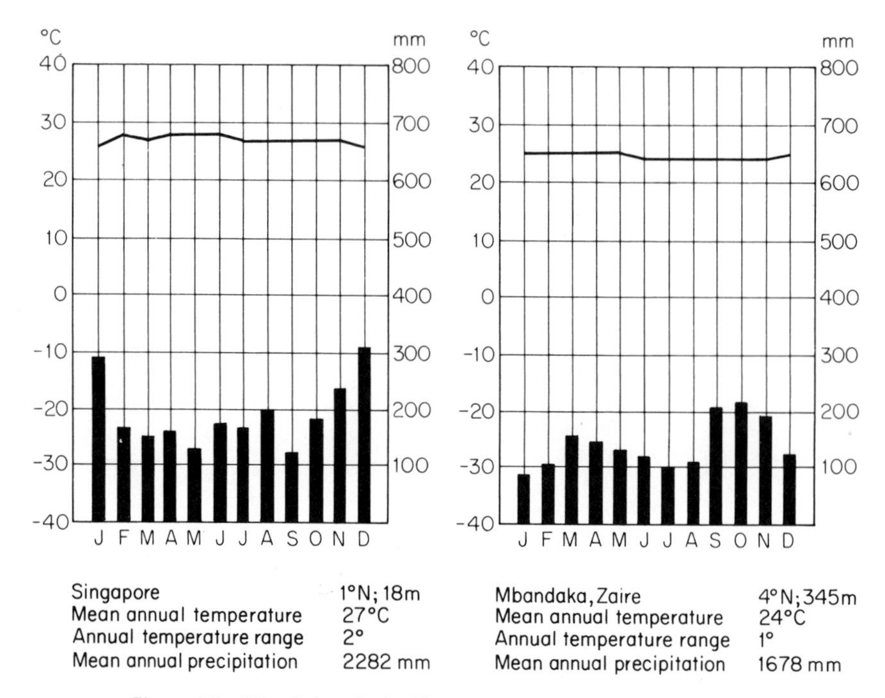

Figure 7.2 Climatic graphs for Singapore and Mbandaka, rainy tropics.

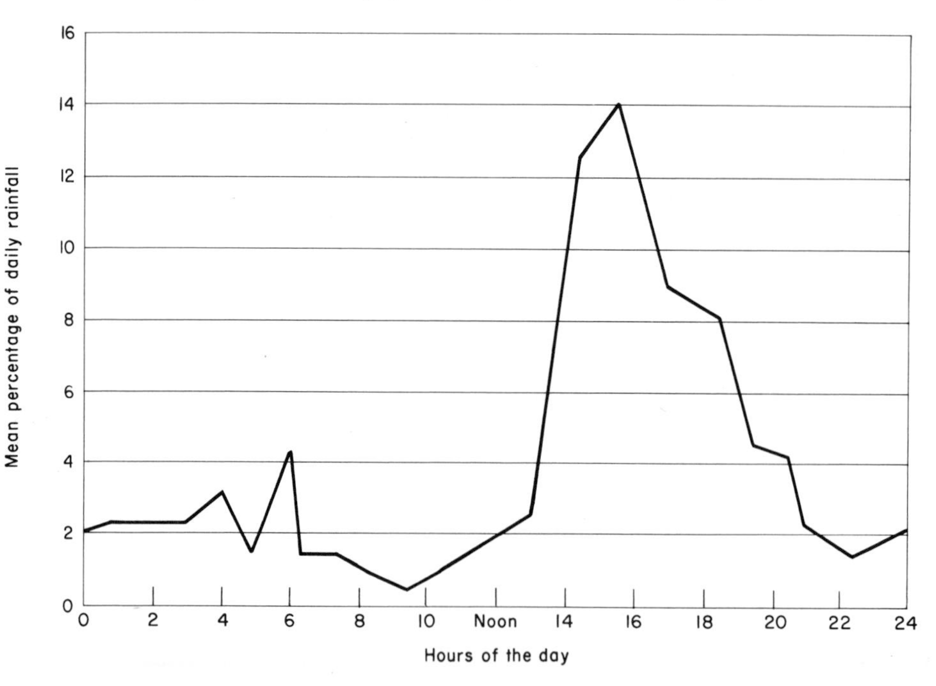

Figure 7.3 Diurnal variation of rainfall at Kuala Lumpur, Malaysia. (After Ooi Jin-bee, "Rural Development in Tropical Areas," *Journ. Trop. Geog.*, Vol. 12, March 1959.)

tropical origin are not likely to differ greatly in temperature characteristics, and zones of convergence accordingly tend to be broad troughs of low pressure accompanied by overcast skies and protracted periods of rain. On rare occasions, air masses from middle latitudes invade the tropics and produce frontal rainfall and lowered temperatures, but, for the most part, the storm lines of the tropics which seem to resemble extratropical fronts are zones of convergence within tropical air masses. In this connection it should be remembered that convergence can result from either different wind directions or wind speeds or both. Therefore, a close study of winds is necessary for prediction of weather in the tropics, for tropical weather is primarily of the air mass variety rather than frontal.

Where mountain ranges lie athwart trade winds which have travelled across warm water, the average annual precipitation of the rainy tropics is greatly exceeded. Examples of these trade-wind coasts are found in the Caribbean Sea, southeastern Brazil, eastern Madagascar, and on the numerous volcanic islands of the tropical Pacific. Wherever the orographic effect is a factor inducing heavy rainfall, leeward sides of the mountains show small annual totals in striking contrast to the heavy rainfall on the windward sides.

Owing to the constantly high temperatures in the equatorial regions, the rates of evaporation and transpiration are high and a correspondingly greater amount of rain is required to maintain satisfactory conditions for plant growth. Thus, the monthly precipitation amounts which might seem adequate in the latitude of the Great Lakes may result in equatorial drought, especially during dry spells of more than a few days. Although there is marked variability of rainfall from year to year in this climate, prolonged droughts of serious proportions are uncommon.

The air masses which dominate the rainy tropical climate are associated with the subtropic high and with moist air in the equatorial convergence (see Fig. 5.4). The trade winds originate in the subtropical cells of high pressure and flow equatorward. They constitute the most dependable wind systems of this type of climate but are found chiefly along its margins on east coasts, where their influence helps extend the rainy tropics farther poleward than in continental interiors. Some authors refer to these areas as the "tropical eastern littorals" or the "tropical windward coasts." Not all trade-wind coasts have the true characteristics of the rainy tropics, however. An example station in the poleward extension of the rainy tropics is Salvador, Brazil. Along the east coast of Brazil, south of Cape São Roque, the onshore flow of the southeast trades is stronger and more nearly perpendicular to the coast in late autumn and early winter. Salvador thus has a distinct autumn maximum of rainfall rather than the dry winter which is more typical of areas under strong trade wind influence. Suva, Fiji, also has an autumn maximum, partly because of its location in a region visited by the South Pacific subtropic high in winter and spring.

The ITCZ dominates by far the greater part of the rainy tropics, and associated weather accounts for most tropical precipitation. Because it is essentially a belt of shallow depressions, convection, and convergence, it does not have strong prevailing winds but rather light, erratic air movement except along some trade-wind coasts. The best developed horizontal temperature gradients in the rainy tropics are between land and sea, where the resulting pressure gradients produce daily land and sea breezes that alleviate daytime heat for many coastal locations.

Salvador, Brazil 13°S; 9 m

	J	F	M	A	M	J	J	A	S	O	N	D	Yr
T (°C)	26	26	26	26	25	24	23	23	24	24	25	26	25
P (mm)	74	78	163	290	298	195	206	112	85	94	143	98	1,837

Suva, Fiji 18°S; 9 m

	J	F	M	A	M	J	J	A	S	O	N	D	Yr
T (°C)	26	26	26	26	25	24	23	23	24	24	25	26	25
P (mm)	321	313	399	385	272	160	162	155	218	216	268	291	3,160

MONSOON TROPICS

The monsoon tropics are found in close association with the rainy tropics, generally along coasts where there is a seasonal onshore flow of moist air. Soil moisture storage during a distinctly wet season of several months duration is adequate to maintain evergreen forests through a shorter dry period. The climate may be regarded as transitional in nature from the rainy tropics to the wet-and-dry tropics, having annual rainfall totals comparable to the former and a regime of precipitation similar to the latter. It is well to remember that the term monsoon tropics does not apply to all climates affected by a monsoonal wind circulation. Its use stems from the characteristic climates of monsoon Asia, but the designation wet-and-dry tropics is applied to those regions with less annual precipitation and a distinct season of drought.

Mean monthly temperatures in the monsoon tropics are not greatly different from those in the rainy tropics—typical values ranging well above 20°C. These climates extend farther poleward, however, and the annual range is greater in some cases. The outstanding feature of the annual march of temperature for a number of stations in the monsoon tropics is the occurrence of the maximum before the high-sun period, that is, in the spring months rather than in July in the Northern Hemisphere. Maximum temperatures usually occur in the period of clearer skies, increasing insolation, and a water budget deficit just before the onset of more persistent cloudiness and heavy rainfall. Saigon is a representative station (Fig. 7.4).

Saigon (Ho Chi Minh City), Vietnam 11°N; 10 m

	J	F	M	A	M	J	J	A	S	O	N	D	Yr
T (°C)	26	26	28	29	28	27	27	27	27	26	26	26	27
P (mm)	6	13	12	65	196	285	242	277	292	259	122	37	1,808

Diurnal variation of temperature in the monsoon tropics is, on the average, slightly greater than in the rainy tropics. It is greatest in the drier months and least

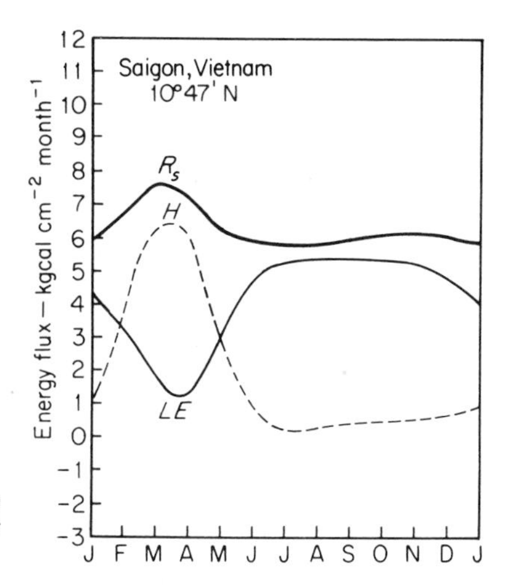

Figure 7.4 Annual march of surface energy balance at Saigon, Vietnam. (After Budyko.)

in the rainy season. During winter there may be some influence from cyclonic disturbances which pass on the poleward margin with resulting short periods of comparatively low temperatures.

Annual precipitation averages above 1,500 mm in most areas of the monsoon tropics. Where a strong onshore flow of moist air meets a coast backed by a mountain barrier exceptional rainfall totals are achieved (see Fig. 7.5).

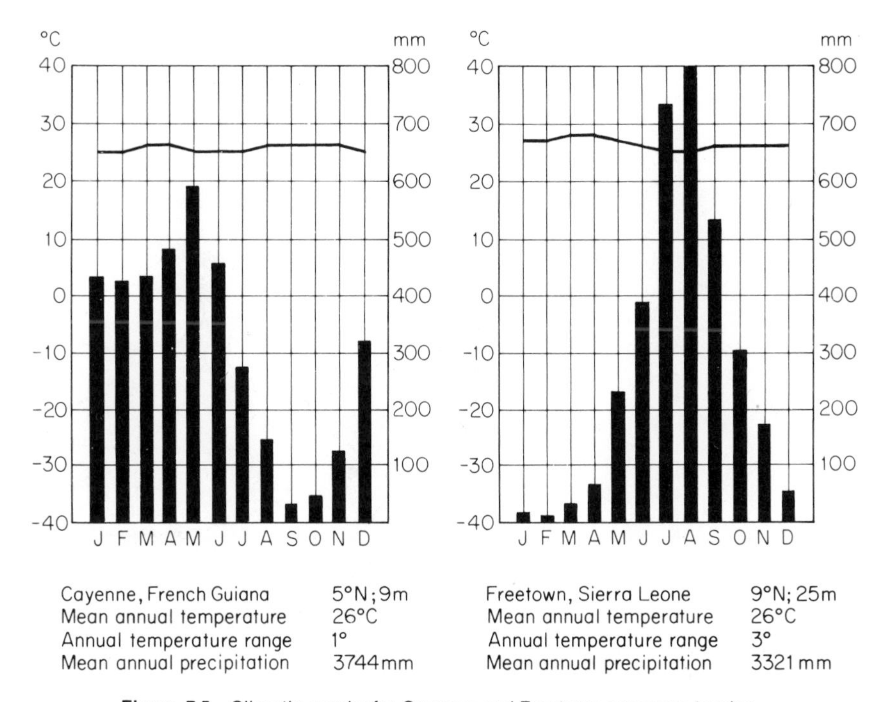

Figure 7.5 Climatic graphs for Cayenne and Freetown, monsoon tropics.

Precipitation which accounts for most of the total in the monsoon tropics is of the showery type, with the orographic effect often playing an important part, but certain areas are visited by tropical depressions. [Summer weather, that is, the rainy season, is similar to that of the rainy tropics, usually with greater monthly rainfall totals. Winter is slightly cooler and somewhat drier, having protracted sunny periods with occasional thundershowers.] Remnants of polar fronts sometimes invade these climates above the level of the trades in winter, producing overcast skies and light rainfall but no very great temperature decreases at the surface.

In southeastern Asia the characteristic pattern of circulation is monsoonal. Air trajectories from the Indian Ocean meet the easterly trades to produce a belt of convergence in summer over the east coast and offshore islands (Fig. 7.6). Much of the moisture precipitated along the east Asian littoral is transported aloft from the Indian Ocean. In winter a southern arm of the westerly jet stream lies south of the Himalayan divide, allowing little moisture into the area. Surface winds are from a northerly direction, and upper-level subsidence suppresses vertical cloud development.

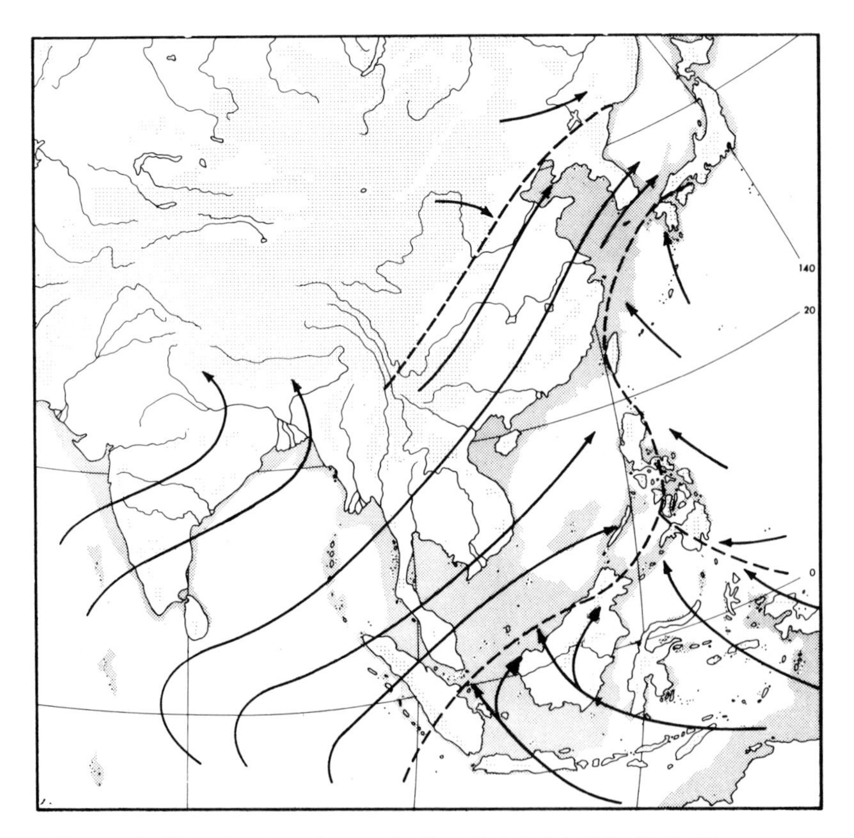

Figure 7.6 Mean air trajectories over Southeast Asia in July. (After Watts, Thompson, and others.)

In other regions where the monsoon tropical climate prevails, the seasonal variation in strength and persistence of the trades combines with the effect of contrasting land and water surfaces to produce a "monsoon tendency." Ordinarily regarded as easterly winds, the trades at times play a part in a shallow tropical flow known as the *equatorial westerlies.* Where the trades blow across the equator, they come under the opposite influence of the Coriolis effect and adopt a westerly component (southwesterly in the Northern Hemisphere or northwesterly in the Southern Hemisphere). For example, when the general circulation migrates northward in the Northern Hemisphere summer, the southeast trades take a position astride or even slightly north of the equator to approach the Guinea Coast of Africa and the Malabar Coast of India as southwest winds. These equatorial westerlies acquire moisture over expanses of warm ocean. As they converge toward the ITCZ and push beneath the upper tropical easterlies, they initiate instability and precipitation. In the case of India they merge into the monsoonal circulation (Fig. 7.6), which is further complicated by heat and moisture exchanges with the land surface, shifts in the upper-air flow patterns, and orographic effects. Occasionally during the summer the processes of convergence at low levels and divergence in the upper easterlies create broad low pressure areas having a cyclonic circulation. These "monsoon depressions" are most common over the Bay of Bengal but also form over the Arabian Sea. They bring increased rainfall and usually weaken as they move onto the Asian continent.

Wherever the trades are found they vary not only with the seasons, migrating to some extent with the general circulation, but also their strength varies periodically in what has been called the "surge of the trades." In the western tropical oceans, that is, along the western margins of oceanic anticyclones, periods of greater velocity in the trades are caused by the incursion of mid-latitude depressions which create a steeper pressure gradient away from the subtropic high centers. When these surges are developing a wind discontinuity, or *shear line,* exists between the strong winds of the surge and the weaker winds of the trades nearer the equator. Convergence along the shear line results in clouds of the cumulus type and showers.

In areas of the monsoon tropics where the monsoon circulation is well defined, winter weather is controlled to a large extent by air moving off the land and hence is somewhat drier. There may be brief local reversals of the typical winter outflow and accompanying showers. Where the winter is not dominated by a strong offshore monsoon, the mere lessening of onshore movement of warm, moist air results in lower monthly precipitation averages (see Fig. 7.7).

Some stations which have the general characteristics of the monsoon tropical climate exhibit asymmetrical regimes of annual rainfall. The northeast coast of South America, for example, has its heaviest rains in December and January, when onshore trades are strongest, and in May through July, when dominated by the ITCZ, but the August-through-October period is relatively dry. Cayenne, French Guiana, with precipitation maximums in January and May, illustrates these features.

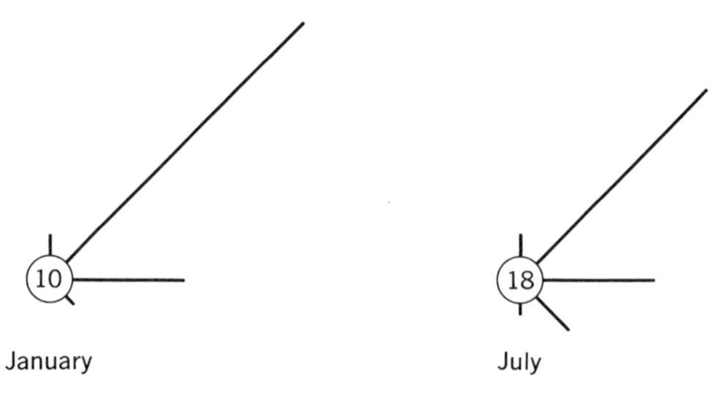

Figure 7.7 Wind roses for Georgetown, Guyana.

WET-AND-DRY TROPICS

As the name implies, the wet-and-dry tropical climate has alternating wet and dry seasons. It is transitional between the rainy and monsoon tropics on the one hand and the tropical arid and semiarid climates on the other. There is a distinctly dry period of two to four months, usually coinciding with the winter, and annual rainfall totals are less than in the rainy tropics. Mean monthly temperatures range from 18°C to above 25°C. The wet-and-dry tropics differ from the monsoon tropics most significantly in the effect of the dry season on vegetation and crops. The characteristic vegetation is savanna, comprising tall grasses or shrubs and scattered taller trees (Fig. 7.8). Latitudinal extent of the climate is from approximately 5° or 10° on the equatorial side to 15° or 20° on the poleward, that is, between the average positions of the intertropical convergence and the subtropic highs.

Owing to a net excess of insolation over outgoing radiation the wet-and-dry tropics have generally high temperatures throughout the year, although on the higher plateaus of South America and eastern Africa altitude produces lower averages. As in the monsoon tropics, there is a tendency toward a maximum temperature in late spring or early summer just before the season of greatest cloudiness and heavy rain. There may be a secondary maximum again just after the rainy season. Along the poleward margins annual ranges are greatest, and, in general, the annual ranges are greater than in the rainy tropics.

Diurnal ranges of temperature are greatest in the dry season and at higher elevations and are somewhat greater than in the rainy tropics. Nighttime temperatures may drop below 15°C in winter. The daytime temperatures of 25° to 30°C in winter are offset to some extent by the lower relative humidity. In summer, on the other hand, the diurnal range is small, temperatures are high, and these factors, combined with the rain and high humidity, produce oppressive conditions similar to the rainy tropics. Relief from the heat and humidity can be found only along coasts with strong sea breezes or at high altitudes.

The outstanding feature of precipitation (indeed of the climate) in the wet-and-dry tropics is the marked seasonal contrast. Many stations have one or more

Figure 7.8 Broadleaf tree and shrub association in the low veld of southeastern Africa. (Photograph by author.)

months with no precipitation recorded over a period of several years, whereas the wettest month has an average of 250 mm or more. Annual totals generally range from 1,000 to 1,500 mm, appreciably lower than in the rainy tropics. On the margins of the climatic region nearest the equator, the dry season is short and it is difficult to define the boundary between the wet-and-dry type and the monsoon tropics or the rainy tropics. Along the poleward margins the dry season is prolonged and conditions grade into the tropical semiarid climate, where potential evapotranspiration exceeds precipitation even in the wet season. Normally winter is the dry season, but along the Coromandel Coast of India spring is the dry period. Madras has a precipitation maximum in late autumn and early winter, when the monsoon circulation brings air across the warm Bay of Bengal from the northeast. The accompanying table shows climatic data for Madras, India, and Cuiaba, Brazil, both in the wet-and-dry tropics.

Madras, India 13°N; 16 m

	J	*F*	*M*	*A*	*M*	*J*	*J*	*A*	*S*	*O*	*N*	*D*	*Yr*
T (°C)	24	26	28	30	33	32	31	30	30	28	26	25	29
P (mm)	24	7	15	25	52	53	83	124	118	267	308	157	1,233

Cuiaba, Brazil 16°; 165 m

T (°C)	26	26	26	26	24	23	23	25	27	27	27	27	26
P (mm)	216	198	232	116	52	13	9	12	37	130	165	195	1,375

Reliability of precipitation is less than in the rainy tropics. A year with destructive floods may also have drought with serious economic consequences. Unseasonable rain or drought can be as disastrous as abnormal rainfall totals. The graph of annual rainfall over a 60-year period at Nagpur, India, illustrates the variability from year to year (Fig. 7.9).

Precipitation in the wet-and-dry tropics is associated with thunderstorms and weak tropical lows. Thundershower activity is greatest at the beginning and end of the rainy season, when violent storms are interspersed with sunny periods. As the rainy season becomes established, shallow tropical lows of the type found in the rainy tropics bring long periods of rain and cloudiness. The dry season weather is essentially like that of the tropical deserts, with only erratic showers.

Circulation patterns associated with tropical air masses exercise the principal

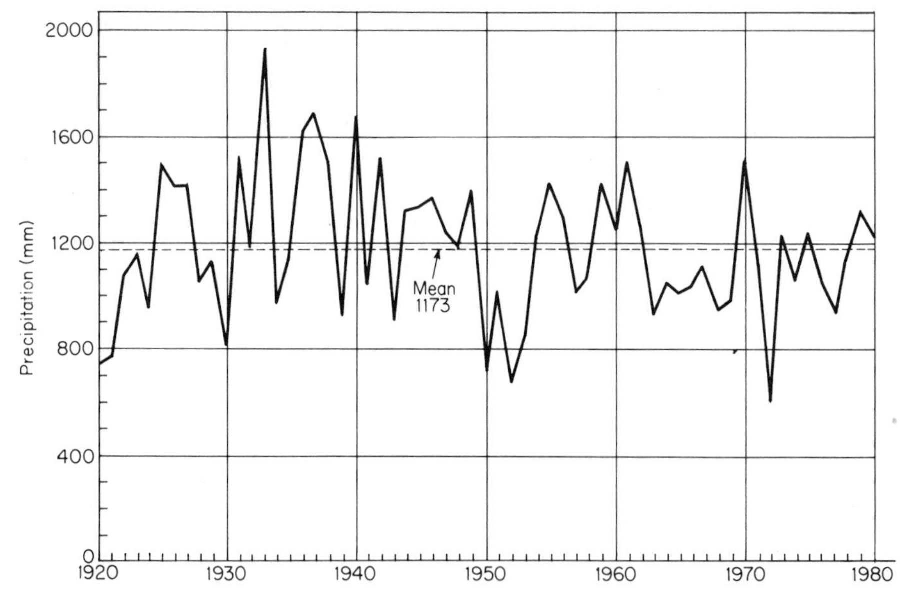

Figure 7.9 Variations in annual rainfall at Nagpur, India. (Data from Government of India Meteorological Department.)

control over the wet-and-dry tropical climate, and a marked seasonal alternation of dominant types is evident. In the wet season the ITCZ and equatorial air masses control the climate, bringing weather not far different from the rainy tropics. This results directly from the poleward migration of the wind and pressure belts of the primary circulation in summer, when the sun is at or near the solstice. Although the effects of the sun's apparent poleward migration lag behind the noon sun's latitudinal position by a month or six weeks, they are nevertheless associated with the summer half of the year and indirectly produce rainy weather which gradually moves poleward until after the reversal of the sun's migration at the solstice. Nephanalyses of satellite photographs have confirmed that intense convective activity indicated by cumulus clouds is at a maximum between 5° and 10°N over the Northern Hemisphere oceans in summer. In India and parts of southeastern Asia, where the monsoon circulation is strong, the ITCZ lies farther north and plays a complementary role to the monsoon in accounting for summer rains. In winter the wet-and-dry tropics are dominated by the subsiding air of the subtropic highs, which have gradually moved toward the equator, bringing arid conditions. In "monsoon Asia" (Fig. 7.10) and in northern Australia this effect is again complementary to the monsoon, which in winter blows off the continent. Thus, the

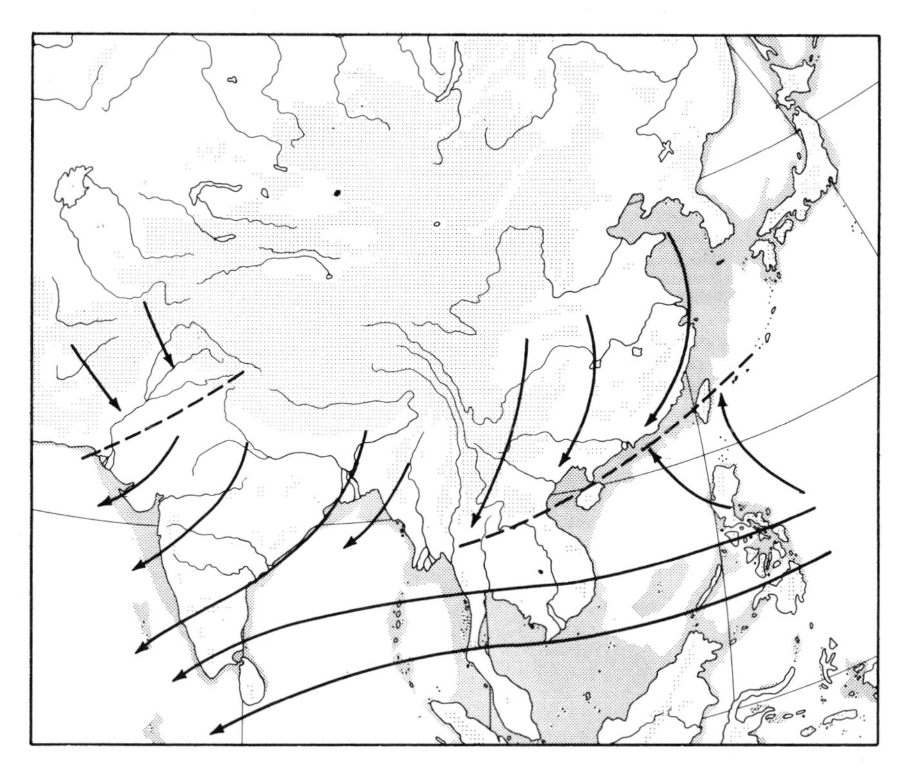

Figure 7.10 Mean air trajectories across southern Asia in January. (After Watts, Thompson, and others.)

climatic region is under the influence of unstable maritime air through most of the rainy season and of stable continental air in the dry season (see Fig. 7.11).

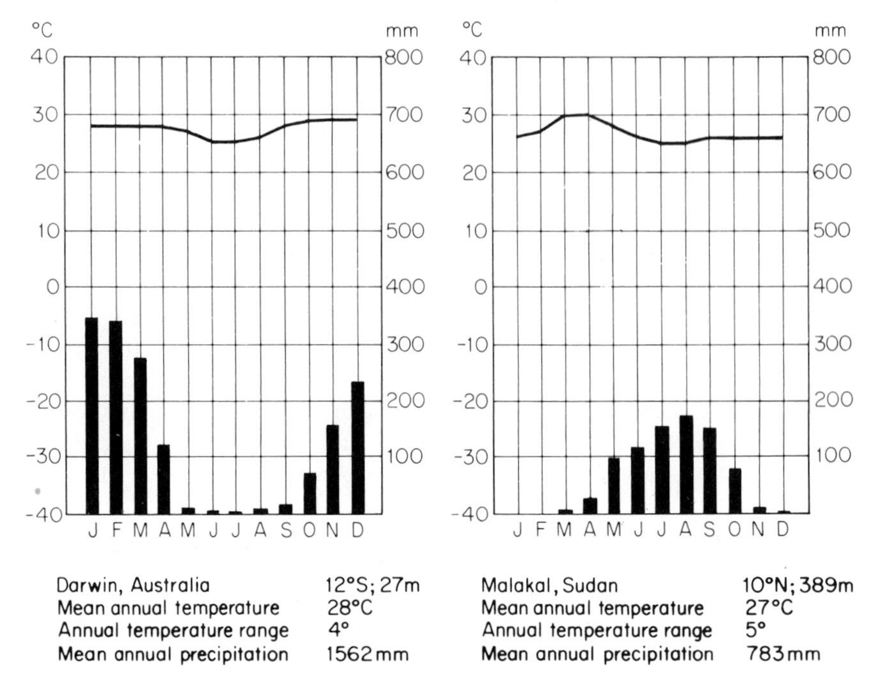

Figure 7.11 Climatic graphs for Darwin and Malakal, wet-and-dry tropics.

TROPICAL ARID AND SEMIARID CLIMATES

The arid and semiarid climates of tropical latitudes have numerous features in common, their differences being more in degree than in kind. The semiarid types are essentially a transition zone from the very dry regions to the bordering moister climates. The distinctive characteristic of these climates is the lack of sufficient rainfall to sustain dense vegetative growth. They are differentiated from the midlatitude counterparts by their higher average temperatures and the consequent need for more rainfall to offset evaporation and transpiration.

The tropical arid and semiarid climates are centered approximately on the latitudes 20° to 25° north and south, where the prevailing air masses are those that subside in the subtropic highs. The process of subsidence results in adiabatic heating and low relative humidity and thus dries the land surfaces. Even where surface heating produces intense low pressure the convection layer is shallow because of upper-air inversions whose effects extend the arid conditions well into the latitudes of the trades. The trade-wind inversion (a subsidence inversion) tends to restrict vertical cloud development. With certain exceptions, notably along west coasts, the core areas of arid climates have transitional semiarid belts situated more or less concentrically around the wetter margins. On the Deccan Plateau of India an elongated area of semiarid climate occurs without a contiguous area of true arid climate. Where tropical arid climates are located along west coasts of continents, ocean currents modify desert conditions. The four principal instances are as follows:

Region	*Ocean Current*	*Desert Name*
Lower California and Sonora	California	Sonoran
Coastal Peru and Chile	Humboldt or Peru	Peru and Atacama
Northwest African Coast	Canaries	Sahara
Southwest Africa	Benguela	Namib

Upwelling associated with these currents reduces temperatures along the immediate coasts and increases stability in the lower air layers, thereby reducing the tendency to cloud formation and precipitation. Advection fogs, stratus, and even drizzle are common along the coasts thus affected, however. In combination with lower temperatures they reduce the water need from precipitation.

Comparison of the climatic data for Walvis Bay, a coastal station of Southwest Africa, and Windhoek, inland but at an elevation of 1,728 m, reveals the influence of the cold Benguela Current (see the accompanying table). In view of its altitude, Windhoek might be expected to have lower summer temperatures. Instead, it is both warmer and wetter than Walvis Bay.

Walvis Bay, Southwest Africa 23°S; 7m

	J	*F*	*M*	*A*	*M*	*J*	*J*	*A*	*S*	*O*	*N*	*D*	*Yr*
T(°C)	19	19	19	18	17	16	14	14	14	15	17	18	17
P(mm)	2	5	8	3	3	0	1	3	1	1	2	1	30

Windhoek, Southwest Africa 23°S; 1,728 m

	J	*F*	*M*	*A*	*M*	*J*	*J*	*A*	*S*	*O*	*N*	*D*	*Yr*
T(°C)	23	21	21	19	16	13	13	16	20	22	23	23	19
P(mm)	77	73	81	38	6	1	1	0	1	12	38	47	370

Another influence on tropical arid and semiarid climates is the effect of mountain barriers. There is no arid area which can be ascribed primarily to mountain barriers, although the Andes in South America and the Atlas Range in northwestern Africa form rather definite rainfall boundaries. A more important effect of mountains in these climates is to induce slightly greater rainfall on their slopes. Examples are to be found in the Ahaggar Moutains in the Sahara, the highlands of Yemen, and along the ranges of Iran and Afghanistan. Not only is the precipitation greater along the mountain slopes, but the temperatures are lower owing to altitude, and potential evapotranspiration is therefore less.

The arid and semiarid areas of East Africa and eastern Brazil must be regarded as atypical locations of these climates in view of their position in latitudes frequently visited by the ITCZ. In parts of northern Kenya average annual rainfall is less than 150 mm. This is presumed to result from subsidence, divergence, and consequent stability of air that reaches the area, especially at upper levels, from the Indian Ocean. Nevertheless, maximum rainfall comes when the ITCZ is in the vicinity on its annual migration. Conditions are even drier farther north along the Somali Coast, where both the summer southwesterlies (recurved southeast trades from the southern Indian Ocean) and the northeasterly winter monsoon have a mainly divergent flow parallel to the coast. This ''monsoon'' reversal also produces a seasonal reversal of the offshore Somali Current, which experiences upwelling during the Northern Hemisphere summer. This upwelling along an east coast is a notable exception to the more common occurrence of the phenomenon along west coast deserts. In far eastern Brazil the stable air of an offshore anticyclone produces a dry summer. In winter the moist southeast trades which bring rainfall to the coast south of Cape São Roque do not penetrate strongly beyond the uplands in most years.

Maximum temperatures in the tropical arid climate are the highest in the world. Extreme maximums above 50°C are common. Azizia, Libya, recorded an official air temperature of 58°C on September 13, 1922. Ground temperatures are much higher under intense insolation. Monthly means of the summer months exceed 30°C at many tropical desert locations. El Golea and Jacobabad are example stations (see accompanying table) that have high summer temperatures, contrasting sharply with Walvis Bay and Windhoek.

El Golea, Algeria 31°N; 398 m

	J	*F*	*M*	*A*	*M*	*J*	*J*	*A*	*S*	*O*	*N*	*D*	*Yr*
T(°C)	9	12	16	21	26	32	34	33	30	22	15	10	22
P(mm)	8	1	6	4	T	1	0	T	2	4	4	14	44

Jacobabad, Pakistan 28°N; 56 m

	J	*F*	*M*	*A*	*M*	*J*	*J*	*A*	*S*	*O*	*N*	*D*	*Yr*
T(°C)	15	18	24	30	35	37	35	34	32	28	22	17	27
P(mm)	8	8	7	2	4	6	37	22	1	0	1	3	99

Annual range of temperature is far greater in the tropical deserts than in the rainy tropics. This is due in part to the generally higher latitudes of the former but results primarily from the clear skies that permit high net radiation in summer and a reduced radiation budget in winter. The annual ranges for El Golea and Jacobabad are 25C° and 22C°, respectively. Compare these with rainy tropical stations, for example, Mbandaka or Singapore (Fig. 7.2). Coastal stations have smaller annual ranges. Just as the annual range increases over dry lands and under clear skies, so is the diurnal range greater. Diurnal ranges of 15° to 25C° are common. On winter nights when long-wave radiational losses are rapid temperatures occasionally drop below freezing. Nocturnal frosts are particularly trying because of the contrast with the warm days.

Temperatures in the semiarid tropical climate are similar to those of the arid type. (See climatic graphs for Lahore and Luxor in Fig. 7.12.) As would be expected, the averages for semiarid climates on the high-latitude borders of the deserts are generally lower and the ranges greater than on the savanna margins nearer the equator.

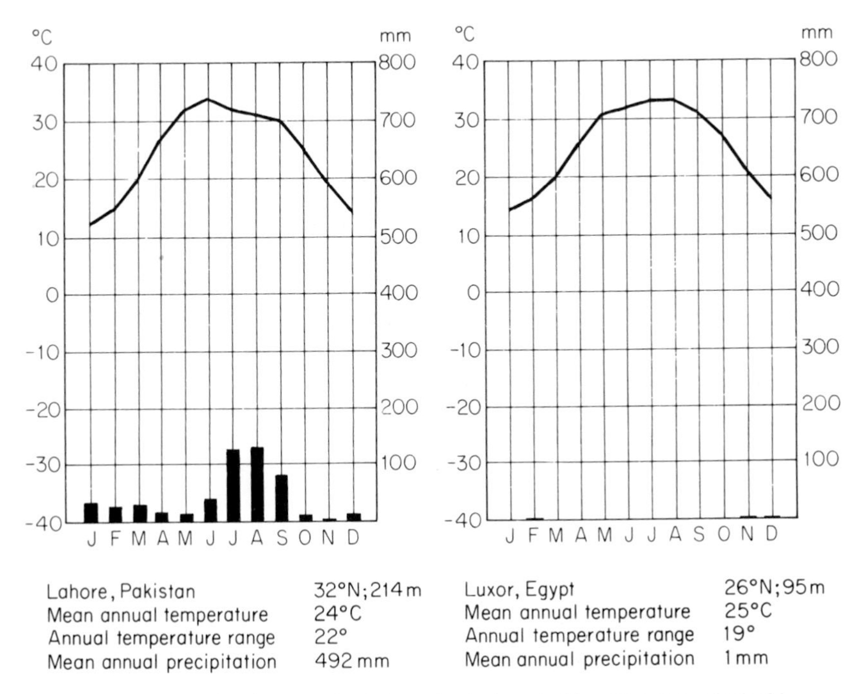

Figure 7.12 Climatic graphs for Lahore, tropical semiarid, and Luxor, tropical arid.

Precipitation in the tropical arid climates is not only low in amount but also erratic. There is no definite seasonal regime that could be termed characteristic of tropical arid stations; the climate is identified by lack of effective precipitation rather than a unique seasonal distribution. For example, stations near the margin of the monsoon tropics have a monsoonal regime, whereas those on west coasts bordering the dry summer subtropics have a winter maximum. The stations with no significant precipitation in any month can hardly be said to have a regime of precipitation. Arica, in northern Chile, has the record minimum average annual rainfall—only 0.5 mm over a 43-year period. In the semiarid climate the seasonal distribution becomes better defined as a rule and is more important in its effect on vegetation and land use. On the other hand, the variability of precipitation is a more critical factor in the semiarid climate than in the true desert. In wetter-than-average years settlers may venture into the semiarid grasslands only to find that the succeeding years are too dry for their type of land use. Figure 7.13 shows the negative departures from mean annual rainfall at the height of a disastrous drought in the Sahel region of northern Africa.

Most of the precipitation of the tropical deserts is of the thunderstorm type. It comes in downpours that rapidly exceed the absorptive capacity of the soil, and because there is little vegetative cover the runoff is great, often of flood proportions. Thus, a relatively small part of the rain is retained as soil moisture. In the semiarid climates there may be longer periods of rain associated with weak lows. In the poleward locations of the semiarid climate, winter rain comes with the mid-latitude cyclones which occasionally reach these areas. On the margins toward the equator, the ITCZ may bring limited summer rain.

The fogs and lower temperatures associated with the cold currents and upwelling along the coastal deserts modify the moisture conditions sufficiently to permit growth of some low forms of vegetation in spite of almost complete lack of rainfall. On the coast of Peru fog is so heavy at times that it is virtually a drizzle and is locally referred to as the *garua*. It is along the Peruvian coast that an interesting and unusual feature of variability in desert precipitation occurs. When the subtropic high moves abnormally seaward, offshore winds and upwelling become weaker and the Peru Current may flow at some distance from the coast. Warm water from the north, known as the *El Niño,* then spreads over the cold water along the shore. Instability and high moisture content above the warm sea surface pro-

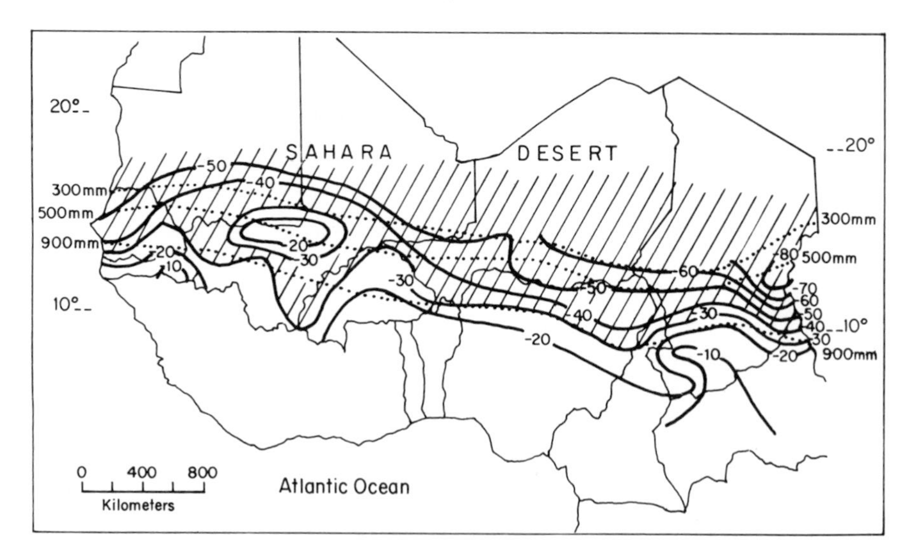

Figure 7.13 Percentage departure of 1973 annual rainfall from normal over the African Sahel. Slant lines indicate area receiving less than 70 percent of normal. Dotted lines are mean annual isohyets. (From Raymond P. Motha et al., "Precipitation Patterns in West Africa," *Monthly Weather Review,* Vol. 108, October 1980, p. 1573.)

duce heavy rains, usually during February and March but as late as May in years of a strong *El Niño*. In 1925 *El Niño* rains caused extensive damage to ruins of sun-dried bricks that had stood near Trujillo for centuries.

QUESTIONS AND PROBLEMS FOR CHAPTER 7

1. How is the monsoon tropical climate distinguished from the rainy tropics? From the wet-and-dry tropics?
2. Explain the strong seasonality of water budget regimes in the wet-and-dry tropics.
3. Why are April and May (rather than July or August) usually the warmest months in the monsoon tropics of Asia?
4. What conditions might lead to a water budget deficit at a place in the rainy tropics?
5. Why is precipitation alone an inadequate criterion for defining arid and semiarid climates?
6. Explain the great interannual variability of rainfall in arid and semiarid climates.
7. Describe and explain the influence of ocean upwelling on the energy and water budgets of tropical arid climates.
8. How are the equatorial westerlies formed? What are their effects on climate?
9. Suggest landscape indicators other than vegetation that might help differentiate the climatic types treated in this chapter.

8

Climates Dominated by Tropical and Polar Air Masses

In the mid-latitude circulation systems of both hemispheres lie the battlegrounds of contrasting air masses, where warm tropical air meets cold polar air along the ever-fluctuating polar fronts. Seasons are primarily warm and cold rather than wet and dry as in the tropics; the changing temperatures from season to season and with the passage of major storms and anticyclones play a much larger part in influencing human activities. Both periodic and erratic weather changes make these latitudes "intemperate" rather than temperate as they are commonly designated. These are the belts of the extratropical cyclones, where most of the precipitation is associated with fronts. All the climates of this group are subject to frost and all experience snow, although the amount and duration of snow cover vary widely. The principal air masses that dominate this group of climates are *cT, mT, cP, mP,* and their modifications. The interactions of these air masses produce the six major climatic types discussed in this chapter.

DRY SUMMER SUBTROPICS

The dry summer subtropical climate prevails on the west coasts of continents in the lower middle latitudes where the controlling air masses are *mT* out of the eastern margins of the subtropic highs. Only in the Mediterranean Basin, where the warm sea promotes cyclonic activity in winter, does this climate extend far eastward from a major ocean. From this region the climate has taken another common name, the Mediterranean type. Other locations are central California, central Chile, the southern tip of Africa, southwestern Australia, and an area in the southeastern part of South Australia and adjacent Victoria.

The chief features of the climate are a hot, dry summer and a mild, rainier winter. During the summer the climate comes under the influence of stable air which flows out of the oceanic subtropic high cells to the west (see Fig. 8.1). (In the

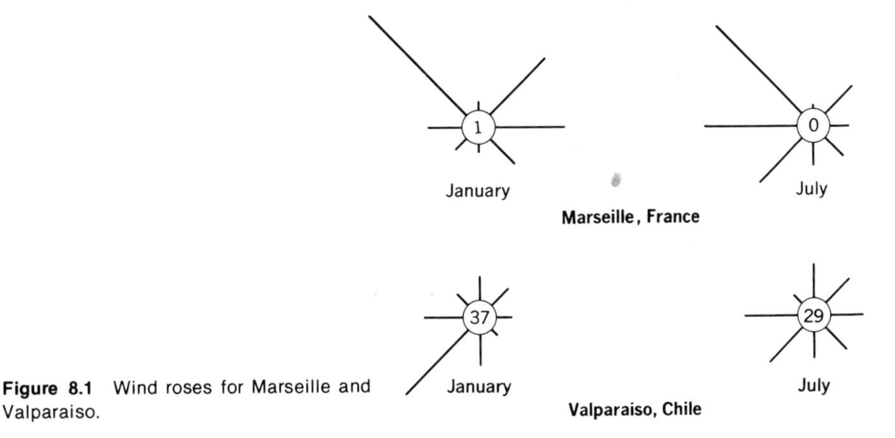

Figure 8.1 Wind roses for Marseille and Valparaiso.

case of the Mediterranean Basin the zone of subsidence in the subtropic high extends far into the eastern end of the Mediterranean Sea.) Under these conditions the climate is very much like that of the tropical arid and semiarid types. In winter the migration of the global circulation brings to these regions the tropical margins of the westerlies with their cyclonic passages and occasional invasions of modified polar air masses. The seasonal contrast reflects the annual march of the energy balance; the example of Lisbon is shown in Fig. 8.2.

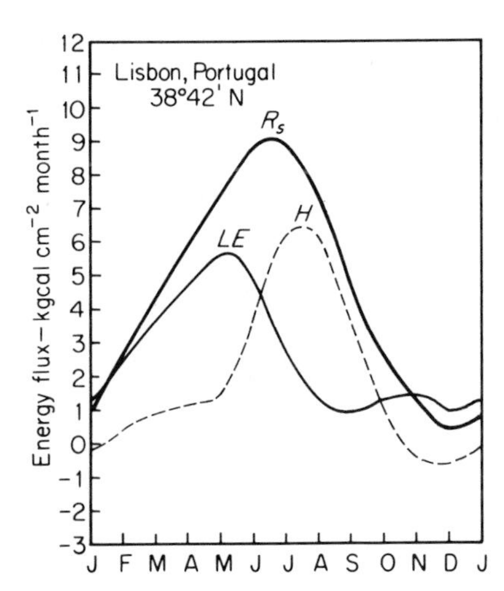

Figure 8.2 Annual march of surface energy balance at Lisbon, Portugal. (After Budyko.)

Annual temperature averages are somewhat lower than in the tropical climates; this is a *sub*tropical climate in terms of latitudinal location and temperature. Monthly averages in summer do not often exceed 27°C, although extreme maximums of over 38°C have been recorded at numerous stations. Coastal locations have much cooler summers because of a marine effect which is increased by cool ocean currents. An exception is found on Mediterranean shores, where the absence of cool currents and the generally warmer water cause higher summer temperatures than in other areas having this type of climate. A comparison of Sacramento and San Francisco affords an illustration of the influence of the ocean (see the accompanying table). Sacramento has a July mean of 25°C, whereas San Francisco on the coast has only 17°C. Note that San Francisco does not achieve its maximum mean monthly temperature until September, when upwelling in the offshore current has ceased, the coastal waters are warmer, and there is less fog.

Sacramento, California 39°N; 13 m

	J	*F*	*M*	*A*	*M*	*J*	*J*	*A*	*S*	*O*	*N*	*D*	*Yr*
T (°C)	8	10	12	16	19	22	25	24	23	18	12	9	17
P (mm)	81	76	60	36	15	3	T	1	5	20	37	82	414

San Francisco, California 38°N; 27 m

	J	*F*	*M*	*A*	*M*	*J*	*J*	*A*	*S*	*O*	*N*	*D*	*Yr*
T (°C)	9	10	12	13	15	16	17	17	18	16	13	10	14
P (mm)	102	88	68	33	12	3	T	1	5	19	40	104	475

The summer temperatures of the interior locations are reminiscent of the subtropical deserts. Daily maximums are high, frequently above 30°C, but the nighttime low may be below 15°C. With clear skies and low relative humidity, daytime heating is intense, but nocturnal cooling is rapid and the development of temperature inversions favors dense concentrations of pollutants in the lower atmosphere. Along the coasts, however, diurnal temperature ranges are much lower. The mean daily maximum for January at Valparaiso on the coast of Chile is about 21°C, whereas at Santiago in the interior it is 30°C. Yet Santiago has a lower mean daily minimum in the same month. Winter is a distinctly cooler season in the dry summer subtropics, the coolest month having a mean temperature usually below 10°C. As in summer, the diurnal ranges are greatest in the interior. Winter frosts occur but they are rarely severe. On the few occasions when night temperatures drop well below the freezing point, there is great damage to citrus and other crops. Freezing in the dry summer subtropics is ordinarily the result of rapid nocturnal radiation and air drainage in the lower layers of a *cP* air mass that has invaded the subtropical latitudes.

Katabatic, or gravity, winds sometimes plague the coastal regions in winter. The *mistral* in southern France and the *bora* along the Adriatic coast of Yugoslavia flow seaward from interior plateaus to create unpleasantly cool, dry conditions. In Southern California a hot, gusty wind known as the *Santa Ana* blows toward the coast in winter when a high pressure center is developed over the western United

States. Wind speeds sometimes reach 45 knots, and because of low relative humidity, the air becomes dusty and forest fire danger is often critical. In South Africa the hot, dry *berg* blows from the plateau as cyclonic circulation replaces high pressure.

Annual precipitation totals in this climate generally fall within the range of 350 to 900 mm, the amount being least on the semiarid margins and increasing poleward toward the marine climate. The summers have little or no rain, and this feature combined with the high temperatures (Fig. 8.3) results in extremely low soil moisture content. The meager summer rainfall consists largely of scattered showers attended by rapid runoff. Typical summer weather is an unbroken series of hot, sunny days. At the coasts, modified temperatures are accompanied by a tendency to fogginess, especially where the effect of cold ocean currents extends poleward from tropical desert coasts. Fog and dew help support plant growth along these coasts. Winter precipitation comes mainly from *mT* or *mP* air in cyclonic storms. Frontal rainfall ordinarily is light and scattered because the low centers normally pass on the high-latitude margins. Strong blocking highs in the westerlies may divert depressions to lower latitudes, especially along the coasts of the northeast Atlantic and northeast Pacific, bringing frontal conditions much like those which are regular occurrences in the marine west-coast climate, but with precipitation concentrated in fewer rainy days. At such times the upper-level Rossby waves have an abnormally great amplitude, and the westerly jet stream "bulges" equatorward.

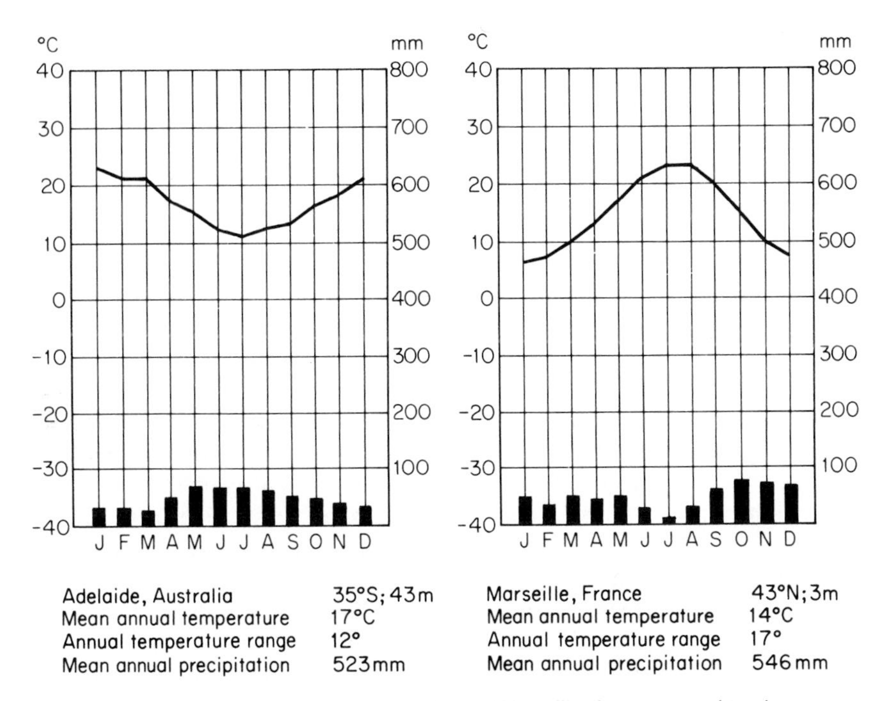

Figure 8.3 Climatic graphs for Adelaide and Marseille, dry summer subtropics.

Stations with little or no rain in the driest summer month commonly have 70 to 100 mm in the wettest winter month. Snow is rare at low elevations. Both snow and total precipitation increase on the slopes of mountains, which at their greater heights are properly classified with the highland climates.

HUMID SUBTROPICS

The humid subtropics lie in approximately the same latitudes as the dry summer subtropics, but they are on the eastern sides of the continents, where they come under the influence of conditionally unstable tropical air out of the western margins of the subtropic highs in summer (see front endpaper). Humid subtropical climates extend farther toward the equator and merge with humid tropical types, for example, in Southeastern Asia. Annual precipitation is greater than in the dry summer subtropics, but summer rainfall may be inadequate to meet the needs of potential evapotranspiration.

Summer temperatures and humidity are uncomfortably like those of the rainy tropics, for the air from the subtropic highs flows onto the land from the lower latitudes with a path over warm water. This is in contrast to the dry summer subtropics, where subsiding air moves out of the high cells from higher latitudes over colder water. Thus, the prevailing air masses over the dry summer subtropics are relatively stable and dry, whereas those of the humid subtropics are moist, warm, and often unstable (see Fig. 8.4). The prevalence of warm currents along east coasts in the subtropics and cold currents along the west coasts accentuates the contrast between the two climatic types. In winter, the humid subtropics of North America are influenced primarily by the belt of mid-latitude cyclones, with the two main air mass types being *cP* and *mT*. In Asia, the monsoon circulation is superimposed upon the global circulation and winter months have less precipitation than summer owing to the prevailing flow of stable air off the continent.

Temperatures in the humid subtropics are similar to those in the dry subtropics, but high relative humidity and warmer water offshore make the summers more like those in the wet tropical climates. The mean temperature of the warmest month is around 27°C for most stations. Mean daily maximums reach 30° to 38°C, and extremes exceed 38°C. Diurnal ranges are small, the nights being much more oppressive than in the dry summer subtropics. Furthermore, temperature variations from day to day are not very great. The growing season is long, but relief from summer conditions of heat and high relative humidity comes with the first salients of polar air in autumn. Then the humid subtropics suddenly lose their similarity to the rainy tropics. Temperatures of the cold month average between 5° and 12°C for most areas in the climatic type. The major exceptions are in monsoon Asia. Hong Kong has, as its lowest monthly mean, 15°C in January. Allahabad, India, has a January average of 16°C. These stations would be classified with the tropical climates but for their distinctly cooler winter seasons. The greater proportion of water in the Southern Hemisphere causes annual ranges to be generally lower than

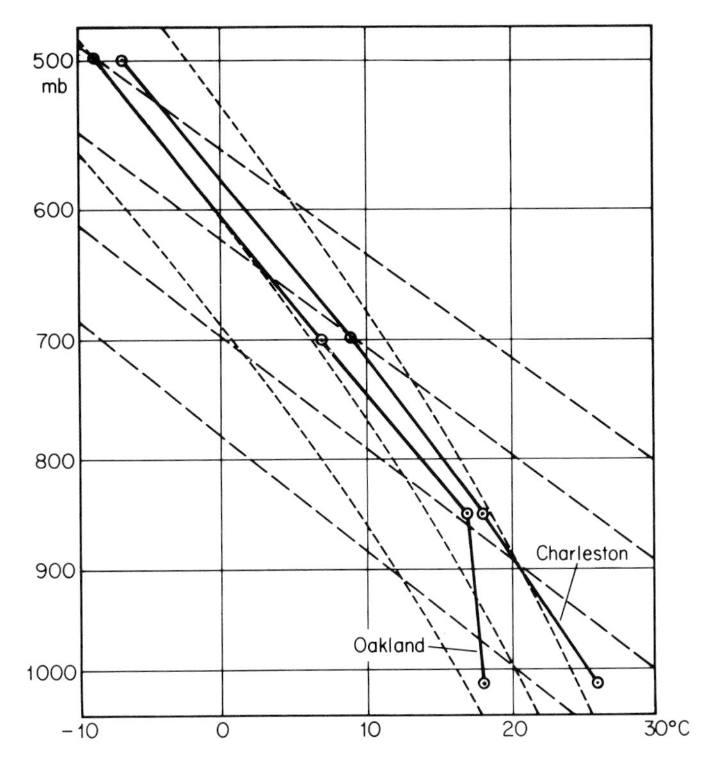

Figure 8.4 Mean vertical temperature soundings for July at Oakland, California, and Charleston, South Carolina. Note the relative stability of the surface layer at Oakland.

in the Northern Hemisphere in this climate. In both cases coastal stations show smaller annual ranges than those in the interior. Compare the temperature data for New Orleans with those for Buenos Aires and Memphis (see the accompanying data). Memphis, an interior station, has an annual range of 22C° compared with 15C° at New Orleans, whereas Buenos Aires has a lower summer maximum and a 14C° annual range.

Buenos Aires, Argentina 35°S; 55 m

	J	*F*	*M*	*A*	*M*	*J*	*J*	*A*	*S*	*O*	*N*	*D*	*Yr*
T (°C)	24	23	21	17	14	11	10	12	14	16	20	22	17
P (mm)	104	82	122	90	79	68	61	68	80	100	90	83	1,027

Memphis, Tennessee 35°N; 86 m

	J	*F*	*M*	*A*	*M*	*J*	*J*	*A*	*S*	*O*	*N*	*D*	*Yr*
T (°C)	6	7	11	17	22	26	28	27	24	18	11	7	17
P (mm)	148	116	124	117	102	92	87	67	71	73	106	119	1,223

New Orleans, Louisiana 30°N; 17 m

	J	*F*	*M*	*A*	*M*	*J*	*J*	*A*	*S*	*O*	*N*	*D*	*Yr*
T (°C)	12	13	16	19	23	26	27	27	25	21	15	13	20
P (mm)	98	101	136	116	111	113	171	136	128	72	85	104	1,369

Day-to-day variation in temperature becomes much greater as polar and tropical air masses alternately advance and retreat in their battle for supremacy in the succession of cyclones. While the winters can best be described as mild, frosts are normal in the higher latitudes and occasionally plague the tropical margins. Extreme minimums of −6° to −12°C have occurred along the United States Gulf Coast. Severe frosts are a hazard to crops and orchards in Florida just as they are in California.

Typical annual averages of precipitation in the humid subtropics vary from 750 to 1,500 mm. The lower averages are found in the regions bordering the semiarid climates, the greater amounts along mountain ranges and in the tropical extensions of the climatic type. Most areas have a fairly uniform distribution of rainfall throughout the year; 80 to 150 mm are typical monthly averages. Nevertheless, great variety can be found in precipitation regimes. In the southeastern United States the primary monthly maximum occurs in March at Vicksburg, in May at Little Rock, and in July at Mobile. Miami has a maximum in September owing to hurricane activity, whereas winter cyclones account for a January maximum at Memphis.

In India and southeastern Asia the monsoon produces a distinctly dry winter as dry continental air flows toward the coasts. Tokyo's autumn maximum represents the blending of a monsoon effect with tropical cyclonic storms (see Fig. 8.5). Its secondary maximum in early summer results from a southwesterly

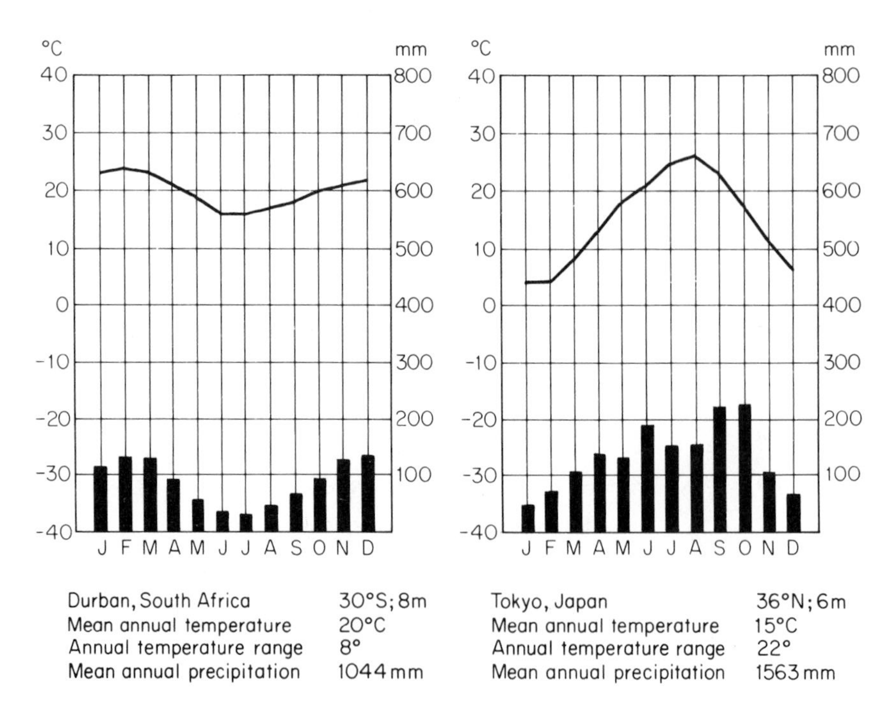

Figure 8.5 Climatic graphs for Durban and Tokyo, humid subtropics.

stream of warm, moist, unstable air of equatorial origin. The rains of this season are known as the *Baiu* in Japan.

For the climate as a whole, summer precipitation is primarily from air mass thunderstorms or line thunderstorms associated with sea breeze fronts and squall lines (see Fig. 8.6). Hurricanes or typhoons (or their remnants) visit the coasts occasionally in late summer and autumn and increase the monthly rainfall totals. As the polar front moves into these regions, frontal precipitation and thunderstorms of the frontal type are common in the converging poleward flow of air drawn into the warm sectors of cyclones.

Winter precipitation is mainly frontal and sometimes comes in the form of snow. Rainfall is lighter and more continuous; rainy days are greater in number, and cool, overcast days are more frequent. Frontal and ground fogs both occur in the humid subtropics, primarily in winter. Coastal fogs of the advection type are generally lacking because of the absence of cold currents in the adjacent oceans, but inland advection fogs may form as air is transported over cooler land surfaces.

Mountain barriers affect the distribution of precipitation in the humid subtropics just as in every other climate. One of the most unusual climatic stations in the world is Cherrapunji, India, to which allusion has already been made in Chapter 3 (see also the accompanying table). Cherrapunji is located at 1,313 m above sea level on the south side of the Khasi Hills in northeastern India. It is more than 300 km from the Bay of Bengal, but when the summer monsoon blows strongly the large area of lowlands allows moist air to pass inland, where it is forced upward

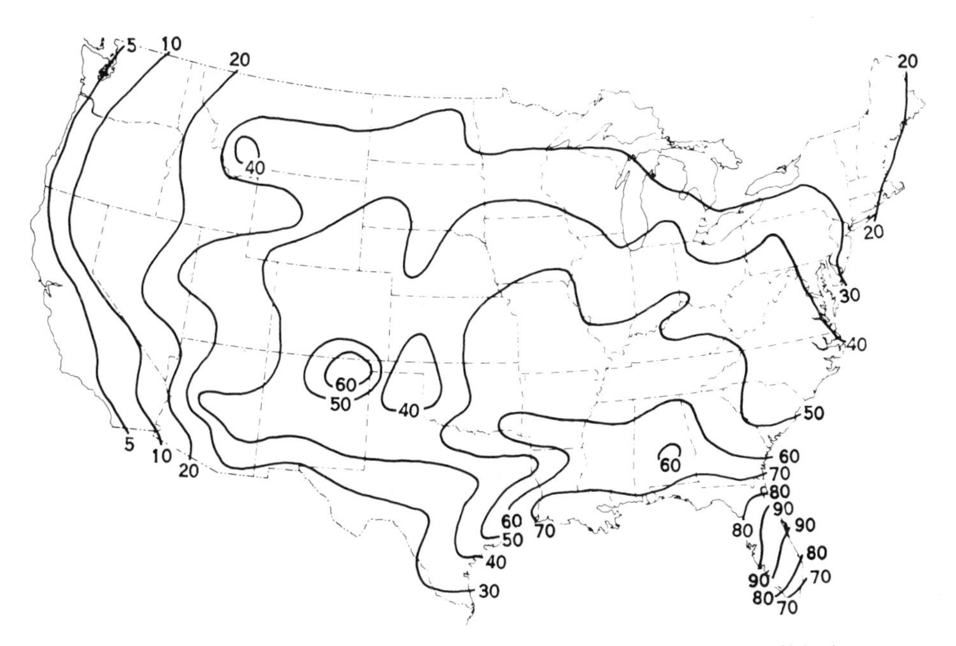

Figure 8.6 Mean annual number of days with thunderstorms in the contiguous United States. (NOAA)

in a funnel-shaped depression. Flood waters in the lowlands are warmer than the sea and probably contribute a share of the moisture that reaches the uplands. The dividing effect of the Khasi Hills acting as a mountain barrier is shown by the fact that Shillong, at 1,500 m but on the northern slope 40 km away, had only 2,410 mm mean annual rainfall in the period 1931–60. Were it not for the lower monthly temperatures and the relatively high annual range of temperature, Cherrapunji would be classified as a monsoon tropical station.

Cherrapunji, India 25°N; 1,313 m

	J	*F*	*M*	*A*	*M*	*J*	*J*	*A*	*S*	*O*	*N*	*D*	*Yr*
T (°C)	12	13	17	19	19	20	20	20	20	19	16	13	17
P (mm)	20	41	179	605	1,705	2,875	2,455	1,827	1,231	447	47	5	11,437

MARINE CLIMATE

The characteristic location of the marine climate is on the west coasts of continents poleward from the dry summer subtropics, where air of oceanic origin flows onshore (see front endpaper). The climate is frequently referred to as the marine west-coast type, although it also occurs along certain southeast coasts and extends well inland in Europe.

Most areas having this type of climate receive ample precipitation in winter, but in summer potential evapotranspiration exceeds rainfall. Winters are mild for the latitude and summers are relatively cool (see data for Portland and Melbourne in the accompanying table). The dominant air masses are of maritime polar origin, but modified *mT* and *cP* also affect the climate. In winter there is a succession of cyclones as the westerlies and the polar front prevail in these latitudes. There are fewer frontal passages in the marine climate in summer, when the subtropical oceanic high pressure cells reach their highest latitudes, diverting cyclonic storms poleward. Along the eastern margin of the Pacific high, an outflow of stable, subsiding air brings distinctly drier summer conditions to the North American Pacific Coast.

Portland, Oregon 46°N; 23 m

	J	*F*	*M*	*A*	*M*	*J*	*J*	*A*	*S*	*O*	*N*	*D*	*Yr*
T (°C)	4	6	8	11	14	17	20	19	17	12	7	5	12
P (mm)	136	107	97	53	51	42	10	17	41	92	135	162	944

Melbourne, Australia 38°S; 44 m

	J	*F*	*M*	*A*	*M*	*J*	*J*	*A*	*S*	*O*	*N*	*D*	*Yr*
T (°C)	20	20	18	15	12	10	10	10	12	14	16	18	15
P (mm)	45	59	50	69	54	52	54	50	58	74	70	68	691

Mean annual temperatures in the marine climate are mostly in the range of 7° to 13°C, unless modified by altitude. The mean monthly temperature of the warmest month is usually 15° to 20°C. Daily maximum temperatures do not often exceed 25°C, but extreme maximums of over 35°C sometimes occur when continental air temporarily invades a coastal area. Nights are cool, partly owing to nocturnal radiation and partly because of the generally low temperatures produced by winds off the sea. The marine influence is seen in low diurnal ranges as well as low annual ranges of temperature, but the effect decreases inland. Horizontal temperature gradients are greater from the coasts toward inland locations than in north-south directions, because the transport of heat from ocean sources is more significant than latitudinal variation of net radiation in this climate. Wintertime anomalies are even greater than in summer, temperatures being from 5° to 15C° higher than the normal for the latitudes. (The hythergraphs in Fig. 8.7 compare the maritimity-continentality of four stations in North America.) Long periods of overcast skies and prevailing onshore flow of maritime air in winter keep diurnal ranges small. Off northwest Europe and the British Isles, the Gulf Stream and the North Atlantic Drift extend the abnormally warm winter conditions into high latitudes. Mean monthly temperatures of the coldest month are above freezing throughout the climatic region. Extreme minimums below -15°C occur at the higher latitudes and at inland and upland stations, but prolonged periods of continuous freezing weather are unusual at the coasts. Outbreaks of continental air are responsible for periods of clear, freezing weather. At such times the wind is from an easterly direction, blowing out of a continental anticyclone (chiefly *cP* or *A* air masses). The North American Pacific Coast is protected to some extent from the cold continental air by the mountain barriers of the Rockies and Cascades. Even when the air pushes to the coast, the stable-air foehn modifies its chilling effects appreciably. In northwest Europe, the marine climate extends farther inland than on

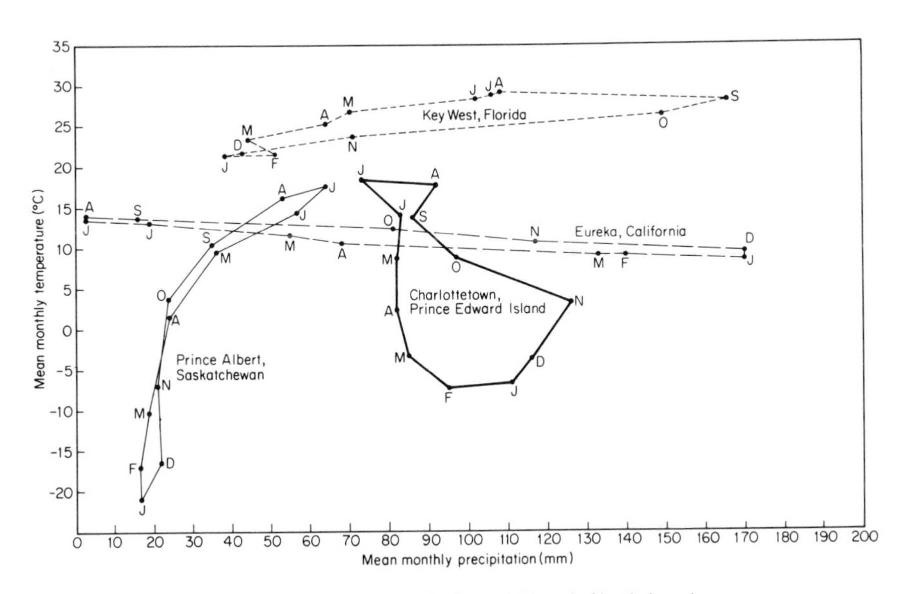

Figure 8.7 Hythergraphs for four stations in North America.

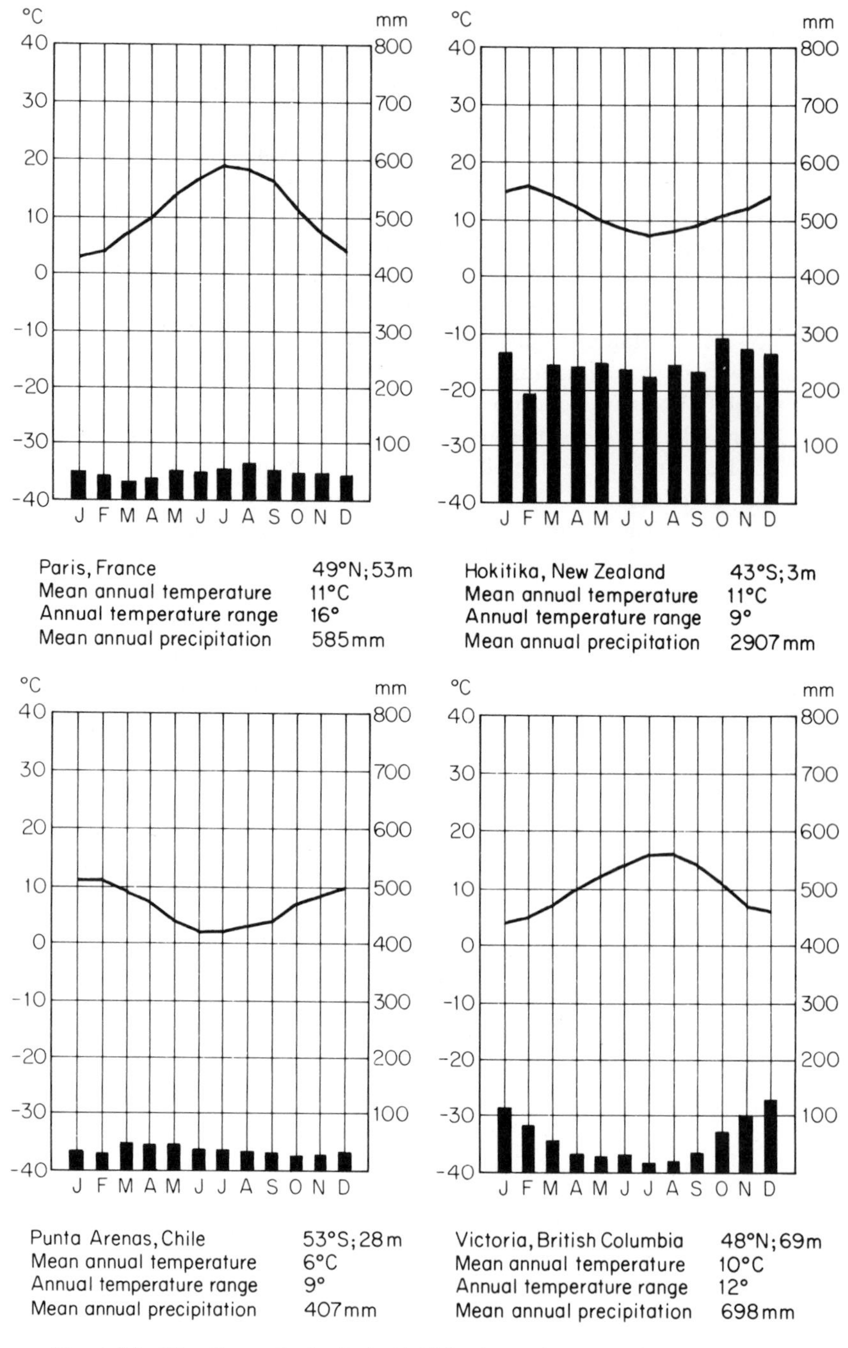

Figure 8.8 Climatic graphs for Paris, Hokitika, Punta Arenas, and Victoria, marine climates.

any other continent, largely because of the absence of an effective mountain barrier transverse to the prevailing westerlies. Not only do the maritime effects reach farther eastward, but the occasional polar outbreaks of northern Europe cover a much larger area, often including the coasts. Boundaries between marine and continental climates are sharply defined by mountains in North America and in southern Chile but are transition zones in Europe.

Annual precipitation ranges from less than 500 to more than 2,500 mm in the marine climate (Fig. 8.8). Most areas experience a water budget deficit in the warmer months, although the dry season is neither so long nor so pronounced as that of the dry summer subtropics. Margins adjacent to the dry summer subtropics have the lowest summer rainfall averages. Less potential evapotranspiration accompanies the lower summer temperatures, and natural vegetation does not show a marked influence of seasonal drought. The water budget graphs for Birmingham and Athens illustrate the contrast (Fig. 8.9).

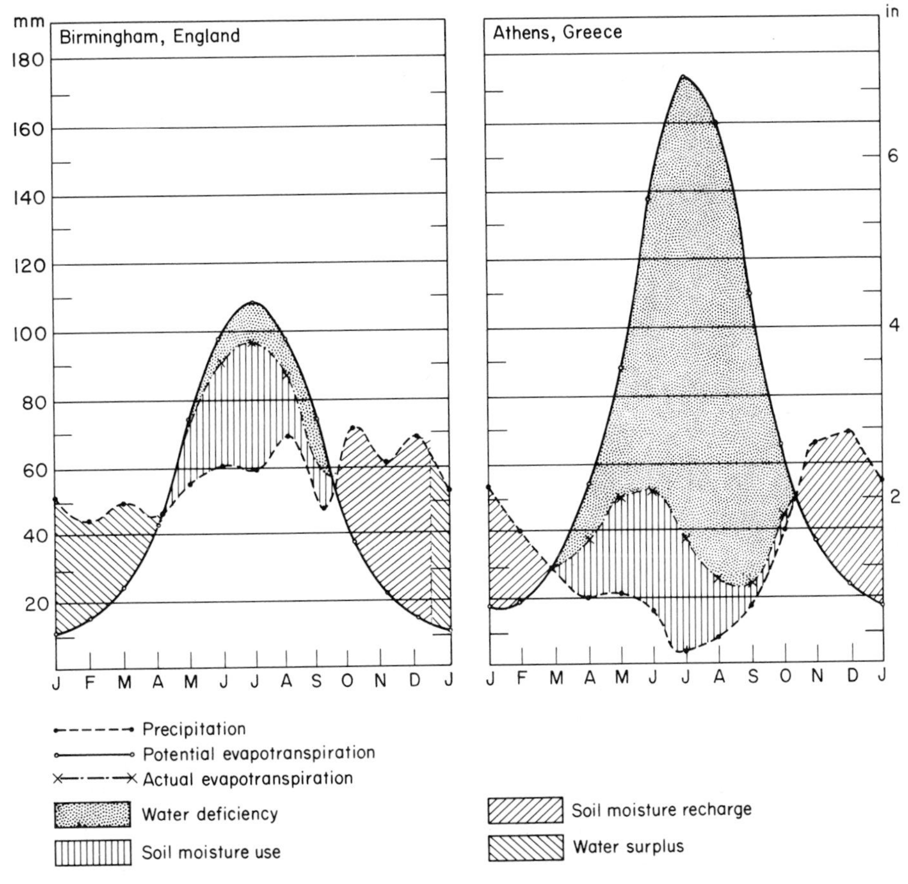

Figure 8.9 Contrasting water budget graphs for Birmingham, marine climate, and Athens, dry summer subtropics. (After method of Thornthwaite, based on data computed by C. W. Thornthwaite Associates, Laboratory of Climatology.)

Leeward of mountain barriers total yearly precipitation may be less than 500 mm. Extreme southern Chile, the Otago Basin of southern New Zealand, and lowlands about the Strait of Juan de Fuca in Washington and British Columbia exhibit this effect. Owing to cold ocean surface temperatures offshore, the dry summer regime along the North American west coast extends farther poleward than in any comparable area of the world. Agricultural lands in the Puget Sound–Willamette Lowland are drained by tiles or ditches in winter and irrigated in summer, suggesting an anomalous subregion of the marine climatic type.

Some areas with a marine climate have a precipitation maximum (or a secondary maximum) in autumn. The graphs for Paris and Punta Arenas illustrate this regime. October is the rainiest month at Juneau and Sitka, Alaska. Autumn maximums are attributable to the warmer source regions of maritime air masses at that season which give the air a higher moisture content than in late winter or spring.

Two features of the marine climate are the reliability of precipitation and the large number of rain-days, especially in winter. Great departures from the annual or monthly means are uncommon. Many stations have more than 150 rain-days a year (see Fig. 8.10). Bahia Felix, Chile, near the Straits of Magellan, has an average of 325 rain-days per year; in 1916, rain fell there on 348 days! In such a climate gray, overcast skies with light rain or drizzle are the rule. There may be a

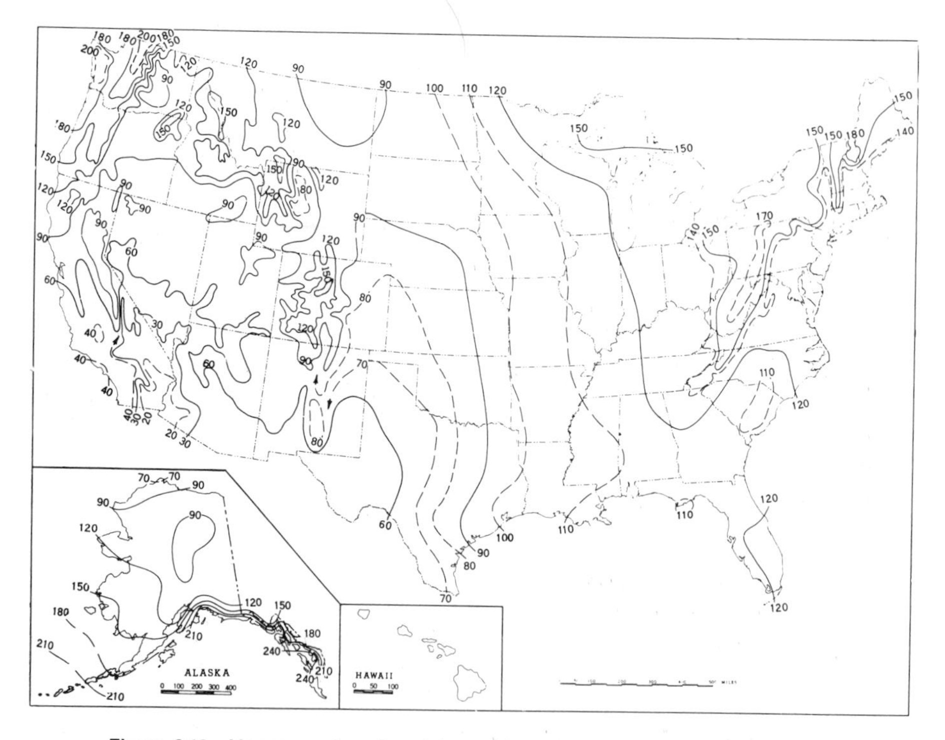

Figure 8.10 Mean annual number of days with precipitation of 0.25 mm or more in the United States. (NOAA)

number of consecutive days with precipitation, but the total rainfall during such periods is small. Dull, drizzly weather is the price paid by inhabitants of the marine climatic regions for mild winters. The parade of cyclonic storms brings relatively warm maritime air, but it also brings moisture, and often strong winds. Stationary and occluded fronts are common, the occlusion process often taking place along coastal mountain ranges. On the other hand, several fronts may pass over a region in rapid succession without a single rainless day. Needless to say, there are few sunny days in winter. In the lowlands, snow may occur on a few days, but ordinarily it remains for only a short time. Frequently, it is "rained off." Only when an invasion of cold polar air follows a snowfall is there a snow cover for more than a day or two.

Summer precipitation is also largely of cyclonic origin. It is more showery and falls on fewer days and yet may reach monthly totals equal to those of winter, the exception being those coasts which come under marked influence of the oceanic highs. Thunderstorms are far less numerous than in the subtropical or the continental climates, for the air which moves in from the high cells tends to be more stable. Most lowland areas average fewer than 10 thunderstorms a year. At higher elevations, in the mountain areas, total precipitation is, of course, greater and thunderstorm activity is more common. Quite often, conditionally unstable maritime air is cooled sufficiently by orographic lifting to set off convective storms along mountain ranges. In company with the showery precipitation and smaller number of rain-days in summer, there is more sunshine.

Fog frequency is high in the marine climate. The maximum incidence of fog is in autumn and winter, when warm, stable maritime air drifts across colder land. The marine climate is remarkably free from violent storms such as hurricanes or tornadoes. However, winter winds of gale force are frequent, and, along the coasts, heavy seas are a hazard to shipping and shore installations. The name "roaring forties," used to indicate the fury of the prevailing westerlies in the Southern Hemisphere, is nearly as apt in the Northern Hemisphere. The North Atlantic, Irish Sea, Straits of Magellan, Australian Bight, and Tasman Sea are all infamous in the annals of merchant shipping.

MID-LATITUDE ARID AND SEMIARID CLIMATES

The dry climates of mid-latitudes differ from the tropical arid and semiarid climates in two important respects: average temperatures are lower, and subsiding air masses are not the chief controlling factors. A major factor influencing the arid and semiarid climates of middle latitudes is their location in the continents, far removed from the windward coasts. Mountain barriers on the windward accentuate the aridity of several areas. The arid and semiarid phases of the dry mid-latitudes differ primarily in the degree of aridity, and they can be treated together conveniently. Differences in water budgets are nonetheless significant in their effects on the associated desert or steppe vegetation and land-use practices (Fig. 8.11). Mid-latitude dry climates of the Northern Hemisphere merge gradually into their

Figure 8.11 Sheep grazing in the mid-latitude steppe. (Wyoming Travel Commission.)

tropical counterparts in several areas (see front endpaper). Temperature is the differentiating criterion through its influence on evapotranspiration.

The lower annual temperature averages are a response to the effects of higher latitude, and in some cases of altitude as well. Appreciable differences occur within the mid-latitude dry climates as a result of the latitudinal spread. The characteristic temperature curve shows a large annual range, a reflection of continental location. In summer the temperatures are high as *cT* air masses develop over the heated land surfaces. Because there is a deficiency of moisture little heat is used in evaporation, most of it being used to further warm the subsiding air mass. In winter the prevalent air masses are again continental, but they are associated primarily with outbreaks of polar air. Denver is representative of the mid-latitude semiarid climate; Lovelock, Balkash, and Santa Cruz typify the arid phase.

Denver, Colorado 40°N; 1,615 m

	J	F	M	A	M	J	J	A	S	O	N	D	Yr
T (°C)	0	1	4	9	14	20	24	23	18	12	5	2	11
P (mm)	12	16	27	47	61	32	31	28	23	24	16	10	327

Lovelock, Nevada 40°N; 1,212 m

	J	F	M	A	M	J	J	A	S	O	N	D	Yr
T (°C)	0	3	6	10	15	19	24	23	18	12	5	1	11
P (mm)	16	16	13	11	16	18	5	6	7	11	14	15	146

Balkash, U.S.S.R. 47°N; 423 m

	J	F	M	A	M	J	J	A	S	O	N	D	Yr
T (°C)	−15	−13	−5	8	16	22	24	22	15	6	−5	−12	5
P (mm)	10	8	10	11	9	14	11	9	4	8	9	12	115

Santa Cruz, Argentina 50°S; 12 m

	J	F	M	A	M	J	J	A	S	O	N	D	Yr
T (°C)	14	14	12	9	5	2	2	4	6	10	12	14	8
P (mm)	21	16	20	17	25	18	16	16	12	7	15	18	201

Note the contrasting temperatures at these representative stations. Deep in the Asian continent, Balkash has an annual range of 39C°, whereas Santa Cruz, on the southern Argentina coast, has a range of only 12C°. Patagonia lies to the leeward of the Andes and has a marine temperature regime, but it receives little moisture from the prevailing westerly flow of air. Only occasionally does cyclonic activity bring easterly winds, and the continent is too narrow to induce a thermal low-pressure center with monsoonal circulation onto the land in summer. The modified marine influence of southern Argentina is unique in the dry mid-latitude climates.

Extreme temperatures range from well below −40°C in winter to above 40°C in summer in these climates. The lowest minimums are recorded in the northern Great Plains and in Siberia. The highest maximums occur at the lower-latitude margins of the climates. With respect to temperature, a summer day in certain parts of the dry mid-latitudes may be uncomfortably like the tropical deserts. Diurnal ranges are great, however, and especially at the higher elevations the nights can be pleasantly cool. Daytime heating and nocturnal cooling effect a large diurnal range, just as in the tropical arid and semiarid climates. Although the same is true to some extent in winter, the mid-latitudes have much lower monthly averages, and invasions of *cP* air bring severe freezing not found in the tropics.

In common with the dry climates of the low latitudes the mid-latitude arid and semiarid climates have deficient precipitation but the water need decreases under lower temperatures. Annual rainfall of 150 to 200 mm may support steppe vegetation in the cool mid-latitudes yet be adequate only for desert scrub in the tropics. Owing to its effect on plant growth, the ratio of precipitation to potential evapotranspiration differentiates the arid from the semiarid phase of the dry mid-latitude climates. Desert vegetation is characteristic of the arid climate and steppe grassland predominates in the semiarid regions. Accordingly, the climates are also well known by their associated vegetation as mid-latitude desert and steppe, respectively.

As in the dry tropical climates, in the mid-latitude counterparts there is no general rule governing the regime of precipitation. The margins near the dry summer subtropics have winter maximums; toward the humid continental climates summer maximums prevail, and where marine influences are marked, as on the east coast of Argentina, there is no great variation from month to month. The graph for San Juan, Argentina, illustrates the winter maximum in the arid type; Dickinson, North Dakota, has an early summer maximum (Fig. 8.12). Absence of a distinct seasonal maximum is shown in the data for Santa Cruz and Balkash.

A characteristic of the precipitation record in all dry climates is the great variability from year to year (see Fig. 8.13, also Fig. 3.20). In general, departures from the long-term mean increase as the mean decreases.

In a study which has become a classic in geographic literature, Henry M. Kendall pointed out that, on the Great Plains in the period 1915–24, humid years were experienced as far west as the Rocky Mountains and dry years occurred as far east as Minnesota. The significance of precipitation variability in the semiarid climate is far greater than in the arid. A 100 percent increase of rainfall in a given

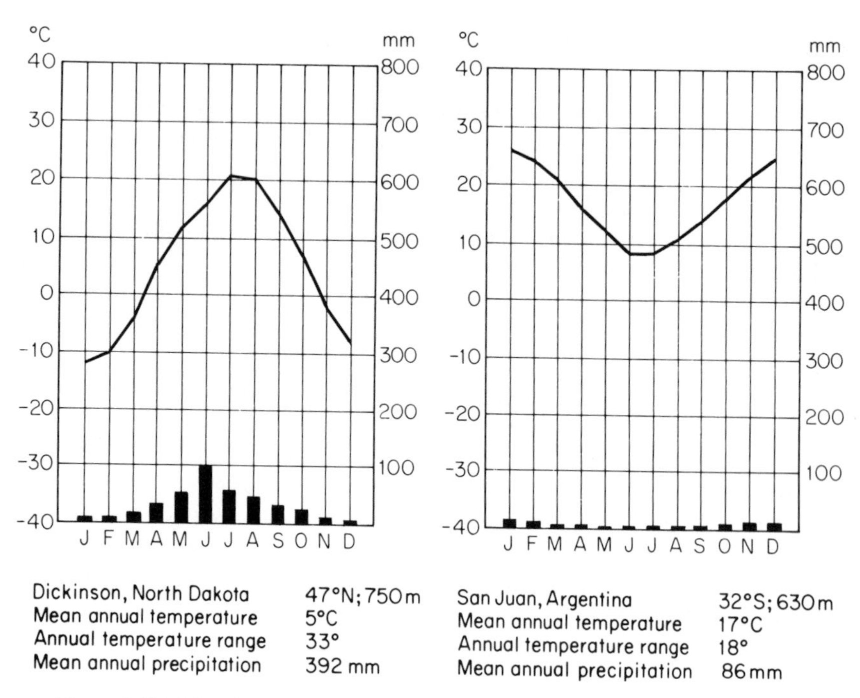

Figure 8.12 Climatic graphs for Dickinson, mid-latitude semiarid, and San Juan, mid-latitude arid.

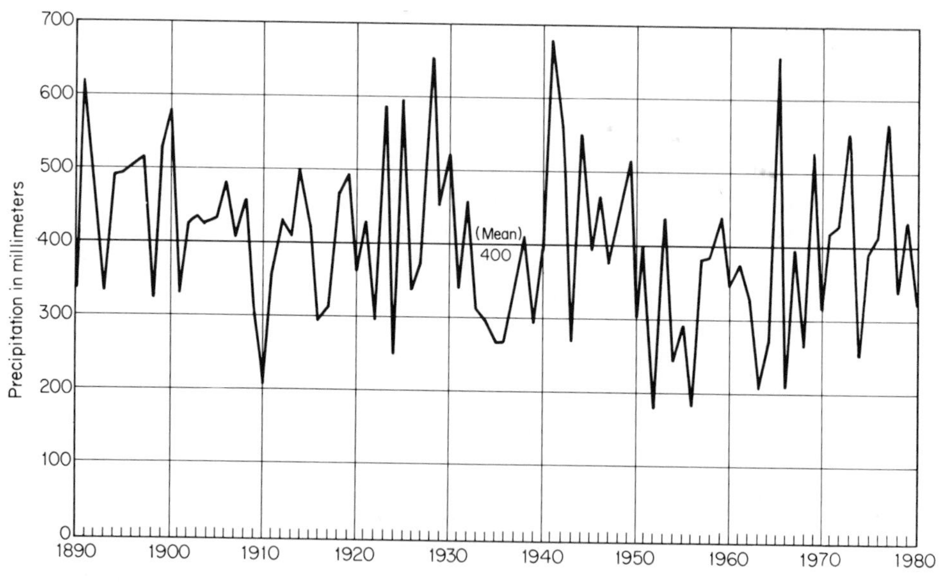

Figure 8.13 Variations in annual precipitation at a station in the mid-latitude semiarid climate near Springfield, Colorado, 1890–1980.

year in an arid climate with an average of 70 mm is not likely to produce great changes in vegetation and human settlement. Agriculture is geared to irrigation, if practiced at all. In the steppe regions, however, a series of wetter-than-average years gives a false impression of agricultural potentialities, and settlers are tempted to institute land uses not actually suited to the long-term climatic pattern. In no climate is there more danger in the use of statistical averages than in the mid-latitude semiarid type. The dry years are as important as the "normal" years, if not more so, in determining the vegetation association and the optimum patterns of land use.

The meager winter precipitation in the dry mid-latitudes is frontal and comes with occasional cyclones, which replace the continental high-pressure centers. It may come as snow, and high winds accompanying polar outbursts in the wake of a cold front produce blinding snowstorms, known as *blizzards* on the Great Plains. The infrequent cyclonic disturbances are sandwiched between periods of cold and generally clear weather. The chinook, or foehn wind, is experienced over the northern Great Plains in winter and spring. A similar wind which descends from the Andes into western Patagonia is known as the *zonda*. In the Soviet Union the *sukhovey* is a warm, dry wind associated with summer anticyclones over southern Siberia. Surface wind speeds are generally high in these climates because of the large expanse of plains and plateaus. Immediately at the surface, the absence of trees is also a factor permitting free flow of air.

Summer weather is influenced by thermally induced low-pressure centers over the continents. Although this may tend to produce monsoon-type winds blowing onto the continent from water bodies, mountain barriers and the distance of the dry climatic regions from the sea limit the amount of moisture that reaches the interior. Summer rainfall comes with scattered thundershowers. All too often a dry period is broken by a violent cloudburst-hailstorm which does as much damage to crops as drought. In the United States, tornadoes sometimes form along belts of frontal thunderstorms, especially in the warmer half of the year.

HUMID CONTINENTAL WARM SUMMER CLIMATE

Extensive areas with humid continental climates occur only in the Northern Hemisphere; the Southern Hemisphere has mostly a water surface at these latitudes. As indicated by their names, the two types of humid continental climate differ primarily with respect to the length and intensity of the summer season. They are considered separately here to facilitate emphasis on their differing temperature characteristics.

The humid continental warm summer climate ranges between approximately 35° and 45° north latitude, encompassing productive agricultural regions of mid-latitude North America and Eurasia. Continental polar air masses dominate the winters, bringing cold weather, but there are intervening surges of *mT* air in the warm sectors of cyclones. In summer *mT* and *cT* air masses bring high temperatures and rainfall maximums (see Fig. 8.7). This is truly the battleground of polar and tropical air, where there are cyclonic "skirmishes" within the larger pattern of seasonal "warfare."

Annual temperature averages mask the large annual ranges in the humid continental warm summer climate. Summer is hot; winter is cold; and even spring and autumn tend to be well-defined seasons. Mean monthly temperatures for June, July, and August are near or above 20°C. The average length of the frost-free season ranges from about 200 days in the south to 150 days along the northern margins. Yet January averages are typically below 0°C. St. Louis, near the boundary of the humid subtropics, has a January average of 0°C but a July average of 26°C. Harbin, near the boundary of the cool summer phase in Manchuria, has −20°C as a January average and 23°C in July (see accompanying table).

St. Louis, Missouri 39°N; 172 m

	J	*F*	*M*	*A*	*M*	*J*	*J*	*A*	*S*	*O*	*N*	*D*	*Yr*
T (°C)	0	2	6	13	19	24	26	25	21	15	7	2	13
P (mm)	50	52	78	94	95	109	84	77	70	73	65	50	897

Harbin, Manchuria 47°N; 143 m

	J	*F*	*M*	*A*	*M*	*J*	*J*	*A*	*S*	*O*	*N*	*D*	*Yr*
T (°C)	−20	−16	−6	6	14	20	23	22	14	6	−7	−17	3
P (mm)	4	6	17	23	44	92	167	119	52	36	12	5	577

Continentality is expressed in the warm summer type not only by high monthly means in summer but also by extreme maximums above 35°C. Furthermore, diurnal ranges are small in summer so that the nights are often uncomfortably warm. Conditions of high humidity and high temperature are similar to those in the humid subtropics. A day and night in central Iowa may be much like the same period in Louisiana or Alabama, for summer temperatures do not decrease rapidly with latitude. In some years warm anticyclones persist for several days or weeks over the Mississippi Valley and Great Plains, causing high temperatures and drought. Approximately 1,300 heat-caused deaths and widespread losses of crops and livestock accompanied a prolonged heat wave over the central and southern United States in the summer of 1980. Figure 8.14 shows the extensive anticyclone which prevailed during the month of July, when temperatures averaged well above normal and no rain fell over the southern Great Plains.

Winter is quite another matter. Not only are the monthly averages and the extreme minimums much lower, but the decrease in these values with increasing latitude is appreciable. The *cPk* outbursts which bring the very low temperatures to the northern parts of the climatic region may not reach the southern margins, or, if they do, they are modified somewhat as they move to lower latitudes. Thus, typical annual minimum temperatures in northern Iowa are on the order of −30°C, whereas in southern Missouri the mean annual minimum is −20°. East of the Great Plains in North America temperatures increase southward at a faster rate in winter than in summer. The influence of the large land mass in Asia results in even

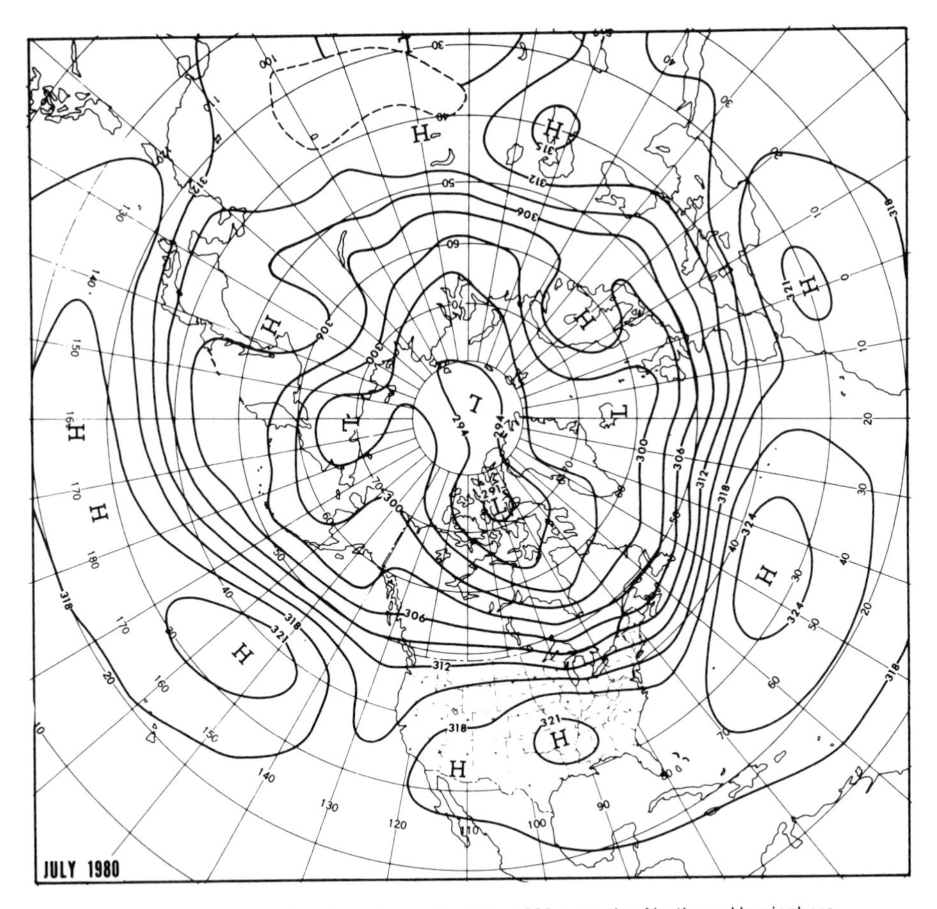

Figure 8.14 Mean 700-mb contours for July 1980 over the Northern Hemisphere. Heights in dekameters. (*Monthly Weather Review*, Vol. 108, p. 1711.)

lower winter temperatures in the humid continental warm summer climate of Manchuria.

Annual precipitation in this climate varies from 500 to 1,250 mm. The amount typically decreases toward the northern latitudes and toward continental interiors. Spring or summer is the season of maximum rainfall. Occasional frontal passages and local squall-line activity account for most of the warm-season rain. Scattered thunderstorms and a high percentage of possible sunshine produce summer weather similar to the humid subtropics. A maximum of rainfall in the summer half of the year is most marked in the interior areas and in eastern Asia, where the monsoonal influence is an important factor. For example, in July, Harbin has 167 mm but in January only 4 mm. A characteristic of the precipitation regimes of interior locations in this climate is the spring or early summer maximum, usually in May or June (see Fig. 8.15).

In spring the air masses are still cold from the north and the contrasting

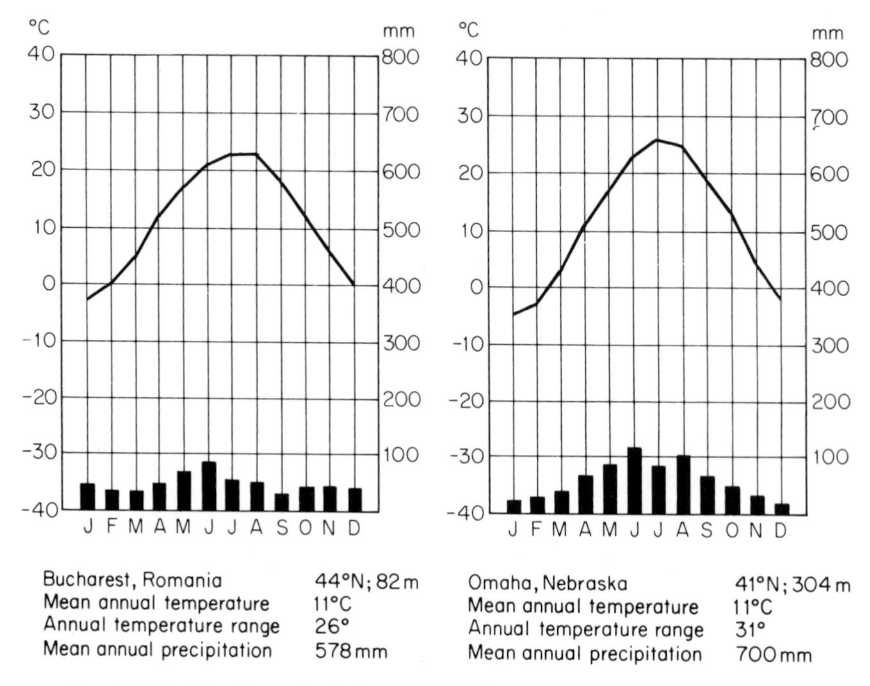

Figure 8.15 Climatic graphs for Bucharest and Omaha, humid continental warm summer climates.

tropical maritime air comes from water that has reached its lowest temperature for the year. The land heats rapidly, however, and the consequent warming of lower layers of advancing maritime air increases instability and leads to numerous convective showers. Although thunderstorms continue throughout the summer, the differences in temperature between the land and the upper air become less and the degree of instability is therefore lessened. Sometimes hail accompanies thunderstorms in this climate. In the Mississippi Valley tornadoes are an additional hazard, primarily in the warm months.

Except on the east coast of the United States, the winter months usually have much less precipitation, a feature which is particularly evident in eastern Manchuria and northern China, where a monsoonal regime prevails. A part of the winter precipitation comes as snow, the proportion increasing with latitude and altitude. It is primarily of frontal origin, and the blizzards behind the cold fronts may extend into the humid continental warm summer climate from the west or north. Average annual snowfall varies from 25 to 100 cm. Frontal weather in winter may also bring rain, occasional fog, and sometimes ice pellets or freezing rain. Thunderstorms can develop in the warm sectors of cyclones whose paths follow the southern margins of the climatic type. In general, winter is much less sunny than summer, a fact taken for granted by residents in the areas with humid continental climate. Nevertheless, conditions are in marked contrast with the wet-and-dry tropics, which also have a precipitation maximum in summer but which have more sunny weather in winter.

Probably the best known feature of winter weather in the humid continental climates is its changeability. Temperatures may drop from above 15°C in the warm sector of a cyclone to −20°C behind the cold front in a period of a day or two, and, although extremes of such magnitude are not the rule in winter cyclones, sudden changes typify most storms. The passage of cyclones becomes more frequent in autumn, but there may be warm, sunny weather between the first two or three cyclones. In the United States the weather associated with stable air in October and November is often called "Indian summer," and since it often follows frosts, the fall colors of deciduous trees add a great deal to the pleasantness of these periods. Just as summer retreats with rearguard delaying actions, so in spring there are late frosts and snowstorms which punctuate the erratic poleward retreat of cyclonic storms. Custom decrees that these late spring storms be termed "unseasonal"; they are in fact a characteristic of the climate.

HUMID CONTINENTAL COOL SUMMER CLIMATE

The most obvious features that distinguish the humid continental cool summer climate from the warm summer type are a cooler summer, a shorter growing season, and a colder winter, reflecting the influence of higher latitude and continentality on the energy and water budgets (see Fig. 8.16). The climatic region lies

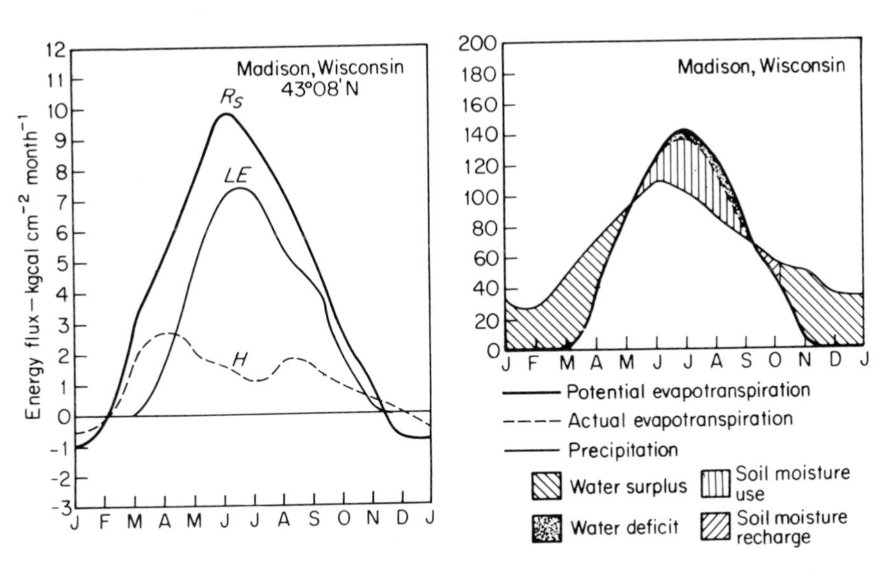

Figure 8.16 Energy and water budgets for Madison, Wisconsin, humid continental cool summer climate. (Data from Sellers and C. W. Thornthwaite Associates, Laboratory of Climatology.)

north of the warm summer type in North America, eastern Europe, and the far east of Asia. In each case it has a wider longitudinal extent than the humid continental warm summer type. Being at higher latitudes, the cool summer type is affected by polar and arctic air for a longer season. The frequency of extratropical cyclones in winter is among the highest in the world. (Figure 8.17 shows the distribution of total January cyclonic passages across Canada and the United States during a 28-year period.) Maritime air often is drawn onto the east coasts of the continents as cyclones approach the sea. In summer incursions of *cT* and *mT* air become more frequent.

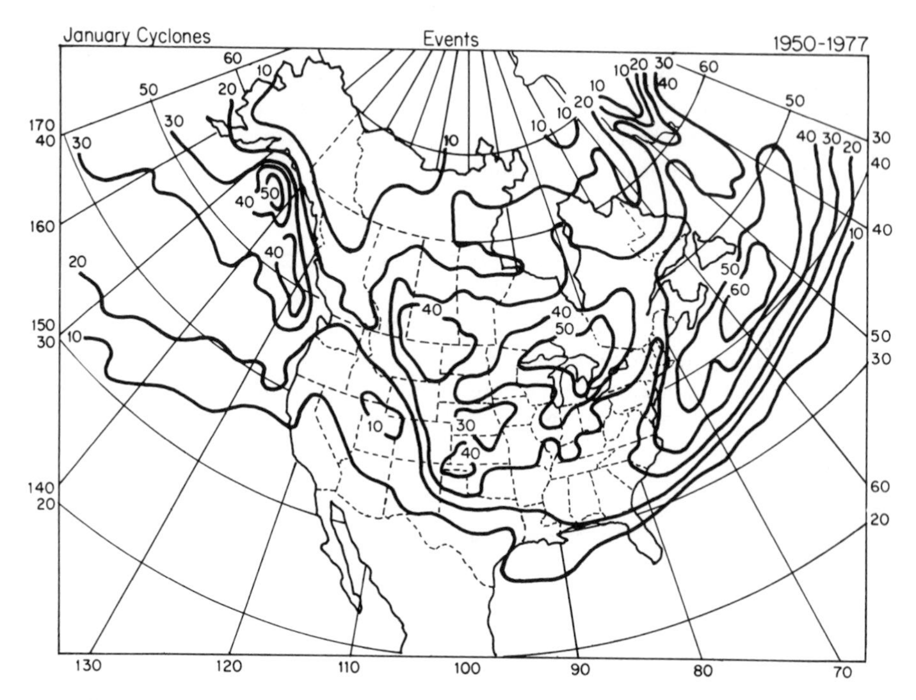

Figure 8.17 Areal distribution of cyclonic passages across Canada and the United States during the month of January. Values are totals for the period 1950-1977. (Zishka and Smith, *Monthly Weather Review*, Vol 108, p. 391; by permission of the American Meteorological Society.)

Temperatures are generally lower both in winter and summer than in the warm summer type. Summer extremes may exceed 35°C in a "heat wave" of tropical continental air, but these are offset by cool periods, often with cloudiness. Winnipeg and Moscow have July means of 20°C and 19°C, respectively (Fig. 8.18), in contrast with July means of 26°C at Omaha and 23°C at Bucharest.

The frost-free season averages less than 150 days, except for coastal locations, and it decreases in the higher latitudes. It is significant, however, that with the increase of latitude the summer daylight periods are longer; this tends to offset the short season to some extent insofar as crop production is concerned. Thus, the

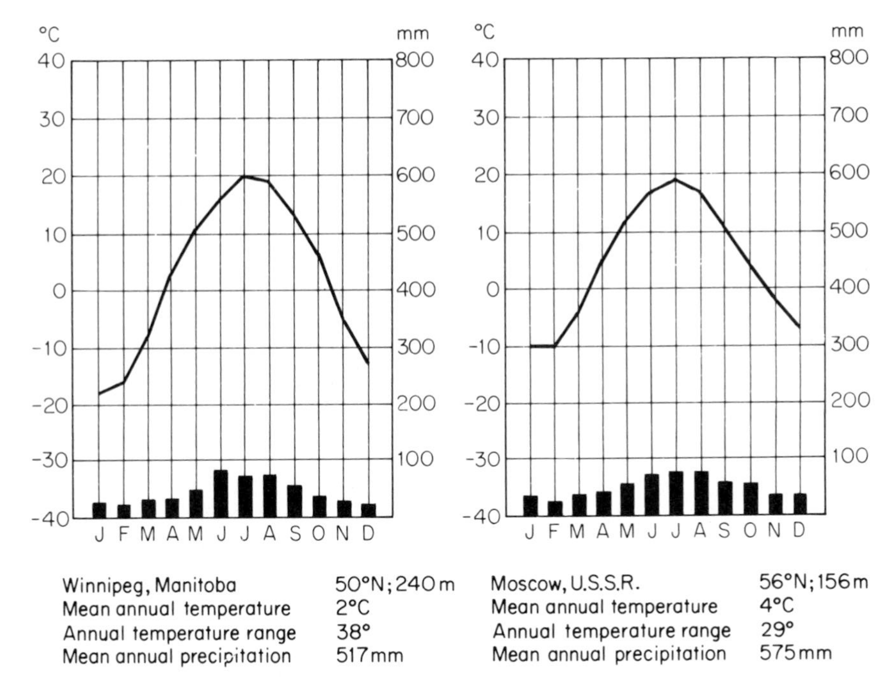

Figure 8.18 Climatic graphs for Winnipeg and Moscow, humid continental cool summer climates.

longest day at Edmonton, Alberta (lat 53°45′ N), has more than 17 hours, compared with less than 15 hours at St. Louis, Missouri (lat. 38°35′ N), in the warm summer climatic type.

In the winter the days are increasingly shorter as one proceeds northward and the contrast between the seasons is correspondingly accentuated. Annual ranges of temperature are even greater than in the humid continental warm summer climate. Winnipeg has a range of 38C°; at Moscow the range is 29C°. These large annual ranges result from a combination of latitude and interior continental location. Winters are severe; typically, the January average is well below 0°C, and for some interior stations it is below −15°C. Note, however, that the climate is identified primarily by its summer rather than is winter temperature characteristics. Harbin, Manchuria, for example, has a January mean temperature of −20°C but a warmer and longer summer than in the humid continental cool summer type of climate. Extreme minimums in the cool summer type fall to −45° or −50°C, but they are not an annual occurrence. A series of several days having temperatures below −15°C can be expected in most winters as polar or arctic air lies over the continents. Short-term changes of temperature are greatest in winter, and they are more marked in connection with the nonperiodic cyclonic disturbances than with the diurnal cycle of insolation.

Annual precipitation in the humid continental cool summer climate is usually less than in the warm summer type. The areas within the climatic region are farther from the sources of maritime tropical air and they are influenced for a greater part of the year by drier polar air masses. Annual amounts for most stations range from 350 to 700 mm, but the temperatures are lower and a given amount of summer precipitation is more effective than in the warm summer type. Regimes of precipitation are similar to those of the warm summer type, the maximum coming in summer in the interiors. Toward the east coasts of North America and Asia and in Japan, the summer maximums become less pronounced and autumn maximums are more characteristic. Summer thunderstorms are not so common, but summer frontal passages occur more frequently. In the winter season the proportion of snow is greater and snow lies for longer periods. Prolonged snow cover is a factor which inhibits the warming of southward-moving air masses, and, consequently, contributes to the severity of winter conditions.

Eastport, Maine 46°N; 24 m

	J	*F*	*M*	*A*	*M*	*J*	*J*	*A*	*S*	*O*	*N*	*D*	*Yr*
T (°C)	−5	−5	−1	5	9	13	16	17	14	9	4	−3	6
P (mm)	104	87	93	89	78	90	78	73	90	92	114	95	1,084

Nemuro, Japan 43°N; 26 m

	J	*F*	*M*	*A*	*M*	*J*	*J*	*A*	*S*	*O*	*N*	*D*	*Yr*
T (°C)	−5	−6	−2	3	7	10	14	18	16	11	5	−1	6
P (mm)	49	40	77	77	99	97	104	106	152	124	92	63	1,081

The effect of ocean water off the east coasts of North America and Asia considerably modifies the humid continental climate. Southerly and easterly winds in the cyclonic circulation bring maritime air to these regions, resulting in increased precipitation and moderated temperatures, especially in winter. Eastport, Maine, has 1,080 mm of precipitation fairly well distributed through the year. Nemuro, on Hokkaido in northern Japan, has the same annual average but a September maximum (see the accompanying table). The peculiar features of the humid continental cool summer climate on the east coasts have led some authors to classify it as a distinctive "modified humid continental" or "New England–Hokkaido" type. On the coasts of New England and the Canadian Atlantic Provinces the precipitation regime varies less than in the interior; Cape Race, Newfoundland, has a November maximum. Mean summer temperatures and annual ranges are both appreciably lower than inland. The Maine coast has an exceedingly steep gradient of annual temperature ranges away from the coast. At Eastport the range is 22C°, whereas at Woodland, only 50 km inland, the range is 30C°. The ocean not only modifies temperature ranges, but also produces a longer frost-free season. A further characteristic of the coast of New England and Canada is the frequency of gales associated with the succession of cyclones. On rare occasions tropical hurricanes reach this area. Off the coast of Canada the tropical water of the Gulf Stream meets the cold Labrador Current and fogs are frequent in the overlying air. Light winds carry the fog onto the coasts, but it rarely penetrates far inland. The greatest fog frequency is in summer months and over the Grand Banks east of Newfoundland.

Maritime effects in northern Japan are analogous to those in New England, but the persistent continental air masses act as a barrier to oceanic air in winter, resulting in an extension of the continental characteristics to the east coast of mainland Asia. The mountainous islands receive moisture in winter from the Sea of Japan and thus do not exhibit the usual dry winter regime of the mainland. Note that Nemuro has a September maximum of rainfall which coincides with the typhoon season. Comparison of the data for Nemuro and Vladivostok reveals the effect of an island location in producing a smaller annual temperature range and a more even distribution of precipitation (see Fig. 8.19). The winter climate of

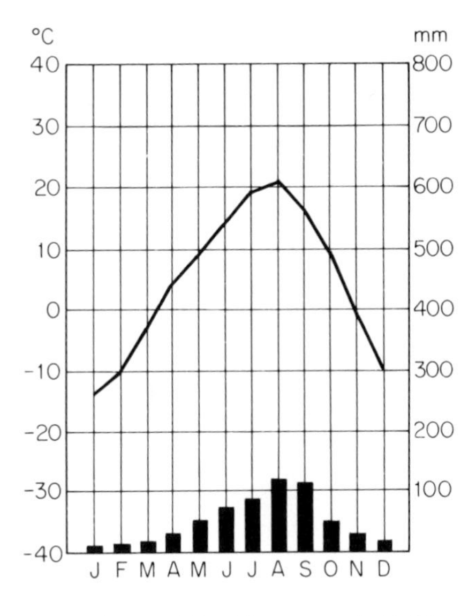

Figure 8.19 Climatic graph for Vladivostok, humid continental cool summer climate.

Vladivostok is essentially continental; in the month of January winds are almost entirely from the north or northwest (Fig. 8.20).

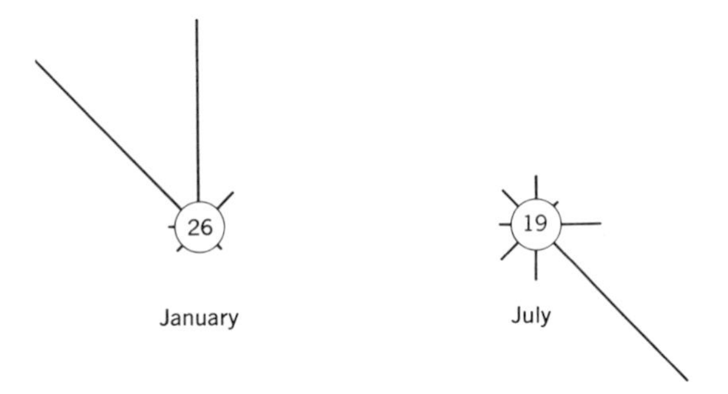

Figure 8.20 Wind roses for Vladivostok, U.S.S.R.

QUESTIONS AND PROBLEMS FOR CHAPTER 8

1. Explain why the dry summer subtropics and humid subtropics have great contrasts in summer rainfall although both are influenced by the oceanic subtropic highs.
2. Why are there more thunderstorms in the humid subtropics than in the marine climate?
3. Compare the effects of mountain barriers on the marine climate of North America with those in Chile.
4. Suppose that the earth were to rotate from east to west. What climatic conditions could be expected in Argentina, the British Isles, and California?
5. Contrast the causes of mid-latitude arid and semiarid climates with those of the low-latitude counterparts.
6. Would ''dry continental'' be an appropriate name for mid-latitude arid and semiarid climates? Explain.
7. Explain the differing precipitation regimes of Bucharest and Vladivostok.
8. Explain the role of climate in creating the fiorded west coasts of the mid-latitudes.

9

Climates Dominated by Polar and Arctic Air Masses; Highland and Ocean Climates

The effects of high latitude, high altitude, and the sea surface on heat and moisture budgets produce three distinctive groups of climates, respectively. At high latitudes, where the dominant air masses are of the polar or arctic types, and at high altitudes decreasing temperature is the principal differentiating criterion. Ocean climates are distinguished by the strong influences of heat storage, circulation, and boundary layer exchanges. Collectively, the wide variety of ocean climates have a greater regional extent than all of the climates of land areas combined.

TAIGA CLIMATE

Taiga is a word of Russian origin and refers to the northern continental forests of Eurasia and North America. Names such as "subarctic" or "boreal forest" have also been applied to the taiga climate. On both continents the climatic type extends farther south on the eastern margins than on the west, reflecting the influence of the large land masses. These regions are the sources of polar air masses, and for most of the year their weather is dominated by cold, dry, stable air. As a result, annual temperature ranges are large and precipitation is meager. There are no areas of taiga climate in the Southern Hemisphere, which consists mainly of water at comparable latitudes.

The mean July temperature at Yakutsk, in east central Siberia, is 20°C, but the January average is −43°C, making the annual range 63C° (see the accompanying table). Ranges of 30C° or more are typical in the interior locations; conditions are moderated near the oceans, particularly on west coasts. Tromso, Norway, has a mean annual range of only 15C°.

Yakutsk, U.S.S.R. 62°N; 103 m

	J	*F*	*M*	*A*	*M*	*J*	*J*	*A*	*S*	*O*	*N*	*D*	*Yr*
T (°C)	−43	−37	−23	−7	7	16	20	16	6	−8	−28	−40	−10
P (mm)	7	6	5	7	16	31	43	38	22	16	13	9	213

Tromso, Norway 70°N; 24 m

	J	*F*	*M*	*A*	*M*	*J*	*J*	*A*	*S*	*O*	*N*	*D*	*Yr*
T (°C)	−3	−3	−2	1	5	9	12	11	8	4	0	−1	3
P (mm)	118	94	113	75	65	57	56	83	115	131	97	115	1,119

Fairbanks, Alaska 65°N; 134 m

	J	*F*	*M*	*A*	*M*	*J*	*J*	*A*	*S*	*O*	*N*	*D*	*Yr*
T (°C)	−24	−19	−13	−1	8	15	15	12	6	−3	−16	−22	−3
P (mm)	23	13	10	6	18	35	47	56	28	22	15	14	287

In the short summer the days are long, and maximum temperatures frequently rise above 25°C. Yakutsk has recorded a temperature of 39°C. Fairbanks has an extreme maximum of 34°C. At such stations a small diurnal range might be expected as a result of the long daylight periods, but even where there are 24 hours of daylight an appreciable difference exists between the amount of effective insolation at noon and that at midnight, when the angle of the sun is low. Thus, diurnal ranges are commonly on the order of 10° or 15C°. Marine influences moderate the diurnal range considerably. The frost-free season in the taiga varies with the latitude and with distance from the sea. Inland it lasts for 50 to 90 days, but there is always the risk of summer frost. In some areas the subsoil is permanently frozen even where summer thawing permits a plant cover (Fig. 9.1). The short duration of the frost-free season is balanced to some extent by the longer days, so that certain hardy crops can mature in a shorter calendar period in the taiga than in lower latitudes. Not only is the summer short, but it also ends abruptly. Autumn is often said to begin with a hard freeze, last a few days, and end with the first snow of winter.

Winter is the dominant season. Except on favored coasts, six to eight months have mean temperatures below 0°C, and many stations have three or four months with means below −15°C (see Fig. 9.2). Furthermore, average daily *maximum* temperatures in winter are generally well below freezing. Under the dominance of the continental anticyclones the weather is usually clear, but the daylight periods are short or entirely absent and outgoing radiation far exceeds insolation. The lowest official surface temperature in the climatic type (−68°C) was recorded at Verkhoyansk in eastern Siberia. It occurred on the 5th and again on the 7th of February 1892. At Oimekon, about 650 km to the southeast, the temperature record is shorter, but in recent years it has been colder than Verkhoyansk. An unofficial −78°C was reported at Oimekon in 1938, but details of the date and exposure

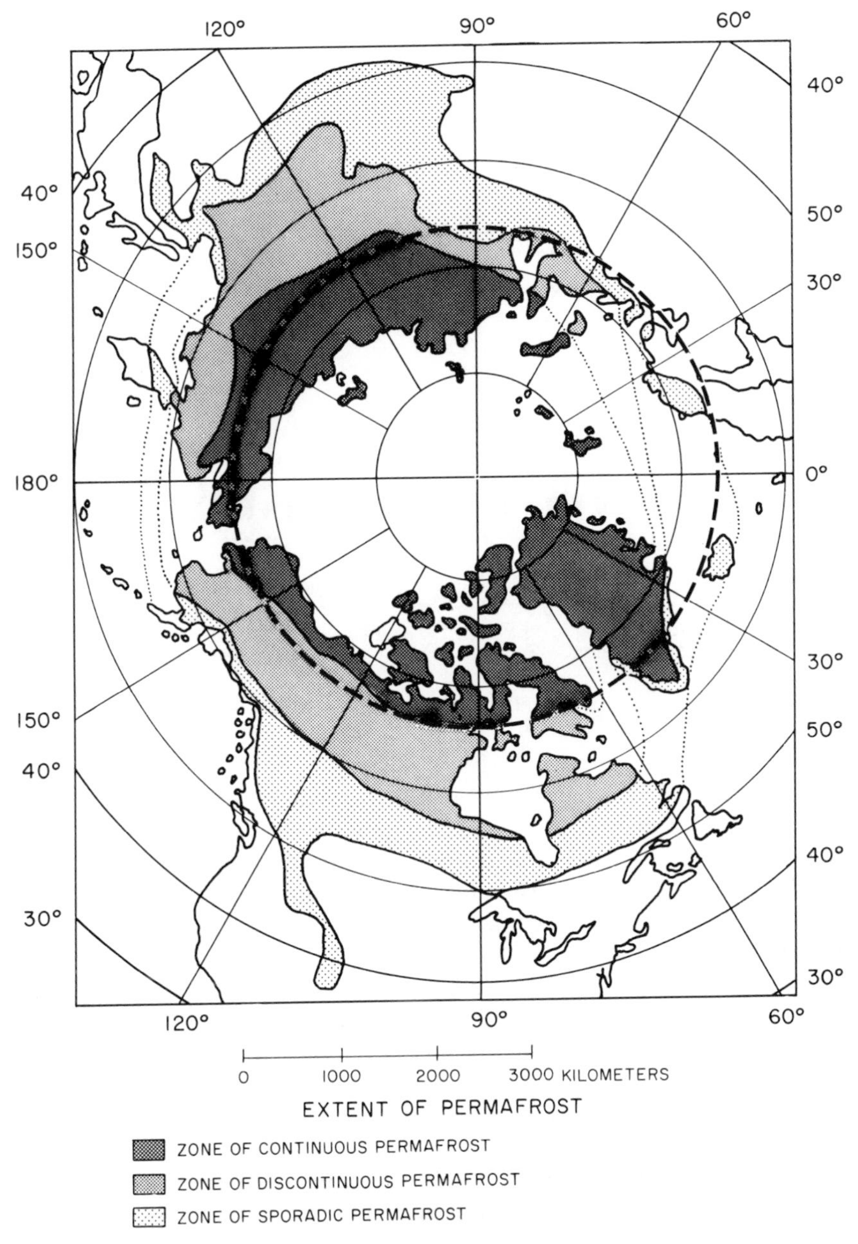

Figure 9.1 Distribution of permafrost conditions in the arctic. (From United States Air Force, Geophysics Research Directorate, *Handbook of Geophysics,* Revised Edition, published by The Macmillan Company, 1960.)

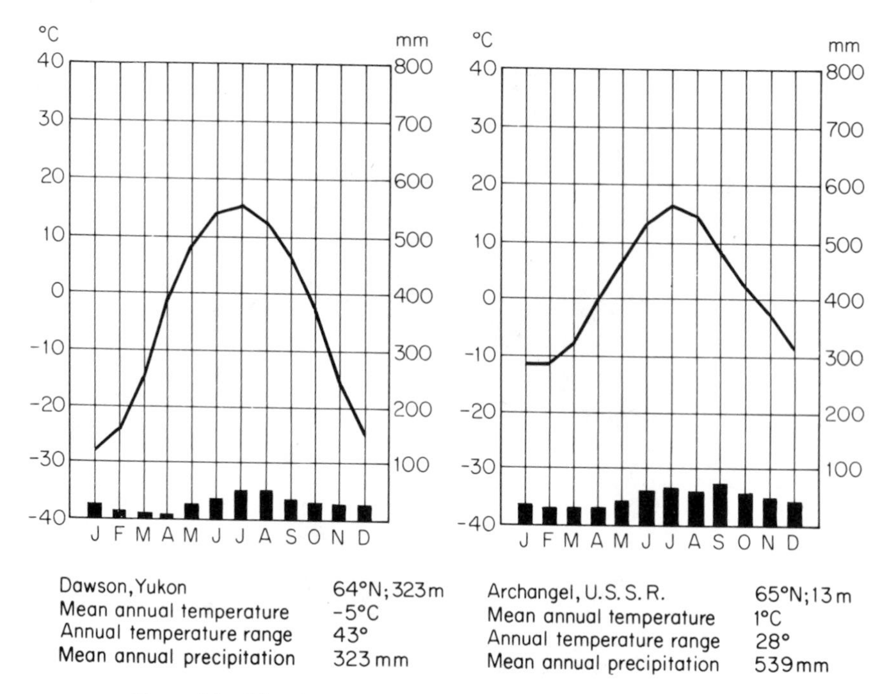

Figure 9.2 Climatic graphs for Dawson and Archangel, taiga climates.

of the thermometer are unknown. In North America the lowest official temperature is −63°C, recorded at Snag, Yukon Territory, on February 3, 1947. Although these are exceptional temperatures, they indicate the extreme cold which frequently envelopes the taiga in winter. Under the calm conditions that sometimes accompany the extremely low temperatures the dry air does not produce the same physiological chilling as moister air would at appreciably higher temperatures. With wind, however, chilling effects rapidly increase and powdered snow may pervade the air, reducing visibility to zero in a "whiteout."

Except along the seaward margins, precipitation in the taiga climate is generally less than 500 mm annually. It is mostly of cyclonic origin and has a definite summer maximum, again with the coasts excepted. (Compare the data for Tromso and Fairbanks.) At interior locations rainfall is often inadequate to meet the demands of potential evapotranspiration during the long daylight periods of summer, so that soils turn to dust and irrigation may even be necessary in the few areas where crops are raised. Winter snow actually accounts for the smaller part of annual precipitation in most of the taiga, although it may accumulate for several months (see Fig. 9.3).

Fog and low stratus clouds are fairly common along the coasts of this climatic region, being especially prevalent in the Aleutian Islands. Ice fogs may form over snow surfaces in winter when temperatures are extremely low.

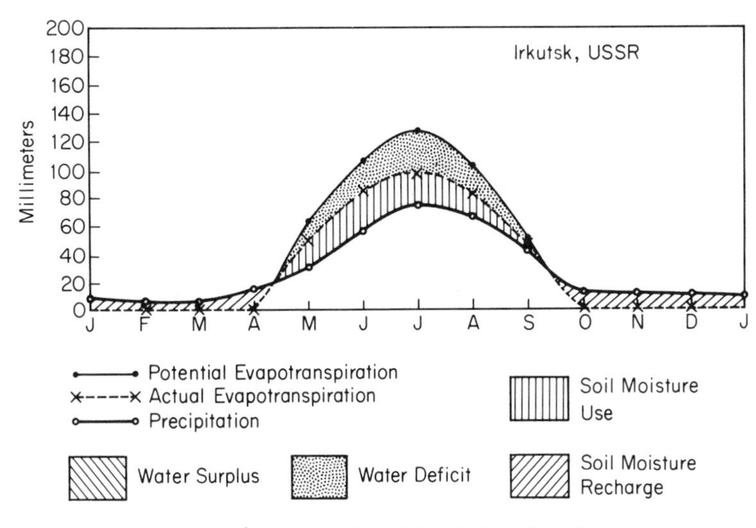

Figure 9.3 Water budget graph for Irkutsk, taiga climate.

TUNDRA CLIMATE

Like taiga, *tundra* is a vegetation term which has also been applied to the associated climate. The tundra climatic region lies north of the taiga and extends along the arctic coasts of North America and Eurasia. For the most part it is found north of the Arctic Circle, but in eastern Canada and in eastern Siberia it extends into lower latitudes in keeping with the continental influence. Northern Iceland, coastal Greenland, and many islands in the Arctic Ocean also have this type of climate. The climate does not occur in the Southern Hemisphere except on a few islands in the Antarctic Ocean and small areas on Antarctica, because of the predominance of water surface in the corresponding latitudes.

For most of the year the tundra climate is under the influence of *cP* or *cA* air masses. In summer the air may have maritime characteristics as a result of the melting of sea ice and more frequent advances of air from oceanic sources in lower latitudes. Even in winter, islands and coasts experience limited marine influences. These effects are the basis for a possible subdivision, the *polar marine climate.* Although the long periods of ice cover on the Arctic Ocean render it less effective as a moderating influence than oceans at lower latitudes, it nevertheless does affect the tundra climate. Even in winter, when entirely frozen over, the Arctic Ocean has enough heat to modify temperatures slightly. For this reason the extreme minimum temperatures occur farther south within the continents, where loss of heat by radiation is great, rather than at the higher latitudes on the coast. The boundary between sea-ice and water is probably one of the most significant factors in the weather of the arctic. Along this boundary the greatest horizontal temperature gradients occur, and it is consequently a zone of frequent storms.

Mean annual temperatures in the tundra are normally below 0°C, and the annual range is large. Barrow, Alaska, has an annual mean of −12°C, but the July average is 4°C and that for February is −28°C (Fig. 9.4). The range of 32C°

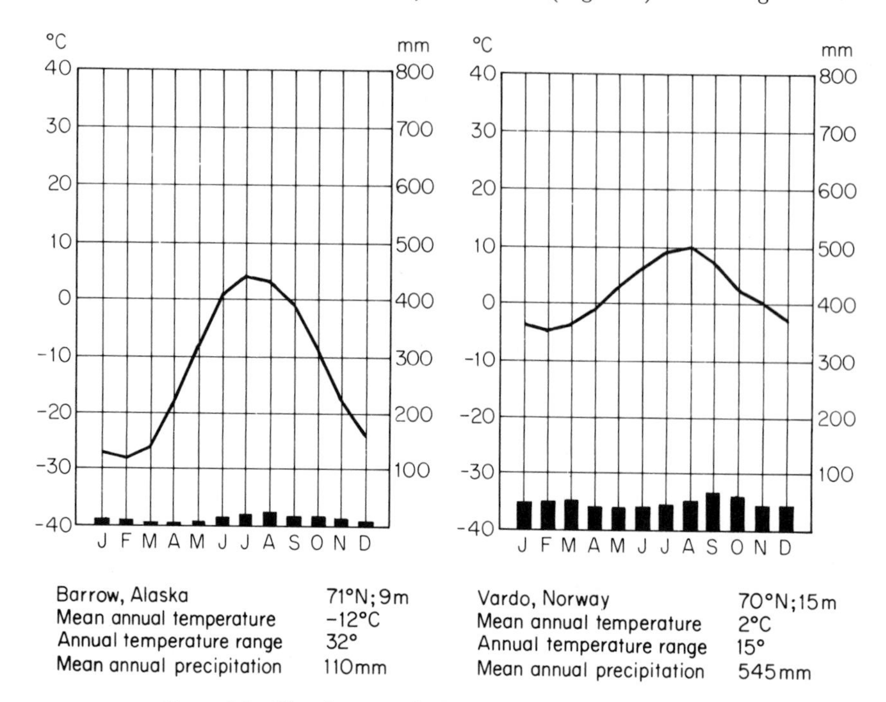

Figure 9.4 Climatic graphs for Barrow and Vardo, tundra climate.

reflects the marked continental nature of the climate in spite of the station's location on a coast. Vardo, Norway, at a comparable latitude, experiences a greater maritime influence, having an August mean of 10°C, a February mean of −5°C, and an annual range of only 15C°. The more moderate temperatures at Vardo are due in part to the effects of the North Atlantic Drift, which creates positive temperature anomalies in the far northern Atlantic. Prevailing southerly and westerly winds in winter augment the maritime influence (Fig. 9.5). There is no comparable movement of warm water through the Aleutian Island chain and Bering Strait into the Arctic Ocean north of Alaska.

Summer temperatures in the tundra are reminiscent of mild winter weather in the mid-latitudes. The average temperature in the warmest month is below 10°C. Because heating is generally inadequate to melt the permafrost, percolation of water is restricted and surface drainage is poor (see Figure 9.1). Even the long daylight period is insufficient to overcome the negative effects of low sun altitude and the ice-choked polar seas. On rare occasions maximum temperatures over 27°C have been recorded, but a more representative maximum is 15° to 18°C. On the other hand, frosts may occur in any month. Most stations record only two to six

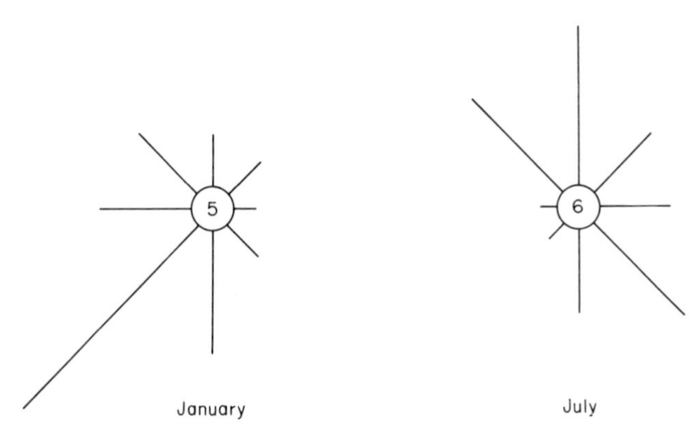

Figure 9.5 Wind roses for Vardo, Norway.

months with average temperatures above 0°C. The exceptions are in especially favored coastal locations near the southern margin of the climatic types. Hardy shallow-rooted sedges, grasses, and woody plants maintain limited growth during the tundra summer by absorbing radiation to create a favorable microclimate.

The long, cold winters of the tundra do not differ greatly from those in the taiga. Six to 10 months have mean monthly temperatures below freezing. The high latitude results in a long period when the sun appears above the horizon each day for only a short time or not at all. During this period, the loss of heat by radiation is far in excess of insolation, but the region is not as dark as might be supposed. Cold weather is accompanied by clear skies, and the long nights are illuminated by the moon, stars, twilight, and displays of the aurora borealis. Extreme minimum temperatures are slightly higher than in the taiga because of the adjacent sea, although the difference between, say, −45°C and −50°C, is of doubtful importance to human beings in the two regions.

Annual precipitation is less than 350 mm over most of the tundra, and it comes largely with cyclonic storms in the warmer half of the year. In the extreme northern reaches of the Atlantic, however, the warmer sea and a persistent upper-level trough of converging air streams combine to produce greater amounts of precipitation (Fig. 3.17). Thus Ivigtut on the southwest coast of Greenland (see accompanying table) and Angmagssalik on the east coast have 1,340 and 780 mm, respectively. On the west coast Godthaab has only 460 mm.

Ivigtut, Greenland 61°N; 5 m

	J	*F*	*M*	*A*	*M*	*J*	*J*	*A*	*S*	*O*	*N*	*D*	*Yr*
T (°C)	−5	−4	−3	0	5	8	10	9	6	2	−2	−4	2
P (mm)	86	125	86	81	99	100	75	109	172	187	144	77	1,340

Snow contributes a greater proportion of the precipitation in the tundra than in the taiga. Wet summer snows are common along the poleward margins. Winter snowfall ordinarily does not exceed 150 to 200 cm and is probably not more than 65 cm in the northern areas. Snow depths in winter are somewhat greater in northeastern Canada and the lands bordering the North Atlantic than in northern Alaska, northwestern Canada, and Siberia. In the former areas, wintertime cyclonic storms are more common. With the low temperatures, the snow is powdery and drifts easily and is consequently difficult to measure accurately. In the regions of scanty snowfall, large areas may be swept bare while others are covered by deep drifts. Blowing snow frequently reduces visibility to a few meters.

Maximum storminess in the tundra occurs in autumn and in spring. In autumn there are appreciable differences between temperatures of the land and the unfrozen sea, and in spring the zone of cyclonic storms is best developed over the region. Some maritime stations have late summer or autumn precipitation maximums that are related to this seasonal distribution of storms. The precipitation graph for Vardo (Fig. 9.4) illustrates this effect.

No general statement can be made about cloudiness and fog in the tundra; the amount varies with storm activity and location with respect to water bodies. Fog and low stratus are prevalent along the seacoasts in the warmer months. Where cold water and drift ice move southward underneath warmer humid air, as along the Labrador Coast and in the Bering Strait, summer fog reaches its maximum. Winter skies are comparatively clear, although ice fog sometimes occurs.

POLAR CLIMATE

In the polar climate mean monthly temperatures are all below 0°C and vegetation is entirely lacking. Snow, ice, or barren rock covers such areas. The polar climate and the associated icecaps predominate over most of Greenland, the permanent ice of the Arctic Ocean, and Antarctica. The widely held concept of a permanent anticyclone at each of the poles has not been borne out by observations, for the high appears to be weakly developed and subject to invasion by cyclones. Nevertheless, the icecaps undoubtedly have a marked influence on the polar climate as well as on climates of adjacent regions. The cold air masses (*A* or *AA*) formed by cooling over the icecaps play an important part in cyclonic activity along the border zone between ice and water and occasionally move into lower latitudes in winter.

The lowest mean annual temperatures on earth are those of the polar icecaps on Greenland and Antarctica. Amundsen–Scott Station at the South Pole recorded a mean daily temperature of −50.0°C during 1976. Annual ranges are large, but not so striking as at some continental locations in lower latitudes. The annual range in Antarctica is greatest over the interior and least at the coasts, responding to effects on the heat balance of latitude, continentality, and the icecap itself. McMurdo Sound, on the Ross Sea coast, has an annual range of 22C° compared with 35C° at Vostok on the interior ice plateau.

Monthly means in the polar summers normally are well below freezing in spite of the continuous daylight. In addition to providing a cold surface the ice reflects much of the solar radiation. Nevertheless, thawing on the arctic ice pack near the North Pole creates mushy ice or pools of water from which limited evaporation occurs (see the heat balance graph, Fig. 9.6). Insolation is exceptionally great during mid-summer in Antarctica owing to the clear atmosphere, but high albedo and heat exhange with the ice result in lower temperatures than those of the North Polar summer. In addition the high elevation of the ice plateau makes it generally inaccessible to advection by warmer air. The "hottest" day recorded at the South Pole was January 12, 1958, when the temperature reached −15°C.

Amundsen-Scott 90°S; 2,800 m (1957-75)

	J	*F*	*M*	*A*	*M*	*J*	*J*	*A*	*S*	*O*	*N*	*D*	*Yr*
T (°C)	−28	−40	−54	−58	−57	−58	−60	−59	−59	−50	−39	−28	−49

McMurdo Sound, Antarctica 78°S; 20 m

T (°C)	−5	−8	−16	−23	−23	−24	−26	−27	−22	−19	−9	−4	−17

Vostok II, Antarctica 78°S, 107°E; 3,420 m (1958-60)

T (°C)	−34	−44	−55	−63	−63	−67	−67	−71	−67	−59	−44	−32	−55

Winters in the polar climates are colder still; monthly means range from −20°C to less than −65°C. A temperature of −66°C was recorded at 3,000 m on the Greenland icecap in 1949. Eismitte, also at an elevation of 3,000 m, had a maximum of −3°C and a minimum of −65°C during a two-year period of observations in 1930–31. An extreme minimum of −88.3°C occurred at the Soviet antarctic base, Vostok II (elevation 3,420 m) on August 24, 1960. A feature of temperature trends in the arctic basin, on Greenland, and in Antarctica is a flat

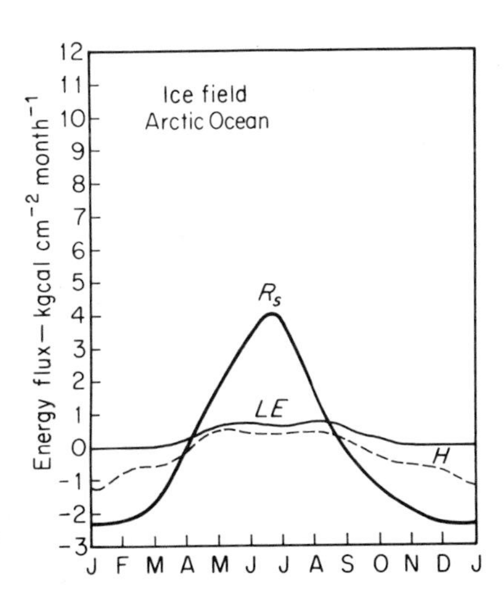

Figure 9.6 Annual march of the energy balance on an ice surface in the arctic. (After Budyko.)

curve during the winter months with the minimum at the end of the dark period. This phenomenon has been called the *kernlose* (or coreless) winter. Its occurrence in Antarctica has been explained as follows: After the sun sets at the end of summer the difference between temperatures of the continent and the surrounding oceans brings about instability and initiates cyclones that move maritime air inland to "ventilate" the lower troposphere above a thin layer of cold air, thus slowing the normal decline of surface temperature. By late winter ice covers a much wider belt around Antarctica and reduces the temperature of southward-moving air masses. The result is a secondary drop in temperature to the seasonal minimum.

Diurnal variations of temperature are small throughout the year in polar climates. In summer they decrease generally toward the poles, where the change in altitude of the sun during the day is least.

Cyclonic storms penetrate the polar icecap regions, bringing cloudiness, some snow, and often fierce winds. Precipitation records are fragmentary, but analyses of satellite photographs reveal the changing boundaries of snowfields and sea ice. Figure 9.7 shows seasonal and interannual variations in the total area of snow and ice cover in the Northern Hemisphere during a nine-year period. The selected monthly values incorporate seasonal snowfall beyond the permanent icecaps and snowfields and are a useful index of albedo changes, which greatly influence the radiation budget in the polar, tundra, and taiga climates. Studies of accumulated snow layers on Greenland have confirmed the short-term records of

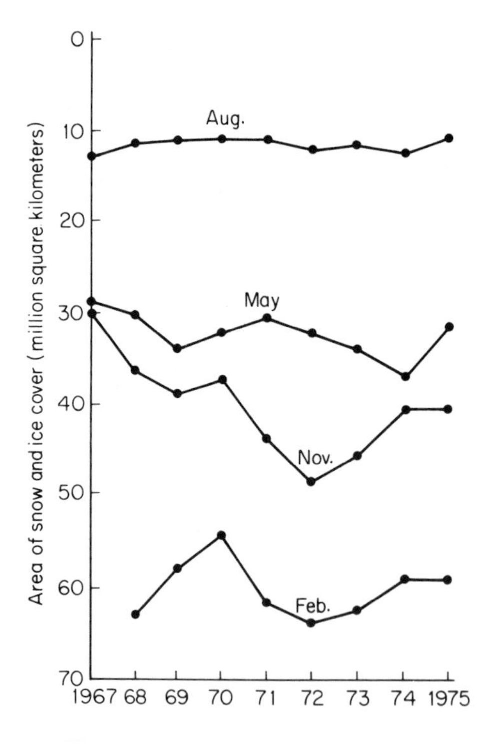

Figure 9.7 Interannual variation of mean snow and ice cover in four selected months, Northern Hemisphere, 1967–1975. Based on analyses of satellite data by National Environmental Satellite Service and G. J. Kukla.

small annual snowfall totals. Estimates based on snow and ice profiles in Antarctica indicate annual values ranging from less than 5 cm on the high interior plateau to more than 50 cm on parts of the coastal periphery. When humid air reaches the maritime fringes of the icecaps, frost and rime may contribute to ice accumulations. One principle is certain: as long as an icecap maintains its depth and extent there must be accretions of snow to compensate for losses by sublimation, melting, and break-off around the margins.

Katabatic winds are a marked feature of weather along icecap coasts. Intense radiational cooling forms a shallow layer of cold surface air that flows downslope, sometimes at great speeds and with local gusts that fan the snow into a blizzard. At Mawson on the antarctic coast (long. 62°53′ E) southeast winds exceeding 50 knots have been recorded in every month, and gusts have exceeded 100 knots on several occasions.

HIGHLAND CLIMATES

The outstanding feature of highland climates is the great diversity of actual climates which prevail, resulting in large differences over short horizontal distances. The principal regions designated as having highland climates are the main mountain chains and high mountain basins of the middle and low latitudes. These are the Cascade–Sierra Nevada and Rockies in North America, the Andes in South America, the Alps in Europe, the Himalayas and associated ranges as well as Tibet in Asia, the Eastern Highlands of Africa, and mountain backbones in Borneo and New Guinea. They do not include all of the world's upland areas. Altitude influences climate everywhere, but where other factors overshadow its influence or where the affected areas are small the highlands have been included in other climatic types for the purposes of regional analysis. Near the poles, for example, climates are cold at both low and high elevations, and a detailed examination of the differences is hardly warranted in a world view of climates. The Ahaggar Mountains in the central Sahara have a modified form of the tropical arid climate, but the differences are primarily of the mesoscale.

There is no single, widespread highland climate with a reasonably distinctive combination of elements analogous to, for example, the rainy tropics or the marine climate. Regions with highland climates are mosaics of innumerable topoclimates fitted intricately into the pattern of relief and altitude. It would be quite impossible to represent them adequately on an ordinary small-scale map of the world or even of a continent. It is necessary, therefore, to indicate the factors which produce the many small climatic differences and to generalize about the characteristics common to the regions included in this climatic type. The regions considered here as having highland climates are those which have considerable areal extent and which have features distinguishing them from other climatic regions at comparable latitudes. The dominant factors which affect highland climates are: altitude, local relief, and the mountain barrier effect.

The normal lapse rate of temperature entails an average decrease of about 6C° per thousand meters increase of elevation. There are important exceptions to this in actual lapse rates (see Chapter 2, "Vertical Distribution of Temperature"), but, on the whole, higher altitudes are associated with significantly lower temperatures and smaller annual ranges. Barometric pressure always decreases with altitude. The atmosphere above areas at high elevations is ordinarily much freer of dust, smoke, and other nongaseous material, and the air is accordingly more transparent to the passage of both incoming and outgoing radiation during cloudless periods.

The influence of local relief is most pronounced in connection with the effectiveness of insolation on slopes having different exposures and with the modification of wind direction and speed. Every variation of slope with respect to the sun's rays produces a different microclimate. In the Northern Hemisphere, the southern slopes of hills and mountains receive more direct insolation, whereas northern slopes are less favored. Many deep valleys and steep northern slopes are exposed to the direct rays of the sun for only a short time each day; in rare cases they may be forever in the shade. Variations in exposure to winds also produce local climates. Wind speeds at high altitudes are generally greater because of the increasing pressure gradient and decreasing friction with the earth's surface; the higher an isolated mountain site is the more nearly wind direction conforms to broad patterns of circulation. Winds tend to be gusty in rugged terrain. Locally, they may be funneled through constrictions in canyons or valleys, forced over ridges, or routed around mountains. Under conditions of fair weather, mountain and valley breezes are generated by the relief itself. Thus the direction and speed of wind in mountain valleys may be quite at variance with the air movement aloft or over adjacent plains. Prevailing winds in mountainous terrain often are conditioned more by the trends of valleys than by global or continental circulation systems.

We have treated the barrier effect of mountains earlier in connection with several of the climatic elements. The barrier effect induced by mountains produces "standing waves" that lie perpendicular to wind direction. Stationary lenticular clouds sometimes form along the crest of the wave. This effect may extend well into the upper troposphere, at the same time generating increased turbulence.

Stable air masses of major proportions can be dammed by a mountain range, creating marked temperature differences between the windward and leeward. In this manner in winter the east–west mountain chains of Europe hold much of the polar air away from the northern Mediterranean coast, the North American Rockies and Cascades protect the Pacific Northwest, and the Himalayas decrease the southward flow of cold air masses from central Asia. Whatever air does overflow these barriers is subject to the stable-air foehn effect and is therefore not so cold when it descends the slopes at the continental margins. (See Chapter 5, "Other Air Mass and Storm Effects.") Mountains in the path of moist winds receive considerably more precipitation on the windward side than on the leeward. The accumulations of snow and ice at the higher elevations affect surface albedo, temperature, and air drainage. In general, mountain barriers tend to differentiate the climates of the windward and leeward wherever they lie in belts of prevailing winds of the major circulation systems.

Since highland climates are distributed through a wide range of latitudes, no specific temperature values characterize all such climates. Under clear skies, insolation is intense at high altitudes and there is a greater proportion of the shorter wave lengths (violet and ultraviolet), which do not penetrate well to lower elevations. As a result, one's skin burns or tans more readily. Temperature differences between daylight and darkness and between sunshine and shade are great. The same clear air that permits easy transmission of insolation also allows rapid loss of heat by radiation and large diurnal ranges of temperature. Night frosts are common even where daytime temperatures rise well above freezing. By contrast, cloudiness reduces the diurnal range and results in a lapse rate approximating the normal both by day and at night. A distinctive feature of highland climates in the tropics is a larger diurnal than annual range of temperature. Monthly means and the annual march of temperature are determined primarily by latitudinal variations in insolation and in the transport of air from land or water source regions. Quito, Ecuador, virtually on the equator, has an insignificant annual range appropriate to its latitude, but the daily range normally exceeds 10C°. Longs Peak has a much greater annual range (18C°) in response to the variation in insolation through the year, and the range is greater still in the high, arid valley at Leh, Kashmir.

Quito, Ecuador O°8′S; 2,811 m

	J	*F*	*M*	*A*	*M*	*J*	*J*	*A*	*S*	*O*	*N*	*D*	*Yr*
T (°C)	13	13	13	13	13	13	13	13	13	13	13	13	13
P (mm)	119	131	154	185	130	54	20	25	81	134	96	104	1,233

Longs Peak, Colorado 40°N; 2,730 m

	J	*F*	*M*	*A*	*M*	*J*	*J*	*A*	*S*	*O*	*N*	*D*	*Yr*
T (°C)	−5	−5	−3	1	5	10	13	13	9	4	−1	−5	3
P (mm)	18	30	52	69	60	42	91	57	44	42	22	23	550

Leh, Kashmir 34°N; 3,514 m

	J	*F*	*M*	*A*	*M*	*J*	*J*	*A*	*S*	*O*	*N*	*D*	*Yr*
T (°C)	−8	−6	0	6	10	14	17	17	13	7	1	−5	6
P (mm)	12	9	12	7	7	4	16	19	12	7	3	8	116

Whereas temperatures are markedly lower with an increase in elevation, precipitation tends to increase, at least up to altitudes of 3,000 to 5,000 m. Above a zone of maximum precipitation in the tropics it again decreases, for the orographic effect has already exhausted most of the precipitable moisture. On mountainous trade-wind islands, there is heavy rainfall on low windward slopes, but at higher elevations an upper-air subsidence inversion (the trade-wind inversion) produces arid conditions. At higher latitudes the increase of precipitation continues upward to mountain crests. As in the case of temperature, no general statement can be made about average amounts of precipitation. Average annual totals are related to the dominant air masses and to the moisture characteristics of prevailing winds (see Fig. 9.8). Fogginess is generally greater in mountains than on nearby plains or valleys. A great deal of mountain fog is simply clouds formed by one or more of the lifting processes, but radiation fogs are common in calm air on clear nights and in the early morning hours.

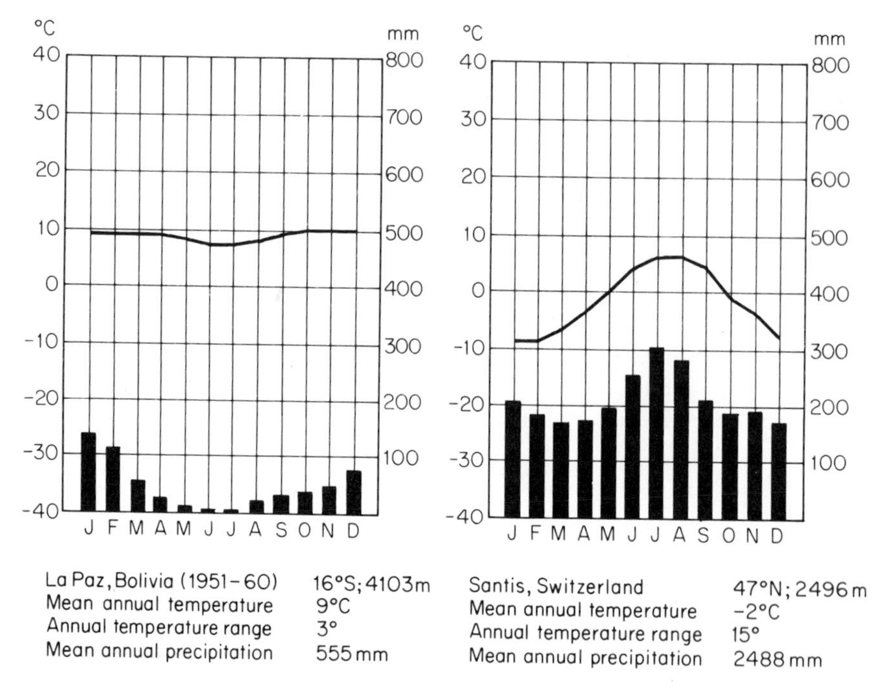

Figure 9.8 Climatic graphs for La Paz and Santis, highland climates.

With increased elevation there is an ever-greater proportion of precipitation in the form of snow, and the snow cover remains for a longer period. The elevation of the lower limit of permanent snow or ice is determined by air temperature, by slope exposure to the sun, and by depth of snowfall. Although the snowline is in general higher in the tropics and becomes lower with increasing latitude, it is not highest at the equator but rather in the two drier belts that correspond roughly with the tropical arid and semiarid climates (see Fig. 9.9). A scanty accumulation of snow obviously can be melted away in a short time under cloudless skies and subsiding air. This also explains the generally higher level of permanent snow on the leeward slopes of mountains and uplands far removed from the oceans. The snowline in the Rockies of Montana is higher than that of the Cascades, which, in turn, is higher than that on the Olympics in the same latitude. Foehn winds account for a more rapid removal of snow on leeward slopes. Other variations from the averages indicated in the figure occur on shaded slopes, where snow remains much

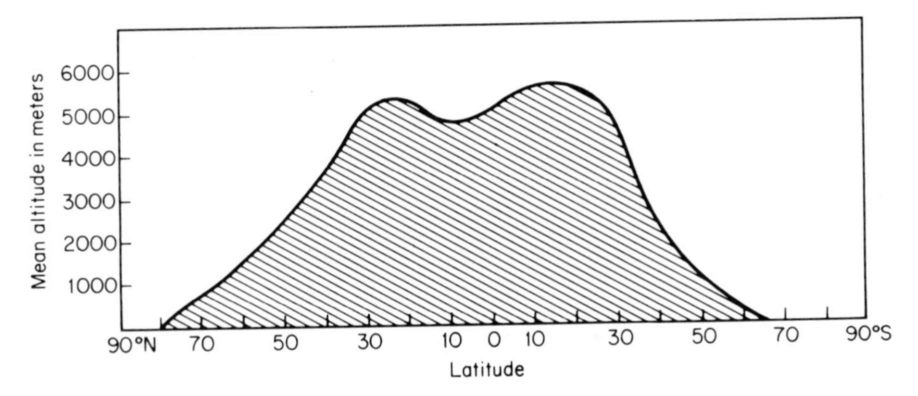

Figure 9.9 Variations of the mean altitude of the permanent snowline with latitude.

longer than on slopes exposed to the sun. Where there is a winter maximum of precipitation, the snowline tends to be lower than in areas at the same latitude with a summer maximum. In any particular location the actual lower limit of snow results from these various factors working together or in opposition.

The causes of precipitation in a highland climate are much the same as elsewhere in the general geographic area in question, except for the intensification by the orographic effect. In the belt of cyclonic storms, for example, the characteristic frontal storms prevail and the annual regime of precipitation is similar to that of the surrounding lowlands, but the annual amount and the total falling in a given storm are both likely to be somewhat greater in the highlands.

OCEAN CLIMATES

Oceans cover 71 percent of the earth, a vast surface for exchange of heat, water, and momentum with the atmosphere. Because they have a thermal capacity more than 1,000 times that of the atmosphere the oceans store huge quantities of heat and help moderate temperatures throughout the climate system. The ability of sea water to transport heat between its surface and lower layers as well as horizontally is the main cause of differences between climates of the oceans and continents. Variations in deep water temperatures are a probable link in the processes of climatic change. As the principal source of evaporated moisture that feeds clouds and precipitation oceans also regulate air temperature by means of latent heat transfer. Their patterns of horizontal and vertical circulation redistribute sensible heat and thereby tend to equalize the global energy budget.

Climates over the oceans reflect not only the influence of the surface water but also that of the atmosphere. As over land, latitude is the basic terrestrial factor that controls the radiation budget, but winds, air masses, and cloudiness act together to produce a variety of ocean climates. Transitions are rarely so abrupt as over the continents, and it is consequently more difficult to define regional boundaries. In addition, the general lack of systematic records for ocean locations remote from islands or coasts hampers detailed analyses. This section, like that on highland climates, will sample the range of conditions and causes rather than attempt a complete spatial differentiation.

The surface energy budgets of two locations illustrate the interaction between sea and atmosphere in creating ocean climates. In the warm equatorial Atlantic northeast of Brazil, where components of the heat balance change relatively little through the year, a large part of the converted net radiation is used in evaporation and only a small amount is transferred from surface to air as sensible heat. (See Fig. 9.10a and compare with the heat balance at Manaus in Fig. 7.1.) During the latter third of the year surplus heat is transferred to lower layers and to higher latitudes.

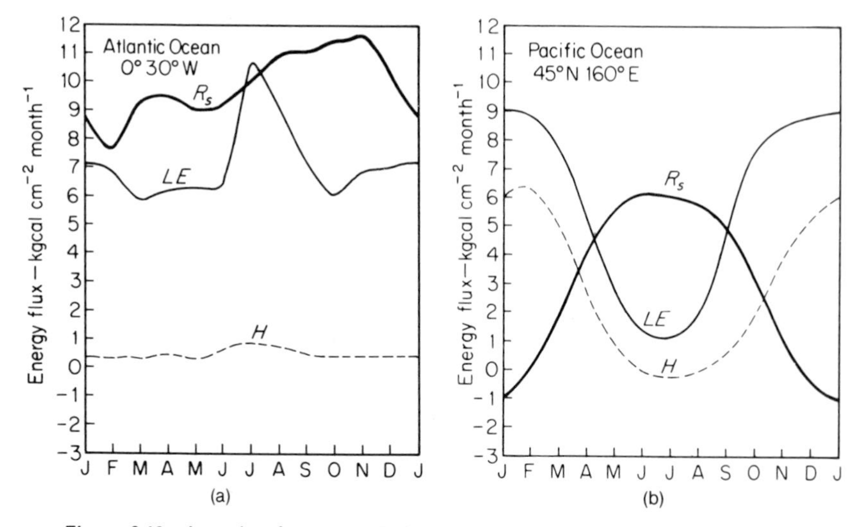

Figure 9.10 Annual surface energy balance graphs for two ocean locations. (a) northeast of Brazil; (b) southeast of the Kurile Islands. (After Budyko.)

In contrast with the equatorial Atlantic, the surface heat balance of the Pacific southeast of the Kurile Islands shows the combined influence of monsoonal atmospheric circulation and the Kuro Shio, or Japan Current (Fig. 9.10b). There is a large net radiation loss in winter and large amounts of latent and sensible heat pass from the warm surface current to the air. Net radiation gain increases in summer, but evaporative and sensible heat is actually transferred from the warm air to the cooler water surface.

Sea surface temperatures are an index of the heat budget. Mean surface water temperatures tend to lag behind the annual maximum and minimum of insolation, reaching their extremes in February and August. The influence of latitude and major ocean currents is evident on the accompanying maps (Figs. 9.11 and 9.12). Air temperatures near the sea surface depend on atmospheric circulation as well as local heat exchanges.

Precipitation, evaporation, and runoff from the land are the main com-

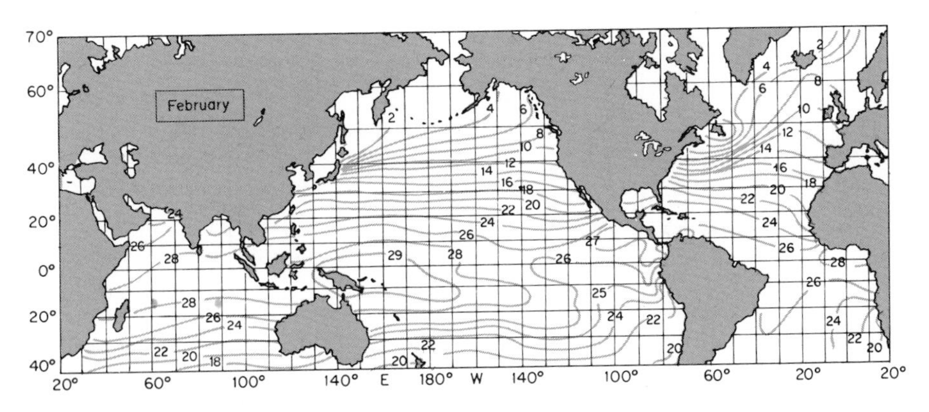

Figure 9.11 Mean sea surface temperatures in February, °C. (After Newell.)

ponents of the oceanic water budget. Estimates based on observations and calculations indicate about 1,140 mm average annual precipitation for the entire world ocean, the equivalent of about 120 mm added by runoff from the continents, and a compensating loss of 1,260 mm by evaporation. Since each of these values varies regionally, some adjustment must be made by horizontal transport of water within and among the oceans. Precipitation over the oceans is generally greatest in zones of atmospheric convergence, especially the intertropical convergence and polar fronts, and it is associated with the storm types common to these areas. It is least in regions dominated by subtropic highs, at high latitudes, and where cold currents prevail. Although isohyetal maps have been drawn for the oceans, they are based on estimates. Orographic effects and daytime convection produce local anomalies in both annual amounts and regimes of precipitation on islands, but indirect evidence suggests that island stations are broadly representative of their regions. Data for Ocean Island near the equator in the western Pacific illustrate the great monthly and annual rainfall variability that causes water supply problems on numerous atolls in the tropical Pacific (Fig. 9.13). Mahe Island, Seychelles, in the Indian

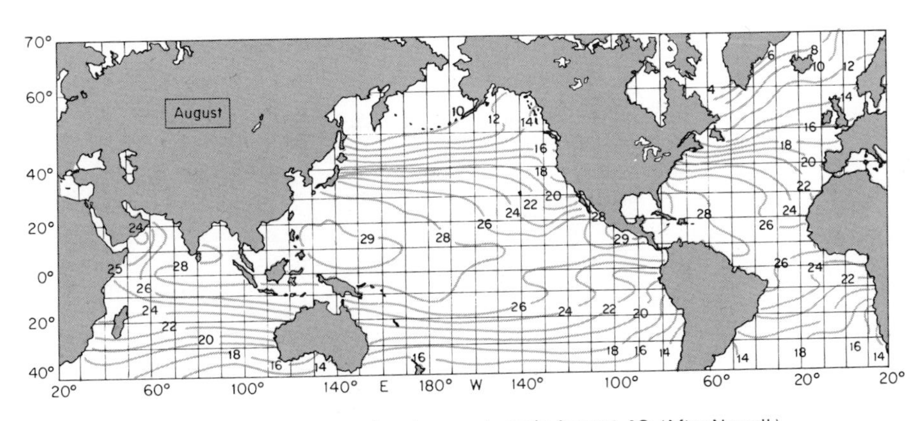

Figure 9.12 Mean sea surface temperatures in August, °C. (After Newell.)

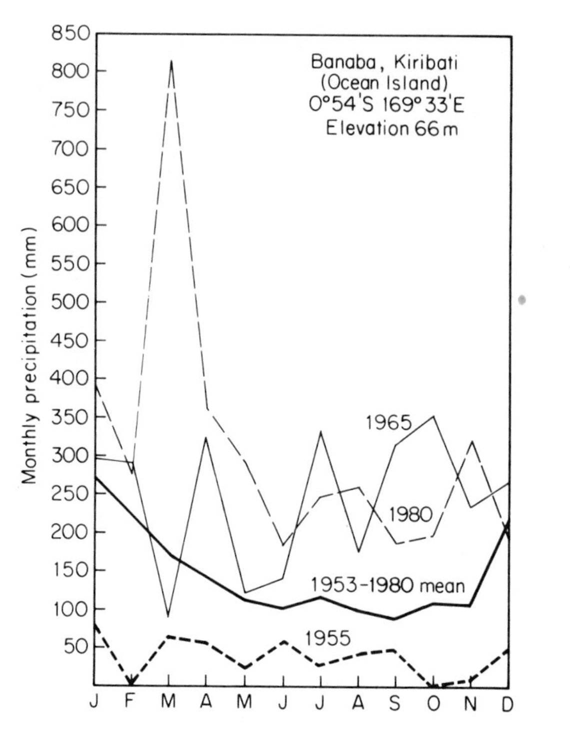

Figure 9.13 Monthy rainfall for selected years on Ocean Island in the western Pacific. (Data from NZ Meteorological Service.)

Ocean, represents a monsoon regime. Both temperature and precipitation at Sal, Cape Verde Islands, show the influence of the subtropic high and the cold Canaries Current. At high latitudes the moisture content and stability of air masses govern precipitation amounts. Climatic data for a station near sea level on mountainous Jan Mayen northeast of Iceland indicate the moderating effect of the North Atlantic

Mahe Island, Seychelles 5°S, 55°E; 3 m

	J	*F*	*M*	*A*	*M*	*J*	*J*	*A*	*S*	*O*	*N*	*D*	*Yr*
T (°C)	27	27	28	28	28	27	26	26	26	26	27	27	27
P (mm)	358	242	204	170	161	85	83	78	116	167	206	332	2,203

Sal, Cape Verde Islands 17°N, 23°W; 55 m

T (°C)	21	21	21	21	22	23	24	26	27	26	24	22	23
P (mm)	2	2	1	0	0	0	18	20	27	15	10	13	108

Jan Mayen 71°N, 8°W; 39 m

T (°C)	−4	−5	−5	−3	−1	2	5	6	4	1	−1	−3	0
P (mm)	79	54	63	58	33	28	36	61	83	93	82	75	735

Drift and an autumn maximum of precipitation coinciding with highest sea temperatures.

Being less affected by topographic irregularities, winds over the oceans conform more closely to the general atmospheric circulation than do those over the continents. In fact, the discovery of prevailing wind patterns in the days of sailing ships laid the foundation for early surface wind models, and the related nautical terms are still widely used. Seasonal changes in surface wind direction are reflected in oceanic circulation and the transfer of heat. Cyclonic systems embedded in the atmospheric circulation often form over water and reach their greatest intensity while still at sea, where essential energy and moisture are available (Fig. 9.14). The divergent flow of oceanic anticyclones and the steering action of upper-level jet streams combine to guide major storms at sea and determine their paths toward continental shores. But the effects are reciprocal. Because the oceans respond slowly to changes in the heat budget their influence on climate over the continents may be delayed for several months. Climatologists have found correlations between anomalies in sea temperature and subsequent weather events thousands of kilometers distant. The intensity and duration of the summer monsoon in India, for example, appears to reflect earlier cooling in the Arabian Sea. Long-range forecasters have had some success in predicting average temperatures and precipitation over North America from sea temperatures in the North Pacific during the previous summer.

Figure 9.14 GOES-East satellite view of Hurricanes *Isis,* off the Pacific Coast of Mexico, and *Allen,* over the Gulf of Mexico, on 8 August 1980. (National Satellite Data Service.)

QUESTIONS AND PROBLEMS FOR CHAPTER 9

1. Why are extreme minimum temperatures often lower in the taiga than in the tundra although the latter is at higher latitudes?
2. Why does the tundra climate extend farther equatorward in eastern North America than in the western part of the continent?
3. Why are there massive ice caps on Greenland and Antarctica in spite of meager precipitation in those areas?
4. What climatic conditions are necessary to maintain permafrost?
5. Why are the mountains and plateaus of Antarctica not included among the highland climates?
6. Why is the mean snowline higher at about 20° latitude than at the equator?
7. What months would be best for climbing in the Himalayas? Why?
8. Why is fog likely to occur more often near mountaintops than in adjacent valleys? Under what conditions might the reverse be true?
9. What special influences on climate result from the mobility of ocean water?
10. Continents experience maritime influences. Are ocean climates subject to continental effects? Explain.

10

Climatic Change

The preceding chapters have examined broad spatial patterns of climate on earth, emphasizing contemporary climates. Weather changes that collectively make up climate are taken for granted; in fact, the only constant feature of weather is change. We are aware of diurnal and seasonal weather fluctuations and the apparently random passage of storm systems. But several kinds of evidence point to variations in climate as well. The implications of a rapid secular change in climate defy the imagination. Agriculture, manufacturing, commerce, and all other human endeavors would require drastic adjustment if the world's climate were to change greatly within a generation. During human history, relatively modest fluctuations on a regional scale have demonstrated the impact of climate on economy and society. Employing the sequence of time categories suggested in Chapter 6, we turn now to the "history of climate"—evidences of changes and theories of their causes.

PROXY EVIDENCE AND GEOCHRONOLOGY

Reconstruction of climates prior to the record of human history depends on "proxy" data, which are inferred from natural climatic indicators. Examples of proxy evidence and the approximate time periods for which they are available are shown in Fig. 10.1. Although abundant, and sometimes contradictory, evidence indicates marked changes in climate since the creation of the earth, the data become increasingly satisfactory through the eons. Plant and animal fossils in various sedimentary deposits are the primary clues to duration and geographical extent of temperature and moisture conditions since the beginning of the Cambrian Period

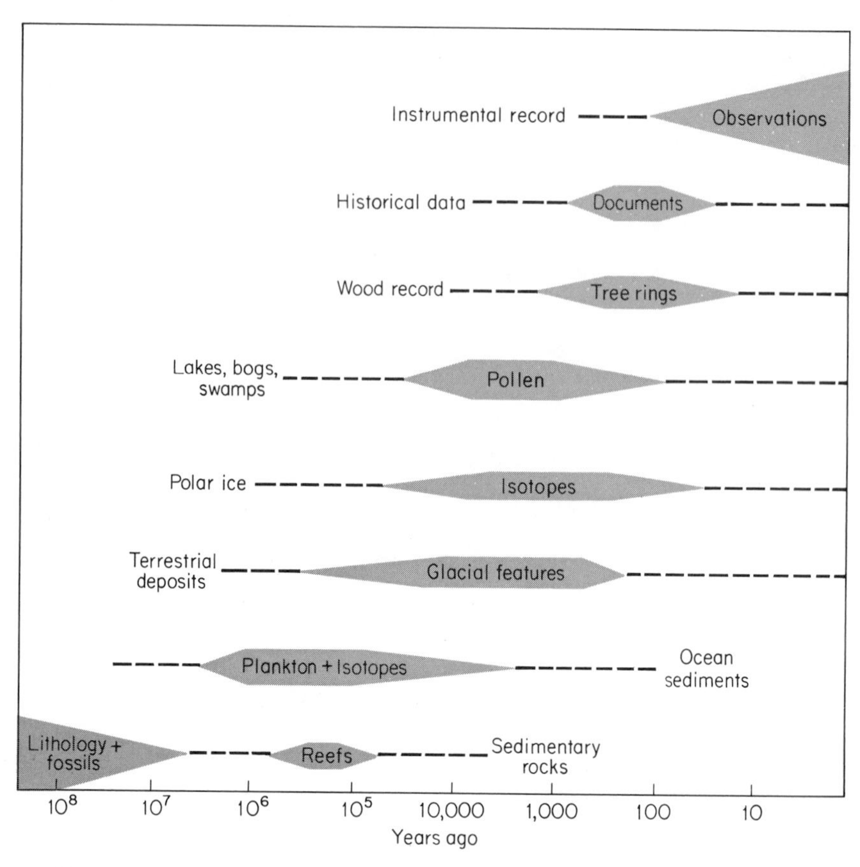

Figure 10.1 Types of evidence for climatic change and their approximate time scales. (After Bernabo, *EDIS*.)

nearly 600 million years ago, when life forms began to develop rapidly and the free oxygen content of the atmosphere increased. Fossils are the basic data of paleoclimatology, which borrows its methods from paleontology. Inasmuch as they depend on past existence of life, it follows that the range of paleoclimatic fluctuations must fall within the limits allowable to support sea life and probably within the limits for plant life on land as well. Thus, sustained global temperatures such as 100°C or −50°C are highly unlikely features of paleoclimates. For estimates of temperature and precipitation from fossils it is assumed that environmental conditions required by a given species today were also favorable for that species in the past. This principle, known as *uniformitarianism,* governs a wide variety of geologic and climatic studies that involve organic residues.

Inorganic deposits of alluvial, aeolian, glacial, or volcanic origin also reflect past climates, although they often contain organic materials that enhance interpretation. Erosional landforms offer corroborative evidence of climatic processes such as flooding, glaciation, or wind action. Shoreline features indicate changing lake and ocean levels in response to climatic fluctuations.

One of the most productive fields for investigation of climatic changes is glaciology. Consistent trends in precipitation and temperature are revealed by the areal extent of glaciers and in the resulting landforms. Most of the direct study of glacier fluctuations and climatic relations has been carried out in the twentieth century. The Swedish glaciologist Hans Ahlmann first correlated measurements of ice ablation (reduction of glacier volume by evaporation, sublimation, and melting) and simultaneous weather observations. The volume of glaciers is affected by annual snowfall, duration and degree of temperatures above the melting point, and the amounts of incoming and outgoing radiation. These factors are affected, in turn, by cloudiness, wind speed, and humidity. The shape of the basin or valley containing the glacier also influences its movement; some glaciers cannot advance until they have increased in volume sufficiently to surmount topographic barriers.

Glacier observations by Ahlmann and his followers have shown that the climate of the arctic and subarctic became warmer in the first part of the twentieth century, the major changes being at the higher latitudes. This trend may have undergone a reversal, however, since mid-century.

Existing ice fields provide proxy data on climates of the distant past as well as for recent decades. Application of the oxygen isotope method of paleotemperature analysis to deep cores from the ice caps of Greenland and Antarctica has yielded temperature estimates for approximately the last 100,000 years. One antarctic core, for example, indicates a relatively rapid cooling on the order of 5C° about 60,000 years ago, followed by slow warming and then another cooling period beginning about 30,000 years ago. The technique of oxygen isotope analysis, discovered by Harold C. Urey, entails determination of the ratio between oxygen-18 and oxygen-16 isotopes in the H_2O molecules by means of mass spectrometry. When air temperatures are high more of the heavier oxygen-18 isotope is evaporated from the ocean and subsequently precipitated on ice caps. Conversely, during cold periods a greater proportion of the lighter oxygen-16 is transferred to polar ice layers. In either case the ocean water undergoes a compensating change in the $^{18}O/^{16}O$ ratio. Calcium carbonate ($CaCO_3$) in the shells of marine microorganisms reflects the ratio, thus making it possible to derive a record of past temperatures and periods of glaciation from deep-sea organic sediments. Oxygen isotope analyses of deep-sea cores provide a continuous temperature record for more than a million years.

Magnitude and sequence of climatic events are fundamental in all studies of climatic change, but chronology is equally essential for correlating different kinds of proxy evidence and for making worldwide comparisons. Whereas oxygen isotope analysis is suitable for estimating temperatures, the stable ^{16}O and ^{18}O isotopes do not provide a direct basis for dating. Geochronology (literally earth dating) of paleoclimates is accomplished primarily with reference to decay rates of radioactive (unstable) isotopes in geological formations and deep-sea deposits. Paleosoils, ancient peat bogs, residues from evaporation of water bodies (*evaporites*), ice cores, and calcium carbonate deposits in caves are other types of proxy records that can be dated by radiometry. Techniques for dating organic remains are especially valuable for correlating the evidences in life forms and geological strata. The carbon-14 method, developed by Willard F. Libby, is applicable to plant and animal matter and even to carbon dioxide in air or water, but its range of accuracy is acceptable for only about the last 50,000 years. Other radiometric methods that span a time scale to include some of the oldest known earth materials complement stratigraphic studies. None is sufficiently precise to identify periods shorter than several thousand years, however. Among the elements whose radioactive isotopes have been found useful in establishing rock chronologies are potassium, rubidium, thorium, and uranium.

Several kinds of proxy data combine climatic evidence with a parallel chronology. A related technique from organic chemistry is based on amino acids, which are essential components of living things. Some amino acids are more stable than others, their rates of decay depending on temperature. Thus, when relative decay rates and temperature responses are calibrated, analyses of organic remains offer the prospect of temperature dating across millions of years.

One of the most detailed chronologies for the past few thousand years lies in *varves,* the annual layers of silt and clay deposited on the bottoms of lakes and ponds that are subject to freezing in winter and thawing in summer (see Fig. 10.2). The only material being deposited in a frozen lake is the fine suspended clay; the surface ice prevents other materials from entering. When thawing begins, fresh water and coarse sediments are introduced. In part because of climatic fluctuations, no two successive years have the same thickness of deposits. Thus, parallel dating of lake beds in the same region is possible through matching of varve thickness. Varves cannot be formed under glaciers because of the necessity for annual freezing and thawing. A comparison of lake beds in once-glaciated areas therefore reveals the dates of ice removal. About 13,700 years are represented in the varve records of Scandinavia, where the oldest varves are in the south and the newest in the north, permitting a chronology of the last glacial recession.

A proxy technique known as *dendrochronology* is based on the annual increment of tree growth, mainly in the middle and high latitudes (see Fig. 10.2). Records in tree rings go back more than 3,000 years in living trees and another 5,000 years or more in fossil wood and ruins. A close relation between annual rainfall and the growth rings of trees under climatic stress was demonstrated by A. E. Douglass in the southwestern United States, where fluctuations in precipitation have been fairly well charted. Subsequent studies at the Laboratory of Tree-Ring Research, University of Arizona, have shown that temperature, pressure, and atmospheric circulation patterns also correlate with tree-ring widths, making it possible to characterize major climatic fluctuations. A 5,405-year record reconstructed from bristlecone pines at the tree line in the White Mountains of California indicates a relatively warm period between 3500 and 1300 B.C. (Fig. 10.3). Cooler and wetter conditions prevailed from 1300 to 200 B.C., and North American glaciers advanced. Predominantly warm summers between 200 B.C. and A.D. 300 gave way to cooling and renewed glacial advances in the Rocky Mountains and Sierra

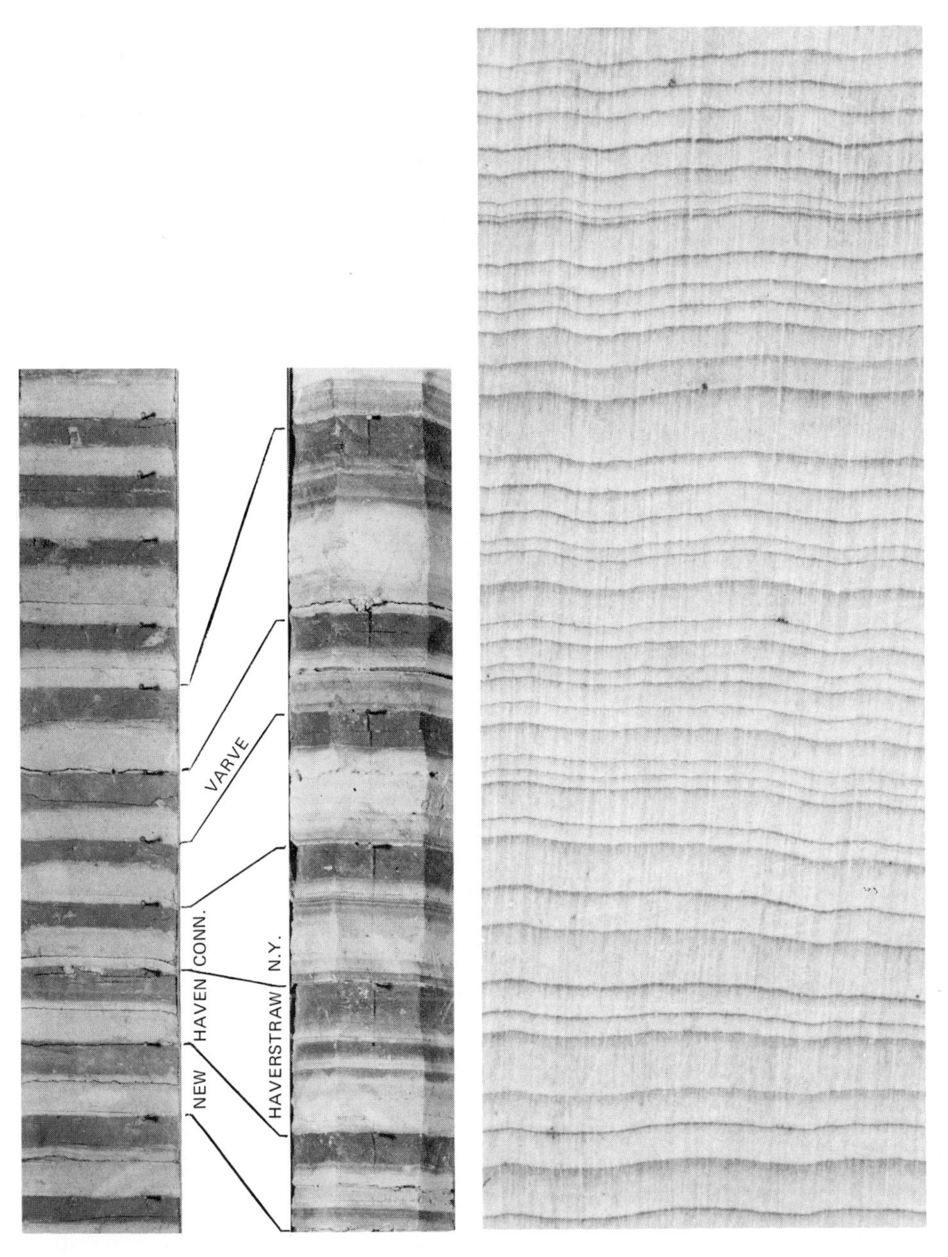

Figure 10.2 Examples of proxy evidence for geochronology. Varve correlations on the left are from sites near New Haven, Connecticut, and Haverstraw, New York. The tree-ring cross section on the right indicates variations in annual growth of a New Zealand conifer. (Varve photograph courtesy The American Museum of Natural History; tree-ring photograph courtesy Peter Dunwiddie, Quaternary Research Center, University of Washington.)

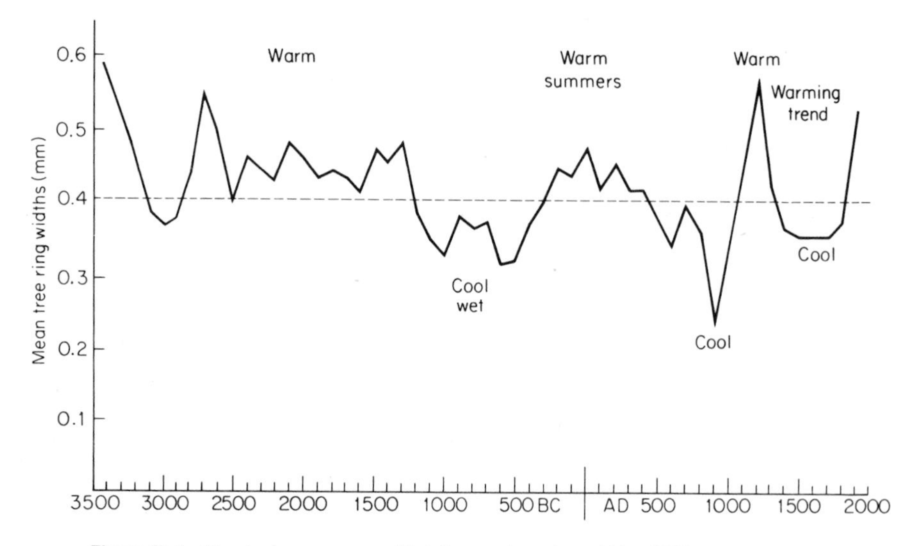

Figure 10.3 Hundred-year means of bristlecone tree ring widths, 3500 B.C. to A.D. 1950, and inferred climatic conditions. (After LaMarche.)

Nevada that lasted for more than 800 years. Around A.D. 1200 there was a temporary warming, followed by another cool period and yet another warming trend during the past century. These inferences coincide remarkably well with other types of proxy evidence.

Dendrochronology has also been employed to date the comparatively recent advance and recession of glaciers. Cross sections of trees partly tilted by glacier ice along the margins of maximum advance but left alive reveal the exact year of the maximum advance of the glacier. While a tree was erect growth rings were concentric, but after tilting by ice pressure the rings lost their symmetry. Narrow rings indicate periods when ice was close enough to a tree which remained erect to affect temperature. A related technique, *lichenometry,* dates the recession of glacier ice or perennial snow fields with reference to the size of lichens, certain species of which have remarkably constant annual growth rates.

Another approach to geochronology uses *pedogenic* criteria, that is, data obtained from the study of soil development. Old soils that have been buried in river flood plains, along fluctuating lake shores, or under windblown deposits furnish some evidence of past climates, for climate is the most active factor in soil formation. Peat bogs also contain evidence for a generalized method of climatic dating. The most effective analyses have been accomplished with the aid of pollen studies, or *palynology.* Under the microscope, pollen grains identify the plant associations that prevailed at various stages during the peat accumulation. The succession of plant types in peat layers shows several long climatic waves during the last 30,000 to 35,000 years. In the British Isles the proxy data from peat correlate with archaeological evidence of technological development.

Fossils of animals as well as plants aid in piecing together the history of climate over the past few thousand years. The bones of arctic mammals have been found far south of their present range, and well-preserved remains of extinct animals occur in glacial ice in the arctic. The mastodon is perhaps the best known example. Remains of desert and steppe animals have been unearthed in western Europe.

A global paleoclimatic calendar, reconstructed from proxy data, shows a sequence of low temperatures and intermittent glaciation from 800 to 600 million years ago, followed by warmer ice-free conditions through the Cambrian, Ordovician, Silurian, and Devonian Periods, that is, between about 550 and 350 million years ago (Fig. 10.4). During the late Carboniferous and early Permian, approximately 300 million years ago, cool-wet climates produced extensive glaciation over Pangaea, a Southern Hemisphere land mass which then comprised much of present day Africa, India, Australia, Antarctica, and South America. Through the Mesozoic Era warm and dry conditions were dominant, and there were no polar ice caps. Beginning in the early Epochs of the Cenozoic Era, polar ice advanced and

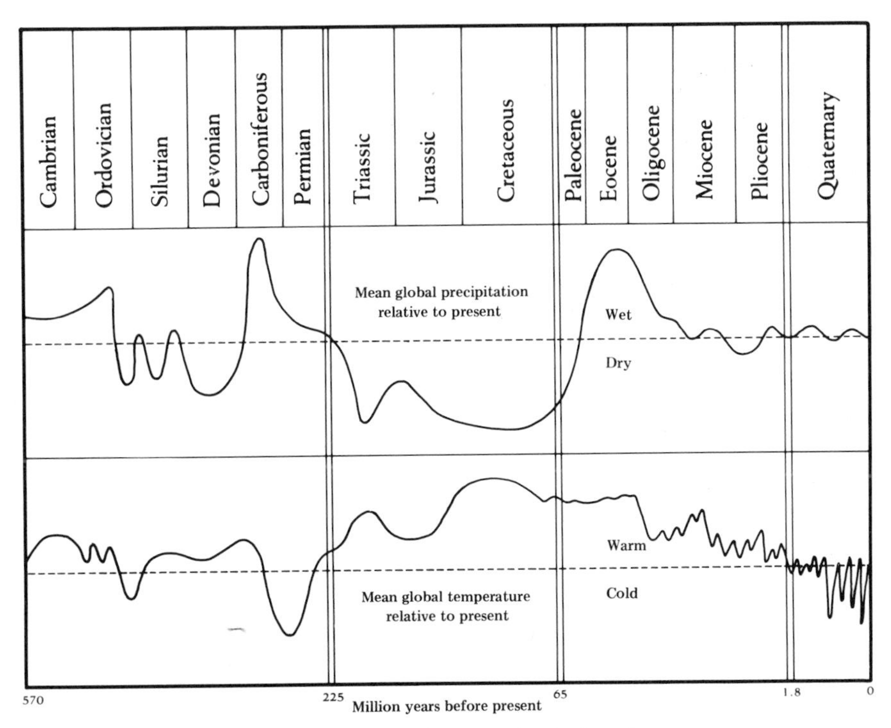

Precipitation and temperature fluctuations since the Precambrian

Figure 10.4 Major fluctuations of global precipitation and temperature from the Cambrian to the present, based on proxy data. Note progressive increase in time scale. (Adapted from L. A. Frakes, *Climates Throughout Geologic Time*, 1979; H. H. Lamb, *Climate: Present, Past and Future*, Vol. 2, 1977.)

receded in concert with a succession of temperature fluctuations, culminating in the glacial ages of the Quaternary. During the Quaternary, the subtropic highs gradually shifted from 50° or 60° latitude toward their present positions. The lowest Northern Hemisphere temperatures since the Precambrian prevailed through most of the Quaternary glaciation. Generalized trends of mid-latitude air temperatures for the past 150,000 years are shown in Fig. 10.5. The last major ice

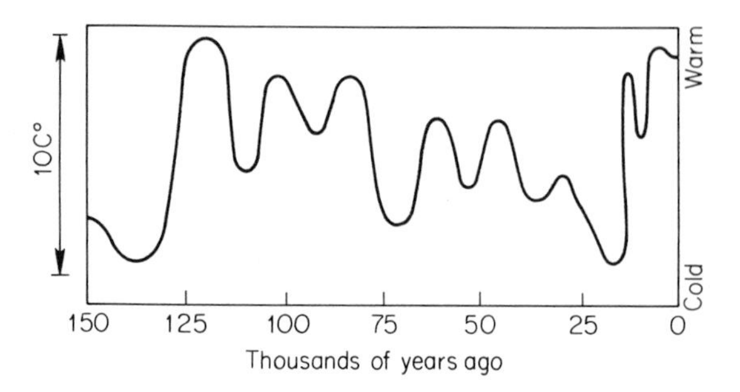

Figure 10.5 Trends in air temperatures of the Northern Hemisphere mid-latitudes during the past 150,000 years. (From *Understanding Climatic Change.* Washington: National Academy of Sciences, 1975.)

advance reached its maximum about 18,000 years ago and was halted by rapid melting and recession that continued until about 6,000 years ago at the height of the Holocene, or most recent interglacial epoch. Concurrently, mean sea level rose approximately 100 m (Fig. 10.6).

CLIMATE DURING RECORDED HISTORY

Throughout the entire period of written and inferred human history proxy data overlap archaeology, documentation, and instrumental records to assist verification of climatic changes. It might be naively assumed that recorded human history would lend itself readily to analysis of past climates. On the contrary, documental climatology deals largely with manuscripts written for purposes other than climatic description, and problems arise from changes in the calendar and lack of continuity of historical records at a given location. Writers have had a tendency to mention droughts, severe storms, or other disasters and generally have neglected relatively stable interludes. Nevertheless, accounts of crop yields or crop failures, of floods, and of migrations of people give evidence of the possible influence of a changing climate. Correlation with various proxy data shows the following pattern since the birth of Christ. As would be expected, there are more data for Europe than for other parts of the world.

During the first century of the Christian era, the pattern of precipitation in Europe and the Mediterranean resembled that of today. This was followed by a wetter period ending about A.D. 350. The fifth century was dry in Europe and prob-

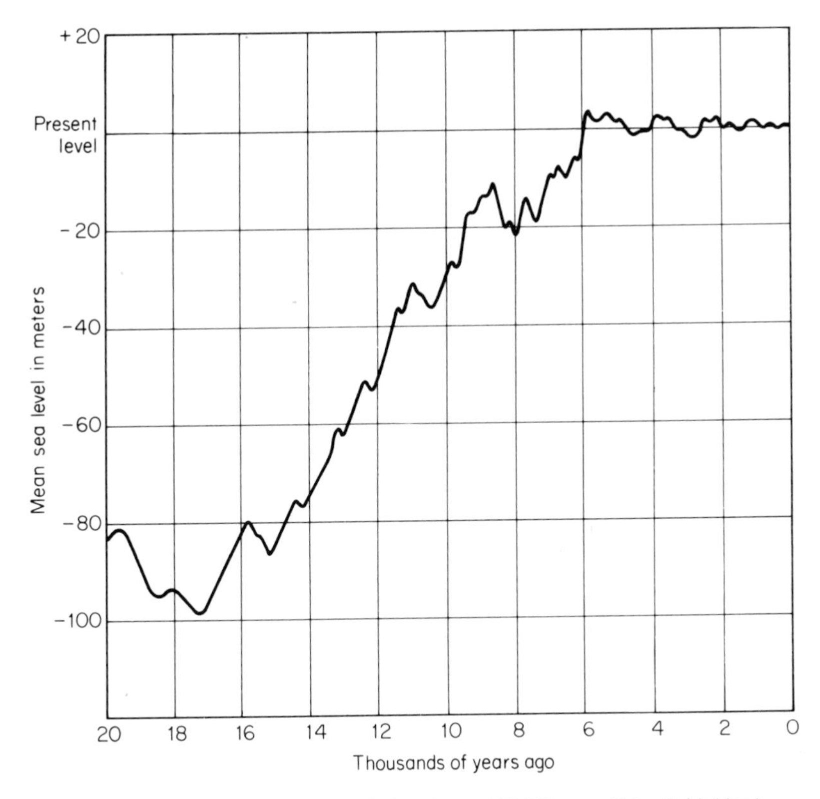

Figure 10.6 Changing sea level during the past 20,000 years. (After Fairbridge.)

ably also in North America, where many western lakes dried up completely. In the seventh century Europe was both warm and dry. Heavy traffic crossed the Alps through passes that are now filled with ice. Tree rings show a dry period in the United States at this time, and the Nile floods were low. Europe was wetter again in the ninth century. During warm, dry conditions in the tenth and eleventh centuries wine production reached its peak in England. Vikings settled Greenland in 984 in a time of mild climate, only to abandon it about A.D. 1410. The first half of the thirteenth century was stormy in the North Sea and North Atlantic. A drought from A.D. 1276 to 1299 is believed to have driven the cliff dwellers from Mesa Verde in Colorado. The fourteenth century was cold and snowy in northern Europe and northern North America. The Aztecs settled Mexico in 1325, when lakes were at higher levels than today. During the so-called "Little Ice Age," extending from 1430 to about 1850, European glaciers advanced farther than at any time since recession of continental ice. Arctic sea ice was also probably much more extensive than at present. Since the middle of the nineteenth century, Northern Hemisphere glaciers have receded to their sixteenth-century positions and sea level has risen by 10 to 12 cm.

CLIMATIC TRENDS SINCE THE ADVENT OF INSTRUMENTS

Finally, let us consider the very short period for which we have instrumental records. Like all forms of historical data relating to weather and climate, records of instrumental observations fade rapidly into the past. Simple rain gauges and wind vanes date from antiquity, yet continuous rainfall records reach back only to about 1700. After the development of reliable thermometers and barometers in the seventeenth century, meteorological observations, mainly at European sites, began to increase slowly. Most early data are fragmentary, although in some cases useful time series have been developed from overlapping records for different places. For example, regional temperature averages have been calculated for central England from 1680 and for the eastern United States from 1738. Mean pressure and wind flow patterns based on barometric records since 1750 have been reconstructed for large portions of the Northern Hemisphere mid-latitudes, and even earlier estimates have been inferred from other weather data. By 1850 a sparse global network of daily surface observations was fairly well established. Not until the twentieth century did upper air data become a significant part of the instrumental record.

One would expect that records of instrumental observations would end speculation about recent climatic changes. Objective analyses of data should provide the facts on trends in temperature and precipitation, the two elements for which records are most abundant. But climatic records are not readily subjected to objective study. In the first place, weather observers have human failings, and even small errors affect calculations that may involve equally small trends. The exposure and height above ground of instruments also affect results substantially. Removal of a weather station to a new site practically destroys the value of its records for purposes of studying climatic change. Even if a station remains at the same site for a century, changes in vegetation, drainage, surrounding buildings, and local atmospheric pollution are likely to produce a greater effect on the climatic record than any macroclimatic change. Continuity of local instrument exposure and observational precision is essential to establish a reliable record of true climatic change. In order to achieve this objective the World Meteorological Organization has recommended that reference climatological (''benchmark'') stations be established permanently at representative sites where local changes in the environment can be kept to a minimum. A properly maintained reference climatological network could provide basic data for the evaluation of apparent trends in other climatic records.

Statistical analyses of temperature records in the United States show a general warming trend during the first half of the twentieth century, but that the change was small. The annual average for many stations increased by less than 1C°, reaching a peak about 1940. Winter temperatures increased slightly more than those for summer months, but the trend was not the same in all parts of the country. In the Pacific Northwest, northern Rockies, and the northern Great Plains mean winter temperatures decreased or showed little change.

World temperature trends were upward from about 1885 to 1940, rising by about 1C° in winter and less than 0.6° for the year. The greatest increases were in winter over the arctic regions, where rises in excess of 3C° characterized the period 1917 to 1937. After 1940 the rate of increase slowed and then began a reversal (see Fig. 10.7). Low-latitude regions experienced comparable trends, but of smaller magnitude.

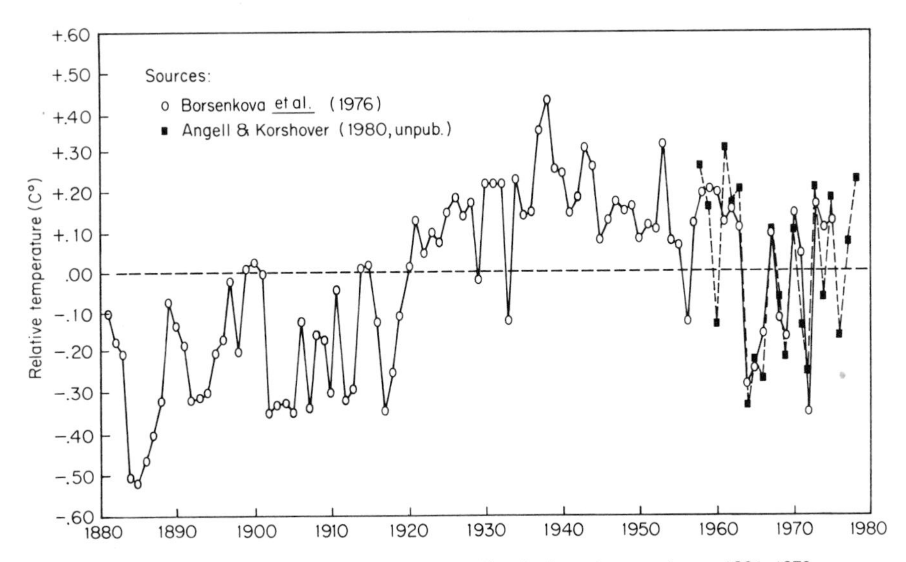

Figure 10.7 Relative mean annual Northern Hemisphere temperatures, 1881–1978. Based on surface records for stations between 15° and 90° N. (Courtesy J. Murray Mitchell, NOAA.)

Additional information on climatic trends, partly inferred from proxy data, indicates a greater warming in eastern Greenland and northern Scandinavia, where mid-century winters averaged 4° to 7C° higher than in 1900. This warming was expressed in the oceans also; codfish were found farther north in the Atlantic than ever before in history.

CLIMATIC CYCLES

It seems clear that there have been climatic fluctuations in historic times, but the regularity of climatic cycles is far less certain. Even the well-recognized daily and seasonal cycles vary in length by large percentages. Daily maximums of temperature, for example, rarely succeed one another at intervals of exactly 24 hours. At high latitudes the daily cycle takes on a character much different in summer from that in winter. The annual cycle often has seasons that are early or late; January and July are by no means always the extreme months in middle and high latitudes. The reversal of the monsoon in southern and eastern Asia is generally regarded as an annual phenomenon, but it occurs at widely differing dates.

Analyses of both proxy and instrumental data have suggested a great number of supposed cycles, few of which have a dependable regularity. Most are barely distinguishable from chance anomalies and properly should be termed rhythms or quasi-periodicities. They range in length from one year to millions of years. The two-year cycle which commonly appears in series of climatic data reflects persistence in the climate system, the heat budget in particular. For example, heat stored by the atmosphere, oceans, and land in one year tends to support higher temperatures well into the next. Evaluations of hypothetical climatic cycles often are more productive when cause and effect are treated together. Even if climatic fluctuations were found to occur randomly it is likely that they are the result of random causes. The relentless search for periodicities in solar activity, earth–sun relations, composition of the atmosphere, and other phenomena that may influence climate will no doubt continue as long as there is even a remote hope of identifying cycles, which would be valuable aids to climate forecasting.

THEORIES OF CLIMATIC CHANGE

Few natural phenomena have attracted the attention of so many scientific fields or elicited so many hypotheses as have climatic changes. The possibility that different causes are active at different time scales adds to the challenge. In view of the fundamental role of insolation in the energy processes that produce weather and climate, it is understandable that most theories have dealt with the possible effects of alterations in the earth's energy budget. One of the simplest and most persistent holds that the sun is a variable star and that changes in the kind and amount of energy emitted alter the solar constant. Increased solar radiation would warm the atmosphere and account for such events as the melting of continental glaciers. A conflicting theory first presented by Sir George Simpson in England during the 1920s suggested that moderately increased insolation would permit higher moisture content of the air, cause stronger meridional transfer of air, and increase precipitation in polar areas. Greater summer cloudiness would inhibit melting of the accumulated snow and ice. Conversely, diminished insolation would weaken the general atmospheric circulation and thereby reduce precipitation at high latitudes. Thus, paradoxically, a lowering of the mean atmospheric temperature might cause a recession of ice sheets, whereas a temperature increase would lead to their advance. Although the Simpson theory appears not to fit recent instrumental evidence, it is a warning against oversimplified explanations of complex processes.

Theories based on a changing sun merge with those concerning sunspot cycles. No completely acceptable theory has been developed to explain how solar variations are translated into climatic fluctuations, if indeed they are. Examination of the number of sunspots during more than two centuries reveals a cycle of about 11.3 years, but the period has been as short as 9 years and as long as 16 (see Fig. 10.8). Multiples of the 11-year cycle and secondary cycles having lengths of 35 years, 80 years, and other periods have also been suggested. Correlations between

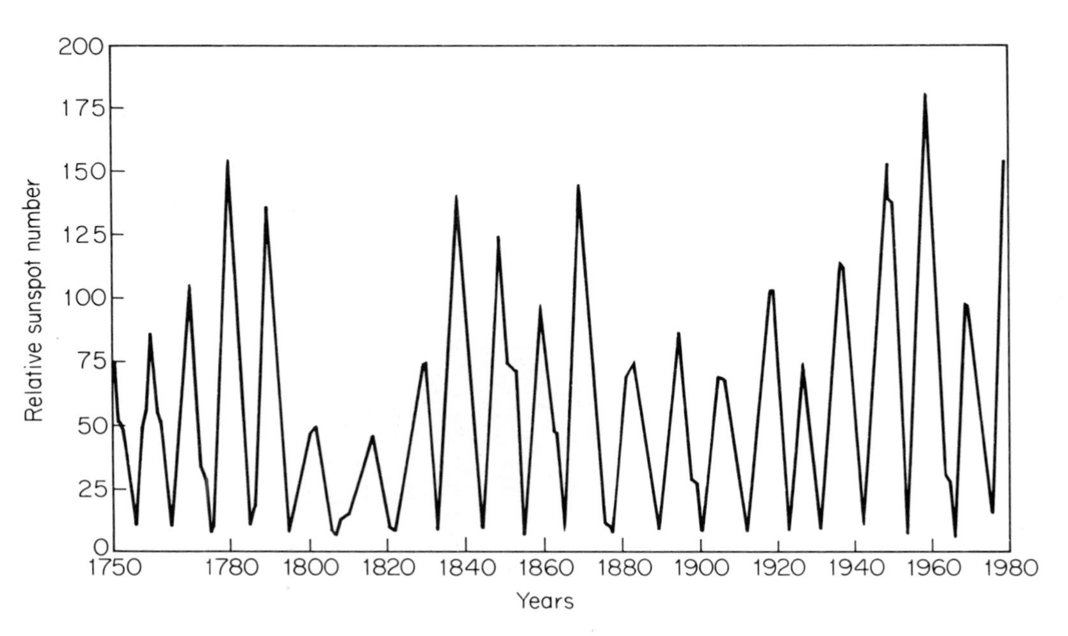

Figure 10.8 Variation of mean annual sunspot activity 1750–1980. (Data from National Geophysical and Solar-Terrestrial Data Center.)

sunspot numbers and weather have been established for specific regions, but a global relationship remains speculative. Increased ratios of the dark areas at centers of sunspots (umbras) to the gray areas around the centers (penumbras) during the period from 1874 to 1970 correlate rather closely with higher mean annual temperatures of the Northern Hemisphere and with drought in the western United States (Fig. 10.9). The effects of sunspot activity may combine with other solar phenomena, such as the sun's rotation and the solar wind, to cause changes in the heat budget, the general circulation, and precipitation patterns on the earth. Emissions of charged protons by solar flares during maximum sunspot periods are known to affect the earth's magnetic field. A strong geomagnetic field deflects the charged particles toward the poles; a weak field allows them to penetrate the atmosphere at lower latitudes. Resulting effects in the ionosphere and stratosphere possibly alter the ozone concentration and, therefore, the transmission of

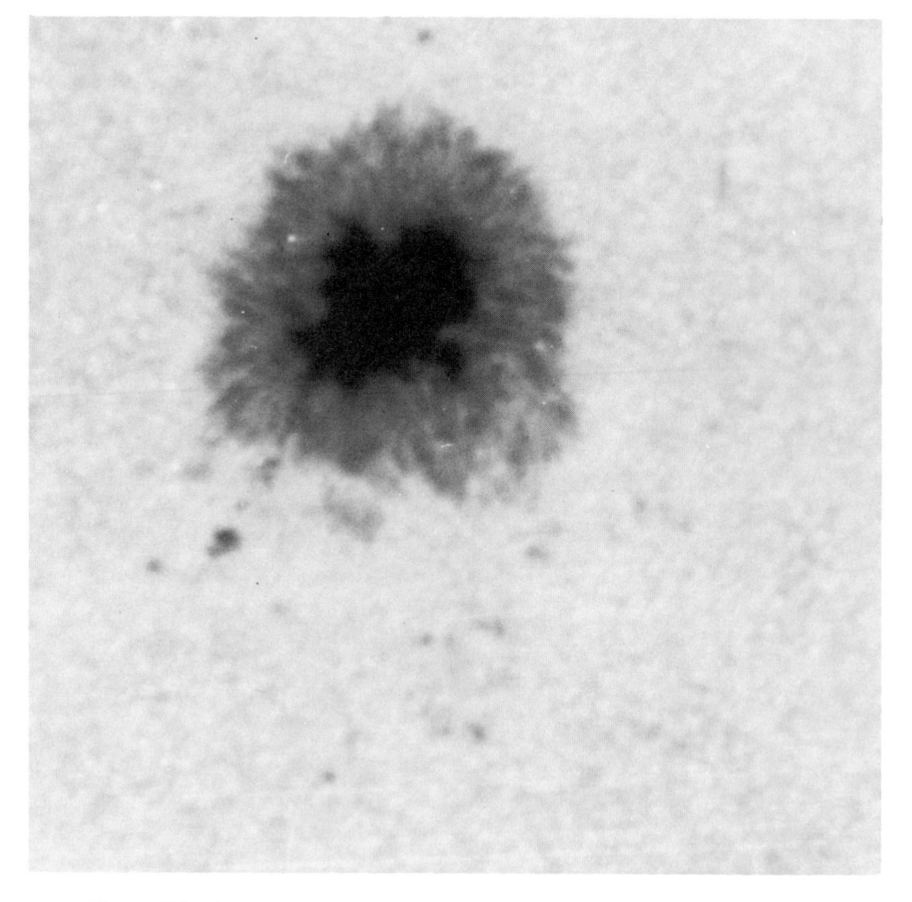

Figure 10.9 Sunspot, showing umbra (dark center) and penumbra (gray area). (NOAA Environmental Research Laboratories, Patrick S. McIntosh.)

ultraviolet radiation. Since the geomagnetic field undergoes a complete reversal at intervals of relative frequency on the geologic time scale, and the magnetic field disappears during reversals, paleoclimatologists have sought correlations between earth magnetism and climatic changes, especially those associated with ice ages and the extinction of species in the fossil record.

Even assuming a constant output from the sun, there are other possible causes of variable insolation. Astronomical theories consider five principal effects. The first three are generally grouped together as the Milankovitch theory, after Milutin Milankovitch, a Yugoslav geophysicist who combined them in a mathematical model to explain expansions and contractions of Pleistocene ice sheets.

1. Changes in the angle which the earth makes with the plane of the ecliptic. The tilt angle varies slowly between 22.1° and 24.5° during a cycle of about 41,000 years, presumably affecting the seasons, temperature distribution, and the general circulation.
2. Changes in the eccentricity of the earth's orbit—period 96,000 years. Resulting variations in the mean distance from earth to sun could affect temperatures on earth.
3. Precession of the equinoxes, the regular change in the time when the earth is a given distance from the sun. At present the earth is closest to the sun in the Northern Hemisphere winter (about 3 January). About 10,500 years ago the Northern Hemisphere winter came at a time of year when the earth was farthest from the sun. Other things being equal (which they never are), winters should have been colder and summers warmer than they are now. In the Southern Hemisphere the reverse applies.
4. Shifting of the earth on its polar axis. This hypothesis, suggested by Robert Hooke in 1686 to explain tropical fossils in England, has been abandoned by most climatologists in favor of theories based on plate tectonics and continental drift, which also could account for apparent "polar wandering."
5. Changes in the rate of the earth's rotation on its axis, affecting the diurnal heat budget and ultimately world climates.

Having reached the outer limits of the atmosphere, insolation is affected by the processes of absorption, reflection, and scattering. Showers of meteoric dust are possible causes of changes in these processes. A more widely accepted theory suggests that from time to time volcanic ash has increased both the albedo of the atmosphere and the proportion of solar radiation absorbed by dust in the stratosphere, reducing insolation at the earth's surface. Ash layers in antarctic ice show a period of intense volcanic activity from about 30,000 to 17,000 years ago, during which temperatures decreased by about 3C°. In the modern era the eruption of Mount Tambora on the Indonesian Island of Sumbawa in 1815 ejected an estimated 150 km^3 of ash into the atmosphere. The following year was known as "the year without a summer" in the United States and Europe, but a direct cause and effect relation is hypothetical. Explosions of ash from Krakatoa in the Strait of Sunda in 1883 produced red sunsets for many months and may have contributed to the severe winters that followed in the Northern Hemisphere. Massive eruptions of other volcanoes in the twentieth century (Katmai, 1912; Agung, 1963; Taal, 1965; Mayon and Fernandina, 1968) forced ash into the stratosphere. Measurements and estimates of direct solar radiation indicate a reduction by 10 to 20 percent for several months following the Katmai eruption and an accompanying decrease of a few tenths of a degree in mean air temperatures. Observed temperatures in the tropical stratosphere increased by 6C° during the year following the 1963 Mt. Agung eruption, whereas surface temperatures decreased by about 0.5C° in the tropics. Figure 10.10 is a satellite photograph of the more recent 1980 eruption of Mount St. Helens in the Cascade Range of Washington State. Effects of this eruption on global climate were undetectable, although local topoclimates were altered by ashfall, destruction of vegetation, and a change in the profile of the volcanic cone.

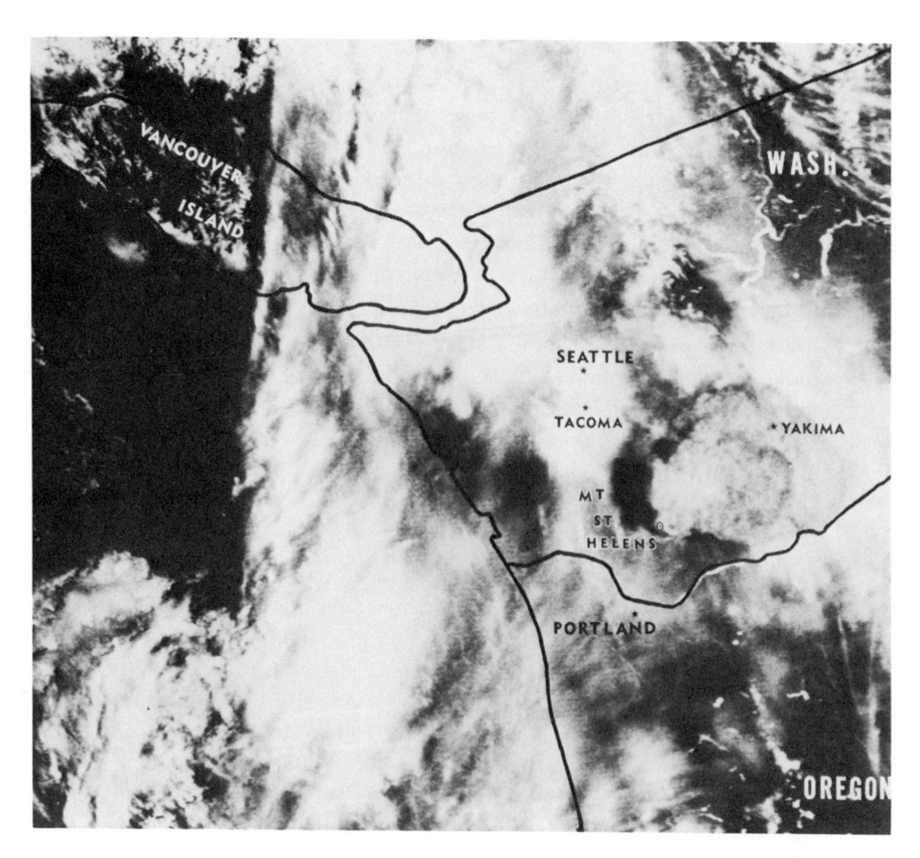

Figure 10.10 TIROS-N view of Mount St. Helens eruption, 18 May 1980. (NOAA, Satellite Data Services.)

Solid particles resulting from wind erosion, burning of vegetation, and pollution by cities and industries also contribute to atmospheric dust. Insofar as their increase is the result of human activities, we may consider culture to be a factor in climatic change (see Chapter 17).

Closely allied to the dust theories are others that postulate variations in the amounts of atmospheric gases, especially those which absorb radiation selectively. Carbon dioxide and water vapor tend to create a greenhouse effect by transmitting short-wave radiation but subsequently absorbing part of the long-wave terrestrial radiation. An increase of CO_2 is thought to produce slightly higher surface temperatures, and it has been proposed as a cause of the warming trend from about 1885 to 1940. (A possible explanation of apparent cooling after 1940 is the overriding effect of rapidly increasing stratospheric dust.) A doubling of atmospheric CO_2 could raise surface air temperature by 2C°. Figure 10.11 shows the trend in measured amounts of CO_2 at Mauna Loa Observatory, Hawaii. Variations in the amount and height of maximum ozone concentration in the upper atmosphere might also affect air temperatures. Sulfur dioxide and chlorine emitted during volcanic eruptions are among the gases that can react chemically, or photochemically, to reduce ozone, which absorbs ultraviolet radiation from the sun as well as a portion of infrared terrestrial radiation. Its increase would lead to a small rise in surface temperatures; a decrease would tend to produce surface cooling.

The influence of the earth's surface on the heat and moisture budgets suggests another category of theories. Continental drift during past geologic eras would account for climatic changes of major proportions as land masses shifted in relation to one another and assumed different latitudinal positions. Refinement of plate tectonics theory by geophysicists since the middle of the twentieth century has given support to explanations of climatic change based on crustal movements. If vertical temperature lapse rates similar to those of the present prevailed in the past, changes in elevation owing to major crustal upheavals might have initiated glacial stages and redistribution of vegetation, both of which are important clues to past climates. Structural warping of ocean basins alters sea level and oceanic circulation, consequently affecting the transport of heat and moisture. Mounting evidence points to a link between climatic fluctuations and circulation systems in the oceans. But the relation is reciprocal. Changing climate may be reflected in water temperatures and circulation, which in turn affect climate. These and other effects that result from interactions within the earth's climate system are termed *autovariations*. For example, changes in surface albedo would require a revised heat budget. It has been calculated that an increase in the mean albedo of the earth (including the atmosphere) by an additional one percent of total insolation would lower average earth temperatures by about 2C°. Freezing of ocean surfaces and formation of extensive ice sheets would greatly increase the proportion of reflected radiation. If no other factors were involved, cooling would appear to beget glaciation, which in turn increases surface albedo, producing further cooling. Such hypotheses of autovariation rapidly lead to circular argument. Obviously, other factors have been at work in the climate system; a single phase of the heat budget should not be examined

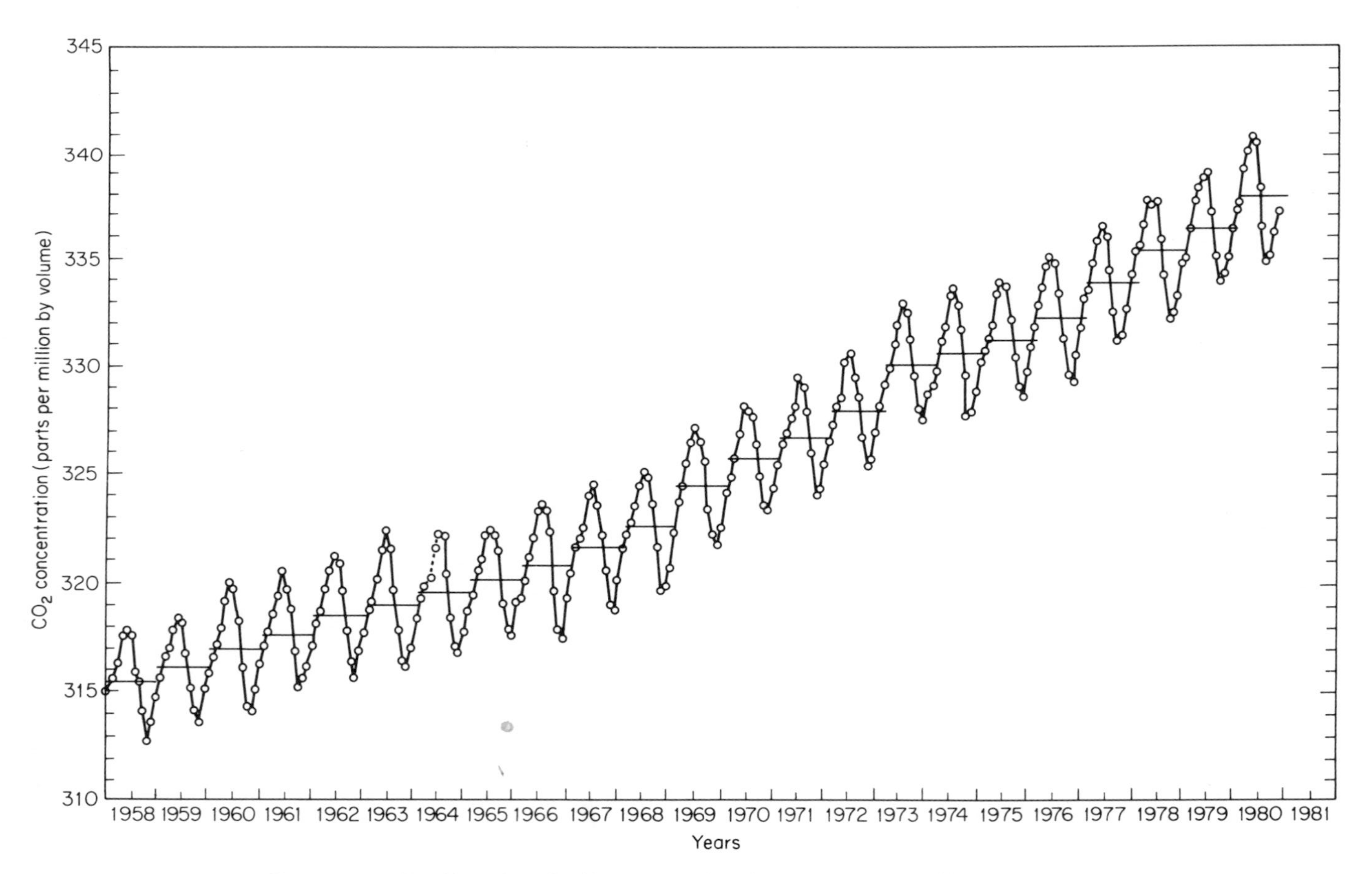

Figure 10.11 Trend in carbon dioxide concentrations in the atmosphere at Mauna Loa Observatory, Hawaii. Annual fluctuations result from removal of carbon dioxide by photosynthesis during the Northern Hemisphere growing season and release during the autumn and winter. (Data after Keeling and NOAA Air Resources Laboratory, courtesy Lester Machta.)

without reference to the others. Nor can changes in the circulation patterns of the atmosphere and oceans be regarded as causes of climatic change, for these motion systems are themselves subject to climatic effects.

FORECASTING CLIMATE

A full understanding of the climate system and explanations of past climates logically lead to prediction of future climates. Extrapolation of past rhythmic changes would be a tempting approach, but thus far no climatic cycles longer than one year have been defined adequately to permit acceptable forecasts for several years in advance. Nevertheless, various statistical probabilities have been calculated for broad applications in long-range planning of human affairs. Most students of climatic change agree that the earth is currently in an interglacial epoch, but few venture predictions of the dates of the next ice age. If apparent trends of the past two centuries repeat themselves, the last half of the twentieth century should continue to be cooler and wetter than the first half in the United States. As always, we should expect years in which the weather departs from the "normal." Forecasts of climate, like weather predictions, cannot be evaluated until the end of the forecast period, but data for checking climatic theories indirectly may be available sooner. For example, meteorological and solar observations beyond the outer limits of the atmosphere increase the store of information on components of both solar and terrestrial radiation.

Meanwhile, if we accept the CO_2 theory, the overall trend in climate during the late twentieth and early twenty-first centuries should be toward higher temperatures. The theory does not rule out other autovariations or astronomical effects that may be superimposed on the long-term trend, however. If we assign major importance to atmospheric dust, the increasingly turbid air should bring a temperature decline. The rate of increase of both CO_2 and dust should be curtailed if "clean" forms of energy replace coal and oil. Any forecast that depends on cultural factors must allow for changing technology, which is fully as difficult to predict as natural phenomena.

Who can guarantee that volcanic eruptions, geomorphic processes, or some other suspected or unknown cause of climatic change will not produce even more marked climatic fluctuations in the next few centuries than have occurred in the past? The uncertainty becomes much greater when extended in time and space. The cover of this book depicts satellite coronagraph images of Comet Howard-Koomen-Michels, entering the sun's coronal field on August 30 and disintegrating August 31, 1979. Whereas minor collisions of this kind appear to have no discernible effect on Earth's climate, they suggest the vulnerability of our tiny climate system to prolonged or intense astronomical events. Ultimately, the forecasting of climate may depend on prediction of changes in and beyond the solar system as much as those which are now considered to be terrestrial variations.

QUESTIONS AND PROBLEMS FOR CHAPTER 10

1. Summarize the kinds of evidence that indicate climatic changes prior to the period of human history.
2. Compare the validity of the following as indicators of climatic change: alpine glaciers; fisheries production; regional drought; wine quality.
3. Outline the theories that attempt to explain climatic changes through geologic time to the present.
4. Explain the carbon dioxide theory of climatic change. Compare the effects of carbon dioxide with those of atmospheric dust.
5. What problems arise in the use of instrumental records as evidence of climatic change?
6. Summarize the evidence for and against the occurrence of climatic cycles.
7. Why might some volcanic eruptions cause observable fluctuations of climate whereas others have a negligible effect?
8. Evaluate the prospects for reliable forecasts of future climate.
9. What changes in the global water budget might result from a long-term surplus or deficit in the earth's radiation budget?

Part III

APPLIED CLIMATOLOGY

11

Climate and Water Resources

Of all the earth's resources, none is more fundamental to life than water. The properties of H_2O in its three physical states make it by far the most useful of compounds. We can breathe it, drink it, bathe in it, travel on it, or see beauty in its different forms. It is a raw material, source of power, waste disposal agent, solvent, medium for heat transfer, or coolant as the needs of modern technology may require. The high specific heat of water, its ability to exist in gaseous, liquid, or solid forms under natural conditions, and its capacity for storing or releasing latent heat with changes of state give it immense influence on atmospheric processes. At the same time, its availability at different times and places is a function of weather and climate. The restless atmosphere is the most active agent in the constant redistribution of water on the earth's surface—a fact that becomes even more striking when we realize that only a minute fraction of 1 percent of the earth's water is contained in the atmosphere at any time. If all atmospheric moisture were precipitated, it would create a layer averaging only about 25 mm deep over the entire globe. The seas and oceans contain about 93 percent of the earth's water; 2 percent is in ice caps and glaciers; fresh-water bodies, ground water, soil moisture, and vegetation account for about 5 percent (see Table 11.1).

THE GLOBAL HYDROLOGIC SYSTEM

The circulation of water from oceans to air and return to the oceans, entailing residence of varying duration in life forms, fresh-water bodies, ice accumulations, or as ground water, comprises the *hydrologic system,* an integral part or subsystem of

TABLE 11.1

World Water Resources *

	Area covered (million km^2)	*Volume (million km^3)*	*Percent of total volume*
World ocean	360	1,370	93
Icecaps	16	24	2
Terrestrial water	134	64	5
Atmosphere	510	0.013	0.001
Total	510	~1,500	100

*Adapted from SMIC, *Inadvertent Climate Modification* (Cambridge, Mass.: MIT Press, 1971), p. 97.

the global climate system that always requires energy for change of state or mass transfer. Figure 11.1 shows that it is an intricate combination of evaporation, transpiration, atmospheric circulation, condensation, precipitation, runoff, infiltration, percolation, and ground-water movements. Whereas the greater part of moisture that eventually falls as precipitation comes from the oceans, some water takes a shortcut in the system and enters the air directly through evaporation and transpiration from soil and vegetation. In middle and high latitudes only a small part of precipitation over land can be traced to evapotranspiration from the same area, but much of the transpiration from rain forests returns as rainfall within the forested region. Precipitation over oceans or lakes is another form of hydrologic shortcut. A slowing of hydrologic activity for various periods occurs when plants and animals use water for cell building, when icecaps and snowfields detain it in a solid state, when chemical action incorporates it in other compounds, or when water is trapped in underground aquifers or deep layers of the oceans. But, sooner or later, most terrestrial water appears elsewhere. Table 11.2 lists average residence times of water in various forms.

TABLE 11.2

Average Residence Time of Water *

Atmosphere	10 days
Terrestrial water	
Rivers	2 weeks
Lakes	10 weeks
Soil	2-50 weeks
Biota	1-20 days
Ground water	1-10,000 years
Oceans	3,600 years
Polar ice	15,000 years

*Adapted from SMIC, *Inadvertent Climate Modification* (Cambridge, Mass.: MIT Press, 1971), p. 96.

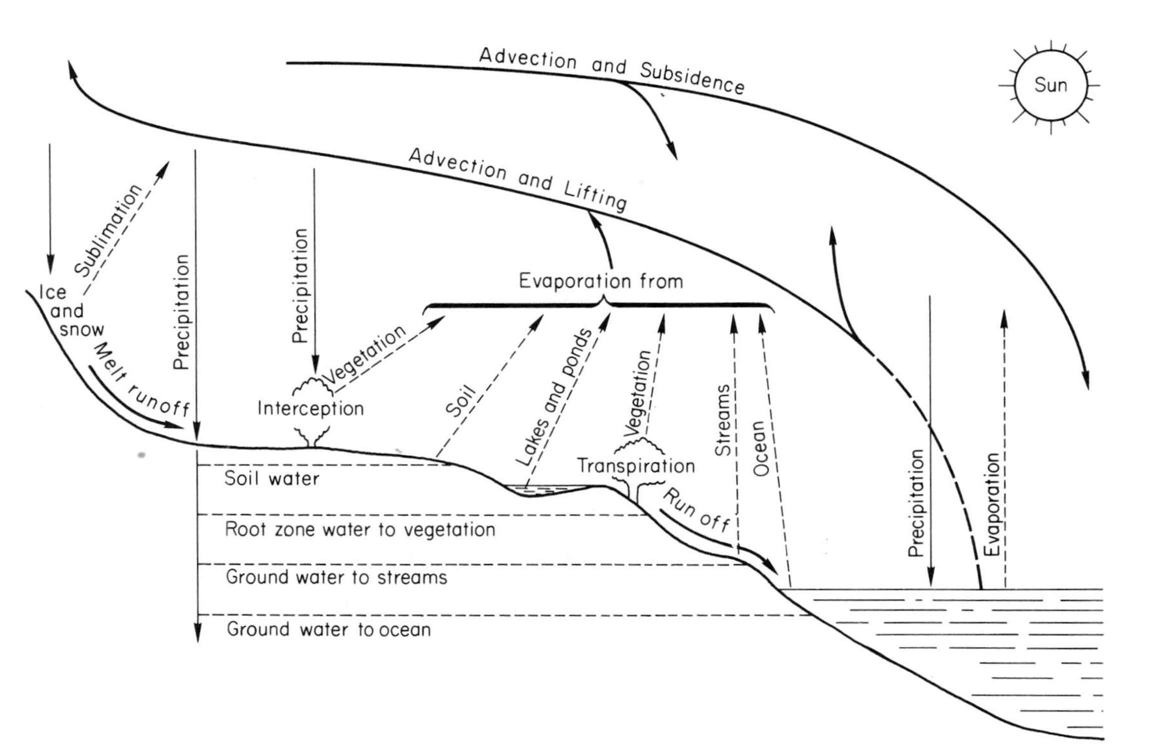

Figure 11.1 Principal features of the hydrologic system.

The atmospheric phases of the hydrologic system have been treated in Chapter 3. The processes of evaporation, condensation, and precipitation are essentially climatic; their functions would be simpler if it were not for constant motion in the atmosphere, which shapes regional patterns of the water budget. For example, maritime air masses carry water onto the land. Wherever they undergo lifting processes—convection, orographic ascent, or convergence—precipitation is the likely result. The processes are active where heat and moisture are abundant, as in the rainy tropics. In dry climates atmospheric moisture and lifting are inadequate, and in very cold climates both heat and moisture are limited. Were it not for the intrusion of warm maritime air masses into high latitudes, polar regions would have even less precipitation. Insolation, through its effects on atmospheric processes, provides the "fuel" to operate a system in which the flow of water vapor across boundaries of watersheds or continents is as important as the flow of rivers to the sea.

THE WATER BUDGET AT THE EARTH'S SURFACE

The major demands for water arise at the earth-atmosphere boundary, which is also the zone of maximum energy and moisture exchange between the surface and the air. Accordingly, we can consider relevant elements of the hydrologic system by means of an accounting procedure—the water budget of the earth's surface. The essential elements of moisture exchange for a given land area and specified period of time are combined in the equation:

$$P = ET + dST + S$$

in which P is the income from precipitation, ET is the loss by evapotranspiration, dST is the gain or loss of storage in the soil, and S is surplus. Thus, all of the income is accounted for by expenditures, an increase or decrease in saving, and a possible surplus that requires wise management. The transfer of moisture to the air by evapotranspiration is a function of both available water and available energy. The amount that could be evaporated and transpired by the available energy is the potential evapotranspiration (PE), or simply the water need (see pp. 148–49). The amount lost upward from the surface is the actual evapotranspiration (AE). It is equivalent to ET in the water budget equation and therefore is limited by precipitation and soil moisture storage. When total precipitation exceeds potential evapotranspiration, part of the water may be needed to restore soil moisture; the remainder, if any, is surplus that may run off, percolate to the ground-water level, or accumulate in ponds. If PE exceeds AE and soil moisture is being depleted, there is a deficit (D), and the equation becomes:

$$P = PE - dST - D$$

Expenditures fail to meet the need imposed by the energy-moisture system. Obviously, the income is inadequate, although savings may help to offset a temporary shortage. Table 11.3 shows an example of water budget accounting on a monthly

TABLE 11.3

Monthly Water Budget—Memphis, Tennessee (all values in millimeters)

	J	F	M	A	M	J	J	A	S	O	N	D	Yr
P	134	112	131	126	106	94	84	80	72	72	107	114	1,232
PE	5	10	31	65	113	152	173	160	114	64	26	10	923
ST*	300	300	300	300	293	241	179	137	119	127	208	300	
dST	0	0	0	0	−7	−52	−62	−42	−18	+8	+81	+92	
AE	5	10	31	65	113	146	146	122	90	64	26	10	828
D	0	0	0	0	0	6	27	38	24	0	0	0	95
S	129	102	100	61	0	0	0	0	0	0	0	12	404

*Amount of soil moisture storage at end of each month, assuming a capacity of 300 mm.

basis for a representative station in a humid subtropical climate. The same data are displayed graphically in Fig. 11.2

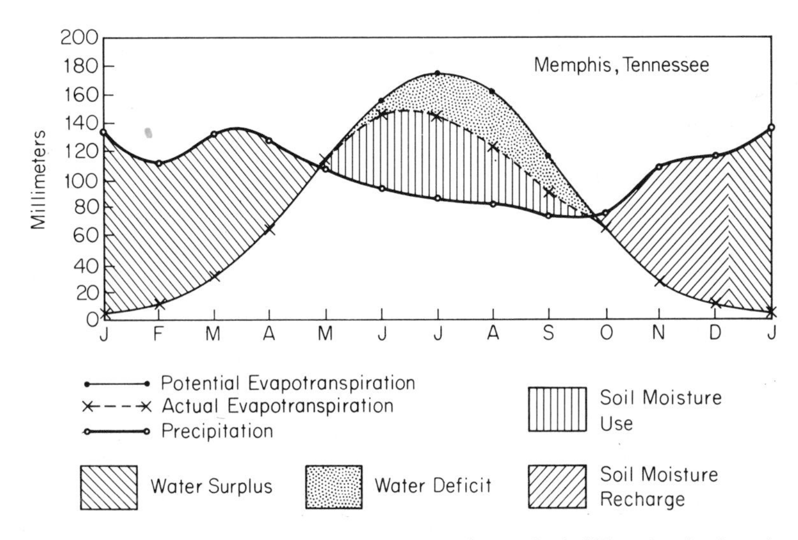

Figure 11.2 Water budget graph for Memphis. (After method of Thornthwaite, based on data computed by C. W. Thornthwaite Associates, Laboratory of Climatology.)

The water budget concept can be applied in the analysis of the moisture exchange at any surface—a plant leaf, a forest canopy, a glacier, an entire continent, or an ocean (see Table 11.4). It is an effective means of determining drought conditions and the proper use of irrigation water. One of the most fruitful applications is in the predictions of runoff (surplus) from river basins under varying climatic conditions. Similarly, ground-water supplies (another form of surplus) may be calculated. Any single element in the budget can be determined mathematically if all the other elements are known.

TABLE 11.4

Estimated Average Water Budgets of Major Continents and Oceans * (cm per year)

Continent	*Precipitation*	*Evaporation*	*Runoff*	*Interocean flow*
Africa	69	43	26	
Asia	60	31	29	
Australia	47	42	5	
Europe	64	39	25	
North America	66	32	34	
South America	163	70	93	
(Total land)	73	42	31	
Ocean				
Atlantic	89	124	23	−12
Indian	117	132	8	−7
Pacific	133	132	7	+8
(World ocean)	114	126	12	0

*Adapted from M. I. Budyko, *Climate and Life,* ed. David H. Miller (New York: Academic Press, Inc., 1974), pp. 227-28 and 238.

EVAPOTRANSPIRATION

The accuracy of water budget accounting depends in turn on the accuracy of the respective values used in the computation. Precipitation and its measurement have been discussed in Chapter 3. It is important to note that precipitation gauges provide reliable totals for large areas (for example, entire river basins) only if placed at representative sampling sites.

The key element in the water budget is evapotranspiration, which is also the link between moisture and energy exchanges. Attempts to measure evapotranspiration directly involve humidity gradients and vapor transport by diffusion or turbulent motion. Because humidity decreases with distance from the evaporating surface, the rate and amount of evapotranspiration can be determined if the difference in water vapor content and the rate of vertical mixing through the surface layer of the air are known. Special instruments and plant chambers assist in obtaining the necessary data.

An indirect method of measuring evapotranspiration depends on the water budget concept itself and employs the drainage lysimeter, or *evapotranspirometer* (Fig. 11.3). The essential features of this device are (1) a sunken field tank filled with soil in which plants like those of the surrounding area are grown and (2) a covered percolation tank sunk into the ground a short distance from the field tank to catch surplus water that drains from the field tank through (3) a connecting underground tube. The field tank receives water only from precipitation or irrigation and loses it by downward percolation or by evapotranspiration. For determination of potential

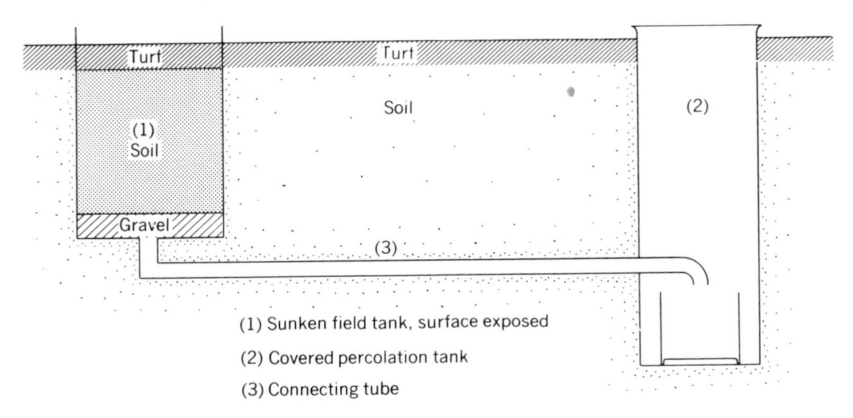

Figure 11.3 Essentials of a field evapotranspirometer. (Adapted from John R. Mather, *The Measurement of Potential Evapotranspiration,* Johns Hopkins Laboratory of Climatology, 1954.)

evapotranspiration, the field tank and the surrounding area are irrigated whenever necessary to maintain optimum soil moisture conditions. When the amounts of water added by rain and auxiliary watering are known, potential evapotranspiration can be found by subtracting the water loss by percolation from the total received in the evapotranspirometer field tank. Soil moisture storage may vary from day to day, and the amount of water collected in the percolation tank varies with it, but over longer periods the fluctuations are not critical.

Expensive but more accurate versions of the evapotranspirometer are weighing lysimeters that provide a constant record of changing water content in the soil and measurements of actual evapotranspiration for periods of only a few minutes. As in the case of precipitation gauges, problems arise in selecting and maintaining representative sites for evapotranspirometers. Standardization of plant cover to provide comparability of transpiration values is also difficult. Artificial watering of sites, especially in dry areas, in order to achieve maximum (potential) evapotranspiration creates an "oasis effect," which might be quite different from the prevailing regional climate. Different slopes and exposures introduce further problems in the interpretation of data.

Because of the rather complex equipment required for measurement of potential evapotranspiration and the attendant difficulties in assuring representative conditions, estimates are often based on the records of actual evaporation from open pans or other types of evaporimeters. Mathematical relationships among moisture, energy, and air movement factors also yield useful approximations to water need. The method developed by Thornthwaite uses temperature and the length of the daylight period as energy expressions. (The *PE* values in Table 11.3 were calculated using his formula.) H. L. Penman in the United Kingdom and M. I. Budyko in the Soviet Union have refined the energy budget approach to potential evapotranspiration by incorporating net radiation as well as factors for temperature, wind speed, and humidity. Although methods which combine the effects of net radiation, temperature, and humidity yield the best theoretical results, often only one of these variables is used to obtain approximate values for practical applications.

SOIL MOISTURE AND GROUND WATER

The role of soil moisture in the water budget is highly variable; it depends not only on the other budget factors but also on the capacity of a soil to hold water gained by infiltration. The amount of water a saturated soil can retain against the pull of gravity is its *field capacity*. It varies mainly with soil texture but also with structure, organic matter content, and the depth of the soil. Fine clays have high field capacities, whereas sandy soils hold little moisture. For agricultural purposes, it is useful to consider the minimum amount of soil water that is necessary in the root zone to allow extraction by plants. This minimum is the *wilting point,* and it depends on the same factors that govern field capacity. Relative field capacities and wilting points for various soil textures are shown in Fig. 11.4.

The water content of soil can be ascertained by several methods, none of which is completely satisfactory for all purposes. A simple but time-consuming procedure is to weigh soil samples before and after drying in an oven. Unfortunately the samples cannot be reused. Weighing lysimeters accurately measure changes of soil moisture in a specific volume of soil. Another device, the *soil moisture tensiometer,* is a porous clay cell filled with water and placed in the soil. A pressure gauge attached to the cell registers changing tension as water moves through the cell toward drier soil or into the cell from wet soil. An electrical method depends on the rela-

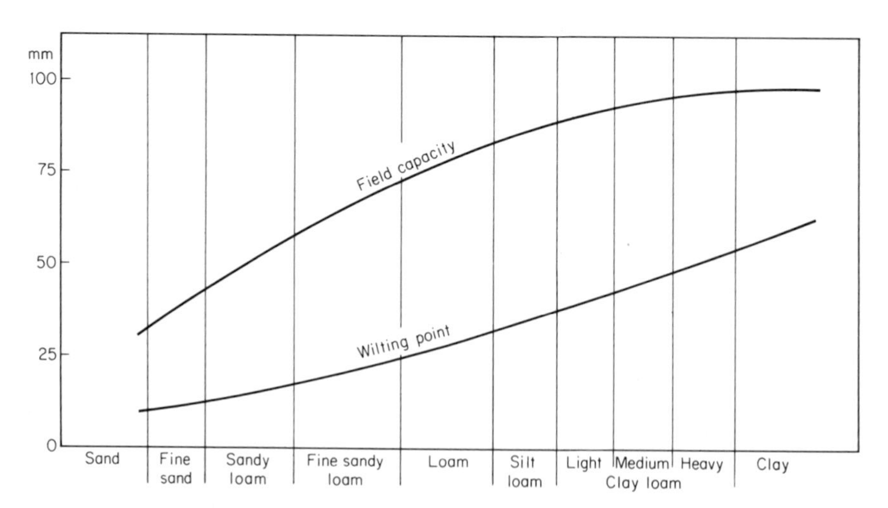

Figure 11.4 Average field capacity and wilting point for different soil textures. Values are in mm per 300 mm depth of soil in the plant root zone. (After Smith and Ruhe, *Water, 1955 Yearbook of Agriculture,* p. 120.)

tionship between soil moisture and the conduction of electricity by buried resistance blocks made of gypsum, nylon, or fiber glass. It is especially useful for remote recording. A neutron scattering method relies on a slowing of fast neutrons by collision with hydrogen atom nuclei. A probe containing a source of fast neutrons and a detector of slow neutrons is lowered into the soil. Since most of the hydrogen is in soil water molecules, the number of slow neutrons detected is an index of soil moisture.

When the soil zone has reached field capacity, excess soil water percolates downward under the force of gravity through a subsurface zone of aeration (vadose water) to a zone saturation known as *ground water.* Depths to the upper surface of ground water, that is, the water table, can be measured in wells and are used to estimate ground-water supplies. Ground water may also be treated as part of the surplus in the water budget, and when other values in the budget (including runoff) are known the amount transported to ground water can be calculated.

RUNOFF AND FLOODS

Runoff is that portion of precipitation that returns to the oceans and other water bodies over the land surface or through the soil and water table. It may involve the direct return of rainfall or the flow from melted snow and ice fields which have temporarily stored the water. Floods differ from simple runoff only in degree, the distinction between the two depending for the most part upon how they affect surface features. River floods result whenever the channel capacity is exceeded by the runoff. Excessive runoff of rainfall or snowmelt is the fundamental cause, but the channel capacity may also be affected by barriers to flow (such as dams or ice jams), sudden changes of direction of the stream, reduced gradient, siltation of the stream bed, or sudden release of water due to a broken dam. On many rivers, floods are defined in relation to arbitrary gauges placed at important points along the stream. Not all floods are "bad." Many of the features of the earth's surface have been sculptured by running water. For centuries agricultural areas in the lower-Nile flood plain and Mesopotamia depended on annual river flooding and the accompanying deposits of fertile silt. What is gained in this way in the lowlands must be lost at higher levels in the watershed.

The amount of runoff in a given region is conditioned by several factors: the amount and intensity of precipitation, temperature, character of the soil, vegetative cover, and slope. If precipitation occurs as rain, the proportion that runs off will depend upon the capacity of soil and vegetation to absorb it. Plants retain some rainfall on their external structures and slow the velocity of raindrops. They also detain water in its horizontal movement. They improve soil structure and their roots provide channels to divert excess soil water into the ground water at greater depths. The high humus content of soils with a dense grass cover enhances absorption, for it acts something like a sponge. Porous soils absorb more water by infiltration than dense clays or rock. If already saturated or if sealed by the intense pounding of raindrops, however, their absorptive capacity is limited. Similarly, impervious subsoil reduces the amount of water that can be stored. When frozen in a wet condition, soils can absorb little rain, but if frozen when porous and nearly dry they have a high capacity for water.

Surface runoff varies with the character of an individual storm. Long-continued rains charge the soil to its full capacity and the proportion which runs off becomes progressively greater. Loose sands and gravels constitute an exception, if they are not underlain with impervious material. In the case of snow or hail, infiltration must await melting. When the soil and plant cover are warm, some melting begins at once. If the surface and the air above are too cold, there is a delay of hours, days, or even months in the runoff. Ultimate surface flow is determined by the various surface and soil characteristics, the rate of melting, and the amount of snow lost by sublimation. Sometimes snow and ice are removed principally by sublimation, or by melting and evaporation, so that the runoff component of the water budget is assumed partly by the air.

CLIMATIC CAUSES OF FLOODS

The predisposition of a climate to storms producing excessive precipitation is the fundamental basis of the flood hazard. In some climates flood-producing storms occur irregularly; in others they follow a seasonal pattern. Two types of storms initiate most rain-caused floods: the violent thundershower, which is of short duration and produces a flash flood, and the prolonged, widespread rain which, through sheer quantity of water, creates extensive flooding over entire watersheds. Flash floods are most common in those regions which experience heavy thunderstorms, but they should be regarded as a potential hazard whenever intense rainfall occurs. The flood-producing storm may be of convective, frontal, or orographic origin; the important feature is its intensity, not its origin. In general the less a flash flood is expected the more disastrous it is likely to be; people continue to settle in narrow valleys and lowland plains in defiance of natural hazards. Although scattered convective storms or a line of frontal thunderstorms contribute to extensive flooding in large river basins, such floods usually follow long-continued rainy weather. This type of weather is most common in the humid climates of middle latitudes when frontal systems occlude or remain stationary for long periods, but it is accentuated dramatically in hurricanes. As long as the storm is fed by moist maritime air, rain can continue. Once the water-holding capacity of soil and vegetation is exceeded, a major flood is imminent. Such conditions are typical of the Ohio Valley in spring, when neither the wet soil, the swollen tributaries, nor the high ground-water table can absorb even a moderate fall of rain.

If the precipitation falls as snow, the flood is not necessarily avoided. It may merely be delayed. The most effective instrument for snow removal is a warm wind. If it is dry (for example, the foehn wind), it carries much of the water away in the air and is not a likely cause of flooding; if it is moist as well as warm, it produces rapid melting and runoff. The effect of sunshine as a factor in melting is confined to the daytime and limited by the albedo of snow. Rainfall on snow does not at once produce runoff of flood intensity, for a snow cover has capacity to hold water. When prolonged rains come in combination with warm winds, however, ideal flood conditions prevail. Another factor in snowmelt runoff is the temperature of the soil under snow cover. If the soil is frozen, there is very little melting and runoff under the snow, but melting due to other causes is likely to produce a high proportion of runoff. Unfrozen soil causes some melting throughout the winter, and a large part of the water enters the soil to become ground water rather than runoff, because the melting process is slow.

The seasonal distribution of precipitation is the principal determining factor in the regimes of rain-caused floods. In the arid climates, the thundershowers and resulting flash floods are both erratic. Concentrations of rainfall in one season (for example, as in the monsoon tropics) are likewise accompanied by seasonal floods. In mountainous areas of mid-latitudes and at high latitudes, flooding due to snowmelt runoff occurs with the onset of the warm season and is not always directly related to the precipitation regime. For example, in the dry summer subtropics the precipitation maximum is in winter, but the maximum runoff of melted snow is in spring or early summer. On the other hand, spring and summer snows in the mountains of marine climates contribute to the runoff in those seasons, often being removed by subsequent rains. In humid continental climates seasonal floods come at the end of the period of soil moisture accumulation, that is, in spring.

Destructive floods occur in every month in the United States, often with loss of life and damage to property running into many millions of dollars. In the United States, as a whole, loss of property and deaths due to floods have been great in all months, with winter and spring the worst seasons. The normal maximum of runoff of streams shows a distinct seasonal pattern. Figure 11.5 shows the normal annual distribution of runoff for representative rivers in the United States and southern Canada. Note the influence of late snowmelt on the regimes of northern rivers. In Florida, snow is not a factor in runoff and the runoff is greatest in autumn in response to the rainier season and decreasing evaporation.

A variation in the normal runoff regime of snow-fed rivers occurs in the Mackenzie, Ob, Yenesei, and other rivers that flow northward at high latitudes in North America and Asia. Melting begins in the headwaters and middle courses of these streams earlier in the spring than it does in the lower flood plains. Because the downstream channels are choked with ice, there is extensive flooding until the thawing season is well underway. A contributory factor is permafrost, which allows little or no infiltration of surface waters.

RUNOFF FORECASTING

Runoff forecasting is concerned with forecasting the amount of rainfall or snowmelt that will be available to run off and the rate of flow in relation to stream capacity. Figure 11.6 summarizes graphically the main factors which are considered in flood forecasting for a particular storm. The techniques of forecasting quantitatively the

Figure 11.5 Normal distribution of runoff by months for representative rivers in the United States and Canada. (From *Water, 1955 Yearbook of Agriculture.*)

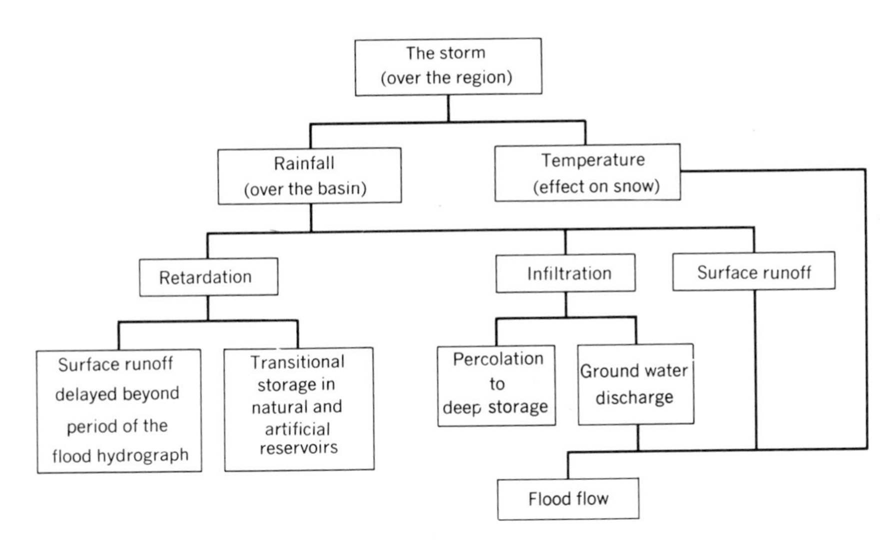

Figure 11.6 Factors considered in runoff forecasting. (From *Climate and Man, 1941 Yearbook of Agriculture.*)

precipitation to be expected from a storm are being refined, but, generally speaking, only qualitative indications of impending disaster due to heavy rainfall can be given in advance of the storm. For example, if a cyclone is drawing large quantities of mT air over a watershed which is already wet, there is a distinct possibility of flooding. Extreme instability of the moist air may presage an even greater flood threat. Temperatures of the conflicting air masses within the storm, the temperature of air expected to follow the storm, and wind are factors which modify the meteorological aspects of the flood hazard. If the precipitation is snow, the threat is deferred until melting begins. Forecasts of snowfall are not of such immediate importance in flood warnings as are forecasts of rainfall and potential snowmelt. When there is some question of whether rain or snow will fall generally over a watershed, the flood forecaster is in a delicate position. In mountainous areas, the lower level of snowfall will determine the area to be affected by rainfall runoff. Thus, a careful assessment of the probable upper-air temperatures is necessary. In regions visited by hurricanes, advance warning of the coming storm is a prerequisite to timely forecasts of "storm surges" along coasts as well as of excessive runoff inland.

Measurements of rain even as it falls provide valuable data for flood prediction. Correlated with subsequent stream flow, these measurements help to establish patterns of rainfall–flood relationships that are the basis for future flood forecasts as well as of immediate utility. *Hydrographs,* that is, graphs on which stream flow is plotted as a function of time, have been useful devices for studying the effects of storms on flooding (see Fig. 11.7). The progress of storm centers across the river basins, the characteristics of air masses, the measured rainfall over the watersheds, and past stream behavior are the principal data employed in making stream-flow forecasts. After precipitated water begins to flow from a watershed more accurate

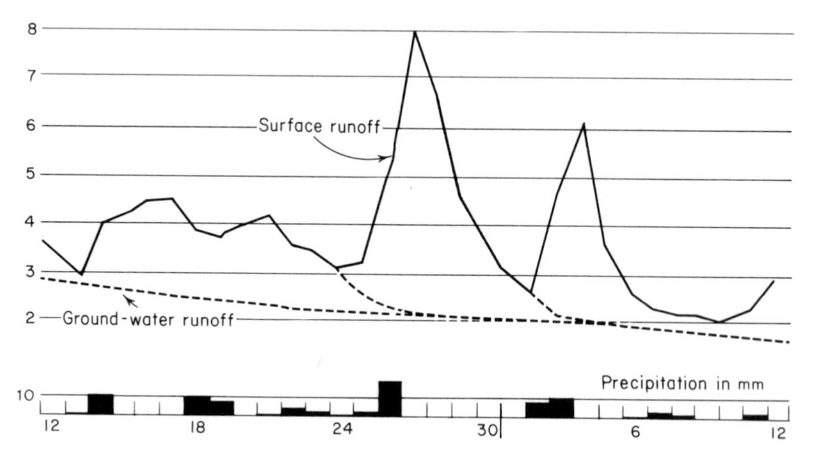

Figure 11.7 Hydrograph of stream flow in the Muskingum River at Dresden, Ohio, during a 31-day period. Values on the left represent relative discharge volume. The lower graph shows daily precipitation in mm. (Adapted from *Water, 1955 Yearbook of Agriculture.*)

predictions of stream discharge can be made. They are based primarily on historical flow patterns of main rivers and their tributaries.

SNOW SURVEYING

In the mountainous regions of middle and high latitudes water is stored in the form of winter snow for release as temperatures rise in spring. This natural reservoir is a vital resource for water users in arid lands adjacent to snow-capped ranges. Efficient planning for water use and runoff control during the season of peak flow depends on prior knowledge of stream discharge. Since it is not possible to predict accurately far in advance how much snow will fall on uplands, the alternative is to survey the snowpack as an indicator of potential runoff. Satellite observations provide data on the changing areal extent of snow cover and augment actual measurements of snow water content.

Each winter several public and private agencies in the western United States cooperate in ground-based snow surveys. The principle of snow surveying involves a sampling technique, since quantitative measurements of all the snow at upper elevations is neither feasible nor necessary. Representative *snow courses* are selected in the snowfields that feed major river systems. Measurements on the snow courses are the bases for runoff forecasts that are prepared early in the winter and revised monthly. Water content is the important factor, for snow densities vary greatly. A snow sampling device permits determination of water content as well as depth. Coupled sections of lightweight aluminum tubing are thrust vertically through the snow to obtain a sample which is weighed to determine its water equivalent. An alternative method uses radioactive isotopes; when the cobalt-60 isotope is inserted to the bottom of a snow layer, it transmits gamma rays at an intensity proportional to snow water content. The *snow pillow* is another snow survey device now widely employed in the western United States (Fig. 11.8). Snow accumulating on a flat, liquid-filled container (pillow) creates internal pressure and activates electrical con-

Figure 11.8 SNOTEL installation. Data registered by the snow pillow (foreground), the standpipe precipitation gauge (right), and a thermistor mounted above the equipment shelter are transmitted by radio to remote receiving stations. (USDA—Soil Conservation Service.)

tacts. A solar cell, battery, and radio assembly at the remote site transmits water content data on command from receiving stations. Known as SNOTEL, this telemetry system depends on backscattering of radio signals by ionized trails of micro-meteorites in the upper atmosphere. SNOTEL installations also incorporate rain gauges and thermometers which are automated for radio transmission of data.

Through the correlation of snow survey information with subsequent runoff from respective watersheds, formulas and graphs to express the relationship have been established. These are fundamental to water supply forecasts which, as a result, have become increasingly accurate. Such forecasts give a good indication of the amount of snowmelt that will enter the streams of a given watershed, but they cannot predict when the runoff will occur or whether there will be destructive floods. When full-scale runoff begins, because of high temperatures or storms, the flood forecaster must resort to the techniques described in the previous section, but information on the amount of water in the snowpack is valuable for assessing the imminence of a major flood. Besides aiding flood forecasting, the snow surveys provide information on the probable water supply for irrigation, hydroelectric power, navigation, fisheries, and a host of other water uses. Snow survey data have been used, with some success, to predict the ground-water supplies of certain river basins.

Special surveys in recreation areas gather supplementary data on conditions for skiing and avalanche forecasts. Unstable snowpacks on steep slopes can be a hazard to transportation and communication facilities, winter sports enthusiasts, and buildings. The danger is so acute each winter in Switzerland that a special Swiss Avalanche Commission has been organized. In the United States, the U.S. Forest Service maintains a Mountain Snow and Avalanche Research Center at Fort Collins, Colorado.

Avalanches are, in a sense, a flash flood of water in the solid form (see Fig. 11.9). While the angle of repose of the snow cover and the underlying surface features are basic considerations in avalanches, they are set off naturally by conditions of snow structure, temperature, humidity, and wind. Any disturbance of an unstable snow cover is likely to cause an avalanche. *Moist* snow is relatively cohesive and does not avalanche readily, but snow *soaked* with water is predisposed to flow or slip. New snow, composed of light, interlocking snow crystals, is stable, as is granular snow, whose particles resist movement. The layering of the snowpack also influences the tendency to avalanche. Generally, the layers reflect weather fluctuations. Some layers may be wet and compact; others are loose and dry. Snow cores provide data on the snowpack strata; meters measure the pressure of the snow, and rams test its resistance and coherence. Avalanche hazard forecasts take into account the amount and characteristics of snow, terrain factors, and the current and predicted weather conditions.

Wind blasts known as *avalanche winds* frequently accompany severe avalanches. They are generated by the mass movement of snow against the air and by greatly reduced pressure behind the moving snow. The more violent blasts have

Figure 11.9 Snow avalanche in a Swiss valley. (Photograph courtesy Institut fédéral pour l'étude de la neige et des avalanches, Suisse.)

been known to break off trees several meters away from the main path of the avalanche.

WATER RESOURCES MANAGEMENT

The earth has a finite amount of water, but fortunately the resource is renewable and reuseable. The fundamental objective of water resources management is to balance the water budget, ensuring adequate quality as well as quantity at a given time and place. Since natural climatic processes often do not create the desired balance locally, various techniques have been developed or proposed to control one or more phases of the hydrologic system. Some examples, summarized in Table

TABLE 11.5

Examples of Methods and Proposals for Altering Natural Hydrologic Processes

Runoff and storage
Dams, weirs, levees, terraces
Ponds, reservoirs, tanks
Drainage, channel dredging
Irrigation, water spreading, flooding
Wells, ground-water recharge
Alter terrain, vegetation, soil
Impervious coverings
Induce melting
Evaporation and transpiration
Combustion, heating and cooling devices
Fans, windbreaks
Alter terrain, vegetation, soil
Reservoir coverings, surface film retardants
Vertical mixing of water bodies
Condensation and precipitation
Fog and dew capture
Industrial cooling towers
Distillation, desalination
Cloud modification
Induce instability of air
Steer storms

11.5, are treated in greater detail in the remaining chapters of Part III, where water and energy budgets are central themes in the application of climatology. It will be evident that acceleration or retardation of one hydrologic process affects others and is likely to cause conflicts among competing water uses.

Simple diversion of surface water (runoff) and withdrawal of underground water (storage) predate written history. Damming and storage in a wet area or during a rainy period for subsequent transfer to meet needs at other places or times has become a widespread practice for redistributing water used in irrigation or hydroelectric generation. Alteration of terrain features, soils, and vegetation also affects runoff and storage. Applications of impervious coverings (for example, urban pavements) greatly increase runoff at the expense of soil moisture and percolation to groundwater.

Other approaches to water management entail modification of atmospheric processes. Suppression of evaporation from reservoirs can be achieved by covers, by vertical mixing to reduce surface temperature, or by windbreaks, whereas heating and mechanical stirring of overlying air accelerate evaporative loss. Thin chemical films (alkanols, for example) that have a low vapor pressure have been used with moderate success to retard evaporation from relatively calm surface waters. Changes in vegetative cover affect evapotranspiration: cultivating practices such as weeding and mulching reduce soil moisture losses; clearing of forests usually results in decreased evapotranspiration and increased runoff.

Distillation for the purpose of desalination or purification of polluted water preempts the normal atmospheric processes of evaporation, condensation, and precipitation to provide water supplies of acceptable quality. Condensation in cooling towers releases latent heat, permitting reuse of water in thermal processes of factories and power plants.

An obvious way to offset a water deficit is to increase precipitation. Capture of fog drip on screens or filaments and collection of dew on cool surfaces are methods of enforcing precipitation to augment meager fresh water supplies along desert coasts, but the amounts of water gained are small. Rainmaking by cloud seeding or other means cannot change the total global water supply, but it might accelerate the hydrologic engine to make more water available at a specific time and place. (See Chapter 17, especially the section "Cloud Seeding.") Schemes to enhance precipitation by inducing atmospheric instability or to steer rain-bearing storms by cloud seeding or other means are somewhat hypothetical at present.

The practical value of understanding the hydrologic system comes from the fact that it provides water on land for the many uses by humans and other life forms. A great deal of mental and physical effort is expended in the attempt to control it for our benefit. Conservation of water resources, indeed of most natural resources, depends on one or more of its phases. Where we cannot control it, we must adjust to its peregrinations.

QUESTIONS AND PROBLEMS FOR CHAPTER 11

1. Summarize the processes that maintain the hydrologic system.
2. Assume that the water data for Memphis, Tennessee (Table 11.3), represent average values for a river basin having an area of 1,000 km^2. Calculate the volume of runoff expected in the month of March.
3. Explain the operating principle of the evapotranspirometer.
4. Define flood in terms of the water budget concept.
5. Referring to Fig. 11.5, explain the differing hydrologic regimes at Banff, Freeport, and Okeechobee.
6. What are the limitations of snow survey data as aids in flood forecasting?
7. What methods are used to determine the amount of water stored in the soil?
8. What are the probable effects of forest removal on the runoff from a watershed?

12

Climate and the Biosphere

The biosphere is the planetary zone of life. It extends well into the atmosphere, to great depths in the oceans, and somewhat less deeply into the earth's crust; but terrestrial life is concentrated near the bottom of the air envelope. Because exchanges of energy and mass are necessary for biological processes at any level, climate ultimately affects all forms of life. Whether the growth medium is air, soil, or water photosynthetic activity is the fundamental process for fixing solar energy in organic matter. Only about one-tenth of one percent of total insolation is converted in the process, but that small fraction supports the earth's life systems on land and in water. In its simplest form, photosynthesis uses carbon dioxide, water, and light to produce organic compounds, store chemical energy, and give off water and oxygen. Along with other chemical processes, it helps to regulate the amounts of atmospheric carbon dioxide, oxygen, and water vapor, which affect energy transfer. Decay, burning, and passage through food chains further complicate the role of biota in the climate system. This chapter examines bioclimatic aspects of natural vegetation, soils, and marine life in the boundary zone between the atmosphere and the earth's surface.

CLIMATIC FACTORS IN PLANT GROWTH

All plants have environmental requirements that must be met if they are to thrive. These may be classified broadly as: (1) climatic, (2) physiographic, (3) edaphic, and (4) biotic. The first is of primary concern here, although the influences of terrain or relief, soils (the edaphic factor), and the interrelations of plants with other plants and with animals cannot be neglected. Climate acts in conjunction with these other factors to set limits to plant growth. Its role is direct in its effects on plants and indirect through its influence on edaphic and biotic factors.

The principal direct effects of climate on plants are exerted by elements of the water and heat budgets: precipitation and soil moisture, humidity, temperature (including soil temperature), sunlight, and wind. Variation in one can change the significance of the others in producing different rates of evapotranspiration and photosynthesis. The moisture factors are the most important over large areas of the earth. Water not only goes into the composition of plant cells but also serves as a medium for transport of nutrients to growing cells and through evapotranspiration acts as a temperature control. For most land plants, the immediate source of moisture is the soil. The amount and availability of soil moisture is not necessarily a simple function of precipitation, but is affected by surface drainage conditions and by the ability of the soil to retain moisture as well as by the losses due to evapotranspiration. Thus, swampy areas occur in the midst of deserts and sandy or gravelly soils in rainy climates may be entirely devoid of vegetation. Just as deficient moisture limits plant growth, so excess amounts restrict certain plants by limiting aeration and the oxygen supply in the soil. Excessive soil moisture tends to develop unfavorable soil characteristics and to increase disease damage.

The humidity of the air in which plants grow has varying significance depending upon the type of plant as well as upon the soil moisture available to it. Low vapor pressure of the air induces increased losses of moisture through transpiration. Many plants can withstand low humidities as long as their roots are supplied with adequate moisture. The xerophytic vegetation of dry climates is adapted to limited moisture in several ways. Thick waxy bark and leaves inhibit loss by transpiration. Certain plants have root systems that extend deep into the soil as well as over a wide radius to gather moisture from a large volume of soil. Some plants of the desert, notably certain species of cacti, store water during relatively wet periods to be used in dry times. Many desert annuals avoid the prolonged dry spells by passing rapidly through the life cycle from seed to seed after a rain. These are better classified as drought-escaping than drought-adapted.

Whereas moisture provides the *medium* for the processes of plant growth, heat and light provide the *energy*. Plants can grow only within certain temperature limits, although the limits are not the same for all plants. Certain algae live in hot springs at more than 90°C, and desert lichens withstand 100°C; arctic mosses and lichens survive −70°C. For each species and each variety, there is a minimum below which growth is not possible, an optimum at which growth is best, and a maximum beyond which growth stops. Most plants cease growth when the soil temperature drops below about 5°C. If the soil temperature is low, the rate of intake of moisture through the roots is decreased and the plant may not be able to replace water lost by transpiration. Freezing temperatures can thus damage the plant cells by producing chemical changes and desiccation. Alternate freezing and thawing are especially damaging. Species differ a great deal in their adaptation to temperature conditions, however, and many plants can endure long periods of below-freezing temperatures although they do not grow. The effect of high temperature is generally to speed up the growth processes. Under natural conditions, high temperatures are rarely the direct cause of death in plants. Rather, the increased evapotranspiration induced by the heat causes dehydration of the plant cells. Up to a point this can be forestalled if the moisture supply is adequate. Consequently, the temperature–moisture relationships are as important as temperature alone.

The moisture requirements of plants become higher as the energy supply increases, and if moisture is available productivity increases. A given amount of precipitation may result in a moisture deficit in a hot climate, but under lower temperatures the same amount may exceed potential evapotranspiration and create a moisture surplus. If the demands of evaporation from soil and transpiration from plants are not met, wilting and eventual death occur. Both evaporation and transpiration are cooling processes that tend to offset the effects of high temperature. Hot winds increase potential water loss and hasten damage to plant tissues; they may be disastrous even when soil moisture is abundant. As growing conditions become cooler, especially at higher altitudes or latitudes, species are successively eliminated from vegetation formations and ultimately all plant life is prohibited regardless of water in snow, ice, or the frozen soil.

Sunlight supplies energy for photosynthesis and is absorbed by chlorophyll. Without adequate light, green plants fail to develop properly, although some species are adapted to shaded conditions. Light regulates the time required for certain species to flower and produce seed. Ordinarily the amount of light available to growing plants is sufficient for normal development so that it is not of major importance in the geographical distribution of vegetation. However, for an individual plant in a specific environment, light conditions can be critical. Sunshine also is closely related to temperature. In combination with atmospheric heat, absorption of direct radiation can drive the temperature above the maximum limit permitted by the moisture supply. In moderate intensity, radiation can stimulate plants to their optimum development. The albedo of plant surfaces aids the control of leaf and stem temperatures.

Wind influences vegetation directly by its physical action upon plants and indirectly by accelerating moisture loss. Through convective heat transfer it affects plant temperatures. Plant leaf temperatures are modified by increases in wind at low speeds. As speed increases, the proportionate effect is diminished. The rending and tearing action of high-speed or gusty winds is a familiar process. Trees are blown over or stripped of leaves and branches; leaves of bushes are shredded; stems of plants are twisted or broken. Abrasion by windblown sand, gravel, or ice particles can also be quite damaging to plants. These are, however, local effects. There is no direct worldwide correlation of wind belts with the vegetation pattern. Extensive wind damage is common on exposed mountains, where it combines with the effects of poor soil, low temperatures, and ice or snow cover to create vertical life zones (see pages 272–74).

Although mean values of the climatic elements may have broad application in determining the suitability of an area for plant growth, the variations from the normal and the extremes are frequently vital considerations. Occasional droughts, floods, heat waves, or frosts can prove fatal to plants otherwise adapted to the normal conditions. A succession of two or three unfavorable years is particularly devastating. Only those plants which escape the disaster or which are able to withstand it survive. Hence climatic anomalies play a leading part in plant adaptation and natural selection.

Duration of minimum conditions for vegetative growth is another consideration which involves the seasonal distribution of the climatic elements. Whether precipitation is well distributed throughout the year, concentrated in a short season, or is erratic sets broad limits for plant associations. Thus, moisture-loving species of the rainy tropics are not adapted to the hot, dry season of the dry summer subtropics. The duration of temperature conditions is likewise restrictive or permissive relative to the plant association in question. Length of growing season is usually defined as the period between the last killing frost of spring and the first killing frost of autumn. It has its widest usage in connection with crop plants, but short growing seasons set limits to natural vegetation as well. As pointed out in the discussion of climates dominated by polar air masses, the longer daily duration of sunlight in summer at high latitudes intensifies the growing season so that plants are able to concentrate their annual growth into a shorter period than at lower latitudes. Furthermore, plants differ in their susceptibility to frost damage; some are simply hardier than others.

WORLD PATTERNS OF VEGETATION

All species have climatic optima for most efficient growth. The fact that these optima as well as extreme climatic limits are different for various plants accounts for the areal distribution of plant associations on earth. Although the other environmental factors may be favorable, they can be negated by an adverse climate. Heat and moisture are the great determiners of where plants will grow naturally, and the energy and water budgets largely set the world pattern of vegetation.

The correlation between climate and vegetation is in broad regional groupings rather than in details and is most evident where undisturbed natural vegetation has reached an approximate equilibrium with the climatic environment (see Fig. 12.1). Constantly frozen conditions restrict the cover to ice and snow or barren rock. Toward progressively warmer climates, tundra and taiga prevail, and since even meager precipitation may equal potential evapotranspiration under the low temperatures, moisture is subordinate to temperature as a limiting factor. In warmer regions, adequacy of moisture becomes critical and a succession of desert, steppe, grassland, forest, and rain forest ranges through arid to very wet climates. Each of these vegetation categories has its cool and hot subdivisions. Thus there are cold deserts and hot deserts, prairie grasslands and savanna, subarctic rain forests and tropical rain forests. In each climatic region, the characteristic vegetation formation consists of associated plant species which, through the centuries, have

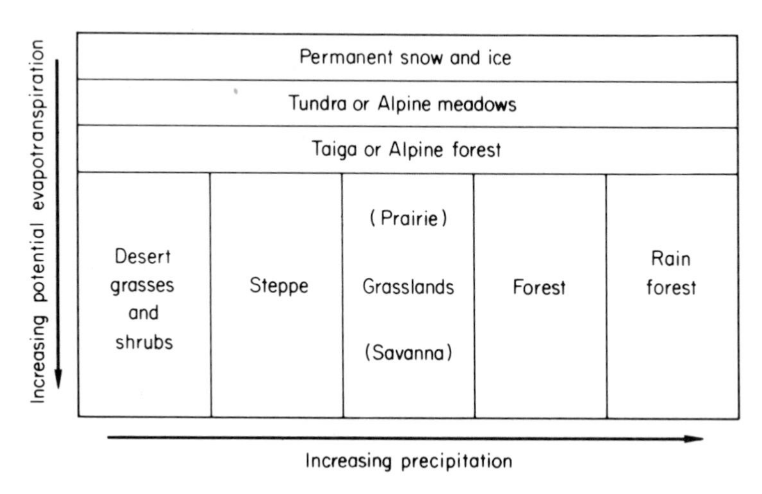

Figure 12.1 Schematic relation between water budget factors and major world vegetation formations.

become adapted to the prevailing climate. The dominant forms are the *climax vegetation.* Other plants may exist and the climax forms may be locally absent because of soils, fire, or human interference, but generally the climax formation reflects climatic influences. Because of their own adaptations to climate and dependence on plant life for both habitat and food, animal species also tend to associate in more or less homogeneous environments. Thus, we recognize "desert animals," "savanna animals," and so on. Many mammals and birds are able to migrate seasonally or in response to climatic anomalies, of course.

VERTICAL DIFFERENTIATION OF VEGETATION

The vertical zonation of vegetation is a common phenomenon of the highland climates. Altitudinal range and variety are greatest in the tropics, but wherever vertical differences of climate occur corresponding changes in vegetation are likely—unless climate prohibits plant growth at all altitudes as it does on the polar ice caps. Altitude duplicates, in some respects, the influences of latitude on the heat budget and on the types of vegetation; but slope, exposure, cloudiness, wind, and orographic effects also condition the climatic environment of highlands. As highland climates are mosaics of many microclimates in a vertical arrangement, so mountain vegetation falls into broad altitudinal belts with many local variations. Thus, the concept of vertical life zones applies regionally, but—in detail—the plant associations vary noticeably. Some of the most striking horizontal variations result from differences in rainfall or insolation on opposite sides of mountains.

Since temperature generally decreases with altitude, whereas precipitation ordinarily becomes greater, the combined result is a rapid increase in the ratio of precipitation to potential evapotranspiration as elevation increases. At the higher levels, the upper limits of plants are more likely to be set by low temperatures than by lack of moisture (Fig. 12.2). On the lower slopes (at least in low and middle latitudes), moisture is the critical factor. On some slopes with relatively sparse rainfall, the moisture demands of a forest are partly met by fog and clouds which reduce evaporation and deposit quantities of water droplets on the plant surfaces.

Although a given level on a tropical mountain may have annual temperature and precipitation averages like those of a place at a higher latitude, it does not have the same climate. Seasonal rhythm in the climatic elements, especially temperature and precipitation, changes with latitude, and the rhythm of plant growth changes with it.

In the tropical highlands of Latin America, four vertical life zones are commonly recognized in relation to vertical differences in temperature and precipitation effectiveness. The lowest is the *tierra caliente* (literally, hot land), extending from sea level up to 750 or 1,000 m and having vegetation typical of the tropical lowlands. Annual temperature averages are generally above 25°C. Above the *tierra caliente,* the *tierra templada* (temperate land) rises to elevations of 2,000 to 2,200 m and embraces a modified from of tropical forest. Temperatures range between 25°C and 18°C on the average. Higher still is the *tierra fria* (cold land) from 2,000 to 3,700 m. There the forest becomes less dense and grades into bushy types. The upper limit of the *tierra fria* coincides with the upper limit of cultivated crops, which is at about the

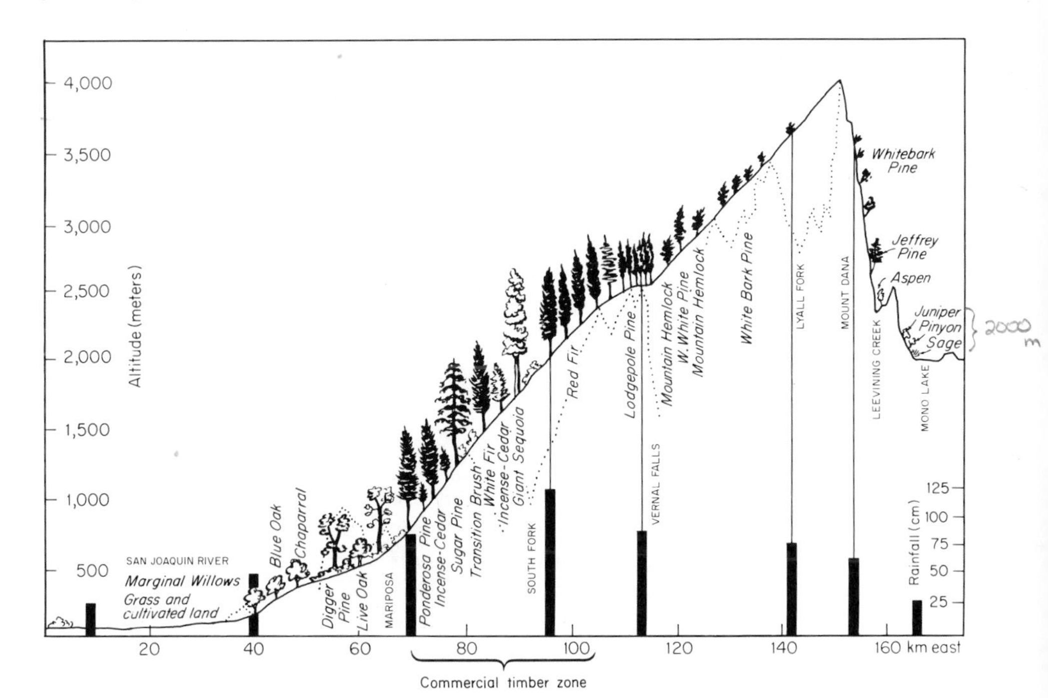

Figure 12.2 Vertical zones of vegetation in the Central Sierra Nevada. (Adapted from *Trees, 1949 Yearbook of Agriculture.)*

12°C annual isotherm. The *tierra helada* (frost land) extends on upward to the permanent snowfield. It is a zone of alpine meadows, whose character depends upon the amount and seasonal distribution of precipitation. Where there is adequate moisture throughout the year, the grasses are relatively dense and perpetually green. Farther from the equator, there is a short dry season and the alpine meadows are dry for a part of the year. Toward the regions of subsidence in the subtropic highs, conditions become drier and the associated vegetation grades into thorny and desert forms.

In the middle latitudes, the *tierra caliente* is absent and there is actually nothing akin to the *tierra templada.* Altitudinal limits to vegetative types are lower, but moisture conditions still have much influence on both horizontal and vertical distribution of plants. In rare cases, where there are small-scale mountain depressions, inversions of the normal vegetation sequence may occur. For example, in a study of vegetation on the Velebit Range of Yugoslavia, F. Kusan found that a persistent air drainage inversion has created a succession of beech and alpine conifers *downward* along steep slopes that drop from 1,600 m to 1,433 m.

CLIMATE AND FORESTRY

Wise management of the world's vegetation resources requires understanding of their ecology, including climatic relations. In view of its economic significance forestry is an appropriate example to illustrate applied bioclimatology. The distribution and growth rates of forest species reflect climatic conditions, and forests, in turn, affect energy and mass exchanges in the climate system. Individual tree species, like other plants, have optimum climatic requirements. Favorable conditions must be met by both the regional macroclimate and the microclimates within the forest itself. The effects of forests on climate are greatest in the area the trees occupy and they are roughly proportional to the density of cover, but forests also modify the climate of adjacent areas. Widespread removal, changes in species composition, or conversion to other types of land use alter surface heat and water budgets and may affect regional mesoclimates (see Table 12.1).

Temperature averages within a forest are slightly lower than in adjacent open areas and the ranges are not so large. In mid-latitude lowlands the mean annual temperature within a forest is about 0.6C° lower than on the outside; at an elevation of 1,000 m the difference is 1C°. The greatest average difference is in summer, when it may be as much as 2C°; in winter it is only about 0.06C°. On the hottest summer days a forest may reduce the surface air temperature by more than 2.5C°, whereas in extremely cold winter weather the difference is only about 1C°.

In low-latitude forests the influence of a tree canopy on air temperature is even greater. In a dense forest the upper canopy shades the ground and acts as the primary absorbing surface during the day, thus retarding the rise of soil temperature. At night the canopy radiates heat more rapidly than the ground, which is slower to cool. The same principle holds true for seasonal variations. In

TABLE 12.1

Changes of Surface Heat and Water Budget Elements after Conversion from Forest to Agricultural Use*

Land use	*Albedo (%)*	*Net radiation (watts/m^2)*	*Sensible heat flux (watts/m^2)*	*Latent heat flux (watts/m^2)*	*Evapo-transpiration (mm per month)*
Coniferous forest	12	60	20	40	41
Deciduous forest	18	53	13	39	40
Farm land, wet	20	50	8	42	43
Farm land, dry	20	50	15	35	36
Grassland	20	50	20	30	31

*Adapted from SMIC, *Inadvertent Climate Modification* (Cambridge, Mass.: MIT Press, 1971), p. 173.

spring and summer, the soil of the forest floor is slow to heat but in autumn and winter it remains slightly warmer than soil outside the forest, though the difference is less. Depending upon the density of the cover, forests may intercept up to 90 percent of the sunlight which is incident at the treetop level. Anyone who has tried to take a photograph inside a forest can attest to this. The reduction in light is the primary cause of paucity of low plants under a dense forest canopy such as is found in the equatorial rain forest.

Clear evidence that total precipitation over a forest differs appreciably from that over adjacent areas is lacking. Rain gauges placed in open spaces within the forest usually indicate greater rainfall than outside the forest, but this is probably due to their being protected from the wind so that they collect a more accurate sample. Tree crowns intercept a part of the precipitation which falls over a forest, and much of it is evaporated. The proportion thus withheld from the soil depends on the density of foliage and also on the duration and type of rainfall. A light rain of short duration may be almost entirely caught in the canopy of a dense forest, whereas in a driving rainstorm accompanied by wind most of the rain reaches the ground. Interception of advection fog may produce fog drip and enhance soil moisture. Snow collected in the crowns of trees may be shaken out by the wind or it may melt and drip to the ground or flow down the trunks. On the other hand, some of this potential soil moisture is lost by sublimation of the snow or by evaporation of melt water (Fig. 12.3). In general, coniferous forests intercept more precipitation than do hardwoods, and deciduous species prevent comparatively little precipitation from reaching the ground in winter. By thinning or clear-cutting portions of evergreen forest, it is possible to reduce the interception of snow in treetops and admit more snow to the ground, where it can augment runoff with smaller losses due to sublimation. Conversely, reforestation may actually result in decreased stream flow from a watershed. Increased transpiration as well as evaporation of water intercepted by the forest canopy reduce surpluses in the water budget. Management of forested

Figure 12.3 Intercepted snow and rime in conifers on a mountain in northern Idaho. Sublimation and evaporation of melt water in the suspended snow reduce the runoff from the area. (U. S. Forest Service.)

watersheds often entails decisions on the relative value of water and biomass production.

Relative humidity is 3 to 10 percent higher within a forest on the average owing to the lower temperatures, lighter air movement, and transpiration from plants. Evaporation from the soil is considerably lessened as a result of the protective influence of the forest, and, if the ground is well covered with plant litter, it is reduced by one-half to two-thirds as compared to evaporation from soil in the open. During the growing season, trunks and branches have lower temperatures than the surrounding air and they are further cooled by radiation at night with the result that dew or fog is formed. The movement of air from a forest to adjacent open country often carries fog with it and aids in preventing frost.

Surface wind speeds are markedly reduced by trees; several meters inside a dense forest wind force is but a small fraction of that on the outside. By restricting air movement, trees aid in reducing evaporation, lowering temperatures, and increasing relative humidity. Within the forest, the reduction in wind results in a more even distribution of snow cover.

The influences of climate on forests, as well as those of forests on climate, must be taken into account in silvicultural practices. As a group, forest species have higher moisture demands than do most other types of plants, and forests are therefore associated with humid climates. On a smaller scale, the variations in climate associated with local differences in slope, exposure, or altitude influence the distribution and rate of growth of trees. In a climax forest dominated by one or more species, other trees equally adapted to the regional climate may not be able to gain a foothold because of shading or other environmental effects created by the existing forest. When burned or cut over, climax forest species may be unable to regenerate without a favorable environment provided by successive plant associations. Some species require protection from direct sunlight and the accompanying high soil temperatures when in the seedling stage but later flourish as the trees that afforded the protection die.

Scientific forestry seeks control of the forest microclimate by logging methods, thinning, and burning in order to maintain a desired forest association. Primarily, this entails control of the amount of light reaching the forest floor. Dense stands tend to develop tall, straight trees of small diameter as the trees fight upward for sunlight. Subsequent natural or artificial thinning allows more light and space for growth as the trees approach maturity. Selective logging of large trees often "releases" smaller trees of the same age, permitting accelerated growth. Too-heavy cutting of certain species may expose the remaining trees to wind damage, however. Windfall is especially common along the margins of clear-cut areas in dense coniferous forests (see Fig. 12.4). Whether the objective is perpetuation of an existing forest, reforestation, or development of a new forest, an optimum climatic

Figure 12.4 Windfall in a coniferous forest.

environment at each stage of growth aids management for timber production, watershed protection, wildlife habitat, grazing, and other forest values.

Since trees have long life spans, they are influenced by climatic fluctuations through the years. During a series of dry years, their growth is limited; in wet years, they grow rapidly. Extreme temperature variations also affect growth; high average temperatures accentuate the water deficits of dry years, and periods of abnormal cold directly retard growth. Increasing evidence of these fluctuations during the life history of trees is being accumulated from examination of their annual growth rings (see Chapter 10, "Proxy Evidence and Geochronology").

FOREST-FIRE WEATHER

Prevention and control of forest fires constitute one of the most expensive, albeit necessary, aspects of forest management. It has been estimated that lightning sets 5,000 to more than 10,000 fires annually in the forests of the western United States alone. Although many are kept from spreading by rain, others are fanned into disasters by wind. Prevention of lightning-caused fires conceivably could begin with control of lightning. To whatever extent cloud seeding or other modification techniques can moderate thunderstorms, lightning suppression is at least a hypothetical solution. Forecasts of potential lightning storms and knowledge of forest conditions resulting from previous weather can be valuable in determining the best techniques for arresting a fire. Because lightning strikes are more common on high elevations, fires thus started are more difficult to reach than most human-caused fires and their control requires special methods and equipment.

The potential occurrence and the rate of spread of forest fires are direct functions of both the local weather and regional climate. Unfortunately, the conditions that create a fire hazard also tend to favor human activities that are responsible for many forest fires. Previous weather is quite as important as that currently prevailing, since it influences the flammability of forest materials. Factors that determine the fire danger include relative humidity, temperature, wind speed and direction, precipitation, and condition of the vegetation and litter. Extreme fire danger develops during long periods of hot, dry weather, especially when associated with the stable air of anticyclones. Katabatic winds such as the mistral in France or the Santa Ana in California frequently create severe fire hazards. Wind hastens drying of the forest and considerably increases the rate of spread of a fire. During the fire season, observations of critical weather elements are taken at stations operated by various public and private agencies. These stations employ standard instruments and usually have in addition a device for determining the moisture content of forest litter. This *fuel moisture indicator* consists of wooden sticks freely exposed to the weather (Fig. 12.5). The fuel sticks are periodically weighed on a scale which is calibrated to give an index of fuel moisture content. Analysis of observational data in the light of general forest conditions is combined with the regional weather forecast to prepare a fire-weather forecast and a fire hazard index. These reports

Figure 12.5 Fuel moisture indicators. The wooden indicators (foreground) are weighed periodically in the shelter to determine an index of forest fire danger. (U. S. Forest Service.)

are the basis for restrictions on logging and travel in forests and for alerting fire-fighting crews. Mobile weather stations provide local information at the scene of a fire. Only with complete knowledge of weather conditions can forest fires be prevented or controlled effectively.

CLIMATE AS A FACTOR IN SOIL FORMATION

Five natural factors influence the formation of soils: climate, plant and animal life (including microorganisms), relief, parent material, and time. Of these climate is the most active. Soils are affected by climate directly throughout their evolution from parent rock to their current state of development. Their character is shaped indirectly by climate acting though vegetation and animal life. Many of the effects of varied relief are ultimately climatic as a result of local microclimates produced by exposure, slope, drainage, or altitudinal differences.

Soil is a layer of earth material which undergoes constant change and development; it is not merely an inert mass of finely divided rock. The dynamic processes which form soils are of three broad classes: physical, chemical, and biological. Physical breakup of parent rock, termed *disintegration,* is induced climatically through the action of rain splash, rainfall and snowmelt runoff, glacier movements, freezing and thawing, and abrasion or transport by wind. Chemical weathering, or *decomposition,* takes place more rapidly under warm, humid conditions than in cold or dry climates. Rocks softened by chemical action become more suitable for plant growth, but they also are more easily worn away by physical processes. Once the weathered parent material begins to support life, plants, bacteria, worms, and so on speed the chemical and physical changes and add organic matter. Within certain limits, chemical and biological activity increase with increasing temperature and moisture; neither is possible without water. Downward transport of colloidal solids is called *eluviation,* whereas *leaching* is the removal of mineral or organic compounds in solution. Both tend to impoverish the top layers of soil. On the other hand, lack of moisture limits infiltration, and rising capillary water tends to leave concentrations of dissolved salts at the surface when it evaporates. The salt flats and alkaline soils of arid climates are examples of soils formed in this way. Broadly speaking, the soils of deserts and steppe grasslands form where evaporation exceeds precipitation, whereas forest soils are associated with the reverse conditions. Thus, total precipitation is not the sole control of moisture in soil formation. Cloudbursts tend to puddle surface soil and seal it against downward water movement, thereby promoting excessive surface runoff. Gentle rains have time to soak into the soil and are more effective in soil-forming processes that require moisture. Obviously the degree of permeability of soil also affects infiltration. Whether the rate of infiltration is rapid, as in sands and gravels, or slow, as in clays, determines the availability of water for plant growth and soil formation. Wind is a drying agent as well as a cause of erosion. By accelerating evaporation from the soil, it increases the water need of plants.

Throughout soil formation the climate within the soil is more important than that of the air above. Temperature changes are conducted downward slowly; at a depth of about 60 to 80 cm diurnal variations are uncommon. Seasonal variation lags considerably at the greater depths, and at about 16 m below the surface there is little or no seasonal temperature change (Fig. 12.6). The structure of the soil and its moisture content help to determine its conductivity; dry, porous soil conducts heat slowly, whereas wet, compact soil and solid rock are much better conductors. Air in the soil may differ in composition from that in the atmosphere, but it is a poor conductor and reduces heat transfer in porous soils. Similarly, snow or a mantle of plant litter inhibits heat conduction at the surface. Rain and snowmelt tend to carry heat downward as they penetrate into soil pores.

Except in the driest deserts, air in the soil below the immediate surface ordinarily is saturated, a fact of considerable importance in both chemical and biological processes so long as temperatures are favorable. The relative humidity of air within the soil is sometimes determined by the relative humidity of the at-

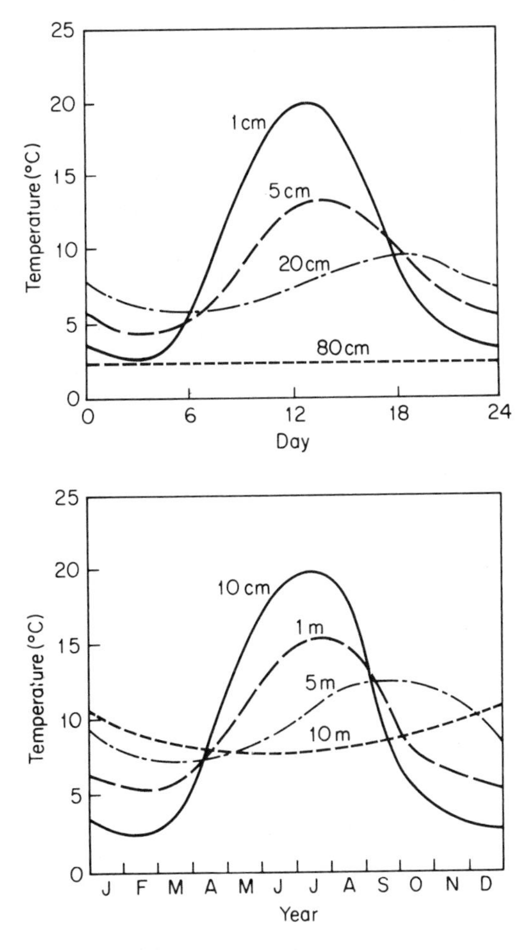

Figure 12.6 Typical daily and annual variations of soil temperature at different depths. (Curves based on data from Geiger, Oke, and others.)

mosphere. If the average relative humidity of the overlying air is near saturation (90 to 100 percent), the soil climate is maintained at a moist level even in the absence of rainfall.

Assuming that the weathered parent material is not removed by erosion (as is likely on a steep slope), the various soil-forming processes combine to produce layers, or *horizons,* in the soil and over a period of time of distinctive *profile* develops. The profile is the most useful feature for soil identification and classification. Characteristics of color, texture, structure (arrangement of soil particles), and the chemical and organic matter content vary downward through the profile, providing a basis for study of soil origin. Figure 12.7 is an idealized soil profile. Not every mature soil exhibits all of the horizons, but it will have some of them.

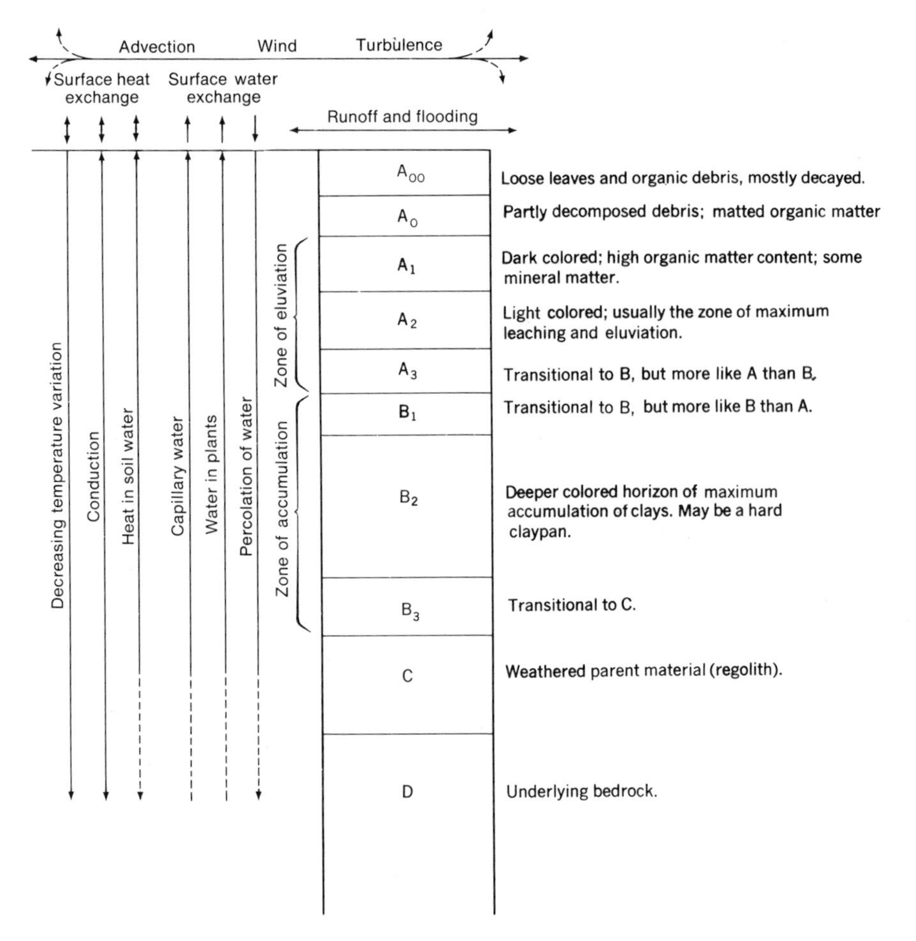

Figure 12.7 Hypothetical soil profile and major exchanges of heat and moisture.

SPATIAL PATTERNS OF SOILS

To the extent that climatic conditions influence the development and properties of soils, major world soil regions correspond in a general way to the world climatic regions and, therefore, to spatial patterns of vegetation. In spite of the dangers that attend prejudgment of soil genesis, we can make some useful generalizations about soil-forming processes in different climates (see Fig. 12.8). Under permanent snow and ice, soils cannot develop. In the tundra, poor vertical drainage and low temperatures result in distinctive shallow, acidic, and often peaty soils of little agricultural value. Infiltration of colloids and bases proceeds at a greater rate under the higher temperatures, longer summers, and moisture surpluses of the taiga climate; this forms characteristic cold-forest soils that have a light-colored (ashy

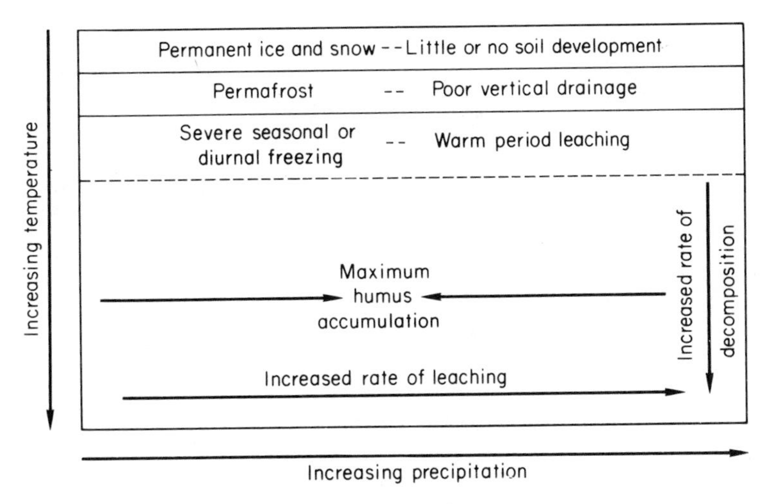

Figure 12.8 Schematic relation of major soil-forming processes to temperature and precipitation.

gray) A_2 horizon. In warmer forest regions, leaching is less significant, and the resulting profiles are neither as acidic nor as well developed. Under deciduous and mixed forests in the humid continental climates and in parts of the marine and dry summer subtropical climates, the soils are not only less acidic but also lack both the concentrated horizon of organic matter at the surface and the gray A_2 horizon (see Fig. 12.9). Where still warmer, humid conditions prevail decomposition of organic matter and leaching of bases and other soluble materials take place even more rapidly to produce acidic soils whose B horizons are stained by iron compounds. Such soils are approximately coextensive with the humid subtropics and are transitional to those of the tropical humid climates. They are often classed with certain tropical soils because the processes that accompany high temperatures and rainfall account for several of their characteristics.

In the humid tropics, bacterial activity and rapid chemical action, combined with excessive leaching, keep humus to a bare minimum. Deep profiles and an abundance of iron and aluminum compounds typify soils of the rainy tropics, monsoon tropics, and wet-and-dry tropics. The native vegetation of subhumid climates is primarily grassland. Sustained surpluses of moisture are uncommon and the processes of leaching and eluviation are slowed accordingly. The abundance of organic matter from grass roots and decayed leaves makes the A horizon much darker than that of forest soils. The change is gradual from the soils of humid areas, which are leached and eluviated, to those of drier regions, which experience far less percolation of salts and colloids. Calcium carbonate accumulates at increasingly high levels in the profile; acidity decreases as precipitation fails to meet the needs of potential evapotranspiration. On the steppe grasslands of semiarid climates alkaline compounds accumulate near the surface, and since the amount of organic matter diminishes progressively as conditions become drier the soils are lighter in color and

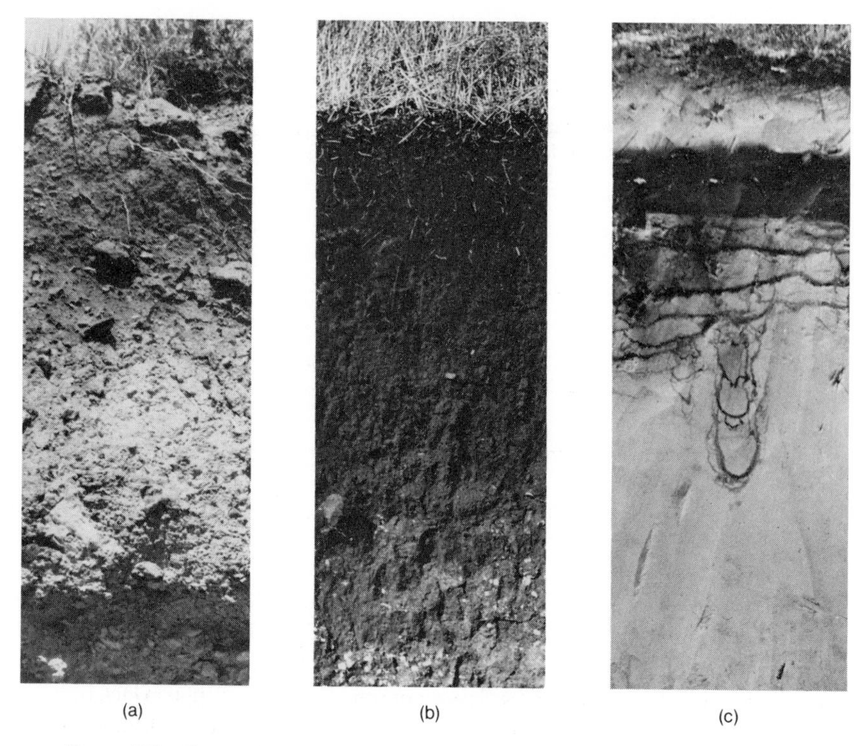

Figure 12.9 Contrasting soil profiles. Each profile represents a depth of one meter. (a) dry climate profile in stony parent soil; (b) subhumid climate and dense grass cover develop a deep organic horizon; (c) an organic layer covered by silt deposits, is leached by percolating water in a humid climate. (Soil Conservation Service—USDA.)

have shallower *A* horizons. Plants that tolerate large amounts of base minerals tend to aid the concentration of calcium salts near the surface.

Desert soils form under extreme water deficits. Their horizons are poorly developed, and they have little or no humus. In the mid-latitude deserts they are typically gray in color; in the tropical deserts a reddish color resulting from certain iron compounds is more typical. The accumulation of soluble salts may be exposed at the surface, especially in poorly drained depressions or flat basins, where surface waters evaporate to leave saline deposits. These *evaporites* are valuable clues to past climates.

The effects of altitude and relief on climate are reflected in the soils of highland regions. Just as highland climates are mosaics of many local climates, so soils of mountains and valleys vary greatly over short distances. Their complex spatial patterns make it impossible to represent them on a small-scale map. Many are shallow and immature because of steep sites and rapid erosional processes. Others form under unique local conditions of climate induced by relief and drainage. Only on gentle slopes, on flat plateaus, or in broad valleys are soils having well-developed profiles likely to be extensive. The altitudinal zonation of mountain soils coincides broadly with highland vegetation—the effect of altitude approximates that of latitude. Alpine soils, for example, are roughly analogous to tundra soils. Acting directly and through the dominant vegetation associations, the changing water budget produces a succession of soils that correspond to the life zones of plants.

CLIMATE AND SOIL EROSION

We have seen that the natural processes of erosion are active in the development of soils and that they are closely associated with climate. Our present concern is accelerated erosion, which tends to destroy soils that are more or less in harmony with natural conditions. The immediate climate-connected causes of accelerated erosion are the same as those which weather and transport parent rock in the early stages of soil formation; consequently, the distinction between "natural" and "accelerated" erosion depends more on the time scale than on erosion processes. But in one way or another, humans are usually indirect agents in accelerated erosion. Where there is excessive precipitation and hence a great potential erosion hazard due to surface runoff, a dense vegetation cover ordinarily exists to inhibit soil erosion. In arid climates vegetation is sparse, but running water is not such a great threat to the soil. Thus, there is a semblance of equilibrium between the forces that form the soil and those that would erode it. The human contribution is essentially negative. Cutting forests, cultivation, grazing of herds, or burning of the plant cover all serve to upset the natural equilibrium and to accelerate erosion. Once the protective cover of vegetation and the organic matter in the topsoil are reduced, erosion proceeds at a more rapid rate. Cultural practices in land use have all too rarely provided for a compensating acceleration of soil-forming processes. Many soil-building practices are known, but they are beyond the scope of this book.

The world distribution of actual accelerated erosion, being in part the result of land-use practices, does not show a close correlation with the patterns of climate and vegetation. Rather it is the potential erosion, or the "erosion hazard," which is closely related to climatic conditions and to the associated soils. The fact that some soils erode more easily than others is traceable, in part, to the effect of climate on their formation (see Fig. 12.10). In the regions of heavy rainfall, running water readily attacks exposed soils. In the humid tropical climates, the soils are generally deficient in humus so that, although the clay subsoils are quite resistant to erosion, the topsoil washes away easily. Where the rainfall is seasonal, where the soil is frozen for a part of the year, or where there is a snow cover for an extended period, the forms and intensity of erosion on exposed soils are altered but may be none the less disastrous. When frozen soil begins to thaw, melting usually takes place at the top first, and, until thawing is complete, vertical drainage of water is impeded.

Figure 12.10 Rainfall and snowmelt rapidly erode fine-textured loess soils on steep slopes of the Palouse region in eastern Washington State. (Soil Conservation Service—USDA.)

Under these conditions, moving surface water can rapidly carry away layers of mud. Snow affords a protective cover until it melts; thereafter the melt water from heavy snowpacks becomes a particularly active agent in sheet and rill erosion. The intensity of rainfall, as well as the amount, is important in determining the degree of erosion due to runoff. Thundershower precipitation does not penetrate the soil as well as light continuous rains, and the resulting runoff and erosion are much greater, especially in dry lands. One cloudburst can do far more damage than several months of gentle rains. Similarly, the concentration of a given amount of rainfall in one season creates an erosion hazard greater than that of an area with the same rainfall total more evenly distributed through the year. Heavy rains which break a period of drought commonly cause considerable loss of soil, for vegetative cover is at a minimum and the loose topsoil is easily removed.

Soils in regions of arid and semiarid climates are the most susceptible to the ravages of wind erosion. Soil water not only adds to the weight of the particles but is a cementing agent which reduces the tendency of soil to be blown about. When exposed to the wind, soils lacking in moisture are readily blown if the particles are small enough. Unfortunately, wind, like water, has a sorting action. It picks up and removes the finer particles which are the basis of soil fertility and leaves the heavier material in drifts or dunes (see Fig. 12.11). Its earthmoving capacity is not as closely related to slope as in the case of erosion by water. In the drought years of the 1930s in the "Dust Bowl" of the United States, the loose topsoil was completely removed from large areas, leaving only the compact clays of the subsoil. Semiarid grasslands are especially critical regions for wind erosion when excessive disturbance of the plant cover (through overgrazing or overcultivation) coincides with unusually dry years.

Temperature has its maximum effect on accelerated erosion indirectly through its influence on plant cover and the weathering processes. Freezing and thawing directly alter the structure of the soil and thus make it more susceptible to the action of wind or running water. When frozen for a continuous period, the soil is largely spared from erosion. The soils of the tundra and taiga are singularly free from accelerated erosion, mostly because of the lack of cultural disturbances but partly owing to the long periods of freezing temperatures.

Figure 12.11 Topsoil eroded by wind in the semiarid climate of western Nebraska. (Soil Conservation Service—USDA.)

The influences of weather and climate on the biological cycles of the oceans are no less complex than those on land but the marine environment changes less rapidly and consequently marine plants and animals usually are slower to reflect atmospheric conditions. Sudden changes in the marine environment are more likely to occur near the surface and in shallow coastal waters, often in response to weather events. Gradual changes with the seasons or longer periods merge into the time scale of climate. Energy and mass exchanges between ocean and air make it difficult to distinguish causes from effects of a changing ocean environment, but important climatic factors include solar radiation, temperature, and wind. Biotic response also depends on the nutrient supply, which is influenced by climate.

As in the case of life on land, photosynthesis is the process by which solar energy enters the cycles of marine life. Availability of light therefore affects distribution of plants and animals in oceans by regulating growth and reproduction. Photosynthetic activity of phytoplankton, the basis of the marine food chain, or pyramid, is limited to the *euphotic,* or illuminated, zone of the sea, where it varies regionally and seasonally. About 50 percent of the light that penetrates the surface of clearest ocean waters reaches a depth of 18 m; about 10 percent reaches 45 m; and only 1 percent reaches 100 m. Dissolved and suspended matter considerably reduce penetration, and selective transmission of the spectrum retards the longer wave lengths while allowing blue light to reach the greatest depths.

Light is a direct factor in the diurnal feeding and schooling patterns of fish and their reactions to fishing gear. Most pelagic (surface swimming) fish rise to the surface before sunset, school, then disperse and sink by sunrise. Demersal (deep swimming) species remain at low levels by day and rise to disperse at night. Light also provides orientation for marine organisms and plays a role in migration.

Ocean water temperatures are determined mainly by absorption of solar radiation, by exchange of heat with the atmosphere, and by mixing. All of these processes depend on atmospheric conditions, but the resulting temperatures tend to change more slowly than those in the air. Sudden variations in temperature are known to be lethal to many species. Entrapment in tide pools or stranding at low tide frequently exposes marine invertebrates to intolerable heat or cold as well as desiccation. Temperature affects the growth rates of marine plants and modifies the nervous, muscular, and metabolic reactions of fish. Spawning, hatching, and growth rates vary not only among species but also with the adequacy of heat and food at each stage of development.

The annual production cycles of marine herbivores vary widely with the availability of algae, which in turn respond to seasonal changes in ocean temperatures, light, and surface stirring by wind (see Fig. 12.12). In tropical waters, where there is a small annual variation in algal growth, herbivore productivity follows a similar curve but with a slight delay in the cycle. In mid-latitudes the seasonal amplitudes of production are greater and there is a longer delay in the response of herbivores, which are able to reduce algae by heavy feeding in mid-

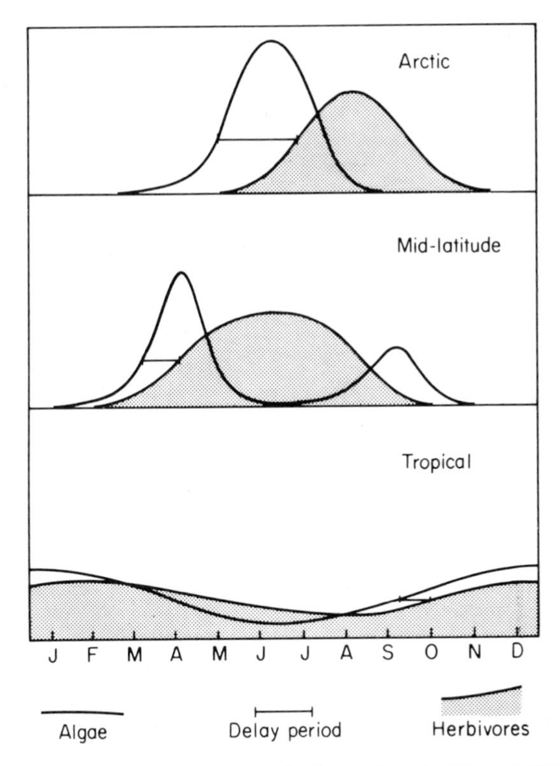

Figure 12.12 Relative annual marine production regimes in different climates. (After D. H. Cushing, *Marine Ecology and Fisheries.* New York: Cambridge University Press, 1975, p. 17.)

summer. The seasonal differences, the delay in herbivore production, and the period of dormancy are greater still at high latitudes.

Fish adapt more rapidly to warming than to cooling, and their tolerance of actual temperatures varies with the temperature range to which they have become acclimated. Figure 12.13 shows the schematic relation between acclimation and actual temperatures for a fish. Optimum temperatures and tolerance ranges differ among species, although exact data are lacking except in a few cases. Other environmental factors modify temperature effects. For example, stresses arising from reduced oxygen can narrow the range of temperature tolerance.

Fish congregate and feed in preferred thermal regions. Some migrate toward the greatest horizontal temperature gradient, where food is more abundant and commercial catches usually are better. Whales often gather in areas of dense advective fog that forms along boundaries between contrasting surface water temperatures. Lower water temperatures following cold winters over eastern North America result in smaller catches of shrimp along the Atlantic coast of the United States. Water temperature differences apparently guide certain anadromous fish toward spawning grounds in streams and lakes. Long-term temperature changes in

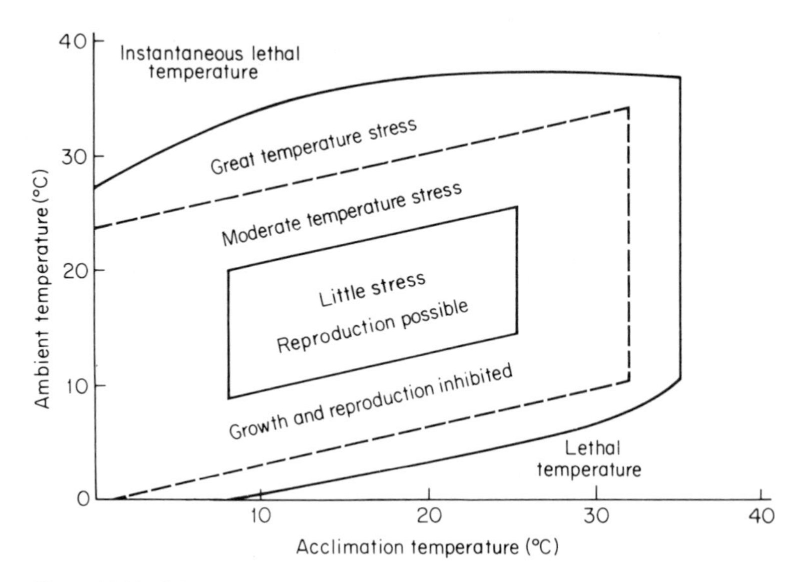

Figure 12.13 Schematic relation of actual temperature to acclimation temperature for a fish. (After Frederick W. Wheaton, *Aquacultural Engineering*. New York: John Wiley & Sons, Inc., 1977, p. 111.)

the oceans, associated with climatic fluctuations, enforce major shifts in fishing activity. The northward migration of cod in the waters off Greenland prior to about 1950 has been cited as evidence of a changing climate; since mid-century both North Atlantic temperatures and cod harvests have declined.

EFFECTS OF WINDS AND CURRENTS ON FISHERIES

Wind is a factor in the motion of ocean currents, drifts, and upwelling. It also is a direct cause of wave action and surface mixing. All of these stirring actions influence thermal conditions, oxygen content, and the nutrient supply. Winds over an oceanic area can disturb local currents and affect the productivity of a fishing ground thousands of kilometers away. Changing courses of ocean currents have been responsible for disastrous declines in commercial fisheries in the past. The herring fishery off Iceland deteriorated catastrophically in the late 1960s as drift ice moved farther south than usual, presumably due to a shift of atmospheric pressure patterns and a stronger outflow of surface winds from the northwest. In years of a strong *El Niño* off the west coast of South America, abnormally high water temperatures reduce the feed for anchovies, which support great numbers of guano birds and constitute the world's largest source of fish meal. The failure of the Peruvian anchoveta fishery in 1972–73 caused economic reverberations in world agriculture, notably in increased prices for soybean meal, which is a substitute for fishmeal. Whereas *El Niño* introduces warm waters that are detrimental to anchovies, it attracts tropical species such as yellowfin tuna.

Mixing by wind, or other dynamic actions in the oceans, redistributes heat, dissolved substances, organic matter, and sediments. Wave action, as during a storm, rapidly alters surface temperature and turbidity, causing most swimming species to descend. Temperature, salinity, acidity, gases (notably oxygen and carbon dioxide), nutrient supplies, suspended sediments, and industrial pollutants are critical factors for marine life. This is especially true along coasts and in estuaries, where mixing and wave action can alter proportions rapidly.

Upwelling of sea water is maintained by steady surface winds, such as those along the eastern margins of the oceanic subtropic highs. The phenomenon also is associated with ocean currents. Its principal relevant effects are on the vertical distribution of oxygen and nutrients. Whereas the ascent of water deficient in oxygen may create a lethal environment, rising organic matter may respond to light and wave-induced oxygenation, thus enhancing biological activity. Cessation of winds over an area of upwelling can disrupt the food chain, however. Several of the world's most productive fisheries are situated in zones of upwelling. Major areas lie along the California, Peru, Canaries, and Benguela Currents.

Fishing is an outdoor activity. It is therefore subject to weather vagaries that determine not only the availability of fish but also the efficiency and safety of fishermen, boats, and equipment. Marine weather forecasts of the "state of the sea" and other conditions that affect the potential catch and operations are a vital service to the fishing industry. The dangers of low temperatures, gales, waves, or sea swell are obvious. Spray and icing are particularly hazardous in middle and high latitudes in winter. Precipitation and fog are critical elements when they freeze on boats and equipment or reduce visibility. The speed and refraction of sound in sea water are functions of temperature as well as salinity and ocean depth, thus affecting acoustical devices that are used for fish detection, depth sounding, and location of gear.

QUESTIONS AND PROBLEMS FOR CHAPTER 12

1. Why are climatic extremes sometimes more important than averages in determining patterns of natural vegetation?
2. Find the meaning of phreatophyte. What effects do phreatophytes have on the water budget?
3. Under what conditions might the vertical differentiation of vegetation on mountain slopes be reversed from the normal pattern?
4. In what ways do desert plants adapt to a restrictive water budget?
5. How do forests influence microclimates? Macroclimates?

6. What criteria are considered in assessing forest fire danger?
7. Explain the relation of the major soil forming processes to climate.
8. Why is water erosion usually more severe in desert climates than in regions that have abundant rainfall?
9. List the environmental factors that affect marine life. Why is wind of greater direct significance to marine life than to plants and animals on land?

13

Climate, Agriculture, and Food

As climate largely defines the global pattern of vegetation, so it sets limits for production of crops and forage, which are the primary bases of today's world food supplies. Domestic animals also respond to climatic differences, both physiologically and through their feed requirements for economic production of food and fiber. In general, crops and animals have their optimum climatic conditions for production, although other variables such as soils, relief, pests, market, and transportation facilities interact to modify the suitability of a particular area for a specific type of cultivation or animal husbandry. The actual distribution of crop plants is determined by the combined influences of physiological, economic, social, technological, and historic forces; but no crop can attain importance in an agricultural system unless it is adapted to prevailing environmental conditions.

General aspects of climatic influences on food production have been known for centuries; yet important details are still being revealed. This chapter examines two broad, overlapping problems of agroclimatology. One is concerned with the influence of climatic elements on specific crops or animals and their productivity. The other pertains to the effect of climate on the spatial distribution of crops and animals. Agriculture is one of the riskiest of all enterprises. Not the least of the hazards are those imposed by seasonal weather vagaries. In the race between population and food supplies even minor fluctuations of climate over a few years could lead to widespread malnutrition and starvation.

CLIMATIC FACTORS IN CROP PRODUCTION

The principal climatic factors affecting crop production are the same as those influencing all vegetation—temperature, length of growing season, moisture conditions, sunlight, and wind—but they must be considered in a different light with respect to crops. Natural vegetation is, axiomatically, adapted to the climatic conditions with which it is associated. For economic reasons, we have domesticated crop plants and cultivated them in environments where they could not survive without our help. In other words, crop plants are less hardy than natural vegetation. Agriculture is essentially a combination of processes designed to promote a favorable environment for growth. With few exceptions the goal is improved yield and quality. Early farmers assessed yield in terms of return per unit of planted seed stock; in modern times the dominant criterion has been production per unit of cultivated area; the trend is toward consideration of yield per unit of water, photosynthetic energy, fertilizer and pesticides, or the energy expended in cultivating, harvesting, and processing. Agricultural practices are so diverse as to include drainage and irrigation to control the moisture factor, artificial sheltering to control temperature, and shading or the use of electric lights to control light. In the extreme, any crop can be grown anywhere if labor and expense are not in question; in the realm of practicability, all crops have their climatic limits for economic production. These limits can be extended by plant breeding and selection as well as by cultivation methods, but secular trends in climate have the potential to disrupt established agricultural systems.

The influences of climatic factors on crops are closely interrelated, and each is modified by the others. Daily, seasonal, or long-term variations in any or all of the climatic elements alter the efficiency of plant growth. The spatial scale of a crop's climatic environment is also significant. Temperatures outside the range of tolerance may prevail at ground level while a short distance above they are well within safe limits. Leaf temperature, which depends on absorption of radiation and the rate of transpiration as well as air temperature, is especially important for growth. Similarly, moisture, light, and wind effects may be quite different in the vicinity of the plant from conditions beyond its ecoclimate.

A device known as the *climatron* has enabled plant physiologists to determine responses of plant materials to climatic factors. Essentially a closed-system growth chamber, the climatron permits controlled experiments under varying combinations of environmental conditions and thus is a valuable aid in research on plant growth and development as well as selection, breeding, and introduction. Its principle also can be applied in studies involving animals, insects, and diseases.

TEMPERATURE AND CROPS

The temperature of the air and of the soil affects all the growth processes of plants. Every variety of every crop plant has minimum, optimum, and maximum temperature limits for each of its stages of growth. Winter rye has relatively low temperature demands and can withstand freezing during a winter period of dormancy. Tropical crops (for example, dates or cacao beans) have high temperature requirements throughout the year. The upper lethal temperatures for active plant cells of most species range from 50° to 60°C but vary with species, stage of growth, and length of exposure to the high temperature. High temperatures are not as serious as low temperatures in arresting plant development—if the moisture supply is adequate and the crop is adapted to the climatic region. Crops ordinarily do not "burn up"; they "dry up." But, under very high temperatures, growth is slowed or even stopped regardless of the moisture supply, and premature loss of leaves or fruit is likely. Disaster to crops usually comes with the combination of dry and hot conditions. Winds that would be expected to produce evaporative cooling often only speed up transpiration and result in dehydration of plant tissues. The optimum temperature for the maximum rate of plant growth is not always the best for crop production. Temperatures that promote quick growth may also result in weak plants which are more easily damaged by wind, hail, insects, or disease. Some crops fail to produce fruit or seed if forced under high temperatures. The highest yields in the mid-latitude grain belts usually occur in years with summers that are cooler than normal for a particular area, permitting increased storage of photosynthetic matter. In addition, cool summers often are accompanied by greater rainfall (see Fig. 13.1).

Within limits, the problem of high temperatures can be solved under field conditions by increasing the moisture supply through irrigation or by moisture-conserving tillage practices. These methods permit successful production of many different crops in hot desert oases. Delicate plants can be protected from the direct rays of the sun by taller tree crops or by cloth or slat shades. Selection of planting sites on the less-exposed slopes of hills also can alleviate high temperatures if the

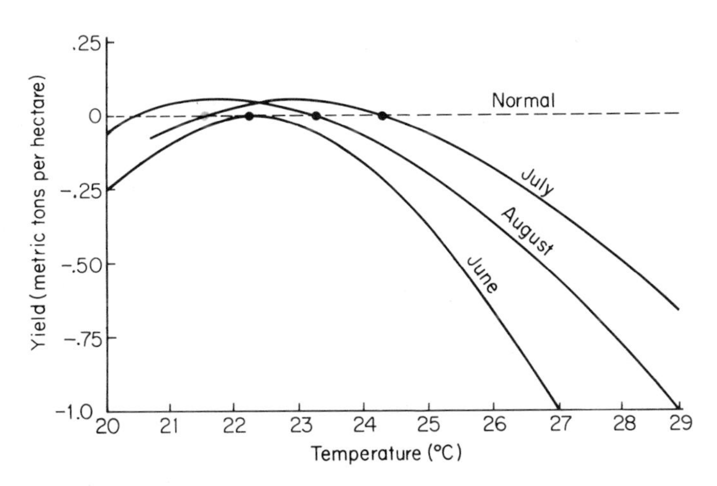

Figure 13.1 Relation of corn production to summer temperatures in the U. S. Corn Belt. (After Thompson.)

crop demands it. Where it is desirable to have low temperatures during the stages of germination or flowering of crops, this can be achieved by planting early in the season or in a cooler location, or, as in the case of winter wheat, by planting before the cold season. Water bodies defer the rise of temperatures to the leeward in spring and thereby help prevent blossoming in fruit orchards until the danger of killing frost is past, for example, in the fruit-producing areas east of the Great Lakes. Citrus fruits take on better coloring and become sweeter when subjected to short periods of near-freezing temperatures. Strawberries develop their best flavor when daytime temperatures are about 10°C, whereas a high sugar content in wine grapes requires abundant heat and sunlight.

Few crops are restricted in their worldwide distribution by the direct effects of high temperature alone, although many yield best when grown near their poleward limits. Agronomists have suggested that a slight decrease of mid-latitude temperatures would result in an actual increase of wheat and corn yields, assuming an unchanged pattern of precipitation. If accompanied by a reduced growing season, lower temperatures could disrupt agricultural practices and cause major shifts of crop zones. Considering that production of many crops has been pushed farther and farther into high latitudes, these effects have increasing significance. In discussing them we must distinguish between chilling and freezing. Prolonged chilling of plants at temperatures above freezing retards growth and can kill those which are adapted only to constantly warm conditions. Chilling may not directly kill plant cells but reduces the vital flow of water from the roots so that transpiration losses cannot be regained. Yellowing of plant leaves sometimes results from this kind of physiological drought. Crops such as rice and cotton are killed by near-freezing temperatures of two or three days' duration; potatoes, maize, and many garden vegetables can "weather" such cold spells with little or no damage, although their growth is slowed. As would be expected, the warm-climate crops are most seriously affected by chilling. Application of cold irrigation water to a field retards growth by reducing temperatures in the soil and at the immediate surface. Evaporation of soil moisture also tends to decrease the temperature. Loss of heat due to evaporation may be serious in the case of flooded rice paddies. In northern Japan chemical films which act as evaporation suppressors are applied to the surface of shallow paddy waters in order to maintain higher daytime temperatures at the level of the growing rice plants (see Fig. 13.2). On the other hand, wet soils do not cool as rapidly at night, and irrigation water can actually help prevent chilling if its temperature is above that of the air.

The influence of low night temperatures varies greatly with different crops. Certain crops, such as potatoes and sugar beets, store carbohydrates more rapidly during periods with cool nights. Cotton, maize, and tobacco require warm nights for maximum development. In high latitudes, the greater length of the daylight period makes it possible for crops otherwise adapted to cool climates to condense their growing season into fewer days. Light as well as heat contributes to speeding growth during the long days. Mid-latitude crops grow structurally but fruit poorly in the tropics.

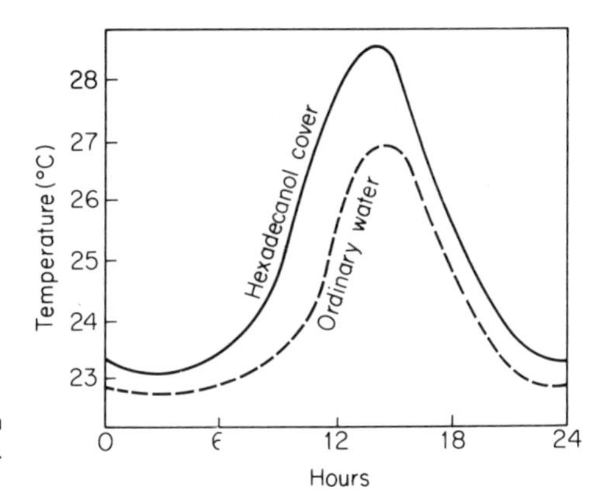

Figure 13.2 Effect of a hexadecanol film on diurnal temperature of rice paddy water. (After Roberts.)

Whenever growing plants are subjected to freezing temperatures, damage or even death is imminent. Some crops, notably fruit trees such as the apple, complete fruiting during the warm season and can withstand below-zero temperatures in the winter. Others have a hardy underground structure in the form of roots, bulbs, tubers, or rhizomes which maintain life while the upper parts of the plant die down during the winter. Annuals simply complete their full life cycle during the growing season, their hardy seeds being the medium for renewed life. Winter freezing can hardly be prevented in climates where it is the normal course of events. Crops adapted to the seasonal cycle of temperature must be selected. Hothouses are economically feasible for only certain high-value crops such as vegetables or flowers.

PHENOLOGY

Although temperature extremes that may damage crops are obvious climatic phenomena which affect agriculture, the rate of plant development at successive stages of growth is of equal importance in determining climatic limits of economic crop production. *Phenology* is the science that relates climate to periodic events in plant and animal life. Phenological data include such facts as dates of germination and emergence of seeds; dates of budding, flowering, and ripening; or seasonal activities of birds and insects. These depend on climatic conditions preceding each event as well as at the time of the event. Their specific relation to climatic elements is not fully understood, but temperature of air and soil, the daily amount and duration of solar radiation, and availability of moisture are major variables. Generations of farmers have kept diaries listing the dates of observed developments in natural vegetation and crops as well as periodic reactions by birds and animals to climate and the seasons. Such information, recorded or not, has been the chief basis for denoting "signs of spring," "signs of autumn," and so on. When correlated with climatic observations, it aids the study of agroclimatology, often revealing influences of latitude, altitude, prevailing winds, or other factors. Maps organize phenological data by using *isophanes* to connect places at which a phenological event took place on the same date (see Fig. 13.3). For example, the winter wheat harvest begins in Texas in June and gradually proceeds northward to Nebraska by early July. The difference in dates of a significant event over a given distance is the *phenological gradient.* Phenological gradients of microclimatic scale are common perpendicular to garden walls, fence rows, or hedges. On the sunny side, plants are

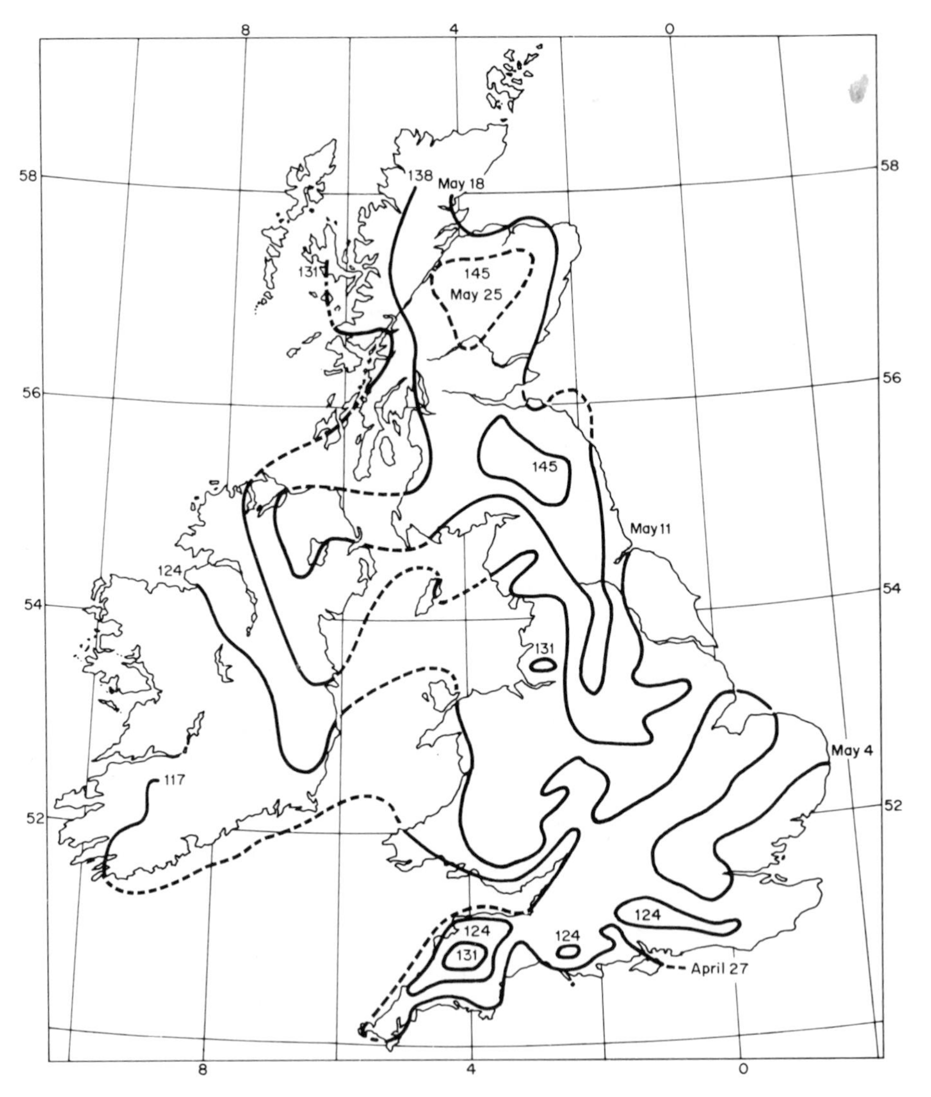

Figure 13.3 Isophanes of mean flowering dates, British Isles. Based on 12 selected plants for the period 1891–1925. (After Bush.)

likely to develop faster near a wall that stores and radiates heat. If shaded by the wall, however, the same variety may mature later. In such cases soil temperature is an important factor.

The rate of development of a plant variety is the resultant of all the environmental factors: climatic, physiographic, edaphic, and biotic. For a particular field under standard cultivation practices, it is primarily a function of climate, heat and light being the most important factors. A close relationship exists, therefore, between plant phenology and both latitude and altitude. It is well recognized that "spring moves northward" in the Northern Hemisphere and that "autumn moves southward." The frost-free season is progressively shorter with increasing latitude. In far northern latitudes the longer days in summer compensate for cooler temperatures to some extent so that certain crops can mature in approximately the same length of time as in mid-latitudes. Figure 13.4 shows graphically the phenology of the *Marquis* variety of wheat in several North American locations. The total time required from seeding to ripening of the wheat was 87 days at Fairbanks, Alaska (lat. 64° N), and 90 days at Lincoln, Nebraska (lat. 41° N). Obviously factors other than monthly temperature averages entered into the development of the wheat plants. Fairbanks has a mean July temperature of 15°C, whereas the comparable value for Lincoln is 26°C. The longer periods of daylight aided the maturity of the crop at the northerly latitudes and offset the lower temperatures. The importance of light in crop development also has been demonstrated in laboratory experiments.

At a given location the period between planting and harvesting is not a specific number of calendar days but rather a summation of energy units, which may be represented as *degree-days.* The duration of a certain temperature is quite as important as temperature averages. A degree-day for a given crop is defined as a day on which the mean daily temperature is one degree above the *threshold temperature* (that is, the minimum temperature for growth) of the plant. Some representative threshold temperatures are:

Spring wheat	0–4°C (depending on variety)
Oats	6°C
Field corn	12–14°C
Sweet corn	10°C
Potatoes	7°C
Peas	4°C
Cotton	17–18°C

Although it has proven to be a useful guide in agriculture, the degree-day concept has shortcomings. It does not differentiate between such combinations as warm spring–cool summer and cold spring–hot summer; it makes no allowance for unfavorably high temperatures nor for diurnal temperature ranges; and it neglects specific temperature values (such as those of leaf surfaces) or trends that may be

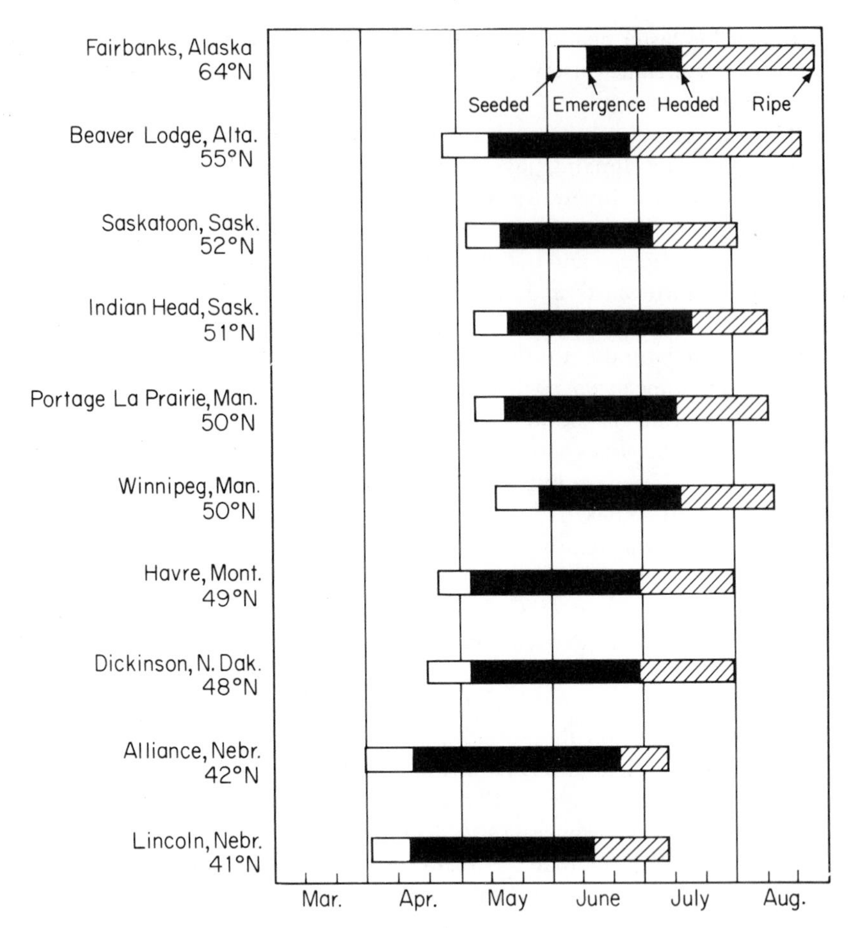

Figure 13.4 Chart of phenology of *Marquis* wheat at selected North American locations. (From M. Y. Nuttonson, "Some Preliminary Observations of Phenological Data as a Tool in the Study of Photoperiodic and Thermal Requirements of Various Plant Material," in A. E. Murneek, R. O. Whyte, et al., *Vernalization and Photoperiodism.* Copyright 1948, The Ronald Press Company.)

essential at various stages of crop development. During germination, for example, soil temperature at the seed depth is more significant than air temperature. Energy units expressed in terms of potential evapotranspiration or potential photosynthesis offer improvements in application of the fundamental concept to plants after they have emerged. Efficiency of growth can then be determined by comparing production of biomass with available energy.

The time required to achieve maturity is also a function of the length of day, or *photoperiod.* In general a crop planted early in the spring requires more calendar days to mature than the same crop planted later. In mid-latitudes the days are shorter in spring and temperatures are lower. For efficiency in the use of labor and equipment at planting and harvesting times it is expedient to make successive plantings of a crop so that the ensuing harvest can be spread across several days or weeks. In this way all the facilities for harvesting, processing, and marketing can be kept to a minimum and yet can be used for a maximum time each season. This is especially significant for vegetable crops that must be harvested immediately at maturity in order to maintain quality. Let us consider green peas as an example. Table 13.1 gives the phenological data for Alaska peas planted on different dates at College Park, Maryland, in 1926. Note that the peas planted on March 29 took 74 days to mature to the harvesting stage, whereas those planted on May 8 required only 51 days. The average length of the photoperiod was 14 hours during the entire period for the first planting and 14.6 hours for the last, and average daily temperatures increased as the season progressed.

Careful studies of crop and animal phenology over a number of years aid in planning management schedules that are best adapted to local climate, availability of labor, and market demands. Sometimes it is also possible to avoid the periods of maximum hazard from insects, diseases, or seasonal weather phenomena. In irrigated areas phenological data indicate the best time to plant in order to make the least demand on water supplies.

TABLE 13.1

Phenology of Alaska Peas at College Park, Maryland, 1926*

Date planted	*Date emerged*	*Date first blossom*	*Date of harvest*	*Total days*
Mar. 29	Apr. 14	May 13	June 11	74
Apr. 3	Apr. 16	May 14	June 11	69
Apr. 9	Apr. 21	May 17	June 14	66
Apr. 16	Apr. 28	May 21	June 17	63
Apr. 24	May 4	May 27	June 22	61
Apr. 30	May 8	May 30	June 22	54
May 8	May 16	June 6	June 28	51

*Data selected from Victor R. Boswell, "Factors Influencing Yield and Quality of Peas," in *Biophysical and Biochemical Studies,* University of Maryland Agricultural Experiment Station, Bulletin No. 306 (College Park: March, 1929).

FROST

The greatest agricultural risk associated with low temperatures is the threat of unseasonable frosts. Two kinds of frosts may be distinguished: (1) advection, or air mass, frost, which results when the temperature at the surface in an air mass is below freezing; and (2) radiation frost, which occurs on clear nights with a temperature inversion and usually results in formation of ice crystals on cold objects. The former, sometimes called *black frosts,* are more properly termed "freezes." Whereas air mass freezes are common in winter in middle and high latitudes, they are an agricultural problem primarily in relation to specific crops which are limited in their winter-hardiness. Actual plant damage may be the result of alternate freezing and thawing, frost-heave in the soil, or desiccation. Winter wheat, for example, can withstand freezing temperatures but is *winter-killed* if the roots are disturbed too much by frost-heave. If unusually severe, an air mass freeze can be disastrous in winter, but it creates a special hazard when it occurs in early autumn before plants have made the necessary physiological adjustments, or in late spring when field crops are in the seedling stage and trees and shrubs are budding or blossoming. Even mature crops can be reduced in quality by subfreezing temperatures. In the subtropics a severe freeze is regarded as unseasonable at any time of year. Millions of dollars in crop damage have been sustained in subtropical climates as a result of cold air masses. The citrus industry is particularly vulnerable. Because the temperature of air masses cannot be controlled on a large scale, not much can be done to forestall the general hazard attending air mass freezes. High-value crops may justify the use of covers or mulches which reduce the loss of soil heat. For production of most field crops, the only satisfactory solution to the problem of freezing is to avoid it as far as possible by planting after the danger is past in spring, and by selecting varieties which will mature before the renewal of the hazard in autumn.

Damage due to radiation frost differs from freeze damage in degree and in its spotty occurrence. Plants which are killed by a general freeze may be only partially damaged by frost, although the economic effects can be just as great. An entire fruit or berry crop may be wiped out by a single hard frost although the plants themselves are not necessarily killed. The hazard is greatest during critical stages of growth; for warm-climate crops, this means the entire growing season. Germinating seeds are not often affected by surface frost, but young seedlings may be killed unless they are of frost-hardy varieties. Crops like potatoes, tomatoes, and melons are vulnerable right up to maturity. The flowering stage is a critical period for most crops of field and orchard. In the spring-wheat belts of North America and Siberia, late summer frosts are destructive even after the kernels of grain have begun to form. Frosty nights followed by warm, sunny days produce a *sun scald* on orchard fruits, considerably reducing their value. Tree trunks are sometimes affected by the alternate freezing and warm sunshine also; orchardists whitewash the trunks to decrease the absorption of heat during the day. In this way the range of temperature experienced by the bark is reduced.

Avoidance of crop damage due to radiation frosts is more feasible than thwarting a general freeze. Preventive measures are based on a knowledge of the conditions for frost. These conditions are (1) a prevailing stable air mass with cool surface temperatures, (2) clear skies to allow loss of heat by radiation, (3) little or no surface wind to mix the cold air near the ground with the warmer air above, (4) a relatively high dew point temperature, and (5) topographic features which induce drainage of cold air into depressions. It is this latter factor which accounts for the spotty distribution of frost. Whether or not frost crystals will form on plant surfaces depends on the dew point of the air. If cooling takes place until the dew point is reached at a temperature above freezing, the formation of dew releases latent heat and retards further cooling. Below 0°C frost will form. This process also releases latent heat, which explains why light frosts of short duration may cause suprisingly little damage. Evaporation of water from soil or plants, especially after a previous rain, also serves to reduce the temperature at ground level, and it is frequently a contributory cause of frost.

As with freezes, a logical adjustment to the frost hazard is to avoid it. Selection of frost-hardy plants is one way. Another is to grow delicate crops on slopes above the level to which below-freezing air is likely to extend or where air drainage or winds can be depended upon to stir the air throughout the night. North-south valleys are in the shade earlier in the evening than those which open to the west and inversions therefore develop sooner and have a longer period in which to concentrate freezing air in the valley bottom. In mid-latitudes mountain slopes frequently exhibit a *thermal belt* of maximum nighttime temperatures between fairly well-defined levels (see Fig. 13.5). Below the belt, frosts developed by radiational cooling and air drainage are a hazard; above the upper limit, low temperatures conform to the normal lapse rate and inhibit plant development by day as well as at night.

Still another practice which has been successful in deciduous fruit orchards of the mid-latitudes entails retarding phenological activity by means of spray irrigation in the late winter. The application of cold water and subsequent evaporative cooling postpone budding and blossoming stages until frost danger is past.

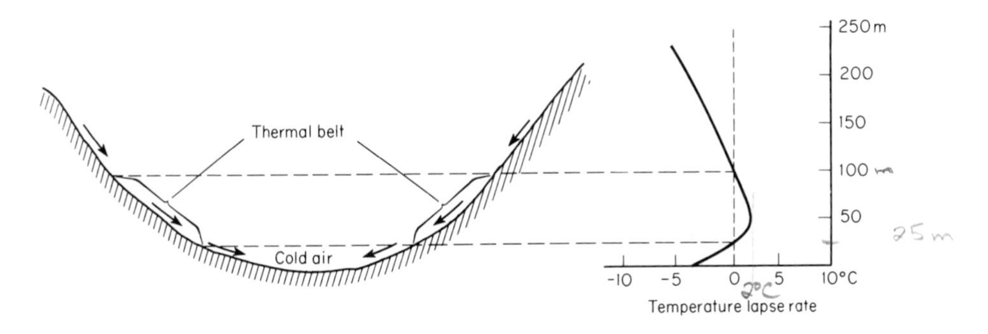

Figure 13.5 Thermal belt characteristic of some mid-latitude valleys.

FROST PREVENTION

Direct frost prevention measures are aimed primarily at breaking up the inversion which accompanies intense nighttime radiation. This is accomplished by stirring the air or heating it, methods which are effective only in stable air. For many years smudge pots were used to combat radiation inversions in citrus groves, supposedly to provide some heat at tree level and form a dense pall of smoke to reduce radiational losses. Cleaner and more efficient heaters are now common. Their direct heating often successfully alleviates the frost danger (see Fig. 13.6).

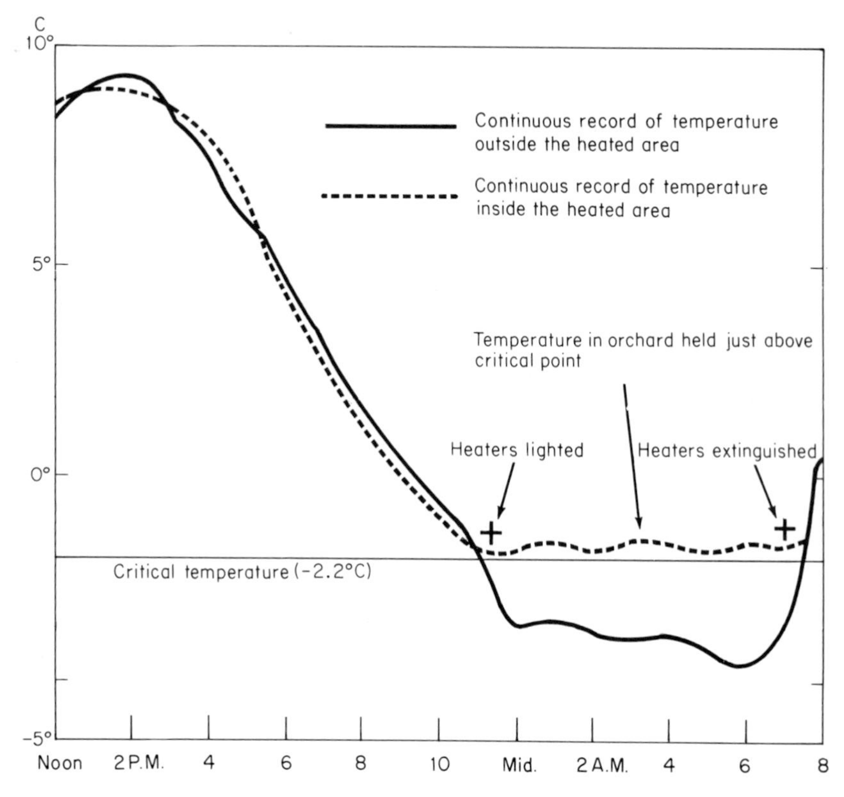

Figure 13.6 Temperatures inside and outside an area using orchard heaters to prevent frost damage. (NOAA)

Another method of mixing stable surface air is by means of huge fans, usually operated by electric or gas-driven motors. Fans cannot prevent damage from an invading air mass with freezing temperatures, whereas heaters may at least reduce the effect of the freeze if it is not too severe (see Fig. 13.7). Airplanes and helicopters have also been used to fan the air above crops, but this is an expensive procedure usually reserved for emergencies. Frost damage to wheat fields in northern states has been successfully prevented by the use of airplanes. The best-known frost prevention activities are those in the citrus areas of California and Florida, where groves are equipped with thermometers and heaters or fans (or both). But these methods are widely employed in fruit orchards throughout the mid-latitudes, especially at the time of spring flowering. Where only a light frost is expected, water sprayed over fruit blossoms has proven effective in ameliorating the hazard. As the water droplets freeze, the release of latent heat tends to maintain the temperature at or near the freezing point. In European orchards and vineyards, fine sprays have achieved protection against temperatures as low as −5°C. Experiments using artificial fogs composed of water droplets and emulsifying agents also show promise of providing a blanket to prevent heat loss.

Figure 13.7 Fan and heater for frost prevention in a lemon grove. (Sunkist Photograph.)

Orchards involve a long-term investment, and expensive measures often prove economically sound. One heavy frost can mean the loss of a whole year's crop; destruction of trees by freezing is an even greater catastrophe. Growers of winter or early vegetables can hardly afford to combat chance frosts by methods justified in orchards, however. Even after a hard spring freeze, vegetables can usually be replanted; if it is too late for one variety, another can be substituted so that the financial loss is not complete. Tender plants are commonly covered with paper or straw for frost protection during the early stages of growth; this is an extension of the principle of cold-frame covers for seedlings. Some plants can be successfully covered with soil by plowing if there is advance warning, although the covering and uncovering processes cause a certain amount of damage. Flooding with water has also been employed as an emergency frost-prevention measure. Cranberry bogs are often saved from frost in this way. In certain circumstances, proper drainage of surface and soil waters can be equally important in frost prevention as a result of the reduction of evaporation. Cultivation practices, such as weeding and mulching, also can inhibit frost.

In order for frost-prevention methods to be effective, it is necessary to assess the risk at the current stage of crop phenology and to have advance warning of the frost danger. Forecasts of general freezing temperatures associated with a cold air mass are based upon analysis of air masses and predictions of their movements and modifications in local areas. While this is by no means a simple task, radiation frost forecasts are even more difficult because of the great local differences in the frost hazard. In the United States federal–state cooperation has made frost warnings especially valuable in the citrus-producing areas, where nightly reports are broadcast during the winter. The frost histories of specific groves aid in making detailed forecasts of radiation frost. In a given locality the threat of damaging frost must be interpreted in relation to soil conditions and heat storage during the previous period, the stage of crop growth, topographic influences, the dew point, and wind, as well as the expected degree and duration of minimum temperatures below freezing. It is as important to know that frost will not occur as to know that it will, if the needless expense of preventive measures is to be avoided. Because of the spotty nature of radiation frost, every efficient farmer who faces a frost hazard has his property equipped with instruments to register local temperature, humidity, and wind so that he can determine the degree of danger for all parts of his orchard or cropland.

THE FROST-FREE SEASON

Climatic records usually derive the frost-free season from the number of days during which the temperature is continuously above 0°C, but for agricultural purposes the period between the last killing frost of spring and the first killing frost of autumn is a more useful index. The effects of freezing temperatures on the principal crops of a locality are the criteria for determining a killing frost, although crops vary in their frost-hardiness.

The dependability of a frost-free season is obviously important in agricultural planning. It is the occasional unexpected frost, especially in late spring or early autumn, that usually causes the greatest damage. Thus, the frost-free season that can be expected in a certain percentage of years offers a basis for choosing crops and calculating risk. For many crops it is better to operate well within the mean frost-free period to avoid risks that attend departures from the normal.

In general, the frost-free season decreases with an increase in latitude. Large areas within the tropics experience no frost except at high altitudes. The other extreme is at the poles, where there is no frost-free season; in the mid-latitudes numerous factors combine to produce wide variations both regionally and from year to year. For example, along windward coasts in the westerlies the frost-free period is longer than in continental interiors or on east coasts in the same latitude. In mountainous regions the seasonal regime of frost is highly complex.

The frost-free period is often equated loosely with the growing season, but if the latter is to have any practical meaning it must incorporate all of the climatic variables that affect crop growth and productivity. A previous section has reviewed temperature effects. In certain climates, especially in the tropics, the length of a growing season may be determined more by the availability of water to meet the needs of potential evapotranspiration than by temperature. The growing season of the dry summer subtropics clearly differs in character from that of the humid subtropics. At high latitudes the length of the daylight period and intensity of summer insolation compensate in large measure for the short frost-free season. The concept of the growing season thus requires careful definition for each crop in terms of its total climatic needs, including freedom from hail, violent winds, or other destructive phenomena.

THE MOISTURE FACTOR

Within rather wide temperature limits, moisture is more important than any other environmental factor in crop production. There are optimum soil moisture conditions for crop development just as there are optimum temperature conditions. Because crop plants obtain their water supplies primarily through their root system, maintenance of soil moisture is the most compelling problem in agriculture. Excessive amounts of water in the soil alter various chemical and biological processes, limiting the amount of oxygen and increasing the formation of compounds that are toxic to plant roots. The underlying cause of inadequate soil aeration may be poor vertical drainage as well as excessive rainfall. Therefore, conditions can be improved to some extent by drainage practices. On the other hand, a high rate of percolation of water through the soil tends to remove plant nutrients and inhibit normal plant growth. Cover crops and addition of humus to the soil help to alleviate this problem. Heavy rainfall may directly damage plants or interfere with flowering and pollination. By packing the surface soil, deluges inhibit normal emergence of tender seedlings. Grain crops are often beaten down, or *lodged,* by rain, making harvest difficult and promoting spoilage or disease. The effect of rain on the harvest and storage of grain and hay is a common problem. Special energy-consuming methods are employed to speed drying and prevent losses in storage. Wet weather during the later stages of maturity and harvest of cotton causes losses of both seed and fiber, often by providing favorable conditions for disease.

Snow and freezing rain are threats to wintering plants. The sheer weight of ice and snow may be sufficient to break limbs on trees and shrubs. A thick ice cover on the ground tends to produce suffocation of crop plants such as winter wheat. Hail is a special case of "excessive moisture" which causes direct damage to plants. Locally it may be a disaster, although over large areas it is a minor hazard compared to drought or the effects of low temperatures. The degree of damage depends on the stage of growth of the crop and upon the intensity of the hailstorm. Because hail is most common in the warm season, it frequently catches crops at a critical stage, pounding young plants into the ground, shredding leaves, or shattering flowers

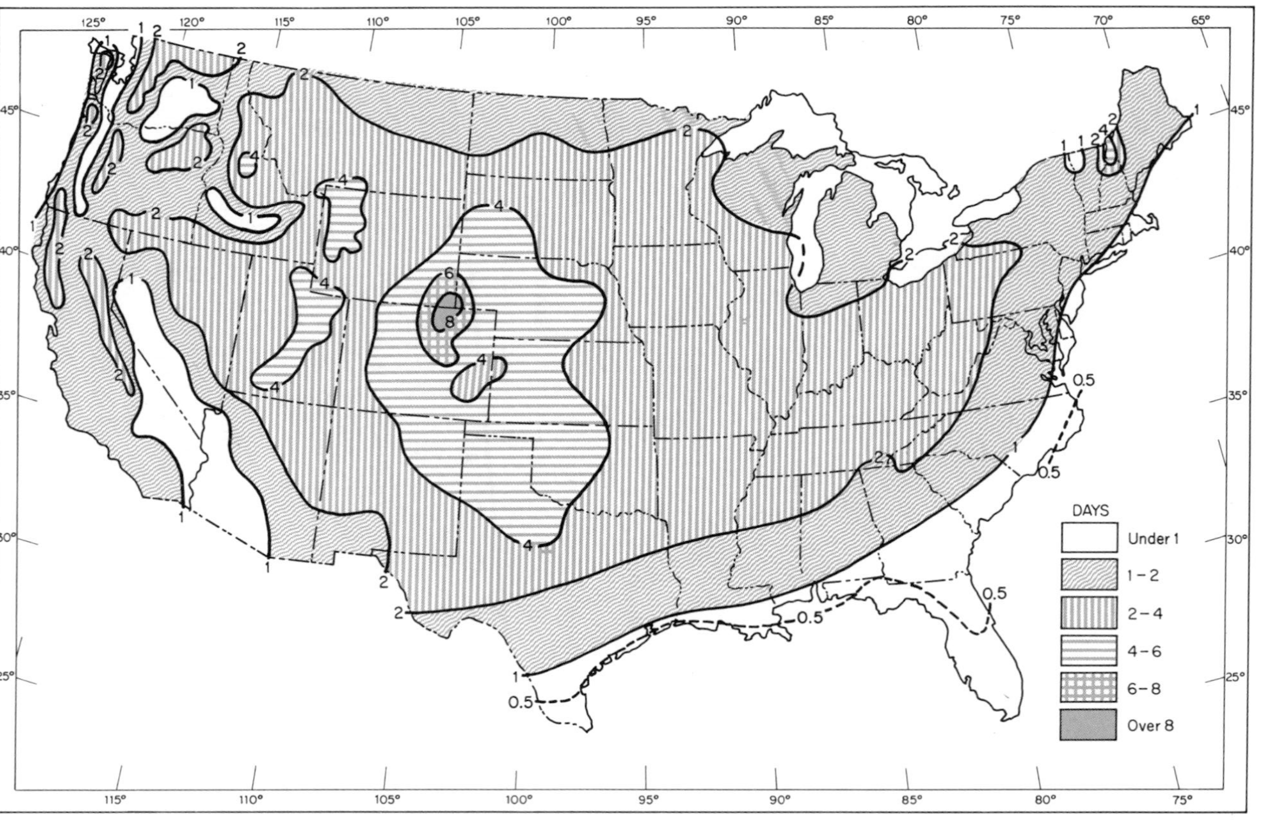

Figure 13.8 Average annual number of days with hail in the contiguous United States. (From John L. Baldwin, *Climates of the United States*. Washington: Environmental Data and Information Service, 1973.)

and seed heads. Figure 13.8 is a map of the average annual number of days on which hail occurs in the United States. The areas in which hail normally causes the most damage are in eastern Wyoming, eastern Colorado, western portions of Nebraska, Kansas, and Oklahoma, and in northern Texas. Many farmers in these and adjacent areas carry hail insurance on their crops. Premium rates are determined on the basis of weather records and hail damage records.

DROUGHT

Under natural conditions excessive moisture is far less an agricultural problem than is drought. *Drought* is the deficit that results when soil moisture is insufficient to meet the demands of potential evapotranspiration. Three classes of drought may be differentiated: (1) permanent drought associated with arid climates; (2) seasonal drought, which occurs in climates with distinct annual periods of dry weather; and (3) drought due to precipitation variability. In every case, the underlying cause of droughts is insufficient rainfall, although any factor that increases water need tends to aid in causing drought. Low relative humidity, wind, and high temperatures are contributory factors because they lead to increased evapotranspiration. Soils which lose their moisture rapidly by evaporation or drainage also augment drought. Drought does not begin with the onset of a dry spell; it occurs when plants are inadequately supplied with moisture from the soil. Thus, crops, growing on soils which have a high capacity for holding water are less susceptible to short periods of dry weather. Land-use practices which tend to increase runoff decrease vital soil moisture storage accordingly.

With these facts in mind, it is obvious that the incidence of drought cannot be determined alone from a map of average precipitation. In addition to the average amount, the seasonal distribution, dependability, intensity, and the form of precipitation must also be known. Furthermore, different crop plants have different moisture requirements. Ultimately, then, drought must be defined in terms of the *water need* of a particular crop growing under a specific combination of environmental conditions. If the minimum water need is not met for these conditions, the plants do not develop and mature properly.

Some climates have rainfall equal to or greater than the potential evapotranspiration in all months. The water budget graph for Zurich shown in Fig. 13.9 illustrates this favorable balance. In most areas monthly precipitation does not regularly exceed water need, however, and for part or all of the year moisture is deficient. In the mid-latitudes the deficiency occurs in summer, when a larger energy supply accelerates evapotranspiration. At Tokyo (Fig. 13.9) precipitation during the first half of the year is in excess of water need. By early summer crops must depend on stored moisture and actually may suffer from a small moisture deficit. As the available energy declines and the autumn rainfall maximum approaches, there is a recharge of soil moisture and a renewal of water surplus. Climates having distinct summer maximums of precipitation may, never-

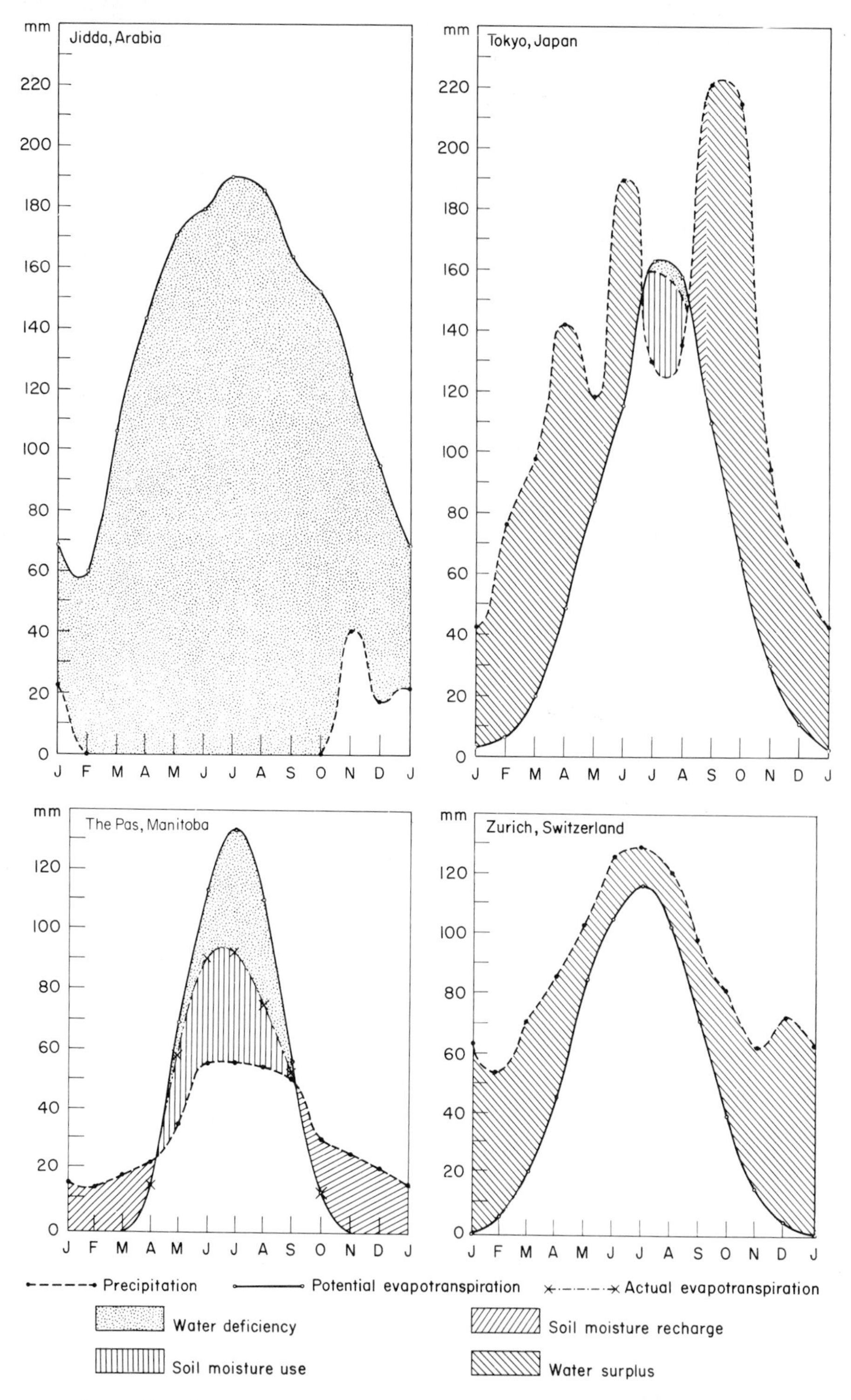

Figure 13.9 Contrasting water budgets for selected stations. (After method of C. W. Thornthwaite, based on data computed by C. W. Thornthwaite Associates, Laboratory of Climatology.)

theless, experience summer water deficits. Such is the case at The Pas, a station on the northern fringes of agricultural production in Manitoba (Fig. 13.9). Although precipitation in summer months is several times that in winter, the longer photoperiod and higher energy income create water need in excess of rainfall and stored soil moisture. The meager winter precipitation, mainly snow, never succeeds in charging the soil to its full capacity. A degree of drought therefore prevails throughout the growing season.

Regions of permanent and seasonal drought coincide broadly with the world pattern of climates. Arid climates experience permanent drought, as the water budget for Jidda, Arabia, indicates (Fig. 13.9). Climates having well-defined seasonal minimums of rainfall are likely to have corresponding seasonal droughts, although their economic significance depends to some extent on whether they coincide with the growing season. In the dry summer subtropics, the dry season comes at the time of maximum energy availability. In the wet-and-dry tropics winter is the dry season, and summer cropping can be carried on if soil moisture is recharged in time. A delay in the onset of the rainy season can be disastrous; some of the worst famines of southern Asia have resulted from a delay of the summer monsoon. Variability of precipitation creates the worst drought hazards, especially in semiarid climates, where a slight departure from the mean may be the critical factor in crop failure. Even in humid climates true drought can reduce crop yields, although it may not wipe out harvests entirely. There are degrees of drought and consequently of drought damage; the effects of moderate drought are not always exhibited in such obvious forms as withering plant leaves but may appear instead in lowered quality or yield.

Prolonged droughts alter the pattern of agricultural land use on a major scale. In the past they have caused migrations from affected areas, for example, the exodus from the Great Plains in the 1930s or from the Sahel region of northern Africa in the 1970s. Several years of abnormally dry weather are usually accompanied by above-average temperatures that accentuate the drought conditions. The water budget graphs in Fig. 13.9 represent average conditions over a number of years. They cannot show the haphazard occurrences of dry spells that plague the farmer in normally wet seasons or in humid climates. Nor do they reveal year-to-year fluctuations in precipitation and therefore the drought hazard. These phenomena are, nonetheless, significant for agriculture. Numerous indices that employ the concept of the water budget have been devised to identify and quantify drought. When applied to daily or weekly values of precipitation, soil moisture, and measured evapotranspiration, the water budget approach provides an accurate running account of real or potential drought for a particular crop (see the discussion of water budgeting on pp. 252–53). Since these data are not easily obtained, it is often expedient to derive an index from more generally available observations of temperature and precipitation. A method developed by Wayne C. Palmer in the United States uses weekly total precipitation and mean temperature to compute evapotranspiration deficits or surpluses, which are compared with data from previous weeks and long-term periods to derive a crop-moisture index. Relative conditions are ex-

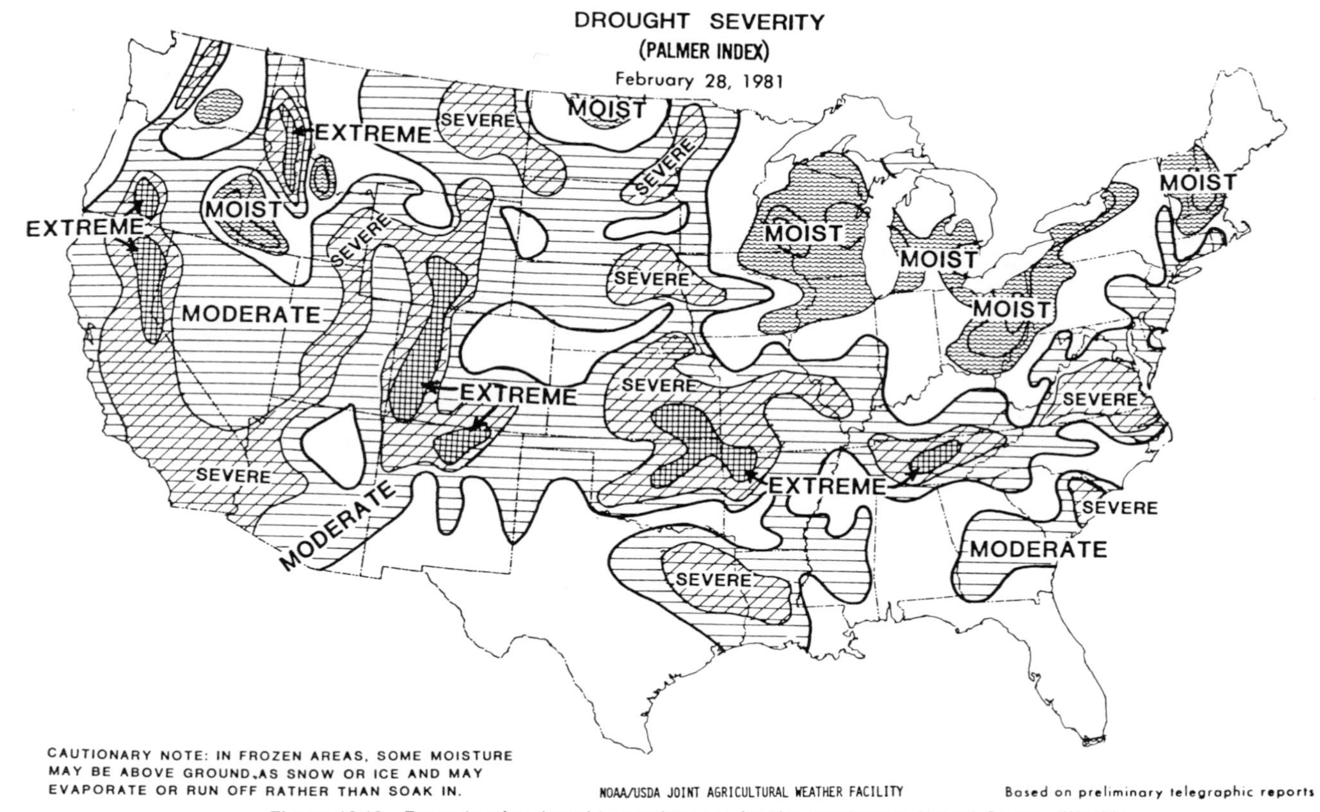

Figure 13.10 Example of a drought severity map for the contiguous United States. *(Weekly Weather and Crop Bulletin,* March 3, 1981, p. 4.)

pressed by index values ranging from + 4 or above (extremely moist) through zero (normal) to −4 or below (extreme drought). An example of a weekly drought severity map for the contiguous United States is shown in Fig. 13.10. Maps of this kind, based on weekly calculations of the Palmer Index, aid interpretation of both abnormally dry and abnormally wet conditions in terms of agricultural practices such as planting, cultivation, and irrigation. They also provide an objective basis for planning drought-relief and governmental proclamations of drought disaster.

COMBATING DROUGHT—IRRIGATION

As we have seen, drought is a condition where water need is in excess of available moisture. Prevention of drought damage to growing crops, then, is a matter of either (1) decreasing the water need of crops or (2) increasing the water supply, or possibly a combination of the two. Planting of crops that have low water demands helps reduce the water need. Cultivation practices which improve the soil structure and inhibit runoff are effective, but limited, drought-prevention measures. Weed control is especially important if the available water is to be used most effectively for crops, for weeds accelerate water loss by transpiration at the expense of soil moisture.

In subhumid and semiarid climates *dry farming* methods depend on the conservation and use of two—or sometimes three—years' rainfall for one year's crop. During the period when a field lies fallow, it is cultivated to kill weeds and to create a soil structure that will retain moisture. Thus, soil moisture stored during the fallow period supplements the meager rainfall in the crop season. In the case of seasonal droughts, the planting schedule often is adjusted to permit maturity and harvest before the effects of the dry season become too great. This is possible only if the temperatures are high enough in the wetter season.

Wherever the water need of crops or grassland cannot be reduced to conform to the moisture supply, the only alternatives are to abandon agriculture or to provide water artificially. At best, the artificial stimulation of rainfall has rather narrow limitations for supplementing natural precipitation. On the other hand, irrigation is a common method for providing all or part of the water need of crops (see Fig. 13.11). In arid regions, or where cropping must be confined to a warm, dry season, agriculture is possible only with irrigation. In semiarid and subhumid climates, irrigation makes possible larger yields and a greater variety of crops. It also lengthens the period during which land can be used productively and makes yields more consistent from year to year. In humid regions, its main value is supplementary in times of drought or to meet the special demands of certain crops, for example, rice. The chief limitations on irrigation are the availability of water from surface- or ground-water sources and the cost of getting it to the fields. Within these limits, irrigation has the advantage that it can be regulated as an element in the water budget to meet the variable demands of different crops, different seasons, or chance droughts.

Figure 13.11 Sprinkler irrigation of alfalfa in western Montana. Evaporation claims part of the water before it can be effective in plant growth. The fractocumulus clouds are unlikely sources of precipitation. (Bureau of Reclamation, U. S. Department of the Interior.)

Ranging between the obvious necessity for irrigation in deserts and the almost total absence of need for it in constantly wet climates are the variable circumstances that determine the amount of irrigation needed and when it should be applied. The total amount of water required for crop production is equal to what is used in transpiration and evaporation, plus what is lost by percolation below the root zone and unavoidable forms of waste incurred in the irrigation process or in rainfall runoff. Evolving techniques of drip (or "trickle") irrigation which deliver water directly to the root zone reduce evaporation and percolation losses markedly. The proportion of water that must be provided by irrigation depends on rainfall. In general, irrigation is desirable whenever the soil moisture storage in the root zone drops to about 40 percent of capacity. In any case, to be effective, water must be applied before conditions reach the wilting point.

The principal advantage in using either soil moisture measurements or evapotranspiration as an index of irrigation needs lies in the resulting efficiency of water use. It is wasteful to irrigate before water is needed or to apply too much water. The absence of rainfall for several days may or may not be an indication of the need for irrigation. Temperature is likely to be a more important factor, since it directly affects evapotranspiration.

In summary, irrigation may be feasible in any climate where water need exceeds precipitation for a long enough period to reduce crop yields. Irrigation is the farmer's best answer to drought. Where irrigation water is made available, temperature becomes the dominant climatic factor controlling crop distribution and yields.

CROPS AND WIND

Wind has its most important effects on crop production indirectly through the transport of moisture and heat in the air. Movement of air increases evapotranspiration, but the effect decreases with increasing wind speed and varies among plant species. Moderate turbulence promotes the consumption of carbon dioxide by photosynthesis. Wind may speed the chilling of plants, or, on occasion, prevent frost by disrupting a temperature inversion. Wind dispersal of pollen and seeds is natural and necessary for native vegetation and may be helpful for certain crops, but it is detrimental when weed seeds are spread or when unwanted cross-fertilization of plants occurs. Continuously strong winds interfere with the pollination activities of insects. Direct mechanical effects are the breaking of plant structures, lodging of hay and cereals, or shattering of seed heads. Fruit and nut crops may be stripped from the trees in high winds. The unfavorable consequences of soil depletion by any form of erosion should be obvious. Low plants are sometimes completely covered by windblown sand or dust; abrasion of plant stems and leaves by sand particles is often associated with wind erosion. Along the shores of salt lakes and oceans, salts transported inland by the wind affect both plants and the soil.

Practices to avert the effects of wind on evapotranspiration are essentially those, including irrigation, employed to combat drought. Mechanical damage due to wind can be lessened somewhat by making use of natural or artificial shelter. Protected valleys and lee slopes are suitable for some types of crops which are easily damaged by wind. Windbreaks composed of trees, shrubs, hedges, or fences are widely used to protect both crops and animals from the wind. Some plants require only temporary protection which can be provided by screens or individual windbreaks. An example of the latter is the use of boards or metal cylinders to shelter young tomato plants. The best permanent windbreaks are rows of trees planted perpendicular to the prevailing winds (see Figs. 13.12 and 13.13). Their moderating effect is felt for a distance equal to several times their height. Where winter snow is common, they have the added advantage of inducing snowdrifts over the adjacent fields. As a widespread measure to combat wind, they have accompanying disadvantages, however. They reduce the area of cultivated land, compete

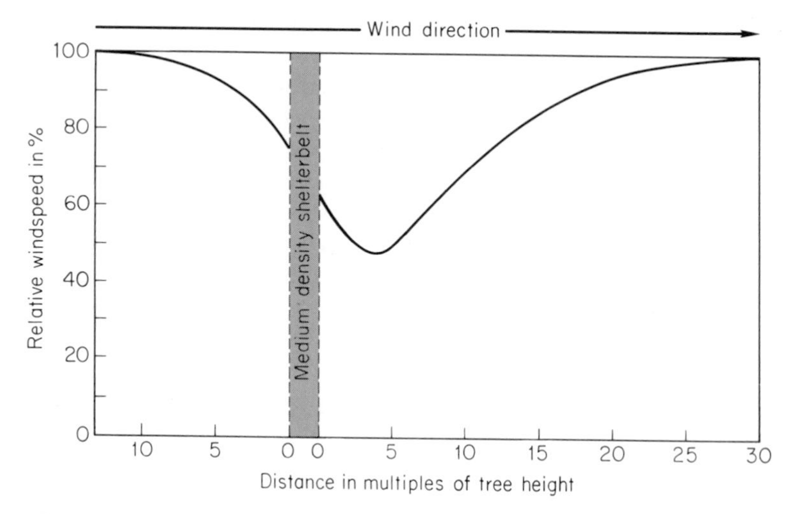

Figure 13.12 Relative wind speeds in the vicinity of a medium-density shelterbelt. (After Staple.)

for soil moisture, and may produce harmful shade. The first of these objections can be overcome if the species can be selected and managed so as to provide a compensation in the form of harvest from the windbreak. If too dense, windbreaks generate turbulence that may nullify their intended effect.

Figure 13.13 Shelterbelts on the Canterbury Plains, New Zealand. The rows of trees, mainly exotic conifers, reduce the effects of a dry northwesterly foehn wind on soil, crops, animals, and farmsteads. Note the alignment perpendicular to the valley. (New Zealand Department of Scientific and Industrial Research.)

Domestic animals are highly dependent on the availability of feed. The climatic factors that affect pastures or feed crops therefore exert an indirect influence on livestock. The suitability of a particular breed of animal to a climate depends on the quality and quantity of feed which is available naturally or which can be grown in that climate quite as much as on the physiological adaptation of the breed to the climate. Generally, it is economically feasible to transport feedstuffs into areas deficient in grain or forage crops only where the demand for an animal product is great or when there is a temporary shortage of feed owing to climatic or other factors. Fluctuations in animal productivity frequently result from variations in the feed supply rather than from the direct effects of climatic elements on the animals. Thus, there may be a decrease in milk or meat production during a drought because the amount and nutritive quality of pastures are impaired. Most animals are adapted to a wide range of climatic conditions provided that their feed and water requirements are met. There are, however, direct climatic effects on the normal body functions of animals. All breeds have optimum ranges of climate for maximum growth and development. When removed to a much different climate they do not ordinarily die as plants often do, but they may fall below minimum economic levels of production.

The climatic elements which affect livestock indirectly through the feed supply are those which influence plant growth or the spread of insects and diseases. Those having direct effects are temperature, light, precipitation, relative humidity, atmospheric pressure, wind, and storms. High temperatures generally reduce production (Fig. 13.14) and reproductive capacity. Dairy cows produce less milk when exposed to high temperatures; the optimum is about 10°C. Hens lay larger eggs in winter and at high latitudes. Many animal breeds consume less feed under hot conditions and therefore do not produce fat and flesh as rapidly. Under extreme heat stress they die. Mid-latitude breeds show a marked decline in fertility when relocated in tropical climates. Extreme cold lowers production also, for too much body energy is required to combat chilling. Where seasonal contrasts are great, most domestic animals adjust to winter by undergoing certain physiological changes, notably the growth of thicker coats. Long exposure to cold, especially if it is accompanied by wind, may cause frostbite or death. Another problem associated with cold weather is the water supply; domestic animals do not satisfy their water requirements from ice or snow.

Sunlight and the duration of daylight influence animals in several ways. Breeds of cattle and pigs with light-colored skins are occasionally sunburned under intense sunshine. The daily feeding period of most classes of livestock is determined in part by the length of day. Cattle normally graze more in shaded locations than in the sun and rest during the midday periods of highest temperature and strongest sunlight. Weight gains of calves and lambs and milk production by dairy cows increase when the winter photoperiod is extended by supplementary lighting, even without increased consumption of feed. Fertility is also affected by the duration of

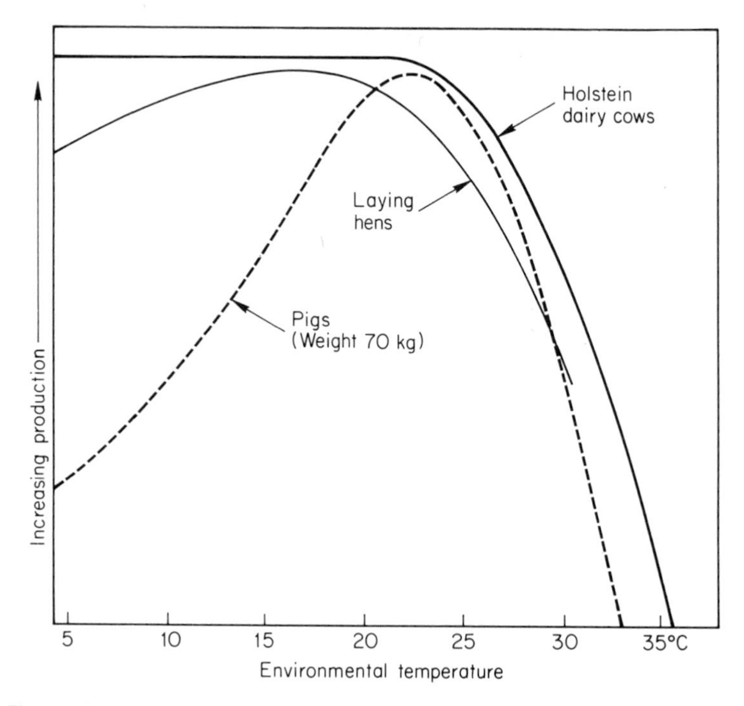

Figure 13.14 Production trends of three farm animals under varying environmental temperatures. (After Bond and Kelly.)

daylight, especially in poultry. Commercial poultry farmers commonly supplement lighting to increase egg production.

The primary influence of precipitation on livestock is through its effect on feed. If it is associated with cold weather, it tends to accentuate the detrimental effects of the low temperatures. Freezing rain can cause much discomfort to say the least. Heavy snow makes moving and grazing difficult or impossible for range animals. Crusted snow or ice may cut their feet and a heavy cover of snow over small animals can cause suffocation. In blinding snowstorms, cattle and sheep often pile up against barriers, where they are trampled or die of suffocation, or they may stampede over precipices. Ranchers who use mountain pastures in summer are well advised to move their livestock to lower elevations before the time of possible autumn snowstorms. For the same reason, too-early grazing of mountain pastures in spring must be avoided.

High relative humidity, whether associated with abundant rainfall or not, influences respiration and perspiration in animals. Very dry air may cause discomfort, but it is ordinarily of less importance than high temperature in a drought provided that feed and water are available. Low relative humidity, wind, and high temperatures all increase the water requirements of animals. Even the camel can go but a few days without water under the extreme heat and dryness of the tropical deserts in summer. Wind may have good or bad effects, depending on wind speed and the accompanying temperatures. Moderate breezes ameliorate discomfort and ill effects caused by high temperatures. On the other hand, high winds increase drying, may fill the air with dust and sand, or intensify the impact of precipitation. Moving cold air chills animals much more rapidly than still cold air of the same temperature.

Changes in atmospheric pressure appear to have some effect on animals. The relatively large pressure differences accompanying changes in altitude are especially significant. For example, the fur-bearing chinchilla had to be "acclimated" to higher pressures in successive stages in order to transfer breeding stock successfully from the high altitudes in the South American Andes. Probably the lesser pressure changes which occur with cyclonic storms are felt to some degree by animals. Decreasing pressure may be one of the stimulants to intensified feeding prior to the arrival of storm centers or fronts.

The climatic elements are combined in storms and they do not act separately on animals. Wind, temperature, and precipitation are the three most important elements. In thunderstorms, lightning may be added to the hazards.

Many of the practices in animal husbandry designed to lessen the negative effects of climate depend on the use of natural or artificial shelters. Specific problems such as the control of temperature can be solved to some extent by heating and air conditioning. Experiments with dairy cattle in the tropics have shown that milk production can be increased by cooling stables with an air-conditioning system. Heating is widely employed in farrowing pens for sows and in brooder houses for poultry. Well-built barns and sheds help to conserve animal body heat in cold climates and in winter; in many instances this is sufficient. Overheating of buildings generally has a detrimental effect. Heating of dairy barns in winter results in reduced milk production. It is usually impractical to provide buildings for large numbers of range livestock, which are best moved to sheltered areas where feed and water are available. Windbreaks of trees and shrubs afford shelter to animals kept in the open.

INSECTS AND DISEASES

Many of the restrictions on productivity and regional distribution of both plants and animals are pathological, that is, they are caused by insects or diseases. Climate nevertheless exerts an indirect influence, for insects and diseases have rather narrow climatic limits for maximum development. A favorable season for enemies of crops can result in a major catastrophe. The infamous potato famine of 1848 in Ireland, resulting from a combination of excessive rainfall and a fungal blight that flourished in the wetness, led to one and a half million deaths and large-scale emigration. Even short-term weather conditions may determine the extent of development and spread of a plague. That the relationships of insects and disease to climate are complex can be seen from the fact that weather conditions sometimes favor parasites or maladies which make inroads on insects. Moreover, the food supply of insects is controlled to varying degrees by weather and climate. Various types of plant enemies, such as mildew, rusts, scabs, and blights, reproduce and spread most rapidly under conditions of warmth and high humidity. Spores of fungus diseases are spread by the wind, making control difficult. Wheat, barley, and certain legumes are largely excluded from humid climates because of disease problems rather than direct climatic factors. Coffee is confined to cooler uplands in the tropics primarily because of disease problems in the hot, wet lowlands. Many mid-latitude crops are unsuited to the tropics for pathological reasons, even though the climate might appear to be directly favorable. A great many diseases and insects which attack plants are kept in check by seasonal fluctuations in temperature or moisture or both. For example, in the areas of the southeastern United States which are infested with the cotton-boll weevil, it has been estimated that 95 percent of hibernating adults die during the winter. Thus, the winter has a marked bearing upon the incidence of boll weevil damage in the summer. The incidence of diseases and insects harmful to plants is considerably less in high latitudes; this is due to the absence of host plants as well as to low temperatures. In other words, the same climatic factors that restrict diseases and insects often limit crop production as well. Wind influences the migration of insects, especially if the air is warm. The direction of migration is with the prevailing wind, and upper winds may be dominant in the distribution, for flights of insects sometimes reach heights above the surface winds.

The density of airborne insects is much greater near the surface when the air is relatively calm, but turbulence carries them upward. In North Africa and Southwest Asia locust swarms are associated with the ITCZ. Rain along the convergence zone provides moisture required for breeding as well as for vegetation upon which the larvae feed. The northerly and southerly winds tend to carry the swarms toward the areas of most favorable ecological conditions (see Fig. 13.15).

Control of insects and diseases which damage plants belongs in the province of plant pathology rather than climatology. Any agricultural practice which provides an unfavorable environment for a crop pest has potential value for controlling it. Forced exposure to extremes of temperature, sunlight, or moisture conditions is one of the most practical approaches to the problem.

Insects and diseases which attack animals have a complicated relationship to climate and other environmental factors. Of the climatic factors, temperature and moisture are the most important, and as a result, there are broad correlations between infestation and both climatic regions and seasons. As in the case of crops, animals generally have fewer pathological enemies in the colder climates. Probably the largest class of malefactors in the livestock industries are internal parasites. Climate and daily weather influence the distribution, rate of spread, and intensity of infestation of parasitic diseases. While developing in that part of the life cycle which is outside the animal, most parasites have limiting ranges of temperature and moisture requirements. Exposure to direct sunlight is usually fatal to eggs and larvae. Thus, control can be achieved to some extent by measures which enhance the destructive effects of climatic conditions on the eggs and larvae of parasites or on their intermediate hosts. Rainfall, wind, and sunlight render valuable assistance in

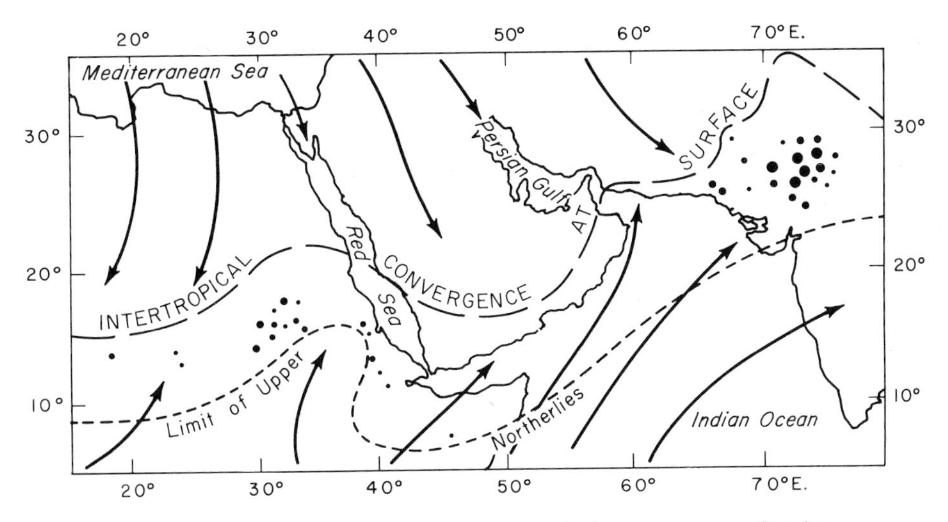

Figure 13.15 Locust swarms in relation to the intertropical convergence zone. Dot size is approximately proportional to the number of swarms reported during a period of 20 days. (After Rainey and Bodenheimer.)

maintaining sanitary conditions in pastures and feedlots. Their aid can be put to the best advantage by rotating animals from one enclosure to another.

It is difficult to separate the discussion of insects from that of diseases because, very often, an insect species is the carrier of a disease. Numerous insects, of course, attack animals directly. One can cite those which sting, chew, suck blood, lay eggs in the hair or skin, or merely swarm menacingly about the animal. Among the many insects that infect animals with disease, an outstanding culprit is the tsetse fly, which is limited to tropical Africa. There are several species of tsetse, none of which can endure low temperature nor extremely hot, dry weather. They live solely on blood, and, since they are disease vectors, they infect the animals on which they feed.

SELECTION AND BREEDING FOR CLIMATIC ADAPTATION

Where a suitable environment cannot be provided economically for a specific plant variety or animal breed, the alternative is to select or breed other types that are adapted to prevailing conditions. This approach has contributed to increased yields and the expansion of agriculture into new regions since the beginning of cultivation and animal husbandry. Modern plant breeding aims to improve yield, quality, resistance to pests, and adaptation to climatic factors, including short-term weather vagaries and longer-term climatic fluctuations. It also increasingly must meet the needs imposed by the changing technology of cultivation, harvesting, and processing; but considerations of climate remain paramount. The difficulty of achieving all objectives simultaneously increases with higher yields, more intensive farming, and the extension of cultivation into areas that are marginal with respect to climate or soil. A genetic strain which is bred to overcome one negative factor may be susceptible to another that was not evident previously. For example, a high-yielding wheat variety may succumb to depredations by fungi, or it may be vulnerable to frost or wind. Nevertheless, the success of hybrid grains, early maturing vegetables, rootstock grafting, and other advances in the "Green Revolution" reflect the benefits of plant selection and breeding.

Food production has also been increased by selection and breeding of domestic animal types that are adpated to a particular climate and the associated disease or insect problem. The Merino sheep is well suited to semiarid climates and tolerates a wide range of temperatures. English mutton breeds, on the other hand, prefer cool, humid climates. Northwest European dairy breeds such as the Holstein and Guernsey produce best in the mid-latitudes; they lack immunity to certain tropical diseases. The Jersey is better adapted to warm conditions than other common dairy breeds. Most large animals make physiological adjustments to prolonged hot weather, but rates of acclimation vary among breeds. Figure 13.16 shows the changes in body temperature and water consumption in Shorthorn cattle subjected to constantly high temperatures for an extended period. By selecting individuals that exhibit desired qualities of tolerance, it frequently is possible to breed lines that are better adapted to a given range of climates.

Crossbreeding of animal types to combine the desirable characteristics of the foundation stock has improved production among several classes of livestock. Since the heat tolerance of European cattle breeds is low, crosses with tropical breeds have been developed in the tropics and subtropics. The Santa Gertrudis is a cross between the Brahman and Shorthorn. Dual-purpose sheep breeds such as the Columbia and Corriedale produce high-grade mutton and good-quality wool under varied climatic conditions. Crossbreeding has also improved the lard-producing characteristics of hogs in hot climates.

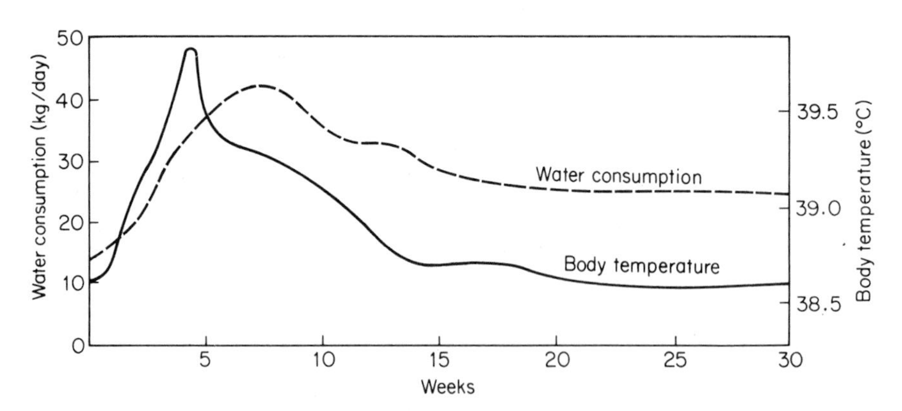

Figure 13.16 Trends of mean body temperature and water consumption by Shorthorn cattle exposed to 32°C temperature and 29 mb vapor pressure for 30 weeks. (Adapted from Dowling, 1968.)

Development of disease resistant types through selection and crossbreeding has been another goal of animal scientists. Some breeds have natural immunities, and it often is possible to combine disease resistance with high productivity. Among tropical breeds of domestic animals, there is commonly some degree of tropical disease immunity as well as heat tolerance, so that crossbreeding has improved both climatic adaptation and disease resistance of high-producing mid-latitude breeds in the tropics and subtropics.

PLANT AND ANIMAL INTRODUCTION

The adaptation of crops and livestock to certain climatic optima is of great significance in connection with the introduction of a particular variety or breed into a new area. To the extent that climate influences the productiveness of crops or livestock, transfer of plants or animals to another region with a similar climate is likely to be successful. Tropical tree crops have been transplanted literally around the world. Coffee, believed to have been native to the highlands of eastern Africa, is now widely grown in tropical Latin America. The native rubber tree of Brazil has met with more favorable circumstances for commercial production in Southeast Asia than in its original habitat. Cacao, also indigenous to the Latin American tropics, is produced in the Guinea Coast region of Africa as well as in South and Central America. The introduction of such Mediterranean crops as citrus fruits, olives, and wine grapes into other dry summer subtropical regions is an example of migration of crops to similar climates. There is a huge list of crops which originated (or at least were acclimated) in Europe, and which have been successfully carried into other mid-latitude climates by European emigrants. Nowhere is there a more striking example of wholesale transfer of plants and animals into a similar climate than the establishment of British agriculture in New Zealand. Breeding and selection play an important part in plant and animal introduction, as do economic and cultural factors. But the suitability of a climate for a particular crop or animal is a primary consideration which is not altered by the fact that many crops and animals have a wide range of adaptability to climate.

Whereas many crop plants have been introduced into new areas in the past largely on a trial-and-error basis, recent efforts have been directed toward detailed study of agroclimatic relationships to improve yields, quality, or disease resistance. As world agriculture has expanded and the demands for food have increased, plant introduction has come to mean more than the establishment of a new crop; it entails new varieties which have better characteristics for productivity in a particular environment. Varieties that have the desired characteristics for a given locality are sought in other areas where the climate is analogous. Regions having climates sufficiently alike to permit equivalent productivity of the same crop variety or crop combination are *agroclimatic analogs*. In the past, simple temperature and rainfall data have dominated studies to determine analogs. As our knowledge of the relationships between plant physiology and climate improves, the analysis of regional climates for the purposes of plant introduction increasingly will use complex heat and water budget approaches as well as single factors such as radiation, light, or wind. In effect, the problem is one of applied climatic classification to achieve a specific objective.

Animals, being far more adaptable to variations in climate than plants, do not present so great a problem in connection with their introduction into a new region. Within broad limits, climate influences animal introduction more through its effects on feed supplies than by direct means. When climates with sharply contrasting temperatures are involved, however, failure to acclimate and to maintain satisfactory production levels may present a serious obstacle to livestock introduction. As stated earlier, crossbreeding and selection have overcome this problem to some extent.

AQUACULTURE

Fish provide only a small fraction of the total world food supply; yet they contribute a significant share of the protein in the diet of many people. The worldwide problem of hunger and malnutrition has stimulated continuing re-examination of aquatic production as a partial solution. The final section of Chapter 12 summarized some aspects of climate in relation to marine life. Rather than review fundamental concepts at length, the emphasis here will be on aquaculture, sometimes termed mariculture when practiced in ocean waters.

Aquaculture, like agriculture, entails maintenance of a special environment for production of organic matter. As in the case of natural aquatic life the principal climatic influences are exerted by light, temperature, and wind. Other factors are water quality, nutrient supply, and undesirable organisms, all of which react to the climate system at least indirectly. The principles of optimum conditions for reproduction and growth, varying among species and different combinations of environmental factors, are applicable. Because the production medium is confined mainly to fresh water bodies or shallow coastal waters, environmental changes may occur rapidly. An objective of aquaculture is to reduce or eliminate harmful effects, while taking advantage of favorable natural conditions. The methods are analogous in many respects to those for crop and animal production, including control of light, temperature, and water movement; aeration and filtration; fertilizing or feeding; selection, breeding, and transplanting; and protection from diseases and preying or competing organisms.

These practices obviously are more feasible in partially enclosed habitats than in the open ocean. They have a long history of success in the rice paddies of southern and eastern Asia and certain other fresh or brackish waters of the tropics, and they have been extended into subtropical and mid-latitude climates (Fig. 13.17). Energy requirements in cooler climates have tended to impose economic restrictions on fish

Figure 13.17 Commercial trout harvest in Michigan. Aquacultural productivity extends beyond the growing season for mid-latitude crops, but it is affected by weather extremes throughout the year. (Soil Conservation Service—USDA.)

farming. Only a massive effort could establish true mariculture in the open ocean. Nevertheless, an intermediate stage of "fish herding" or transplanting offers potential for increasing oceanic production if due attention is given to climatic and other limitations. For example, the Pacific oyster reproduces in the coastal waters of Japan but not in the cooler Puget Sound of Washington State, where commercial oyster beds depend on "seed" imported from Japan.

Because bivalve molluscs are sedentary, they are more easily cultivated than most fish, but they are also less able to avoid desiccation or changes in temperature and water quality. Several species of oysters, clams, and mussels are adapted to mid-latitude marine environments. Proposals to extend their production have included the use of waste industrial heat to warm shallow coastal waters. Swimming species such as carp and catfish readily adapt to pond culture and are most productive in warm humid climates.

QUESTIONS AND PROBLEMS FOR CHAPTER 13

1. How do the factors that influence crop production differ from those which affect natural vegetation?
2. Cite three examples to illustrate the application of phenological data in agriculture.
3. Account for the fact that certain crops mature in fewer days in Finland than in Bulgaria.
4. Explain the practicable methods of frost prevention and the physical processes involved in each.
5. As a consultant to a hail insurance company, how would you adjust premium rates to compensate for risk in different parts of the United States?
6. How might the energy budget concept be applied to estimate feed, water, and housing requirements of domestic animals?
7. What criteria would be appropriate for a climatic classification that is intended for use in plant and animal introduction?
8. Referring to the water budget for Memphis (Table 11.3), determine the amount of irrigation water that should be applied in an "average" July to restore soil moisture to the minimum desirable percentage of capacity.
9. Why are precipitation data inadequate for defining degrees of drought? What other factors must be considered?
10. Explain the role of climate in the world distribution of wheat, coffee, and dairy production.
11. Compare the utility of energy and water budgets in aquaculture, noting variations among climatic types and different species.

14

Climate, Energy, and Industrial Technology

Every economic activity depends on energy, and the production or consumption of energy in any form is subject to the effects of weather and climate. Aspects of these fundamental relations with respect to food were presented in Chapter 13. Construction, transportation, and manufacturing suggest other problems associated with energy resource development and energy use. Application of climatology and meteorology for more efficient management of energy entails decisions at both the planning and operational stages of industrial technology. Whereas the time scale of climate is appropriate in long-range planning, weather is dominant in daily operations. Wise decisions clearly rely on an understanding of relevant processes in the climate system as well as the requirements of specific economic enterprises.

CLIMATIC FACTORS IN ENERGY MANAGEMENT

The energy budget of any subsystem on earth is part of the global energy budget. Solar radiation is not only the ultimate source of most of our energy supplies; it also drives the exchange processes that determine climate. Eventually, it returns to extraterrestrial space as unrecoverable waste heat. Thus, any climatic effect on the radiation budget influences energy demand and availability, however remote the connection. Climate and weather impinge on the entire range of energy uses, altering the efficiency and consequences of conversion. In turn, technological manipulation of energy may modify climate.

Whether the objective is development of an energy resource or more efficient use of energy, there are several considerations that merit attention at successive

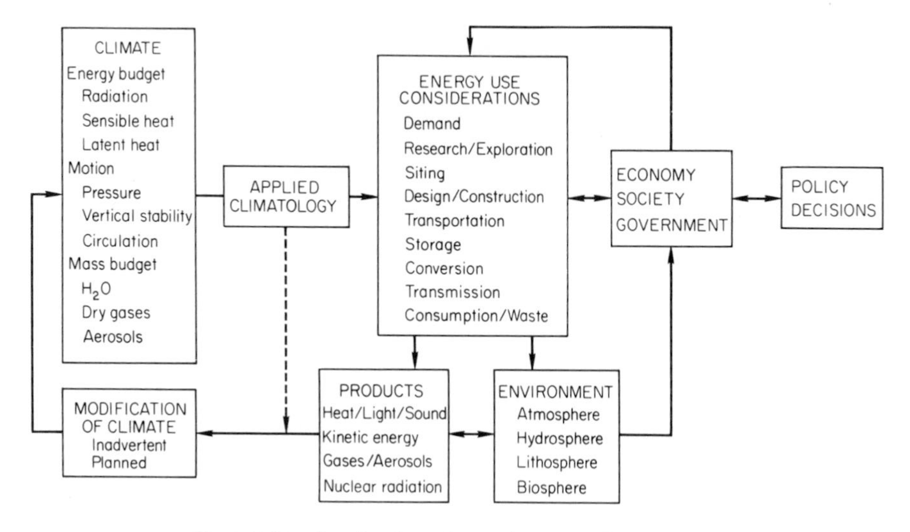

Figure 14.1 The climate system and energy policy decisions.

stages of planning and operation. These are indicated in Fig. 14.1. All respond to the vagaries of weather and climate and therefore can benefit from climatology in a framework of time and space. Their relative significance varies among different energy resources. They are interdependent but lead to consumption or waste, which always affects the environment in some manner. (The terms "consumption" and "waste" here imply functional abandonment. In the strict physical sense energy cannot be created nor consumed.) While recognizing that final decisions are economic, social, and political, the following sections emphasize the contribution of climatological information in improving those decisions.

Actual or potential energy resources that depend directly on climate include solar power, wind, water power and hydroelectricity, biomass (food and fuel), ocean waves and currents, and the temperature and salinity differential of ocean waters. Fossil fuels are linked to solar radiation more remotely through time. Climate is not a genetic factor in such energy resources as tidal action, geothermal heat, or nuclear reactions, but as in the case of all energy conversion it influences their exploitation.

The most useful climatic information for solar power technology obviously is the temporal and spatial distribution of insolation (Fig. 14.2; see also Fig. 2.15). Daily and seasonal variations and the consequent need for a system of energy storage appropriate to the local climate are major deterrents to complete reliance on an otherwise "free," inexhaustible resource. Owing to the unreliability and sparse coverage of solar radiation measurements, it is often necessary to derive estimates from other variables, among which latitude, altitude, and time of the year are basic. Mean values of daily duration of sunshine, cloud cover, relative humidity, and maximum daily temperature also have been incorporated in formulas that attempt to express their correlation with solar radiation. The complex relations of these fac-

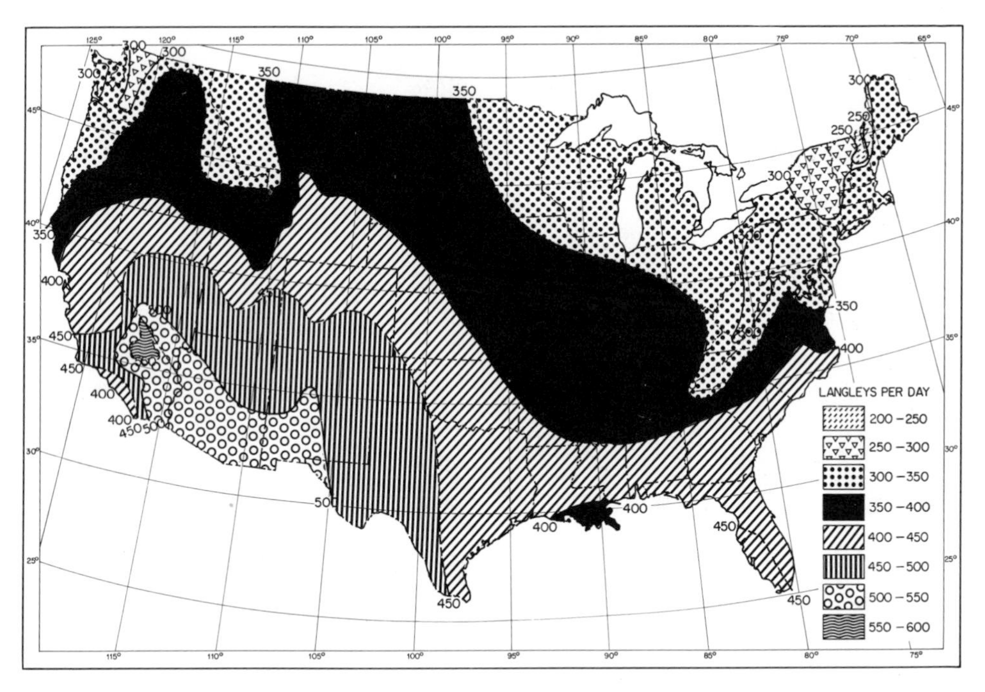

Figure 14.2 Mean annual distribution of daily solar radiation in the contiguous United States in langleys. (NOAA)

tors make difficult the derivation of a universal equation suited for all locations and seasons.

Since maximum intensity of direct solar radiation is achieved on a surface perpendicular to the solar beam, measurements (as by a pyranometer) or estimates of insolation on a horizontal plane must be adjusted for slope angles of both fixed and variable solar collectors. This is accomplished by calculations involving principles of trigonometry. Absorption by blackened receiving surfaces decreases slowly for sun angles from 90° to about 30° and then much more rapidly to zero on a flat surface parallel to the solar beam.

Whereas flat plates exposed to the sky collect both direct and diffuse radiation, focusing–concentrating collectors that "track the sun" depend primarily on direct solar radiation (Fig. 14.3). Measurements of the direct solar beam by a pyrheliometer provide data for calculating energy potential at a site. Whether fixed or tracking collectors are used it is necessary to consider not only the design and orientation of receiving apparatus but also astronomical and atmospheric factors that affect availability of solar energy at the earth's surface (see p. 23). Sun-path diagrams which chart the hourly position of the sun in the sky on any date make it possible to determine the most effective orientation for a fixed plate or to regulate the movement of a direct beam collector. These diagrams are also useful in the design and orientation of buildings for passive solar heating or for maximum shading (Fig. 14.4).

Figure 14.3 Solar energy test facility near Albuquerque, New Mexico. Arrays of adjustable mirrors focus reflected sunlight on receivers mounted on the tower (left) to heat water or another medium which can be used to generate electricity. (Photograph courtesy Sandia National Laboratories.)

The key consideration in development of wind power is choice of a site having winds of adequate speed and dependability for economic operation. A practical threshold for large windmills that generate electricity is 30 km per hour for 40 percent or more of the time. Small windmills such as those used to pump water on farms convert wind force to mechanical energy and usually can meet needs by harnessing less reliable air flow. Wind power increases as the cube of the wind speed; a doubling of wind speed results in an eightfold increase in power. Whereas light breezes do not afford sufficient force, high velocities and violent gusts are hazards to equipment, especially when accompanied by icing conditions. Sites in wind funnels between mountains or on summits usually have higher wind speeds but also may experience gustiness and unreliable flow as regional wind patterns change seasonally or with the passage of pressure systems. Although high altitude sites may benefit from greater wind speeds, the potential wind power is reduced in direct proportion to decreasing air density. At the elevation of Denver, Colorado (1,732 m), wind power is diminished by about 15 percent in comparison with the same speed at sea level. Air density also decreases with increasing temperature and with decreasing pressure and therefore varies with changing weather patterns. Available power from a wind blowing perpendicular to a plane is expressed by the equation:

$$P = \tfrac{1}{2} D V^3$$

where P is the power in watts per m^2, D is the air density in kg per m^3, and V is the

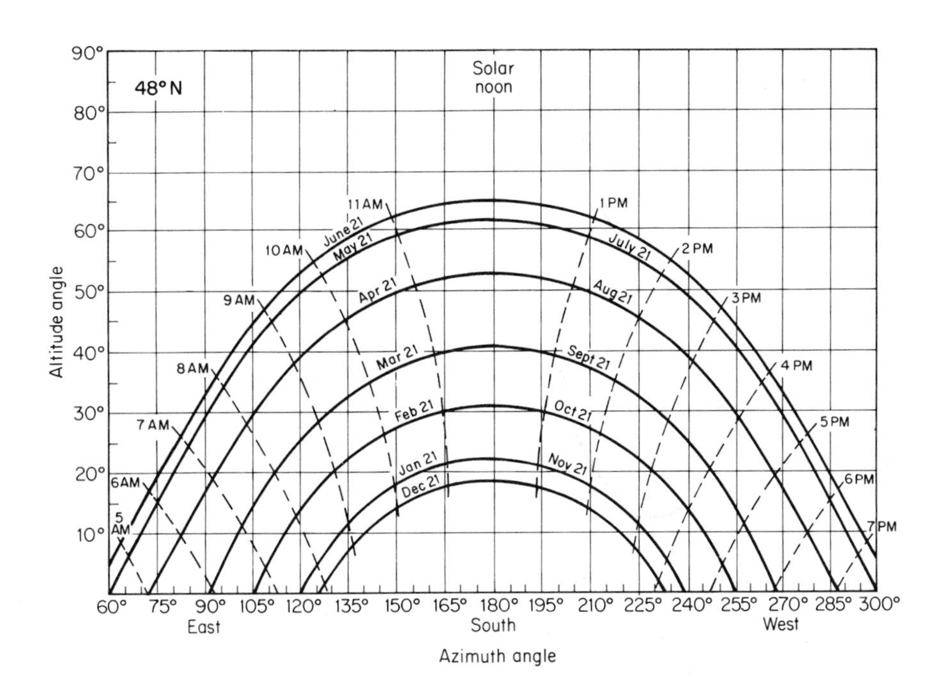

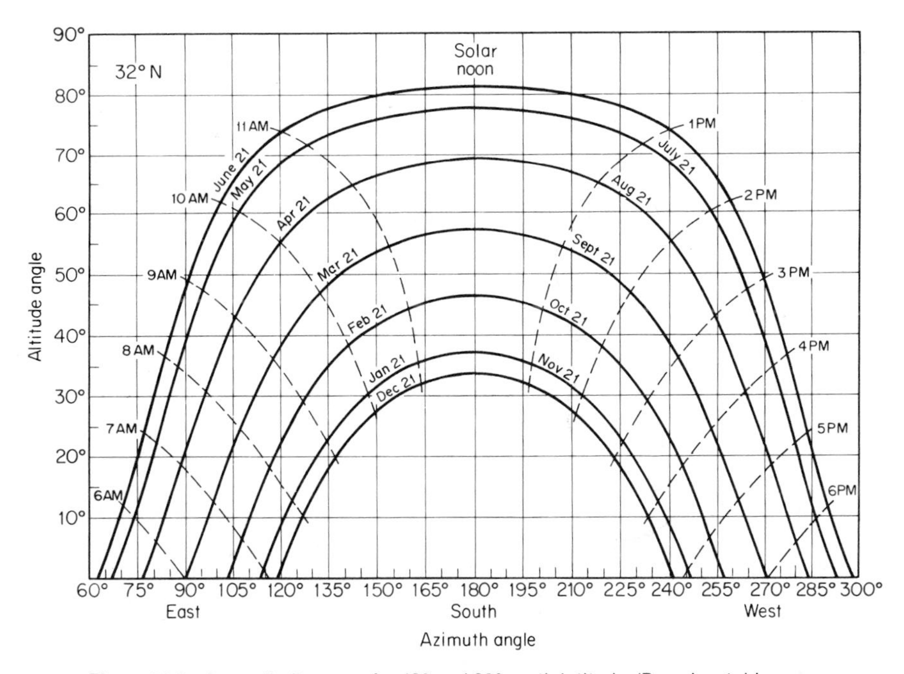

Figure 14.4 Sun-path diagrams for 48° and 32° north latitude. (Based on tables computed by the U. S. Navy Hydrographic Office.)

wind speed in m per sec. It is not feasible to extract all of the wind's energy. Ordinary propeller-type electric generators capture about 35 percent of potential wind power. Because wind speed along the ground is diminished by friction, wind power devices ideally are mounted at the greatest structurally feasible height (Fig. 14.5).

Figure 14.5 Wind turbines in the Columbia Gorge "wind funnel" near Goldendale, Washington. Spanning nearly 100 meters, each rotor can generate 2.5 megawatts at wind speeds of 47 km per hour. (Bonneville Power Administration.)

Estimates of wind speeds at levels above the normal height of anemometers (10 m) are based on surface wind speeds and rawinsonde data. A common height for planning commercial developments is 50 m. Figure 14.6 shows estimates of mean annual wind power for the contiguous United States at 50 m above exposed areas.

Wind power has the advantage of minimal effects on the environment through waste heat, noise, or alteration of local climate. Its disadvantage is the natural fluctuation of wind force and the difficulty of storage from periods of high wind speeds. Once wind power is converted to another form, usually electricity, the problems of efficient transmission and use arise.

The fundamental climatic factor affecting the supply of hydroelectric power is precipitation on the watershed. Volume and fluctuation of runoff are essential information in siting, design, and operation of a generating station (see "Runoff and Floods," Chapter 11). One of the practical objectives of attempts to induce precipitation through cloud modification is to augment surface water supplies, particularly during periods of depleted reservoir storage. Unfortunately, such periods often coincide with minimum cloudiness and regional drought.

Regardless of the resource employed in power generation, electrical trans-

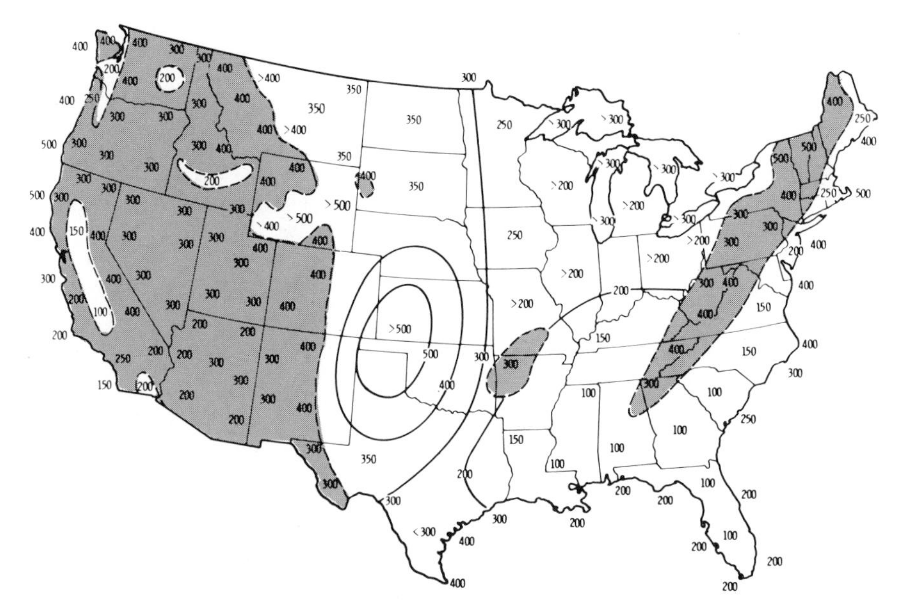

Figure 14.6 Mean annual wind power potential at a height of 50 m in the contiguous United States. Values in watts/m^2. (Courtesy Dennis L. Elliott, Battelle Pacific Northwest Laboratories.)

mission invokes the effects of weather and climate. Wind causes the greatest damage to power poles and lines (Fig. 14.7); strong, gusty winds blow over poles and snap lines, or blow trees and other debris into the lines. Hurricanes are by far the worst storms for overhead power lines, but even good winds for kite-flying are bad for power lines, if kites are not kept well in the open. Moisture conditions affect electrical transmission and the functioning of insulators. Wet snow or freezing rain adds weight to lines and poles and makes them more vulnerable to wind. To remove ice from lines, power companies employ various methods of temporarily disconnecting a line from the main circuit and heating it with an abnormally high power load. Temperature fluctuations influence the operation of switches, transformers, and other equipment. In hot weather, lines expand and sag and are more easily damaged by high winds. Thunderstorms bring the dangers not only of wind and precipitation but also of lightning, which causes at least a temporary power failure if it strikes an unprotected line. Weather forecasts aid in anticipating heavy power demands and extra maintenance problems. As a storm or cold wave moves across the country, the load is often sustained by rerouting electricity around affected areas or by bringing emergency generators into use at a moment's notice. Unfortunately, demand for heat and light often increases during periods of most severe weather effects on transmission. In view of the many problems which weather imposes on overhead lines, it may well be asked if it wouldn't be wise to put the lines underground. From a meteorological point of view it would, but the cost would be exorbitant.

Figure 14.7 Electric power transmission towers damaged by wind. (Bureau of Reclamation, U. S. Department of the Interior.)

As sources of energy, the fossil fuels and biomass differ from the so-called "clean" forms of solar, wind, and water power in their greater yields of waste heat, gases, and solid particles. Ordinary combustion of coal and petroleum derivatives produces not only heat but also varying amounts of aerosols and water vapor, CO_2, SO_2, or other gases, which may modify climate to some degree. Nuclear power production results in waste heat (thermal pollution) and entails the potential hazard of escaping radiation.

Whereas coal and oil ideally are transported during favorable weather and stored near the point of use, natural gas usually is delivered by pipeline to meet demands that are likely to change with weather conditions. Because the dispatch of gas cannot respond as rapidly as can electrical service, anticipation of future demand in the light of weather forecasts is even more desirable. Temperature, wind, cloudiness, and precipitation all affect the use of gas for heating. Of these the most important is temperature (see Fig. 14.8). The demand for gas during a cold wave must be met by dispatch from a remote supply *before* the low temperatures occur. For the most efficient and economical service, the gas dispatcher constantly refers to meteorological conditions and forecasts for the area being served. Several gas companies employ meteorologists who can interpret weather events specifically in terms of gas distribution problems.

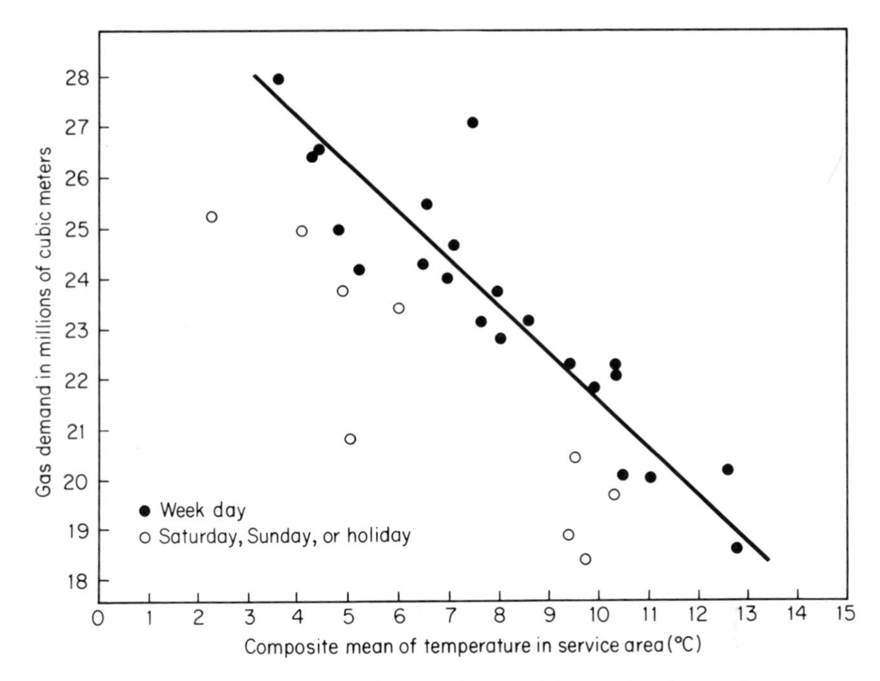

Figure 14.8 Relation of demand for natural gas to air temperature in a service area. Data are for a single winter month. The solid line represents estimated demand for gas at different mean daily temperatures, averaged as a composite for the entire service area. (Data from Pacific Gas and Electric Company.)

DESIGN AND CONSTRUCTION

The objectives of structural design include efficient use of energy and space, safety, durability, esthetic quality, and minimal impact on the environment; each has climatic implications, whether for an individual structure or an entire city. Local weather variables affect decisions on method, scheduling, and costs of construction. Thus, there are two broad aspects of the relation between outdoor construction projects and climate or weather: (1) influences on design and choice of materials and (2) direct effects on construction activities. The design of a given structure must take into account potential stresses imposed by temperature extremes and fluctuations, wind force, humidity changes, rainfall, or loads of snow and ice. These factors also help determine the selection of materials best suited to climatic extremes because of their inherent strength or their resistance to damage by physical or chemical weathering. Qualities such as heat conductivity, flexibility, or permeability may also be important in a particular structure. The rate of progress and efficiency of actual construction depend on conditions that favor all phases of day-to-day operations.

Temperature is one of the most important climatic elements to be considered in structural design. Metals, in particular, expand and contract with changing temperature, and allowance must be made to prevent buckling of girders and supports during warm periods or contraction and possible breaks in cold weather. Vertical metal towers expand more on the sunny side than on the shaded side, thus undergoing a diurnal bending that creates stresses. Paints which reflect a large part of insolation help to lessen this problem. Reflecting paints and safety devices on storage tanks minimize the danger of overflow or bursting due to expansion of the contents. In general, extreme temperatures and sudden changes of temperature bring on the greatest troubles. Concrete, stone, and other types of masonry are subject to damage when temperatures drop rapidly to below freezing, especially if they are wet. The expansion of ice in cracks or in damp earth adjacent to masonry walls commonly causes structures to break if they are not properly drained and reinforced.

Wind exerts a direct force on any structure in its path, and the maximum expected wind speeds are among the stress factors considered in designing bridges, towers, and buildings. Calculations based on climatic records and wind-tunnel experiments assist structural engineering. Bridges often are located across gorges or in river valleys where they are exposed to unusually strong, gusty winds. Suspension bridges are particularly vulnerable to wind, for they tend to sway and to develop destructive undulations. A notable example was the first Tacoma Narrows Bridge at Tacoma, Washington, which was destroyed by wind on November 7, 1940. On December 28, 1879, sections of the Tay Bridge at Dundee, Scotland, were blown down by westerly gales. A train that was crossing the bridge at the time fell into the water, killing all 75 persons aboard.

High towers, even if supported by cables, sometimes develop a "whipping action" and collapse under the force of strong winds. The total force of wind on an object in its path results from the impact of air on the exposed side and the lower pressure on the leeward. It depends on the size and shape of the object and is proportional to the square of the wind velocity. The full force of wind is never felt, even by flat surfaces, because the air tends to flow around them. Streamlining of surfaces can appreciably reduce the effect of wind pressure. Venting in broad structural members allows the air to pass through with less impact also. Solid structures incorporate considerably more strength in order to withstand wind. Additional stress on hollow structures results from the difference in pressure between the air inside and the wind outside. Vents help to equalize the pressure.

Relative humidity affects construction materials, preservatives, and paints. Metals are generally more susceptible to corrosion under conditions of high relative humidity. Chemical and physical deterioration of paints is speeded by moist air, and masonry construction also disintegrates more rapidly under moist conditions. Rainfall not only enhances the destructive influences of moisture but also produces direct physical effects on exposed surfaces. Heavy rains erode earthen structures and cause floods which undermine foundations or carry away ground structures. Dams, canals, piers, and bridges must withstand maximum flood stages. Accumulations of snow and ice create extra weight on horizontal surfaces; additional supports may be required to prevent collapse (Fig. 14.9). Potential accretions of ice

Figure 14.9 Snow and ice load on houses following a winter storm. (NOAA)

on towers, bridges, and cables also call for safety margins in design. When heavily loaded with ice, such structures are more vulnerable to the wind.

Much construction is seasonal, especially in climates where there are marked seasonal weather fluctuations. Economic efficiency dictates that work be planned carefully to take advantage of favorable weather and to avoid loss of time or damage to materials and equipment in bad weather. Climatic records of precipitation and temperature therefore aid in scheduling various types of work. Concrete and mortar are damaged if frozen before they set thoroughly. If the amount of newly poured concrete is small, it can be protected by coverings or artificial heating. Frozen earth impedes some types of construction. On the other hand, in the tundra or muskeg, heavy loads can be moved over the frozen ground with less difficulty. Most machinery does not function well in temperatures below freezing, and special problems arise from the necessity for antifreeze, heaters, and closed cabs on mobile equipment. Metal parts break more readily in extremely low temperatures. Temperature also affects the efficiency of workers on a construction project. Very high temperatures appreciably slow the pace of manual work among the unacclimated and may lead to illness among men doing heavy physical work; low temperatures make more clothing necessary and thereby complicate manual tasks. Accidents occur more frequently during periods of freezing temperatures or excessive heat than under normal temperatures. Icy conditions are especially hazardous. Housing and feeding of construction crews are also influenced by temperature.

Precipitation generally has an adverse effect on construction. Heavy rains pit newly laid concrete, erode earth embankments, and slow down or stop most earth-moving activities. A deep snow cover usually brings work to a halt. Heavy precipitation and fog reduce visibility and thereby increase the danger of accidents. Special facilities have to be provided for protection of materials and equipment subject to damage by moisture. Winds tend to increase the unfavorable effects of precipitation as well as the accident hazard on outdoor construction work, especially on tall structures. Partially completed structures are generally more easily damaged by strong or gusty winds. Certain kinds of construction activities are specifically affected by wind. Spray-painting is an obvious example. Drying winds prevent concrete and mortar from setting properly unless their surfaces are specially dampened.

AVIATION

Among all modes of transport, aviation engenders the most complex and far-ranging relations between energy use and atmospheric conditions. Efficiency, safety, and economy of operation are directly dependent on climate and weather as well as wise use of energy. Ventures farther into, and through, the upper atmosphere have improved our understanding of weather processes and at the same time made that knowledge more necessary in planning and executing air or space travel. Flight in the stratosphere has raised new questions about the effects of engine emissions on climate. Since aviation is based on the earth's surface, weather at the lower levels is as important as that in the upper atmosphere. The probability of certain conditions has a bearing on many decisions that concern both the functioning of air terminals and flight.

The use of climatic information in aviation begins with the selection of an airport site. Level land, absence of surrounding obstructions, and access to centers of population may be factors affecting wind or visibility. Radiation fogs form more often in broad valleys, and if the site is to the leeward of industrial centers, smoke and other pollutants accentuate reduced visibility. Climatological studies assist selection of a site that experiences a minimum of weather hazards such as poor visibility, low clouds, sudden wind shifts, dangerous turbulence, and thunderstorms. The ideal site is windward of air pollution sources, on a modest upland that provides air drainage into adjacent valleys while lying below mean prevailing cloud bases, and well removed from marshes, lakes, or other moisture sources. Prevailing winds determine the best orientation of runways to permit takeoff and landing into the wind. Wind speeds as well as directions are taken into account; wind from some directions may be so light as to have little influence on traffic. Modern jet transports can land on dry runways in cross-wind speeds as great as 30 knots, which are unsafe for small aircraft.

A flight cannot be judged by its speed and economy of operation in the air alone; low clouds, fog, or other weather conditions at takeoff or landing can affect the entire flight. Even in good weather the length of the takeoff run for conventional airplanes is greater at high-altitude airports, where the air is less dense. Piston aircraft require a 5 percent longer run and jet aircraft about 10 percent for each 500-m increase in altitude of the airport. If unfavorable weather is confined to a particular locality, landings may be possible at an alternate field; or the conditions may be temporary, allowing the aircraft to circle the airport while awaiting improvement. Either case requires additional expenditure of energy. Development of so-called "blind landing" devices is gradually resolving the problems of low ceilings and poor visibility.

To provide continuous information on airport weather, the U.S. National Weather Service and the Federal Aviation Administration issue aviation weather reports throughout the United States. Similar agencies serve aviation in other countries. Hourly reports transmitted to other terminals and to aircraft in flight contain data on cloud height and coverage, visibility, sea-level barometric pressure, temperature, dew point, wind direction and speed, altimeter setting, precipitation, and hazardous conditions. Some airports are equipped to observe and report both horizontal and landing-path angular visibility range (see Fig. 14.10). Special observations of weather elements that are significant to operations are reported when sudden changes occur.

In the United States the Federal Aviation Administration prescribes minimum conditions for airport safety. In addition to low ceilings and restricted visibility the pilot must, of course, consider wind, icing, and other potential dangers in relation to aircraft capabilities and navigation equipment. Most major airlines provide supplementary weather service for their pilots and dispatchers.

Knowledge of weather and climate also enters into the design of aircraft and engines. Stresses likely to be encountered in severe storms, the possibility of damage by hail or lightning, and icing must be taken into account in aeronautical engineering. Temperature variations, precipitation, and icing affect power plant efficiency. Engines must function through a wide horizontal and vertical range of pressure and temperature. At the same time, it is necessary to consider the potential effects of aircraft operation on the atmospheric environment.

Figure 14.10 Forward scatter meter. This instrument measures the reduction in transmitted light due to scattering by pollutants and supplements aircraft runway visibility observations. (Photograph by Hutchins Photography, Inc., courtesy EG & G, Environmental Equipment Division.)

WEATHER IN FLIGHT

In planning a flight the pilot has the aid of weather maps, upper-air data, and forecasts. Except for very low flights, pressure distribution is a guide to route planning; gradient winds blow along isobars, but it may be expedient to avoid the worst frontal weather in an extratropical cyclone rather than seek tailwinds. Air-traffic density often restricts aircraft to predetermined routes regardless of headwinds. In the case of long-range flight across entire continents or oceans the route can be adjusted more easily to the pressure pattern, and therefore the winds, at upper levels. The westerly jet streams have made it possible to achieve remarkable records on west-to-east flights even though an airplane may have deviated appreciably from a great circle route. Overseas flights often follow routes in winter different from those taken in summer to allow for latitudinal shifts in wind and pressure systems.

Winds are less significant for supersonic aircraft on high-altitude flights shorter than about 6,000 km. But the efficiency of supersonic jet engines is greater at lower air temperatures (and greater densities at a given level), and it may be feasible to deviate from the great circle route or a particular pressure level in order to travel through colder air, thereby reducing fuel consumption.

The climatic probabilities of various weather conditions at air terminals and in flight are among the criteria for determining routes and scheduling flights. Synoptic climatology aids the identification of both favorable and hazardous conditions that are likely to be encountered in different atmospheric flow patterns. Flight through storms entails all the hazards of wind, turbulence, precipitation, low temperatures, poor visibility, and icing. Entry into a tornado is out of the question. Hurricanes are certainly not suitable for routine air travel, although they have been traversed for carefully controlled experiments and observations. They usually are shallow enough to allow safe flight at upper levels. Next in order of violence are thunderstorms and squall lines, which because of severe turbulence, icing, and possible damage by lightning or hail are best avoided. In a mid-latitude cyclone a variety of weather conditions may affect a flight. Low ceilings and reduced visibility accompany the typical warm front, and there may be hidden thunderstorms. Icing is likely in the cool upper levels, and in winter ice pellets or freezing rain are hazards. Air mass weather in the warm sector of a cyclone may include fair weather cumulus, daytime convective turbulence, or scattered thunderstorms. The cold front is usually the most dangerous portion of the mid-latitude cyclone. Frontal thunderstorms and a sudden change of wind direction and temperature are common features that bear on safety and navigation. Occluded fronts often combine the flight conditions of the warm and cold fronts. Both warm- and cold-front occlusions have warm-front weather in advance of the occlusion, and both have upper-front weather like that of the cold front (see Chapter 5, "Fronts").

TURBULENCE

Atmospheric turbulence creates discomfort, increases stresses on the aircraft, and reduces the efficiency of operation. In its most violent forms it can be catastrophic. Four principal conditions lead to turbulence: thermal, frictional, frontal, and that associated with upper-level waves. Thermal turbulence is produced by vertical convection above a heated surface or by advection of cold air over a warm surface and is accentuated in an air mass that already has an unstable lapse rate. Frictional (sometimes called mechanical) turbulence results from flow across rough terrain. Its effects often combine with convection currents when an irregular surface is heated or cooled unevenly. A strong flow of stable air across a mountain range frequently develops wave action with updrafts and downdrafts that extend to several times the height of the mountains and for a distance of 150 km or more downwind. Turbulence in such a *mountain wave* is usually greatest at low levels and again near the tropopause to the lee of the mountains (see Fig. 14.11). It can be a hazard in any of the world's climatic regions where there are uplands.

Along a cold front, the wind shift and the lifting of warm air above the cold may cause wind shear, that is, pronounced differences in wind velocity in adjacent air streams. Aircraft sometimes encounter layers of turbulent wind shear when climbing or descending through a temperature inversion at any altitude up to the tropopause. Warm air moving over a ground inversion layer, as in a valley at night, creates a thin zone of wind shear that can be especially dangerous. Although tur-

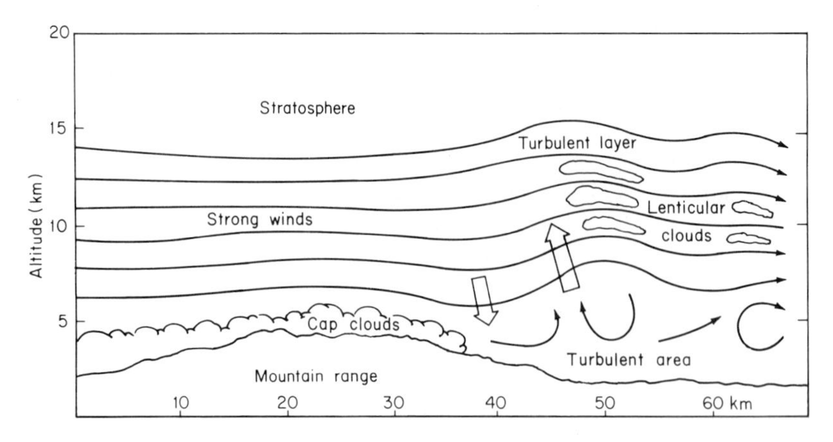

Figure 14.11 Mountain wave turbulence in the lee of a mountain range.

bulent conditions may have accompanying clouds as warning signs, they also disturb clear air. Wave-induced turbulence (WIT, also known as clear air turbulence or CAT in clear air) is common in the vicinity of a jet stream, where both horizontal and vertical shear are generated (see Fig. 14.12). Refinements in the techniques of observation by Doppler radar and infrared detection of water vapor assist detection of wind shear at considerable distances.

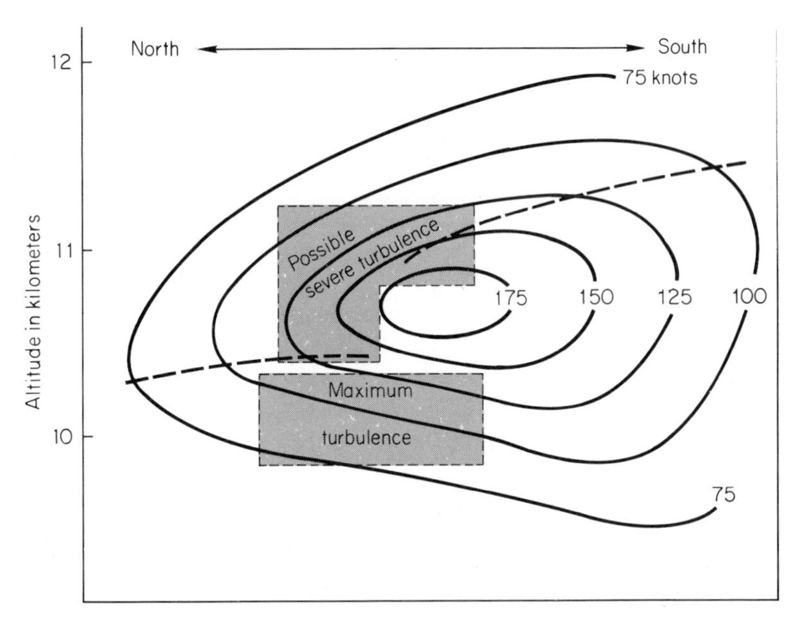

Figure 14.12 Wave-induced turbulence in the vicinity of the polar front jet stream. Solid lines are isotachs in the jet-stream cross section.

AIRCRAFT ICING

Ice formation on aircraft is one of the great hazards to flight in the troposphere, especially in the climates of mid-latitudes. Icing may result when water droplets are present and the temperature of the exposed aircraft surface is at or below freezing. It increases with air speed up to about 400 knots and then decreases under heating by friction along the fuselage and projections. Clouds containing supercooled droplets are breeding places for icing. In turbulent air the drops are large, and when they strike the cold airplane surface they spread and freeze to form *clear ice.* In smoother air the small droplets freeze immediately upon striking the leading edges and take the granular or crystalline form of *rime ice.* Often the two types occur together, but clear ice is more tenacious than rime and prevalent in turbulent clouds. It forms most commonly at temperatures between −25° and 0°C and can result from the fall of rain through cold air. Clear ice creates blunt leading edges and thus alters their aerodynamic properties. It also adds weight to the aircraft. Icing can be expected in clouds from which snow or ice pellets fall. Freezing rain, on the other hand, falls from warmer levels and can be avoided by climbing into the warmer air. Rime icing is most frequent in stratiform clouds at temperatures of −20° to −10°C and tends to accumulate as sharp projections from leading edges. It does not disrupt air flow as seriously as clear ice but is, nevertheless, a cause for caution.

In turbojet engines ice may form on guide vanes or other structures at the air intake, or in fuel systems, and lead to reduced engine thrust and eventual turbine failure. Icing of piston engine carburetors can occur in clear air at temperatures above freezing if the relative humidity is high, but it also forms in clouds. The expansion of air and vaporization of fuel at the intake causes an adiabatic decrease in temperature and condensation of water vapor. If the temperature drops below 0°C, the moisture deposited as frost in the carburetor can reduce power or possibly cause engine failure. Carburetor icing is most common at air temperatures of −10° to 25°C in moist air.

WATER TRANSPORT

The pattern of general circulation and availability of wind energy over the oceans determined, to a large extent, the major trade routes of sailing days. Even today, when steam and motor vessels dominate ocean shipping, wind direction and force and "state of the sea" are important in the safe and economic operation of ships. Navigation is especially hazardous in coastal waters when strong winds are blowing. On the open sea the efficiency of fuel consumption depends on winds and the waves and swell they generate. The pressure of wind on a ship varies as the square of the wind speed. A doubling of the speed increases wind pressure four times. Ships moving against headwinds must counter the relative wind speeds which represent the true wind speed plus the speed of the ship. Wind from the stern increases a ship's speed by about 1 percent, but headwinds decrease its speed by from 3 to 13 percent, depending on the size of the vessel and its load. A longer route, planned in relation to winds and storms, can result in greater economy of operation, less time at sea, and a smoother voyage. Marine climatology and route forecasts support these objectives.

In rough seas it is difficult to hold a course and may be prudent to change the course to reduce the effect of waves and swell (see Fig. 14.13). Harbors in sheltered coastal sites, such as bays and estuaries, have always been desirable because of the relative absence of high winds and rough water, but even in the snuggest harbor damage can occur to vessels that are not properly moored. High or gusty winds make a landing dangerous. In cold climates, wind often concentrates ice in harbors, increasing the possibility of damage due to collision.

Next to wind, visibility is the most significant weather element in water transportation. Fog is the most common obstruction to visibility over water, although industrial smoke may intensify it in harbors. In the open sea, fog interferes with the accurate determination of the ship's position and increases the danger of hitting another ship, rocks, or icebergs. Along coasts where upwelling of cold water occurs, fog is common, and particular care must be exercised to avoid shallows or other danger points. Beacon lights, radio, radar, and electronic devices for sounding ocean depths have considerably lessened the danger of collision or running aground in fog. Even so, dense fogs can bring harbor traffic to a standstill and many marine accidents due to fog occur every year.

Extreme temperatures affect wharf activities and must be met with extra protection for cargo, equipment, and personnel. Modern ships are equipped for control of temperature, humidity, gases, and even barometric pressure in holds that contain perishable or weather-sensitive cargo. If the water temperature is low

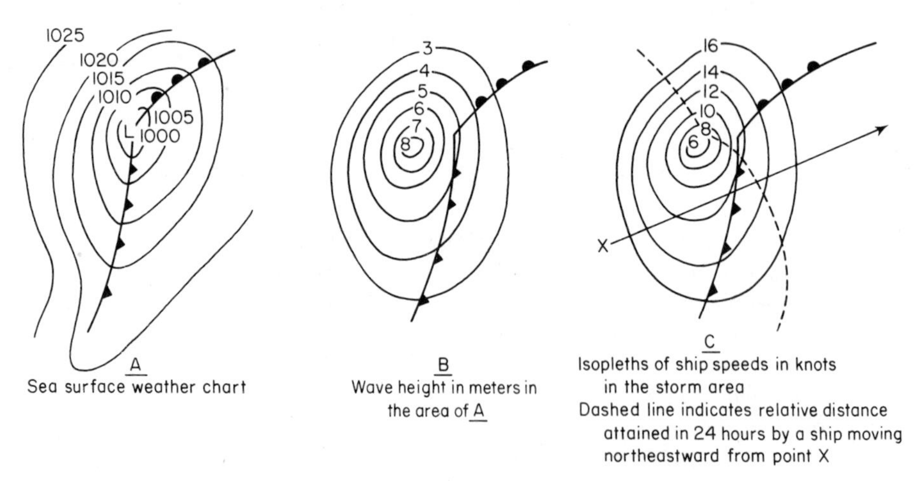

Figure 14.13 Relation of ship speed to wave height in the area of an extratropical cyclone. (After J. J. Schule, "Weather Routing of Ships," in *Meteorology as Applied to Navigation of Ships,* Technical Note No. 23. Geneva: WMO, 1958.)

enough to form ice, otherwise navigable waters may become impassable. In arctic seas, traffic is suspended for several months each year. At lower latitudes, icebreakers open vital channels (see Fig. 14.14). On the Great Lakes the ice cover grows from the shoreline to disrupt shipping even though the lakes are rarely frozen over their entire area. The ice is generally thickest in protected bays; straits become congested with ice carried by wind and currents. Broken ice, or "slush," can damage steering gear and propellers and is not moved effectively by an icebreaker, for it tends to close around the ship. Heavy shipping moves on the Great Lakes for only eight or nine months a year, so that it is necessary to stockpile materials such as coal and iron ore in the navigation season to support industry during the winter. Dates of the annual freeze-up vary by as much as three weeks. Forecasts based on water temperatures at key sites and predictions of heat loss assist shippers in preparing for the halt in traffic. Ice thickness and calculations of heat budget components are variables considered in estimating dates of break-up as a guide to cargo scheduling.

Similar problems relating to ice on rivers, canals, lakes, and harbors occur throughout the higher middle latitudes. The net effect is a seasonal rhythm of

Figure 14.14 Coast Guard icebreaker *Mackinaw* opening a passage in the Great Lakes. (U. S. Coast Guard Official Photograph.)

waterborne traffic with variations in the length of the shipping period from year to year and place to place.

Storms incorporate several adverse weather phenomena, and, depending on their severity, it is sometimes desirable to avoid them. Precipitation is not ordinarily a major factor affecting shipping unless it results in icing. Loading and unloading may be disrupted if perishable cargo is involved, and special protective measures are necessary for loaded cargo. Heavy rain or snow can affect visibility. Lightning is a serious hazard to non-metallic craft unless they are fitted with conductors to carry the discharge to the water. Hurricanes call for special "evasive action." Wind speeds and sea swell are greatest along the leading edge of a hurricane, making travel to the rear of the storm advisable. Where forecasts are available, it is often worthwhile to alter a ship's speed to arrive in port before or after a storm reaches the port.

Radio weather reports and marine forecasts are broadcast from many coastal stations as an aid to ocean navigation. Much of the interpretation of these reports, however, is left to the mariner; indeed, valuable weather data for both marine and land forecasts are contributed by ships at sea.

RAILWAYS AND HIGHWAYS

Railroads would seem to be little affected by weather because of their permanent all-weather track systems. In fact, trains are likely to be on the move in storms which have brought automobile traffic to a halt and grounded all airplanes. But railroads are far from immune to weather hazards, and their operation must be constantly geared to weather changes. Severe storms may damage tracks, bridges, signals, and communication lines. Floods, heavy snow and avalanches are particularly troublesome in mountains, and it has been expedient in many cases to build expensive tunnels to avoid high-altitude weather as well as steep grades. Low visibility due to fog or precipitation calls for extra caution and decreased speeds.

In the movement of freight, special attention must be given to perishable goods and livestock to prevent losses during unfavorable weather. There are seasonal variations in the types and quality of freight. Fresh fruits and vegetables must be moved immediately after harvest, the time of which is determined in part by weather. Freight lines serving ports which are icebound in winter necessarily experience much of the seasonal variation in traffic imposed on water transport. Perishable freight for export must be routed to other ports; nonperishable goods may be stockpiled.

The major consideration in the relation of weather to the use of highways is safety. Poor visibility, slippery surfaces, and gusty winds are the primary driving hazards. Good weather, especially on weekends, brings large numbers of pleasure drivers onto the highways and traffic problems and accidents increase (see Fig. 14.15). Highway patrols experience maximums of activity in fine holiday weather and in bad weather.

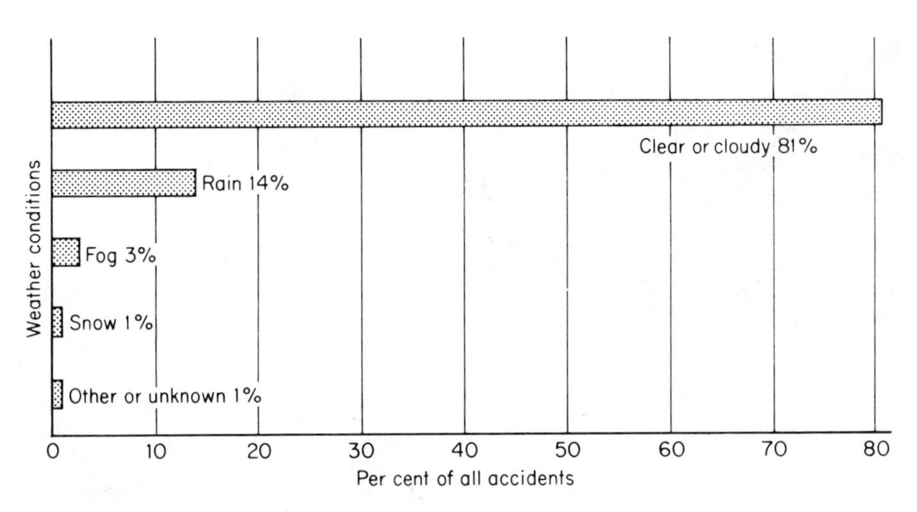

Figure 14.15 Percentage of motor vehicle accidents in different weather conditions. Based on total accident reports for 30 states from 1913 to 1961. (Data from National Safety Council.)

Many of the maintenance problems on highways and roads result from weather and climate. The expense of erecting signs to warn motorists of hazards associated with weather is great. Unsurfaced roads suffer heavy damage from precipitation, and, when wet, are damaged by traffic. Even paved highways undergo erosion at the shoulders, which may lead to collapse of the roadbed. Floods can wash out bridges, underpasses, and even entire sections of road, or they may carry debris onto the highway. Gales blow trees, utility poles, and other obstructions onto the roadway. Alternate freezing and thawing causes frost-heave in the road surface and subsequent breakup of the roadbed under heavy traffic. It is often necessary to prohibit heavy vehicles from wet or thawing roads to reduce surface damage. Applications of sand increase traction, and salt speeds melting when roads are icy. Bridges tend to develop icy surfaces sooner than roadways because they are exposed on all sides to cold air. Snow is one of the most expensive highway maintenance problems in the middle and high latitudes and in mountains (see Fig. 14.16). Where drifting is common, snow fences or windbreaks along the windward sides of highways help induce drifts away from the road. Tunnels and snow sheds are built in mountain passes to avoid deep snow as well as avalanches. Because of the excessive cost of keeping them open with snowplows, some passes are closed for all or part of the winter. Snow removal crews and equipment must be ready for immediate action in snow areas; ironically, great quantities of energy are expended in order to keep roads open so that vehicles can use more energy.

The need to keep city streets clear of ice and snow is even more urgent than on the highways. Sewers and gutters easily become clogged by heavy accumulations of snow. Snow of great depth cannot be merely plowed to the side but must be hauled

Figure 14.16 Snow removal on Chinook Pass, Cascade Mountains, Washington. (Washington State Department of Transportation.)

away. If the temperature is not too low, it may be removed by melting with water. Installation of hot-water or steam pipes under streets is a possible solution to the snow problem.

MANUFACTURING

The influences of weather and climate on manufacturing may be divided into two general categories: those that influence siting and those affecting operations once an industry has been established. Climatic factors influence the location of a factory because of their effects on transportation, raw materials, labor supply, energy conversion, and waste disposal. The relative importance of climatic factors depends on the type of manufacturing in question, and climate is frequently overshadowed by more pertinent economic decisions. Food-processing plants are likely to be located where the agricultural produce for manufacture is available. We can say that the availability of raw material is the primary factor, but certainly climate plays a fundamental role. Similarly, pulp and paper mills are located where there are abundant water supplies, which again presumes favorable climatic conditions. A factory demanding year-round access to water transportation would hardly be located on a harbor that freezes for part of the year. The attractiveness of a climate to a large labor force may induce workers to move to a new area; at least it will be considered in the category of amenities. The cost of heating, or air conditioning, to provide satisfactory conditions for work and for certain manufacturing processes bears a direct relation to climate (see Table 14.1). Factory buildings, warehouses, and other structures should be designed with climate in mind. Factories which dispose of wastes through stacks must be located to minimize air pollution over settled areas.

Weather conditions are reflected in almost every phase of factory operation. Storms cause workers to arrive late for work, hamper essential outdoor activities, cause damage to goods and equipment, or interrupt power. Some processes, notably in the chemical and related industries, have a narrow range of temperature requirements for efficient production. Wind and low relative humidity increase the

TABLE 14.1

Optimum Indoor Temperature and Humidity for Selected Industrial Operations *

	Temperature (°C)	*Relative humidity (%)*
Food processing:		
Milling	18-20	60-80
Flour storage	15	50-60
Baking	25-27	60-75
Candy making	18-20	40-50
Process cheese mfg.	15	90
Textile manufacturing:		
Cotton	20-25	60
Wool	20-25	70
Silk	22-25	75
Synthetics	21-29	55-60
Miscellaneous:		
Cosmetics mfg.	20	55-60
Cosmetics storage	10-15	50
Drug mfg.	20-24	60-70
Electrical equip. mfg.	21	60-65
Paper mfg.	20-24	65
Paper storage	15-21	40-50
Photo film mfg.	20	60
Printing	20	50
Rubber mfg.	21-24	50-70

*After G. Grundke, adapted from Helmut E. Landsberg, *Physical Climatology,* 2nd ed. (DuBois, Pa.: Gray Printing Company, Inc., 1958), p. 392.

fire hazard at manufacturing plants, especially if combustible or explosive materials are being processed or stored.

PRODUCTS OF ENERGY CONSUMPTION: ATMOSPHERIC POLLUTION

A widespread problem resulting from energy use is air pollution, which is simply an unwanted concentration of any phenomenon in the atmosphere. It is not the problem of atmospheric scientists alone. Weather and climate are not the sources of air pollution, but atmospheric conditions do greatly affect the rate of diffusion of contaminating agents both horizontally and vertically. Since air pollution is far more common around cities than elsewhere and has increased with urban growth, it is evidently a human-caused problem. There are always some foreign materials in the air—in liquid, solid, and gaseous forms. Their increase stems from a multitude of sources. Industry, agriculture, heating systems, incinerators, motor vehicle exhausts, and evaporation of volatile liquids account for a large share of the smoke, dust, gases, and vapors that may be factors in climatic change. More than 1,600 chemical compounds of natural or cultural origin have been identified in the atmosphere. Heat, noise, light, and various forms of radiation are additional pollutants that demand attention.

The immediate meteorological effects of pollution are concerned mainly with visibility and solar radiation. One often can locate a distant city by its cap of smoke or haze. Reduced visibility due to pollution correlates with the greater urban activity on weekdays. Since many contaminants are hygroscopic they act as nuclei of condensation and create a haze or fog, further reducing visibility. Moreover, it appears that the water droplets thus formed are more stable than normal cloud droplets and do not evaporate so readily upon being heated. Oily substances, especially, tend to form a protective coating around a droplet, making it difficult to disperse. Originally, the term smog denoted a combination of fog and pollutants. Solar radiation is appreciably reduced by polluted air in the daytime, and outgoing radiation is reduced at night; the net effect is a lowered diurnal range of temperature. Parts of the solar spectrum are transmitted selectively by different gaseous constituents in the air so that the quality as well as the quantity of insolation is impaired. Chemical reactions among various types of contaminants in the air produce new compounds that in some cases are more damaging than the original wastes. Certain of these reactions are photochemical, that is, they take place under the effects of sunlight, especially the ultraviolet wave-lengths.

Stable air and light winds or calms are conducive to concentration of pollutants at or near the source of contamination. For numerous reasons industries commonly locate in valleys and depressions for which stable air has an affinity and where temperature inversions are common. Temperature inversions are particularly suited to the formation of palls of smoke and industrial haze. As the warm fumes, gases, or airborne solids rise, they cool adiabatically as well as by radiation and mixing. Since the air is warmer overhead in an inversion, the pollutants are soon at a temperature equal to that of the surrounding air. Therefore, they do not rise further. Cooling by radiation at night from the top of a smoke layer induces subsidence, and the concentration of pollutants increases at and below the top of the inversion. Stable air, especially with an inverted lapse rate, often is accompanied by radiation fog, which combines with the pollutants to form smog. As already indicated, hygroscopic particles may hasten the condensation process. If the inversion layer is well above the surface, a pall will form with less drastic effects, but it will nevertheless inhibit the passage of radiation and may eventually "build down" to the surface. The most intense smogs usually develop when the top of the inversion is within 500 m of the ground. Pollutants tend to concentrate near the ground in a nocturnal surface inversion. By day, heating of the ground may create a layer of turbulent mixing which does not completely overcome the inversion (see Fig. 14.17). Under such conditions the pollutants mix more uniformly through the layer below the inversion. Sodar systems, which detect sonic waves reflected from dust and air layers of different density, are used to determine the particulate concentration and height of the mixing layer.

Unstable air and strong winds are inimical to formation of dense smogs. Rising air currents carry wastes upward and winds disperse them through a large volume of air. Prevailing winds carry pollutants away from single sources and pro-

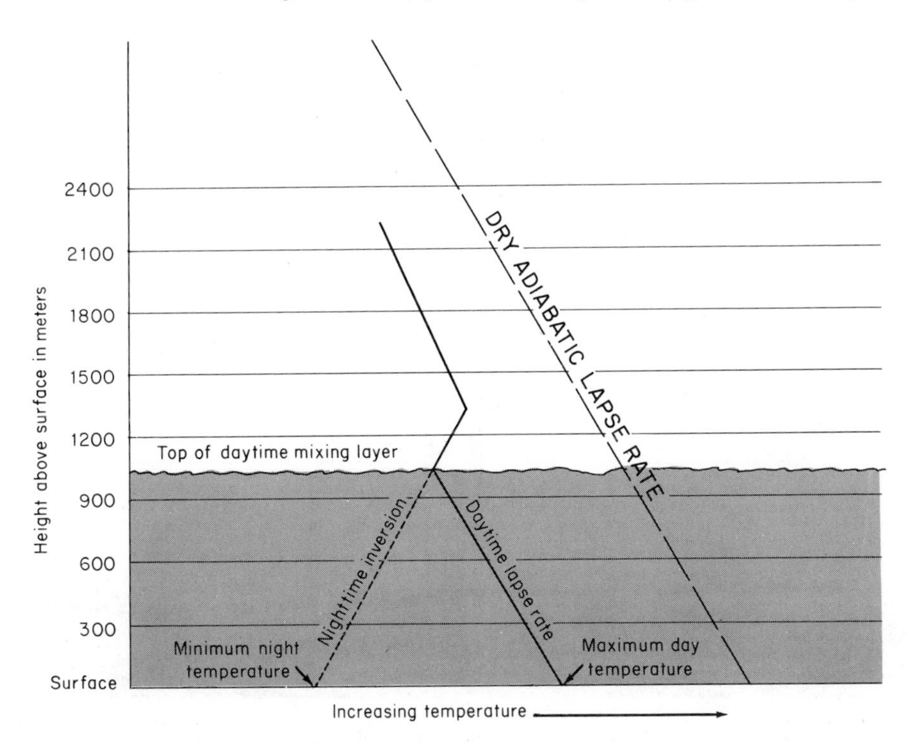

Figure 14.17 Height of daytime surface-mixing layer in relation to temperature lapse rate.

duce leeward plumes that experience a progressive decrease in intensity of pollution (see Fig. 14.18). But local winds, such as sea or valley breezes, may act perversely, even returning polluted air that has drifted away from its source.

The wash-out effect of precipitation on local air pollution is not as great as might be supposed. Although some materials are removed by rain, dilution by wind and unstable air is more effective in reducing concentrations during a storm. On a larger scale, however, most pollutants that are dispersed to high levels return to the surface in precipitation, often after intervening processes have altered their physical or chemical properties. For example, oxides of sulfur and nitrogen emitted during burning of fossil fuels undergo chemical changes to produce sulfuric and nitric acid. Small amounts of these or other acidic compounds in rain or snow result in *acid precipitation,* which can harm plant and animal life, pollute water bodies, and corrode structural surfaces at great distances from the polluting source. Damaging acids also may be contained in fog, dew, or frost. Deposition of dry particles accounts for an estimated 10 to 30 percent of acids reaching the surface, the propor-

Figure 14.18 Smoke released from a meteorological tower indicates winds moving from opposite directions. Towers such as this aid studies of atmospheric pollution and other industrial problems. (Brookhaven National Laboratory.)

tion depending on distance from the source and weather conditions along the dispersion path.

The atmosphere has been regarded all too often as a sewage disposal plant, but it can be overtaxed locally, regionally, and potentially on a global scale. Areas having a high frequency of low-level inversions are particularly unsuitable for factories that eject large quantities of airborne wastes. Siting of a factory to take advantage of variations in the ability of the atmosphere to disperse pollutants merely transfers the problem. Tall chimneys help to disperse effluents above the surface level, adding to the total atmospheric load. Local air pollution concentrations can be alleviated by restricting the release of wastes to periods of favorable meteorological conditions, that is, strong winds and unstable air. Forecasts of *air pollution potential* greatly aid in controlling emissions during critical periods (see Fig. 14.19). Mean depth of the surface mixing layer and mean wind speed within the layer are major criteria for such forecasts. Areas of greatest pollution potential coincide with regions of frequent stagnating anticyclones and low-level inversions.

In spite of procedures that have been developed to enhance dilution of industrial and domestic pollutants, the reduction of emissions remains the ultimate solution to the problem. Enforcement of clean air standards in the urban areas of Great Britain helped decrease smoke pollution by 80 percent between 1960 and 1980. Sulfur dioxide decreased by about 50 percent during the same period, owing mainly to controls on the burning of coal.

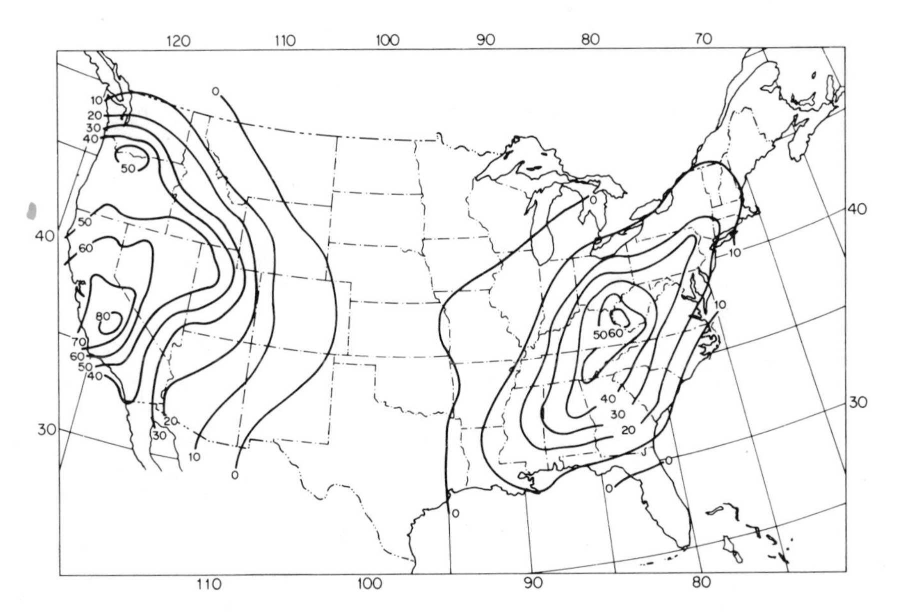

Figure 14.19 Predicted high pollution potential days per year in the contiguous United States. Data are derived from meteorological conditions and do not indicate actual pollution which may occur locally in urban environments. (After McCormick, Leighton, and M. Smith.)

QUESTIONS AND PROBLEMS FOR CHAPTER 14

1. What natural energy resources are available in the climate system? How is each affected by climate?
2. Summarize the weather-related problems of power transmission, suggesting possible solutions for each.
3. Explain how water budget data can be used in the design and operation of a municipal sewer system.
4. An engineer is designing a bridge at a site 30 km from the nearest climatic station. How can a knowledge of climatology help ensure allowance for potential weather damage to the finished structure?
5. Explain the origin of the mountain wave and suggest safety precautions for pilots who fly over mountainous areas.
6. Referring to Fig. 14.13, explain why the highest ocean waves occur west of the cyclone center rather than in the area of lowest pressure.
7. Rank according to relative importance the *climatic* factors that should be considered in selecting a site for each of the following types of manufacturing: ship building, oil refining, copper smelting, whisky distilling, perfume making, meat packing.
8. Compare the climatic problems of a natural gas company with those faced by a producer of hydroelectric power.
9. What are the features of a climate that favor high air pollution potential?
10. Explain why a dense concentration of pollutants might form near the ground even though there is no temperature inversion at any level in the overlying troposphere.

15

Human Bioclimatology

Human health, energy, and comfort are affected more by climate than by any other element of the physical environment. Physiological functions of the human body respond to changes in the weather, and the incidence of certain diseases varies with climate and the seasons. Our selection of amounts and types of food and clothing also tends to reflect weather and climate. The state of the atmosphere even influences our mental and emotional outlook.

Different human beings do not react to identical climates in the same way, however; the relationship is complicated by individual physical differences, age, diet, past climatic experience, and cultural influences. Nor are all individuals equally adaptable to a change of climate. Of all life forms, nevertheless, humans are the most adaptable to varying atmospheric conditions. Climate has been blamed for human failure more often than the facts justify, especially in the tropics, but it is unquestionably a factor in efficiency.

Among the climatic elements that affect the human body, the more important are temperature, sunshine, and humidity. Wind exerts an influence largely through its effects on skin temperature and body moisture; the circulatory, respiratory, and nervous systems register changes in atmospheric pressure. Such elements as cloudiness, visibility, and storms induce psychological reactions that range from positive reinforcement to physiological disturbances. Acting together the climatic elements constitute the climatic environment that directly influences our comfort and well-being.

HEAT BUDGET OF THE BODY

The human body maintains a balance between incoming and outgoing heat by means of the chemical process of metabolism and the physiological processes of thermoregulation in response to external factors of radiation, temperature, moisture, and air movement. Each of the physical processes of heat transfer is represented in the exchange equation:

$$M \pm Cd \pm Cv \pm R - E = 0$$

in which M is the heat of metabolism, Cd the gain or loss by conduction, Cv the gain or loss by convection, R the gain or loss by radiation, and E the loss due to evaporation. If the metabolic heat exceeds the sum of the other budget elements, body temperature will rise; if the resultant is less than zero, body temperature will decrease. Normal internal temperature is about 37°C; the temperature of the skin, which is the primary surface for heat exchange, varies around a mean of about 33°C. Metabolic heat depends ultimately on the intake and digestion of food. It is increased by muscular activity in the form of exercise or shivering. Excess heat is lost by radiation to surroundings, evaporation from the skin and respiratory passages, conduction to air and cold objects, convective transport in moving air (including cold air taken into the diaphragm and lungs), and minor losses to cold food and drink.

Within the normal range of air temperatures experienced by most people, the most effective cooling is achieved by evaporation of moisture from the skin, and this is the type of cooling over which the body has maximum control. As activity is increased, so is the metabolic rate. The body adjusts by increasing the circulation of blood near the skin and by perspiration, although people differ greatly in their ability to sweat. As long as evaporation can remove the secreted moisture, it will have a cooling effect, but if the relative humidity is high and the sweating profuse, a feeling of discomfort develops. Sweat which drops from the body merely represents a loss of water and is not effective in cooling the skin. The most comfortable relative humidity values lie in the range of 30 to 70 percent. Increased speed of air movement along the skin surfaces aids evaporation. At temperatures below about 20°C, evaporation from the skin loses importance as a cooling factor, for the rate of perspiration is reduced. Radiation, conduction, and convection serve to reduce the temperature of exposed skin, but these processes are restricted by clothing. Heat loss from the lungs increases with rapid breathing (panting) and becomes critical as the air temperature drops to extreme minima. At −40°C the dissipation of body heat through the lungs may account for one-fifth of the total loss. As activity increases the volume of air inhaled also increases, and at very low temperatures damage to the lungs and a serious lowering of body temperature may result.

Air temperature as measured by a thermometer is not, in itself, a reliable index of the temperature one feels. Individual responses depend not only on the elements of the heat exchange equation but also on subjective factors that vary widely among different persons and from time to time for the same person. The temperature the body actually feels, or senses, is the *sensible temperature.* Also sometimes termed the apparent temperature, it is not measurable by any instrument. Sensible temperature and wet-bulb temperature are approximately the same when the skin is moist and exposed to normal dry-bulb temperatures, however, since evaporation is then the principal cooling process. Under warm conditions, low relative humidity tends to reduce the sensible temperature because the rate of evaporation is greater.

High relative humidity, in combination with high air temperature, results in a lowered rate of evaporation and consequently a high sensible temperature. The familiar saying, "It's not the heat; it's the humidity," is not strictly true, for the feeling of oppression which we associate with hot, humid weather is dependent fully as much on the high temperatures as on the high relative humidity.

When air temperatures are low, evaporation from the skin becomes secondary to conduction as a cooling process. Because water vapor is a better conductor than the dry gases of air, high specific humidity promotes rapid conductive loss from the body. This explains why we find "dry cold" more bearable than "damp cold" so long as the air is calm. Thus, the sensible temperature is increased in summer but decreased in winter by high specific humidity. Owing to convective loss, it decreases in all seasons with increasing wind speed unless the air temperature is well above that of the body.

Air temperature and a moisture factor have been the primary bases for an objective index of sensible temperature. When the rate of air movement can be determined, a close approximation is the *effective temperature,* the temperature of nearly calm, saturated air which would elicit the same thermal sensation for a normally clad, sedentary person as that produced by the actual dry-bulb temperature, relative humidity, and air movement. Mean wet-bulb temperatures or mean relative humidities show a close correlation with subjective body responses under a given combination of dry-bulb temperature and wind speed. They therefore permit useful estimates of the range of conditions that are comfortable to most people. Although the perception of comfort varies with individuals, living habits, physical activity, the seasons, and climates, nearly everyone finds conditions uncomfortable or oppressive when wet-bulb values exceed 29°C. Air at dry-bulb temperatures below 16° to 18°C seems cool or cold to everyone regardless of the relative humidity. E. C. Thom has suggested a further refinement of temperature-moisture effects at the warm margin of the comfort zone. In his formula:

$$TH = 0.4(t_d + t_w) + 4.8$$

TH is the temperature–humidity index, t_d the dry-bulb temperature in °C, and t_w the simultaneous wet-bulb temperature. For conditions of little air movement, the index approximates the effective temperature (base at saturation). Most mid-latitude residents feel discomfort as the index rises above 21; everyone is uncomfortable when it reaches 26. The discomfort threshold is generally higher among people in the humid tropics and lower in cold climates.

The comfort zone for a group of people is a generalization derived from the individual comfort zones of the people in the group. Individual differences arise from variations in age, state of health, physical activity, type and amount of clothing, psychological factors, and past climatic experience. The normal, or group, reaction reflects not only the composite of individual responses but also the local climate and the season; hence the comfort zone (that is, the range of comfortable sensible temperatures for the group) depends on climate as well as the characteristics of the group. Because it is based on generalized statistics, it can tell little of the differing reactions of persons in different stages of acclimation or of the effects of short-term weather changes.

COOLING POWER

Since the great variety of individual differences makes it impracticable, if not impossible, to obtain measurements of the sensible temperature, the concept of *cooling power* has been developed to express the combined effects of air temperature and air movement in more objective terms. With respect to the human body, cooling power may be defined simply as the ability of the air to enhance the loss of body heat.

Several formulas have been devised to express cooling power in terms of observed values of temperature, humidity, and wind speed. The most-used indices of cooling power are derived from special instruments, however. Measurements of skin and body temperatures of living subjects under controlled conditions yield more realistic results and promise, eventually, to furnish the basis for sound conclusions on the reaction of the human body to varying conditions.

As we have seen, the body loses heat through a combination of several processes. Temperature and wind speed, which are the primary bases of theoretical cooling power, are only two of the factors in actual sensible temperature, but they are the major factors when the air is cold. Cooling effects due to low temperatures and wind have been called the dry convective cooling power of the atmosphere, or simply *wind chill,* a concept developed from experiments in Antarctica by Paul A. Siple. (When air temperatures are above the minimum that the body can maintain by radiation and evaporation losses, moving air heats rather than cools the body.) The colder the air and the higher the wind speeds, the greater is the loss by dry convective cooling. A direct index of the rate of loss is the quantity of heat transferred from a unit area during a given period (see Fig. 15.1). An expression of the relative cooling power is the *wind chill equivalent temperature,* that is, the temperature which at one wind speed would produce the same cooling effect on bare skin as the actual temperature and wind combination. For example, the wind chill temperature of a 70-km per hr wind at −12°C is equivalent to a 6-km per hr wind at −40°C (see Table 15.1). These relationships do not imply that a body will cool below the air temperature; rather they indicate a rate of cooling, which increases rapidly at low wind speeds and then more slowly at higher speeds.

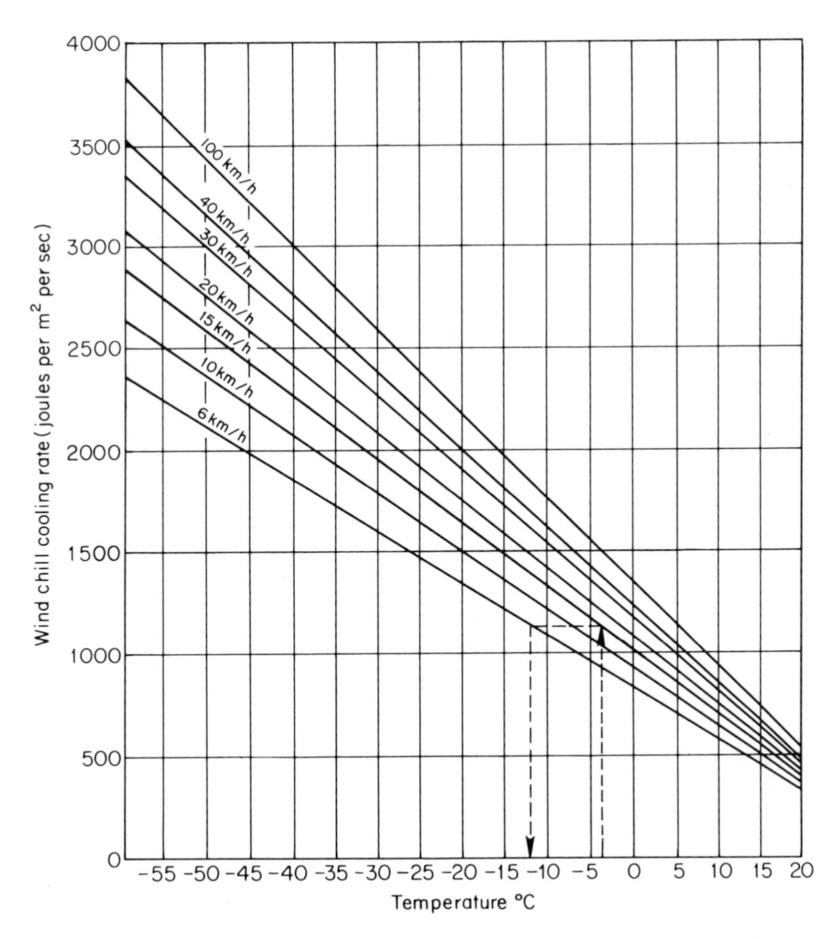

Figure 15.1 Rate of cooling produced by wind at different air temperatures. (NOAA, National Weather Service.)

CLOTHING AND CLIMATE

With respect to climate, the main purposes served by clothing are: protection against temperature changes and extremes, protection from excessive sunshine, and protection from precipitation. Clothing is also worn to protect the body against physical damage such as abrasion, cutting, or burning. With respect to culture, clothing is worn for adornment, prestige, fashion, custom, and other social reasons which vary from group to group as well as among individuals. Very often, the design and selection of clothing is dictated more by cultural than by climatic factors.

Clothing protects against the cold by trapping still air within its open spaces and in the layer next to the skin. Layers of ''dead air'' are good insulators because

TABLE 15.1

Wind Chill Equivalent Temperature as a Function of Wind Speed and Air Temperature*

Dry-Bulb Tempera-ture (°C)	*Wind Speed (km per hr.)*										
	6	10	20	30	40	50	60	70	80	90	100
20	20	18	16	14	13	13	12	12	12	12	12
16	16	14	11	9	7	7	6	6	5	5	5
12	12	9	5	3	1	0	-0	-1	-1	-1	-1
8	8	5	0	-3	-5	-6	-7	-7	-8	-8	-8
4	4	0	-5	-8	-11	-12	-13	-14	-14	-14	-14
0	0	-4	-10	-14	-17	-18	-19	-20	-21	-21	-21
-4	-4	-8	-15	-20	-23	-25	-26	-27	-27	-27	-27
-8	-8	-13	-21	-25	-29	-31	-32	-33	-34	-34	-34
-12	-12	-17	-26	-31	-35	-37	-39	-40	-40	-40	-40
-16	-16	-22	-31	-37	-41	-43	-45	-46	-47	-47	-47
-20	-20	-26	-36	-43	-47	-49	-51	-52	-53	-53	-53
-24	-24	-31	-42	-48	-53	-56	-58	-59	-60	-60	-60
-28	-28	-35	-47	-54	-59	-62	-64	-65	-66	-66	-66
-32	-32	-40	-52	-60	-65	-68	-70	-72	-73	-73	-73
-36	-36	-44	-57	-65	-71	-74	-77	-78	-79	-79	-79
-40	-40	-49	-63	-71	-77	-80	-83	-85	-86	-86	-86
-44	-44	-53	-68	-77	-83	-87	-89	-91	-92	-92	-92
-48	-48	-58	-73	-82	-89	-93	-96	-98	-99	-99	-99
-52	-52	-62	-78	-88	-95	-99	-102	-104	-105	-105	-105
-56	-56	-67	-84	-94	-101	-105	-109	-111	-112	-112	-112
-60	-60	-71	-89	-99	-107	-112	-115	-117	-118	-118	-118

*NOAA, National Weather Service.

they do not readily conduct heat away from the body. Therefore, the fundamental aim in design of effective clothing for cold weather is to provide insulation. Knit or loosely woven woolens are better than cotton cloth with a hard weave, and several thin layers of clothing are better than one heavy layer.

In order to be effective against wind chill, clothing must be reasonably impervious to the passage of air. Wind decreases the insulating capacity of clothing by forcing air into the openings and disturbing the still air. It also reduces the air layers between successive garments and between body and clothing by pressing the clothes against the body. Completely airtight materials are unsatisfactory because they do not provide for ventilation to remove moist air from near the skin. When clothing becomes wet from perspiration, it loses much of its insulating power. In cold weather, wet socks are one of the first danger signs of possible freezing. This is one of the reasons for avoiding strenuous exercise in polar climates. Another is the fact that exercise sets up a bellows action in clothing, thus increasing convective heat loss. The ideal material for an outer garment in cold winds is one that will largely restrict the passage of wind but that will "breathe," that is, allow the escape of water vapor. Undergarments should be resilient enough to maintain their insulating properties against the pressure of the wind.

A rather obvious method of combating the effects of low temperatures and wind is the use of an electrically heated suit regulated by a thermostat. Heated suits have been used with some success by high-altitude fliers and certain sedentary workers, but they are generally impractical for persons active in the outdoors.

For protection against heat, clothing should be loose and allow free transfer of heat away from the body. One of the primary functions is to shade the skin from direct rays of the sun. Lightweight and light-colored materials that reflect insolation are best, and they need to be porous to allow maximum air movement adjacent to the skin. Wearing coats and ties in hot summer weather is justified only by fashion; in terms of applied climatology it is indefensible. For temperatures above that of the body, insulating materials are of some value in preventing overheating. The desert Bedouin finds the same clothing that shields him from extreme insolation in the daytime useful against the cold night air. But in highly humid air, such clothing would be unsuitable. In any case, water vapor in the air absorbs a large proportion of ultraviolet rays so that the danger from that component of the solar spectrum is not so great in humid climates. Head covers in hot weather are especially important to prevent illness due to excessive heat and sunlight. They must permit free ventilation of the head, shade the eyes, and reflect sunlight. Footwear to insulate against hot ground is also necessary in hot climates, especially if the ground is dry. Warm-weather shoes should be porous to allow as much air movement as possible.

Clothing to ward off precipitation is usually of special design and not necessarily effective against accompanying temperatures, which must be regulated by undergarments. Completely waterproof garments made of rubber or plastic do not allow the escape of body moisture, and therefore become uncomfortable at any temperature. Ideally, the material and the design of rainwear should prevent all rain or melted snow from passing through to the undergarments or body regardless of wind force and at the same time they should allow free escape of evaporated perspiration. For practical purposes, it is necessary to strike a compromise between the ideal and a utilitarian garment that will permit the maximum efficiency of physical activity. Rain clothes suitable for a policeman or sentry may be quite unsatisfactory for more active persons such as loggers. Although experiments in clothing design must take into account the variations and extremes of the weather elements, the type of activity performed by the wearer is also of major importance. There is no such thing as "all-weather" clothing. A great deal of discomfort is endured by persons who fail to modify the amount and type of their apparel to conform to changes in the weather or tendencies in their own heat budgets.

INDOOR COMFORT

Temperature is unquestionably the most important element of indoor climate. Merely raising the air temperature in a cold room does not meet all the requirements for a comfortable environment, however. As outdoors, moisture content and motion of the air are additional factors affecting the human response. The concept of effective temperature combines the effects of temperature, humidity, and moving air. Early studies of human comfort assumed a reference base at saturation in calm air. Because indoor air rarely is saturated, an average relative humidity value affords a more realistic standard for comparison. The American Society of Heating, Refrigerating and Air-Conditioning Engineers has adopted a modified definition that incorporates 50 percent relative humidity rather than saturation as the basic humidity value. Figure 15.2 is the resulting ASHRAE Comfort Chart. It shows combinations of wet-bulb and dry-bulb temperatures, as well as relative humidity values, that elicit corresponding effective temperatures for persons who are adjusted to indoor conditions where the speed of air movement is less than 0.23 meters per second. An increase in air movement lowers the effective temperature; an increase in relative humidity raises it.

Because the effective temperature varies among different people, a composite is derived from the individual reactions of persons in a group. The shaded comfort zone in Fig. 15.2 indicates the range of temperature and humidity conditions which most office workers find comfortable when wearing normal indoor clothing. The partially overlapping diamond-shaped area is the comfort zone for lightly clad persons at rest. The effective temperature lines and comfort zones are based on the responses of American subjects in controlled laboratory experiments. They do not necessarily apply in all climatic regions nor among groups of people who are accustomed to other indoor conditions. For example, in some mid-latitude countries where room temperatures are generally lower, the optimum effective temperature for comparable levels of activity may be appreciably lower than in the United States. In the tropics comfortable effective temperatures are higher owing to different patterns of climatic experience and customs in dress and diet.

Another factor affecting the feeling of comfort in a room is the transfer of heat

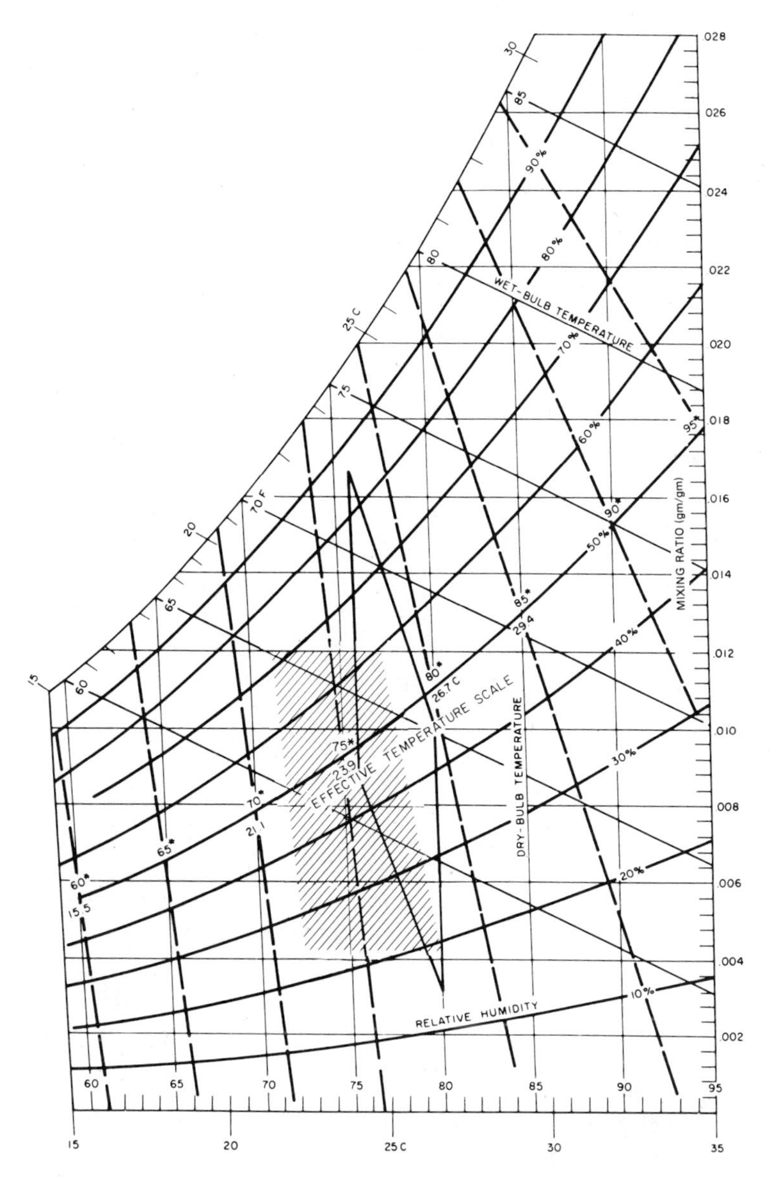

Figure 15.2 ASHRAE Comfort Chart. Dashed lines indicate effective temperatures based on 50 percent relative humidity and air movement less than 0.23 m per sec. Shaded comfort zone is for normally clad office workers; diamond-shaped zone is for lightly clad persons at rest indoors. (Reprinted with permission from the 1977 Fundamentals Volume, ASHRAE HANDBOOK & Product Directory.)

by radiation between the body and walls, floor, ceiling, and solid objects such as furniture. Radiational loss to cold walls may create a sensation of discomfort even though air temperature is at the theoretical optimum. The body may experience simultaneous gains and losses of radiant heat on different sides. The average effect is expressed as the *mean radiant temperature,* an index combining air speed and *globe temperature.* The latter is measured by an ordinary thermometer inserted in a blackened, air-filled, copper sphere about 15 cm in diameter. When wall temperatures are low due to cold weather, it is necessary to keep the indoor air temperature a few degrees higher than the standard in order to compensate for the radiative loss. Conversely, warm surfaces radiate heat to the body independently of the temperature of the intervening air and can maintain comfort, even though air temperature is two or three degrees below the standard.

WEATHER AND HEALTH

Weather changes and extremes produce a variety of influences on human health, some of which result in illnesses arising from direct effects of atmospheric conditions on the body. Temperature extremes are the most common causes of illness related to weather. *Heat stroke,* or *hyperthermia,* develops when the body is unable to maintain its heat balance at high relative humidity and air temperature above that of the body; it may lead to death if deep body temperature rises above the critical level of 42°C. Symptoms are fever, nausea, dizziness, and headache. Treatment entails reducing body temperature by means of cold baths. *Heat exhaustion* is a milder form of hyperthermia identified by dizziness, lassitude, and perhaps fainting. It is more common in crowded rooms than in the outdoors. When the body suffers an excessive loss of salts and water in perspiration, *heat cramps* may result. Adequate liquid and salt intake help to prevent heat cramps and also relieve the condition if it develops. Because of changes in metabolism and blood circulation, the appetite and digestion are impaired in hot weather; digestive disorders are more common in summer and in the tropics.

The summer heat waves of mid-latitudes create especially trying conditions in terms of health and comfort. The more disastrous heat waves accompany subsiding continental air; they affect large regions for several consecutive days and cause numerous heat deaths. Figure 15.3 shows graphically the correlation between average July temperatures and heat deaths in the United States during the drought years of the 1930s, when nearly 15,000 people died as a result of failure to withstand heat stress.

A common direct effect of low temperatures is *frostbite.* The extremities and exposed portions of the body are most likely to suffer, and the danger is heightened by increased movement. Perspiration, especially from the feet, accelerates conductive heat loss, as does the accumulation of condensed moisture about the face. The precautions against wind chill, namely, adequate clothing and moderation of physical activity, help to prevent frostbite. Prolonged exposure to low air tem-

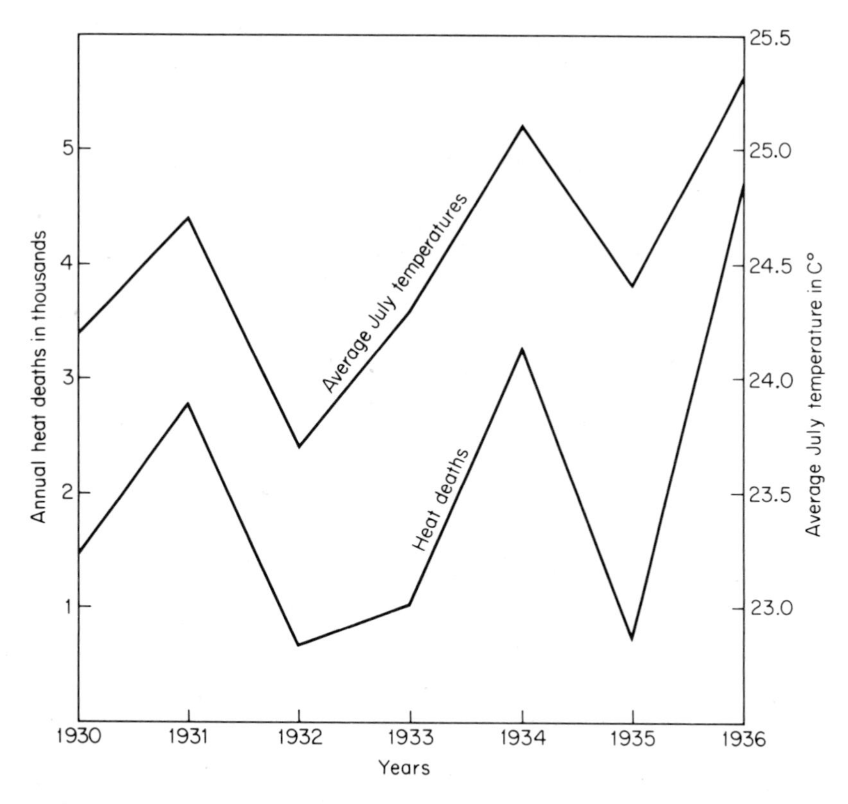

Figure 15.3 Mean July temperatures and number of heat deaths per year in the United States 1930–1936. (NOAA, Environmental Data and Information Service.)

peratures can alter the body's heat balance, producing *hypothermia* and causing eventual death. The development of hypothermia is hastened by physical exhaustion. Low temperatures are an aggravating factor in a number of ailments such as arthritis, swollen sinuses, chillblains, and "stiff joints." Sudden lowering of temperature puts a severe strain on persons with cardiac disorders, although there is no evidence that climate directly causes such ailments. Indeed, virtually all physiological functions react in some manner to temperature changes.

Atmospheric pressure and relative humidity changes appear to bear a relation to certain kinds of pains, notably those associated with respiratory infection and muscular aches. The irritation of respiratory passages, which is commonly brought about by dry air, is further enhanced by wind and dust. Temperature and humidity are significant factors in the release of pollens and consequently affect the incidence of allergies. Very dry air is a major contributory cause of chapped skin and it inhibits the healing of sores and wounds.

To the extent that health is a function of nutrition, the influences of weather and climate upon diet indirectly affect human physiology. One facet of the influence results from the availability of different types of food in different climates. More important is the effect on appetite and selection of food. Temperature is the most significant consideration in this respect. Under cold conditions, the body requires a greater food intake to maintain heat; increased amounts of fats and carbohydrates are needed. Vitamins and minerals are, nonetheless, essential. Primitive peoples in the arctic achieve a balanced diet by eating virtually all parts of animals and fish. People of more sophisticated cultures secure a proper balance through a variety of foods. Nutritional diseases may result from insufficient intake of calories, vitamins, or minerals. Adequate water must also be provided, and where it is obtained from melted snow, the deficiency in minerals has to be made up through foods.

Nutritional requirements in hot climates differ from those in the midlatitudes in that more salt, water, and certain vitamins are needed. Whether these requirements are met is another question. Suppression of appetite may be a major factor in malnutrition in the tropics. There is some evidence that meat and eggs produced in the tropics are deficient in certain nutrients, notably vitamin B_1, presumably as a result of the effect of heat on the living animals. Still other deficiencies in food can be traced to leaching of tropical soils by heavy rains. Whether humans can improve their heat tolerance by controlling their diet is as yet an unsolved problem.

SUNSHINE AND HEALTH

Besides the obvious relation between sunshine and air temperature, the solar spectrum produces several effects on the human body. Infrared rays are absorbed by the body or clothing and converted to heat, thus offsetting much of the cooling power of the air. It is therefore perfectly natural to seek the shade in hot climates and the sunny exposures in cool climates. The visible part of the spectrum (light) affects mainly the eyes. The intense sunlight of the arid tropics or that reflected off snowfields can cause forms of blindness, headaches, and related discomforts.

Ultraviolet rays are valuable for their ability to form vitamin D in the skin and to devitalize bacteria and germs. These qualities explain in part why many health resorts are in sunny locations. On the other hand, ultraviolet radiation can cause premature aging of the skin and sunburn (erythema) to the point of illness. When the skin becomes pigmented (tanned), the pigment affords some protection against further inflammation but does not eliminate the risk of skin cancer, which can be fatal. Blonde persons are generally more susceptible to sunburn than those having darker skins. A "healthy suntan" is commonly regarded as the mark of a successful vacation, although it is probably less effective and potentially more hazardous than fresh air, a change of scene, and relaxation in promoting physical and mental well-being.

The apparent increase in the incidence of skin cancer with decreasing latitude is caused by the greater amounts of ultraviolet radiation where the sun angle is higher and the ozone content of the atmosphere is less. Combined with intense heat, ultraviolet rays are also a factor causing cataract of the eye. These maladies are among the reasons for concern about possible depletion of ozone at high levels in the atmosphere.

Tourist centers and resort areas are always proud to advertise whatever favorable sunshine data they can glean from climatic records. It is true that there are some large regional variations in duration of sunshine. Moreover, the intensity of sunlight is of primary significance in its effect on the body. At high latitudes and in the morning and evening hours, the low angle of the sun's rays makes sunlight much less potent. Smog and haze also detract from its power. At high altitudes, the air is much clearer so that solar radiation is more intense. Possible daily sunshine decreases with latitude in winter and increases in summer; the actual average duration varies with the climate, specifically with cloudiness. Figures 15.4 and 15.5 show the distribution of percentage of possible sunshine in the contiguous United States in winter and summer, respectively. It is evident that the drier climates are generally favored with more sunshine, but the claims of Florida are fairly well substantiated, especially in winter.

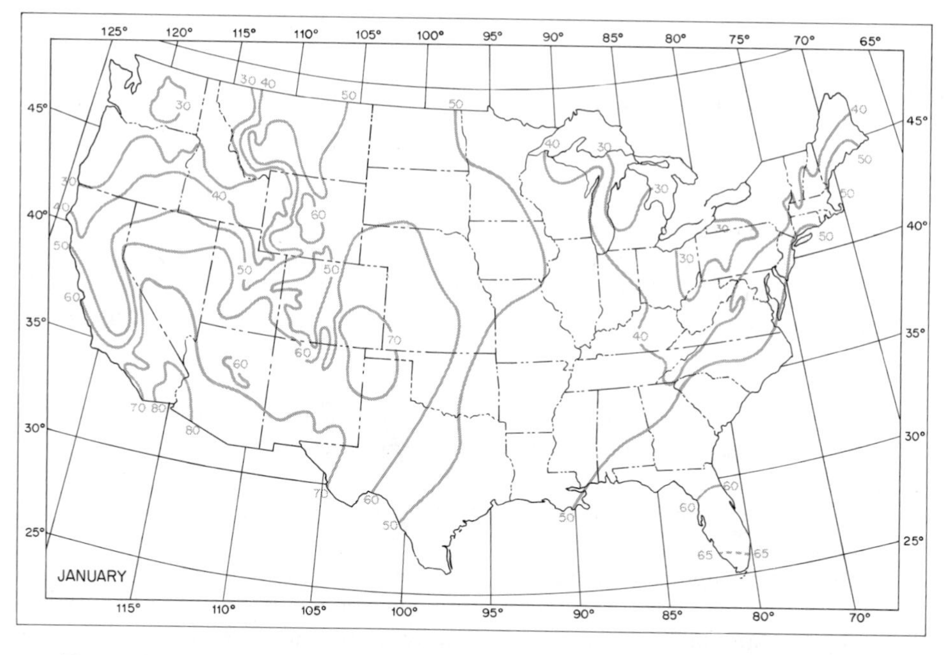

Figure 15.4 Mean percentage of possible sunshine in January in the contiguous United States. (NOAA, Environmental Data and Information Service.)

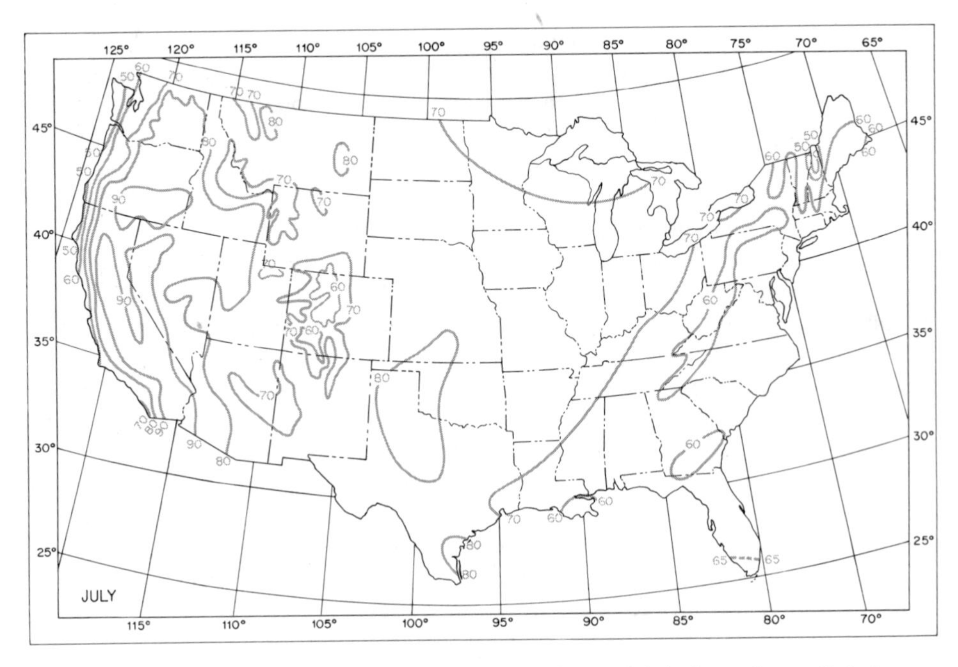

Figure 15.5 Mean percentage of possible sunshine in July in the contiguous United States. (NOAA, Environmental Data and Information Service.)

AIR POLLUTION AND HEALTH

The kinds of atmospheric pollution that may affect human health and comfort include gases (and their odors), liquids, inorganic solids, pollens and organic dusts, heat, light, noise, turbulence, and radioactivity. Their great variety requires a broad definition of atmospheric pollution: an undesirable concentration of any phenomenon in the air. Obviously pollution is a matter of degree as well as kind. Much less is known about the cumulative and long-term influence of moderate pollution on health than about specific disasters arising from abnormal concentrations, yet the former are equally important to human welfare. Moreover, the effects often are difficult to distinguish from other ailments. The incidence of harm from pollution has been highest among infants, the aged, and persons who were already ill.

Chemical effluents and solid particles are responsible for irritation of eyes and mucous membranes, intensification (if not the cause) of respiratory disorders, and pathological conditions leading to death. The more common chemical pollutants are sulfur dioxide, carbon monoxide, nitrogen oxides, ozone, and various hydrocarbons. Many airborne pollutants ordinarily occur in minute quantities, but some (for example, lead and arsenic) tend to accumulate in the body.

Although air pollution has no strict climatic boundaries, hazardous concentrations are more likely to develop in stable air. From a review of the physical causes of atmospheric stability it is evident that the conditions for a potential hazard are present in topographic depressions that are subject to air drainage and radiation inversions and in regions of stagnating anticyclones. These have been the circumstances of major air pollution disasters since the beginning of the industrial revolution. In December 1930, 63 persons died as a result of intense industrial pollution in the Meuse Valley of Belgium. Several others suffered respiratory ailments, and many animals died. At Donora, Pennsylvania, in the Monongahela Valley, 21 people died during a pollution episode in late October 1948. Thousands experienced respiratory and related disorders. The worst air pollution disaster on record occurred in London during the period December 5 to December 9, 1952, when a subsiding continental polar air mass trapped a shallow layer of smoke over the Thames Valley. In the Greater London area nearly 4,000 deaths were attributed primarily to sulfur dioxide effluents from burning coal.

Events such as the foregoing have awakened government agencies to the need for control measures as well as for an improved understanding of the atmospheric conditions that enhance pollution. Urban and regional planning increasingly recognizes the topographic and climatic aspects of air pollution. In the United States the federal Air Quality Act of 1967 provided, among other things, for the designation of air quality control regions as a geographic framework for a systematic approach to air pollution problems. A Clean Air Act took effect in Britain in 1956. To facilitate public health warnings and the regulation of contaminating sources, the relative concentrations of one or more pollutants are used increasingly as bases for an *air pollution index.* Combined with weather forecasts and data on local pollution climatology, the index is an aid in averting disasters.

Natural aeroallergens are a form of pollution having close relations with weather and climate. Many plants have a narrow range of temperature and moisture requirements for the production, release, and dispersal of allergenic substances. Hay fever, asthma, and certain other respiratory ailments are initiated or aggravated as pollens and organic dusts are spread, often for great distances, by the wind. The end of the growing season for plant sources normally brings a halt to the "hay fever season."

As a type of natural pollution, noise may be generated directly by wind and precipitation. Certain human-caused noises originate more often in one kind of weather than another. The roar of power mowers in the suburbs has a close correlation with sunny weekends. Weather conditions also influence the propagation of noise. Unstable air and turbulence in strong winds dissipate sound rapidly. Stable air associated with surface inversions of temperature and light, steady winds tend to concentrate sound near the ground, where it projects for great distances.

CLIMATES AND DISEASE

Many factors, such as cleanliness, nutrition, physical activity, and social contacts, act together in determining the incidence, severity, and spread of a disease. Climate is another factor, one which varies in importance according to the disease

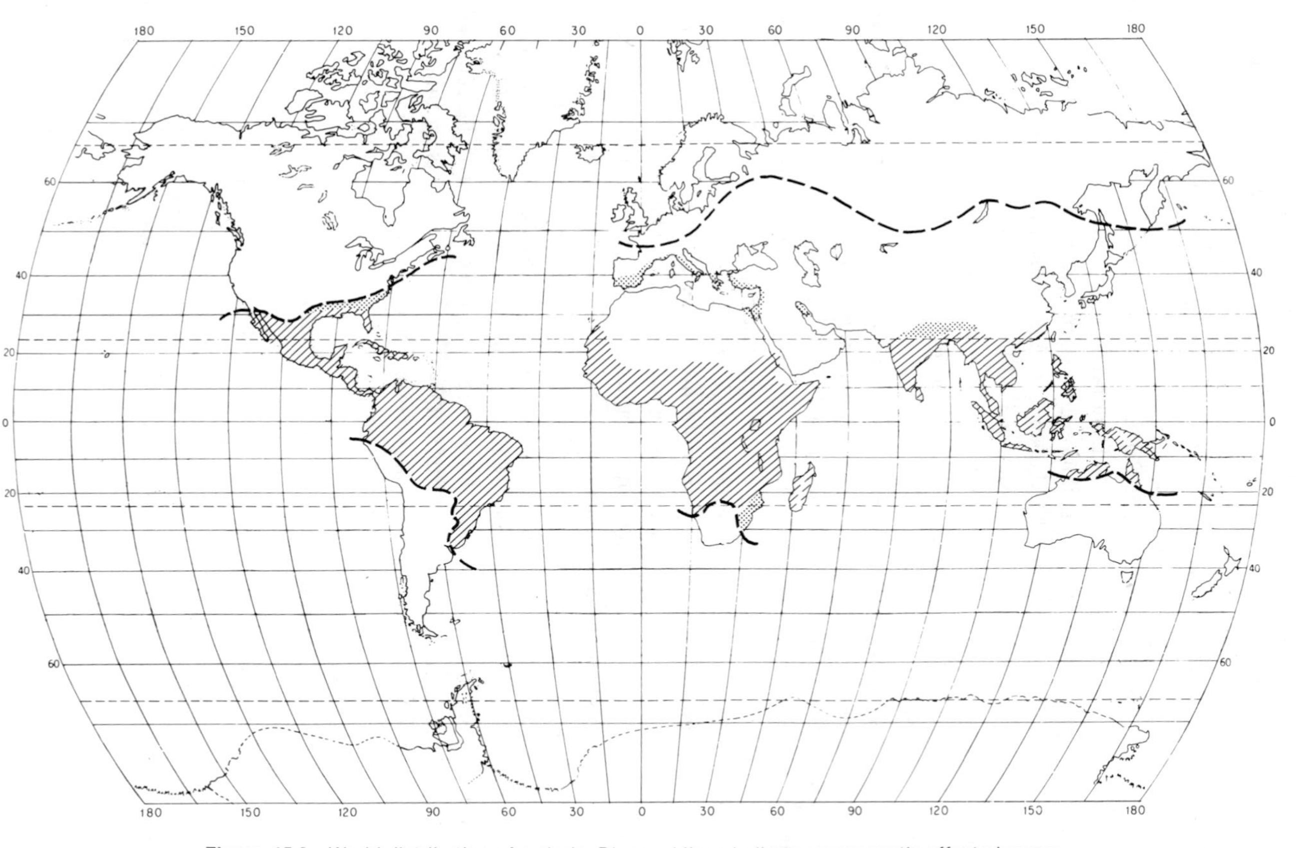

Figure 15.6 World distribution of malaria. Diagonal lines indicate permanently affected areas; shaded areas have histories of malaria; dashed line is the approximate poleward limit of malaria-carrying mosquitoes. (Adapted from American Geographical Society *Atlas of Diseases.)*

in question and the physiological and cultural characteristics of potential or actual victims. The climatic relations of all the common diseases cannot be discussed here; rather the aim will be to outline briefly the broad influences which weather and climate may exert on disease. The relations are extremely complex. Much cooperation in research among specialists in branches of climatology and of medicine is necessary to determine the role of climate in causing, modifying, or facilitating recovery from a specific disease.

There are two basic aspects of climatic influence on disease: the relationship of climatic factors to disease organisms or their carriers and the effects of weather and climate on the body's resistance. Many diseases are associated primarily with certain climates or with a season because of the temperature, moisture, and other requirements of the microscopic organisms that cause them. A number of parasites which attack humans are confined to the tropics and subtropics, where they find suitable conditions of warmth and moisture. Scarlet fever is virtually unknown in the tropics, whereas leprosy flourishes there. Some diseases depend on intermediate carriers and are restricted to environments favorable to those carriers. Yellow fever and malaria, for example, are spread by certain species of mosquitoes that thrive in warm humid climates. (Figure 15.6 is a map of world areas where climatic conditions are favorable for malaria-carrying species of anopheles mosquitoes.) Rocky Mountain spotted fever occurs in the summer when the tick carriers are active.

Many diseases follow a distinct seasonal pattern (see Fig. 15.7). Pneumonia and influenza are common seasonal diseases of the mid-latitudes; their greater incidence in winter is probably due to the lowered resistance in the upper respiratory tract and closer group confinement at that season. Measles and scarlet fever cases are most numerous in spring. Infectious diseases occurring chiefly in winter and spring are much more widespread among the population than those that have their maximum frequency in summer and autumn.

Few diseases are caused directly by climate. A given combination of climatic elements may modify the metabolic rate, respiration, circulation, and the mental outlook of the individual so as to either strengthen or weaken his resistance to disease. Chilling, for example, lowers body resistance to most illnesses. Even in the tropics, sudden decreases in air temperature may be followed by outbreaks of sickness. Persons who fail to adjust their rate of physical activity in high or low temperatures often suffer some degree of exhaustion and a predisposition to disease. The stress imposed by low barometric pressures at high altitudes also enhances development of a number of diseases. As we have seen, weather conditions can effect a concentration of pollutants and thereby increase their potential influences on health.

Favorable atmospheric conditions can assist the body in warding off disease and in promoting recovery if the disease is contracted. Fresh air, sunshine, mild temperatures, and moderate relative humidity all have therapeutic values. Fresh air and sunlight have long been recognized in the treatment of tuberculosis. Rickets and certain skin diseases respond to sunshine. The change of climate often prescribed for various kinds of illness is beneficial only if it is accompanied by rest, im-

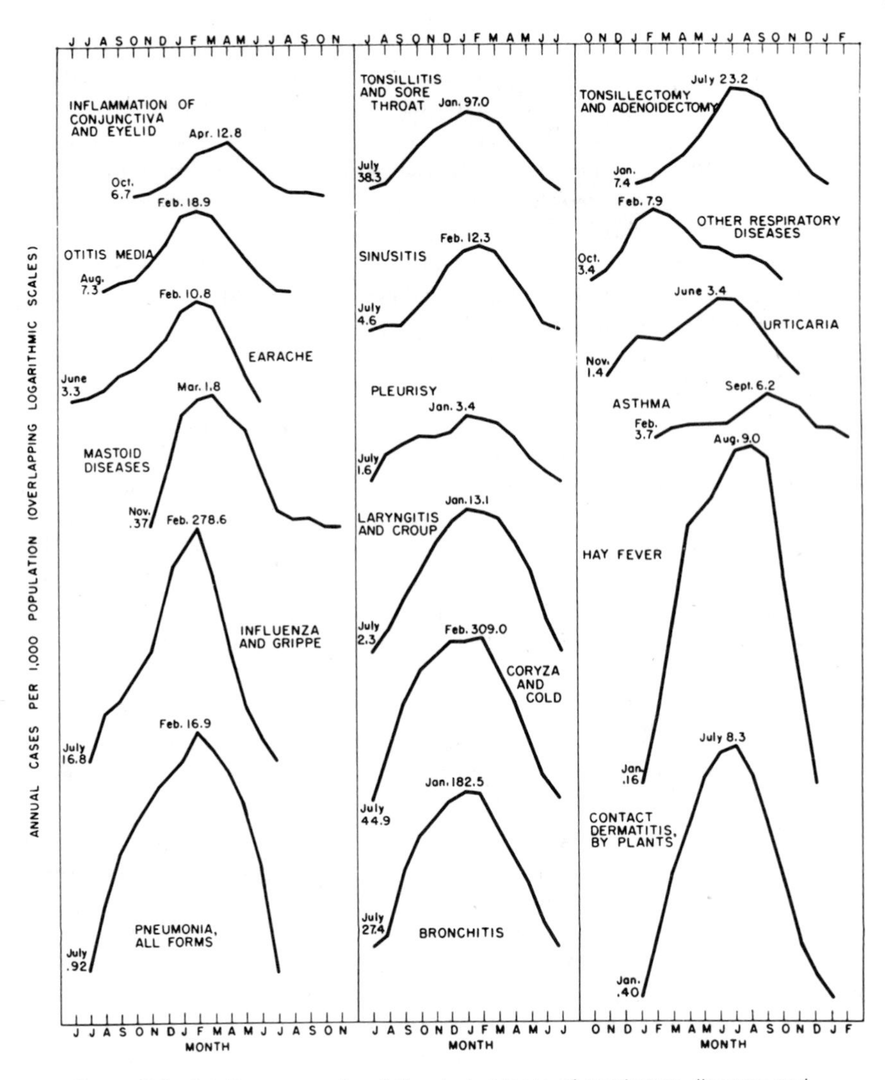

Figure 15.7 Relative seasonal variation in incidence of respiratory diseases and allergies in the United States. (Data from U. S. Department of Health and Human Services.)

proved mental outlook, proper medical care, and good food. Moving to a location with a quite different climate may introduce problems that offset any actual curative powers of the new climate. Functional ailments which are obviously psychosomatic often are alleviated by moving to a different climate if emotional stress on the patient is thereby reduced. In such cases the change of scenery and social environment may be far more important than the change in atmospheric conditions.

ACCLIMATION

Acclimation is the process by which human beings (and other animals) become adapted to an unfamiliar set of climatic conditions. In the broad, popular sense it implies adjustment to all phases of a new physical and cultural environment, and indeed it is difficult to distinguish the purely climatic phenomena from other factors in strange surroundings. The chief problems in adjustment to a new climate frequently are not climatic at all but arise from homesickness, boredom, or social incompatibility.

In the narrower sense of physiological climatology, acclimation entails actual changes in the human body brought about by climatic influences. It connotes a decrease in physiological stress as the body continues to be exposed to the new conditions. Temporary adjustments are made to daily and seasonal weather changes. But, when a person moves to a different climate, a more permanent adaptation gradually takes place. As would be expected from the foregoing sections, temperature is the element of greatest significance in acclimation. Table 15.2 summarizes major physiological responses to high and low temperatures. In a hot climate the cutaneous blood vessels dilate so that more blood will be exposed to the low cooling power of the air and thereby maintain normal body heat. The blood supply is gradually increased in volume in order to meet the great capacity of the capillary vessels, and it undergoes chemical changes. A larger number of sweat glands become active and there is a concomitant increase in thirst. Oxygen consumption declines. During the period of adjustment, there may be physical discomfort, lack of energy, and an indifferent appetite. Persons in poor health may find their condition worsened by the stress. They are well advised to make the shift between two widely differing climates in gradual stages. In some cases a satisfactory stage of acclimation may never be achieved.

Acclimation is not so clearly defined in a cold climate as in a warm one. Residents of cold climates cannot withstand continuous exposure to extreme cold but they are less affected than nonacclimated persons. Repeated contact with cold air leads to a decrease in the flow of blood through constricted capillaries near the skin. Blood viscosity and the number of circulating white cells increase; the liver enlarges; appetite and oxygen consumption both increase in a natural attempt to provide more body heat. Sweating does not occur so readily and tends to be confined to certain skin areas, especially the palms, soles of the feet, and the axillae.

At high altitudes, adjustment must be made to both the lower temperatures and low pressure. Increased pulse and respiratory rates and greater production of red blood cells are the principal modifications which occur to offset the reduced oxygen supply. People who live at high elevations develop greater lung capacity and larger chest cavities than residents of lowlands.

Although the human body is capable of withstanding remarkable extremes of climate, it functions best under circumstances where there is a reasonable degree of comfort and freedom from stress on vital organs. In general, children and healthy

TABLE 15.2

Responses of the Human Body to Thermal Stress *

At low temperatures	*At high temperatures*
TEMPERATURE REGULATION	
Constriction of skin blood vessels	Dilation of skin blood vessels
Concentration of blood	Dilution of blood
Flection to reduce surface exposure	Extension to increase exposure
Increased muscle tone	Decreased muscle tone
Shivering	Sweating
Inclination to increase activity	Inclination to decrease activity
CONSEQUENT DISTURBANCES	
Increased urine volume	Decreased urine volume
	Mobilization of tissue fluid
	Thirst and dehydration
Danger of inadequate blood supply to exposed parts; frostbite	Reduced blood supply to brain; dizziness; nausea; fainting
	Reduced chloride balance; heat cramps
Discomfort leading to neuroses	Discomfort leading to neuroses
Increased appetite	Decreased appetite
FAILURE OF REGULATION	
Falling body temperature	Rising body temperature
Drowsiness	Impaired heat regulating center
Cessation of heartbeat and respiration	Failure of nervous regulation; cessation of breathing

*Adapted from *Arid Zone Research, X – Climatology, Reviews of Research,* Propioclimates of Man and Domestic Animals. Reproduced by permission of UNESCO. © UNESCO, 1958.

persons are more adaptable to a change of climate than the aged or infirm. Fortunately, it is possible to lighten stress during acclimation by proper diet, clothing and housing, or by control of physical activity. Many questions concerning these and other factors which modify climatic influences still remain to be answered by research in physiological climatology and related fields.

PERCEPTION OF CLIMATE AND CLIMATIC HAZARDS

In addition to the daily adjustments people make to the changing weather, they face decisions involving climate when planning a vacation or when they contemplate a change of residence for any reason, be it improved economic or social status, retirement, health, or climate itself. It has become increasingly practicable to move to places perceived to have a "good" climate. Fortunately climate and other considerations do not lead to the same choice for everyone. The dry summer subtropics have a high rating throughout the world. Arid and semiarid climates having moderate temperatures are also attractive to many, and a climate with warm winters usually is preferred to a cold-winter type, hence the annual migration between the mid-latitudes and the subtropical "Sun Belts" in Europe, North America, and Australia. For many people, moderate temperatures and abundant sunshine, combined with freedom from weather hazards, are the cardinal attributes of a pleasant climate. The cheering effect of a sunny day following a surfeit of gray skies is no mere illusion. Yet long periods of clear skies and glaring sunshine can induce ennui. Change, too, has its advocates if the extremes are muted. Clearly, cultural as well as physical factors influence the human response. It may never be possible to identify, let alone quantify, all the variables that underlie perception of climate.

Human perception of climate and its hazards embraces the criteria of cause, magnitude, time and duration, spatial arrangement, uncertainty, and resultant effects. Rarely do different individuals or social groups apply these criteria in the same way. At one extreme a climatic event is regarded as a supernatural act; at another it is analyzed in mathematical terms as a system of physical processes. Climatic hazards such as droughts, floods, severe storms, or heat waves thus evoke a broad spectrum of cultural perspectives that vary not only among individuals but also with time and distance. Whereas popular accounts of past climatic disasters may exaggerate the facts, it is common for people to minimize the threat of predicted hazards and to resist procedures that might help to avert a catastrophe. Remoteness from an affected area also influences the level of concern.

Beliefs relating to climatic change and human ability to modify climate are particularly intriguing aspects of the perception of climate. The layman's interest in a possibly changing climate usually is aroused by extreme weather events rather than by gradual trends. Even where the average temperature may have risen as much as 2° or 3C° the change would go unnoticed by most people, unless it is manifested in extremes. The increased use of cooling systems in homes and buildings is more the result of their greater availability at moderate cost than any real effect of a warming climate. Air conditioning, transportation, eating habits, clothing, agricultural practices, and many other attributes of culture have changed to such a degree that we cannot rely on simple personal assessments of climatic change. In spite of advances in technology, human societies have become more vulnerable to the hazards of weather and climate because of growing pressures on food supplies, energy resources, and living space. Acceptance of greater risks in order to relieve these pressures has led to an increase of property losses and human suffering in the wake of severe weather episodes. Whatever the actual trends in climate may be, the perception of the climate system determines social, economic, and political decisions that are related to climate.

An improved, but still incomplete, understanding of the climate system also has altered perceptions of the human impact on climate. In little more than a century, hopes of controlling climate in desert and semiarid regions by means of cultivation, irrigation, or afforestation have given way to apprehension about the effects of energy consumption, pollution, or desertification. Concurrently, human beings have relentlessly pursued mastery of their environment, even experimenting with techniques for direct modification of climate.

QUESTIONS AND PROBLEMS FOR CHAPTER 15

1. List the processes that maintain the heat balance of the human body and explain how each depends on the climatic environment.
2. Why do most people find very high and very low temperatures even less comfortable when the vapor pressure is high?
3. If the dry-bulb temperature is 24°C and the relative humidity is 30 percent, what is the corresponding temperature–humidity index? (See Table B.2, Appendix.)
4. To what extent might individual perceptions of thermal comfort be changed in order to promote energy conservation?
5. Explain how each of the following affects human health and comfort: ultraviolet radiation, stable air, wind chill.
6. Assuming favorable social, economic, and political conditions, which of the world's climatic types would you choose for permanent residence? Justify your choice.
7. Harmful effects of atmospheric aerosols on human health are known to medical science. What would be the result if there were no solid particles in the air?
8. Explain why respiratory infections are more widespread during winter months in the mid-latitudes whereas intestinal maladies have a peak in summer.

16

Climate and Housing

Successful control of the environment within buildings achieves the dual goal of enhanced human comfort and modification of climate on a small scale. Human dwellings have reflected influences of weather and climate since prehistoric times for the excellent reason that the primary function of housing is to shelter the inhabitants and material contents against weather and climate. Whether one has in mind a cottage, a mansion, or a skyscraper, atmospheric conditions are relevant factors in efficient siting, choice of materials, design, and air conditioning of the structure. The task of "building climatology" is to analyze outdoor climatic factors that influence structural integrity and indoor climates. Temperature, radiation, precipitation, and wind are the chief elements requiring attention. Humidity, cloudiness, and visibility may be criteria for certain decisions. Economic and social factors normally have a major influence on housing, but the principles of applied climatology afford a basis for many refinements in location and construction of buildings as well as improvements in the use of energy for heating or cooling. Indeed, that which is economically feasible and socially acceptable in housing is determined in part by climate.

CLIMATIC ASPECTS OF SITE

Assuming it has been decided to construct a building in a given city or rural area, local microclimatic conditions weigh heavily in the selection of the best site. A good building site may possess microclimatic advantages which offset some of the disadvantages of the prevailing regional climate, that is, the macroclimate. Variable microclimatic conditions are commonly induced by local relief, but they may also be brought about by landscaping, adjacent buildings, water bodies, and industrial wastes. Since some of these controls of microclimate are likely to be altered by expanding settlement, especially in a city, it is advisable to anticipate possible major changes in an environment which might drastically affect the microclimate. Consider, for example, the shading effect of a tall building on a solar-heated home. Ideally, a complete site study should include the microclimate along with such matters as bedrock, drainage, land cost, or proximity to services. Unfortunately, detailed instrumental observation of the microclimate requires time and is not economically feasible except for the most expensive structures. The alternative is to apply known principles of microclimatology and to make reasoned inferences with respect to the climatic conditions at a particular site.

Wind is an important climatic element in site selection, for it produces direct effects on a building and modifies temperature and moisture effects. How much shelter from the wind is desirable depends on the regional climate and on the type of building on the site. In warm and humid climates, free circulation of air helps to lessen excessive humidity and high temperatures; in cold climates, wind may have value in moisture control, but it also increases convective cooling. In either case, high-velocity winds are to be avoided if practicable, otherwise they must be allowed for in building design. Local relief is the most significant as well as the most permanent of the common controls of wind conditions at a specific site. Windward slopes, summits, and plains are likely to receive the full force of surface winds. Sites where topographic barriers produce a constriction in air flow often have strong and gusty winds. Valleys and slopes are subjected to the effects of air drainage and, possibly, mountain and valley breezes. These local movements of air can be of immense value in warm climates and in summer, but they may be annoying in winter. Selection of a site should take into account the directions of prevailing winds and their temperature and moisture characteristics, but a wind which occurs only a small percentage of the time may be more objectionable than all others combined. An outstanding example is a wind from the direction of sources of air pollution. The problem is not confined to cities. Few farmers would want their homes to the leeward of livestock pens or barns. Objectionable odors, pollens, dust, and even insects are carried by the wind.

Types of surface and ground cover to the windward modify wind force, temperature, and humidity. A forest, orchard, or park has a moderating effect on summer temperatures, but pavements and other bare surfaces magnify temperature extremes. Buildings in the vicinity modify both direction and speed of the wind and may create undesirable local currents. A cluster of large buildings absorbs heat during a hot day and radiates it at night. Summer nights may be abnormally warm in the area of its influence. Water bodies aid in reducing diurnal and seasonal temperature ranges and generate breezes. If they are too shallow, however, they warm readily and thus create oppressive relative humidities in their vicinities in summer. In winter in the middle and high latitudes, shallow lakes and ponds may freeze, rapidly losing their moderating effect.

Temperature conditions, other than those directly influenced by the exchange of air between the site and its surroundings, are controlled by sunshine and by elevation. Since the approximate location of the housing site is assumed to have been decided, latitude and its effects on duration of possible sunshine and noon angle of the sun are not variable factors. Average cloudiness may differ within a given area because of relief features, and industrial smoke may greatly decrease the effective solar energy, another reason for selecting a site to the windward of industrial establishments. Maximum benefit from insolation is achieved by choosing a site on the slopes facing toward the equator, where the sun's rays are more nearly perpendicular to the ground surface. In contrast, poleward slopes receive the sun at low angles and may actually be in the shade for part of the day. In the Northern Hemisphere, slopes on the east and southeast are sunniest in the morning; west and southwest slopes are sunniest in the afternoon. Where maximum solar heating is desired, westerly exposures are better than those on the east because insolation is received at the time of highest air temperatures. An easterly exposure provides heating in the early morning; direct insolation is less in the afternoon, and cooling proceeds faster in the evening. Sites in deep valleys have a shortened period of possible sunshine, which may be a distinct disadvantage in winter. In a built-up area the difference in sunniness on opposite sides of a street is worth considering. Nearby tall buildings may shut out most of the sunlight.

The effect of elevation on temperature-design decisions is closely related to relief. Except where differences of several tens or hundreds of meters in elevation are involved, higher sites are cooled more by the freer circulation of air than by the effects of the normal lapse rate of temperature in the daytime. Air drainage and temperature inversions commonly disrupt the normal lapse rate at night. Under clear skies, a broad valley has greater extremes of temperature than adjacent slopes, where a temperature inversion provides warmer night air. Slope sites at or above the average level of inversions have the advantage of being above most of the associated evils such as smoke, haze, and radiation fog.

Precipitation differences will not be greatly significant in the normal range of site possibilities, unless there are wide variations in elevation and orographic conditions, in which case lee slopes usually are drier. If there is considerable wind, some knolls and ridges produce the opposite effect as rain or snow is swept upward on the windward and deposited on the lee. An examination of local vegetation often aids in determining whether this is a common occurrence. Sites exposed to the wind will be most affected by driving rains. The pattern of snowdrifts depends on local eddies produced by obstructions. If deep snow is a winter problem, a site near a highway or street which is regularly cleared is better than an isolated one, unless one prefers to be snowbound. Where heavy showers occur or where there is a heavy runoff of snowmelt, flooding can necessitate excessive costs in design and maintenance of a building and its site. It is never wise to build on or near a river bank without thorough knowledge of the precipitation characteristics of the watershed and the resulting flood regime.

CLIMATE CONDITIONING

Even the best building sites are subject to weather changes, and other factors may make it necessary to use a site which is poor from the point of view of applied climatology. Nevertheless, proper orientation, design, and choice of materials for a house can overcome many of the climatic disadvantages of a site, the primary objective being to create a suitable climate in and around the house, that is, *climate conditioning.* Climate conditioning is concerned with landscaping and the placement of other buildings as well as with the design of the house itself. In its broadest sense, it includes certain aspects of the field of urban planning, for groups of buildings and the associated streets and parks create their own microclimate. For many householders, convenience, architectural harmony, view, and cost of maintenance offset the minor advantages of a slightly better microclimate, but it is well at least to recognize the climatic aspects of site, design, and construction.

BUILDING ORIENTATION

The effects of weather elements are strongly directional. Orientation of a building with respect to wind and sun consequently influence wind force, precipitation, temperature, and light on different external surfaces. Narrow building lots in a residential area preclude wide choice in orientation of a dwelling. Adverse climate requires compensation in design and in the mutual protection afforded by closely spaced houses. Where there is greater freedom, it is expedient to orient buildings to achieve the maximum control over the microclimate. The broad face of a house obviously endures greater wind pressure than does the narrow dimension under the same wind speeds. This may be desirable if high temperatures are prevalent, but it becomes a problem if winds are exceptionally vigorous or cold. Strong, cold winds on the side of the main entrance are particularly objectionable. Orientation with respect to local and prevailing winds as well as to buildings or other windbreaks in the vicinity determines the relative effect of wind on different parts of a house. A wind rose showing average windspeeds and frequencies from different directions is especially useful if it is interpreted in connection with temperature and moisture data. The possibility of wind-driven rain merits special investigation.

The rules which govern the relation of sunshine to site exposure apply equally to building orientation. In the Northern Hemisphere, the greatest amount of insolation is received by the east and southeast sides of a house in the morning and by the west and southwest sides in the afternoon. In summer, with increasing latitude, the path of the sun in the sky describes more complete circles so that the sun rises north of east and sets north of west; in winter the arc of the sun's visible path is shortened, and east and west exposures receive very little sunlight. Ordinarily, an east–west alignment of a rectangular house provides the maximum gain of solar energy in winter and minimum in summer.

CLIMATE CONDITIONING THROUGH DESIGN

The climatic factors to consider in housing design are insolation, temperature, wind, and moisture. Insolation provides heat and light and has certain health-giving powers. Climate conditioning seeks to control its effects through such design features as shape, layout of rooms, placement of openings, insulating materials, overhangs, or roof orientation and slope. The functions of various rooms determine the most efficient layout with respect to the sun. For example, a kitchen and breakfast nook are best placed where they will receive morning sun, whereas the living room may benefit from afternoon sun, especially in winter. Bright sunlight can sometimes be a nuisance; the study and certain kinds of workrooms are better located on the shady side of the house, where lighting is more even throughout the day.

By varying the ground plan from the traditional rectangle, it is possible to obtain considerably more sun control for various rooms. An L-shaped house can be oriented to provide a greater variety of sunny exposures than a square one covering the same ground area and thereby capture more solar energy. Conversely, its orientation can increase shading if desired. Admission of sunlight to the interior need not be confined to windows in vertical walls on the sunny exposures. Skylights offer one method of increasing the light in a one-story house or on the top floor of a multistoried building. They are difficult to maintain, however. Clerestories are suited to rooms on the shady side of a building where sunlight is desired. The principle of the clerestory is illustrated in Fig. 16.1. In combination with ceilings of the

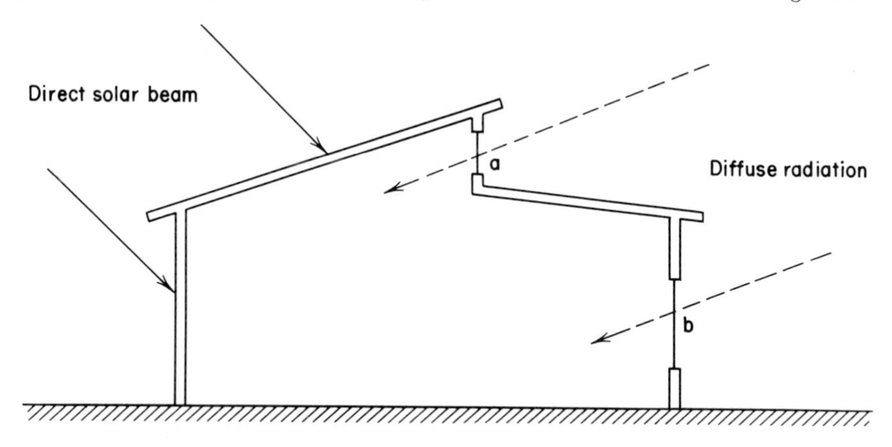

Figure 16.1 Principle of the clerestory. The clerestory window at "a" and the wall window at "b" admit scattered and reflected light. The opaque roof and wall on the sunny exposure keep out direct sunlight.

proper color and texture, it can greatly improve lighting of rooms that face away from the sun. When oriented toward direct sunlight, a clerestory is an effective device for passive collection of solar energy.

Unwanted bright sunlight is controlled from the interior by means of curtains or blinds over openings. Glass bricks and translucent windows also reduce glare while admitting light. Overhangs, awnings, and screens serve much the same purpose on the outside and have the added advantage of reducing heating of walls. In sunny climates of the middle latitudes overhangs can be designed to expose walls and windows in winter when the sun is at a low angle but to provide noonday shade in summer (see Fig. 16.2). The exact size and design of overhangs for the degree of shading desired can be calculated from the sun path data for different times of the year at the latitude of the site (see Fig. 14.4). In low latitudes, overhangs shade walkways and the ground around the house, thereby helping to reduce the temperature of the walls. Overhangs would have to be inordinately large in high latitudes to afford shade which is not usually necessary anyway.

Design features in a well-built house take into account possible extremes in outside temperature and aim to maintain indoor temperatures at a comfortable level with a minimum of artificial heating or cooling. Basically, this entails control of the processes of heat transfer: radiation, conduction, and convection. These processes are affected by precipitation, humidity, and wind as well as by insolation and air temperature.

Solar heat is a blessing or a problem in relation to a specific house design depending on the climate and the season, and it is sometimes difficult to separate the benefits from the disadvantages. Well-insulated floors, walls, and roofs inhibit the passage of heat by conduction to or from the interior of a house. Light-colored and smooth surfaces reflect a greater percentage of sunlight than do dark and rough surfaces and consequently do not heat so rapidly under intense insolation. Through

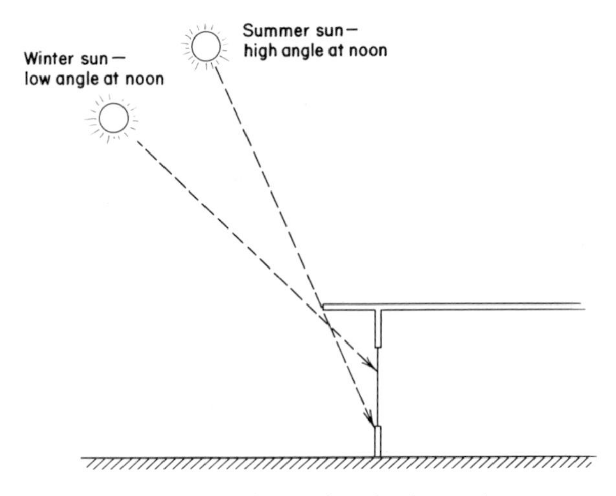

Figure 16.2 Effects of an overhang in winter and summer.

selection of materials for walls and roofs or the paint applied to them, it is possible to control solar heating to some extent. A light-colored surface will also reflect insolation in winter, but since there is less sunshine at that time this is generally not so important as summer cooling. Double roofs between which air can move freely are effective in reducing the flow of absorbed solar heat into a house in hot climates. Venting of attics has a similar though less marked effect. In semiarid and arid climates, flat roofs can be kept covered with water and thus aid cooling by evaporation. Short-wave solar radiation passes through clear air and windows with relative ease (even at low air temperatures) and upon being absorbed is converted into long-wave radiation which is partially trapped. Single-pane windows conduct and radiate a great deal more heat than insulated walls, however, and the gain in solar heat through large and numerous windows during short, sunny winter days may be more than balanced by losses at night or on cloudy days. Storm sash or double-pane glazing with the intervening space sealed airtight will reduce conductive loss. Cooling due to evaporation of water from the exterior window surfaces can also be reduced in this way. Drapes drawn over the interior of windows partially curb radiation losses. In general, the less sunshine at a particular site, the more glass is needed to overcome dark interiors, but it must be remembered that glass expedites the passage of heat in both directions.

Wind causes convective cooling, intensifies precipitation effects, and exerts direct force upon a house. Therefore, house design should provide for insulation against too much cooling, tight construction to prevent the entrance of cold air or wind-driven precipitation, and overall strength to withstand wind pressure (see Fig. 16.3). Under hot conditions, a movement of air through a building is useful in

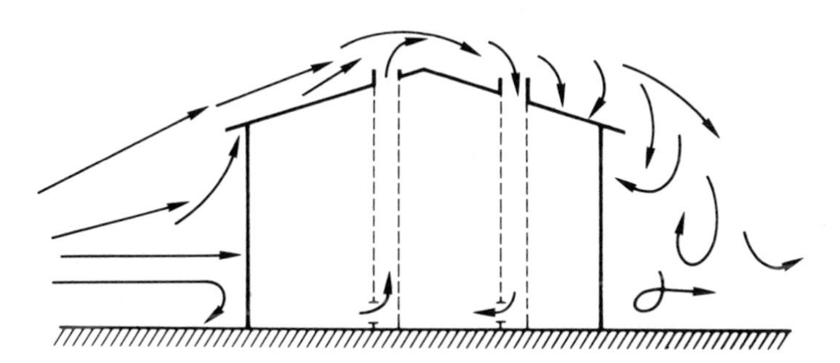

Figure 16.3 Typical wind effects on a house. Eddies generated across the roof create unequal pressures and gustiness; they may cause reverse flow in chimneys or ventilation ducts. Open windows further modify air flow both inside and in the vicinity of the building.

controlling temperature and humidity. Tropical houses have numerous openings to permit free flow of air. Native dwellings are commonly designed so that walls can be propped up or rolled up all around in a manner to provide shade but allow breezes to circulate. The tent of the desert nomad is a similar example of climatic conditioning through design. The flaps can easily be arranged to create shade and

Figure 16.4 Climate conditioning through design in a tropical desert. (An Exxon photograph.)

admit air by day and to restrict cool breezes by night (Fig. 16.4). Movement of air through a house is modified by the openings and obstacles as well as by overhangs and the shape of the exterior. Various kinds of deflectors aid in directing air currents. For maximum airiness, floor-to-ceiling walls may need to be replaced by partial dividers or screens, for to have its maximum effect a breeze must pass through the house without being trapped in rooms on the windward. Some architects experiment with scale models of buildings in a wind tunnel before making final decisions on design details.

It was pointed out in the discussion of orientation of a house that the larger the surface exposed to the wind, the greater will be the total wind pressure upon it. Long eaves and overhangs tend to catch the wind and therefore increase the danger of wind damage. Loosely attached roof coverings of any type can be lifted and swept away by a fierce wind. Low, streamlined buildings of simple design are best for regions with violent winds. Departures from this principle call for compensating strength in structural members and walls. Possible destructive effects of flying debris to roofs, walls, and windows must also be taken into consideration. A part of the stress resulting from gale force winds is due to the difference in pressure between the outside and the interior. This can be reduced to some extent by louvered venting or by opening windows on the lee side. A well-built house can withstand hurricane winds, but the cost of "tornado-proofing" is normally beyond the range of the homeowner, who shares a calculated risk with his insurance company.

The most obvious aspect of moisture problems in housing is precipitation. Roofs and drains should permit rapid removal of heavy rains and runoff without damage to surrounding grounds. Structural design should accommodate possible extreme rainfall intensities and the weight of water that could accumulate if drains become clogged. If not properly diverted, dirty runoff may stain walls. Where there is danger of seepage along basement walls and foundations tile drains facilitate drainage away from the building. Walls need to be of tight construction to prevent precipitation from being blown into small openings. Water which enters crevices and subsequently freezes can be especially destructive. Overhangs protect entrances, paths, and windows as well as walls, but they must be designed with sun and wind also in mind. A canopy over an entrance shields the door and the opening when it is in use. Vestibules that extend beyond the main wall serve the same purpose and are more efficient in cold climates. Low windows are subject to rain splatter and covering by drifted snow; doorsills need to be high enough to dam a shallow flow of water from the doorstep in a deluge.

In regions of heavy snowfall, roofs must be strong enough to hold the weight of several tons of snow. Some indication of the possible snow load can be obtained from climatic records (see Fig. 16.5). The stress is augmented when rain falls onto a snow-covered roof. A method for computing maximum snow loads as a basis for building codes in Canada uses maximum anticipated snow depths, an assumed density of 0.2 g per cm^3, and an added weight representing a one-day maximum fall

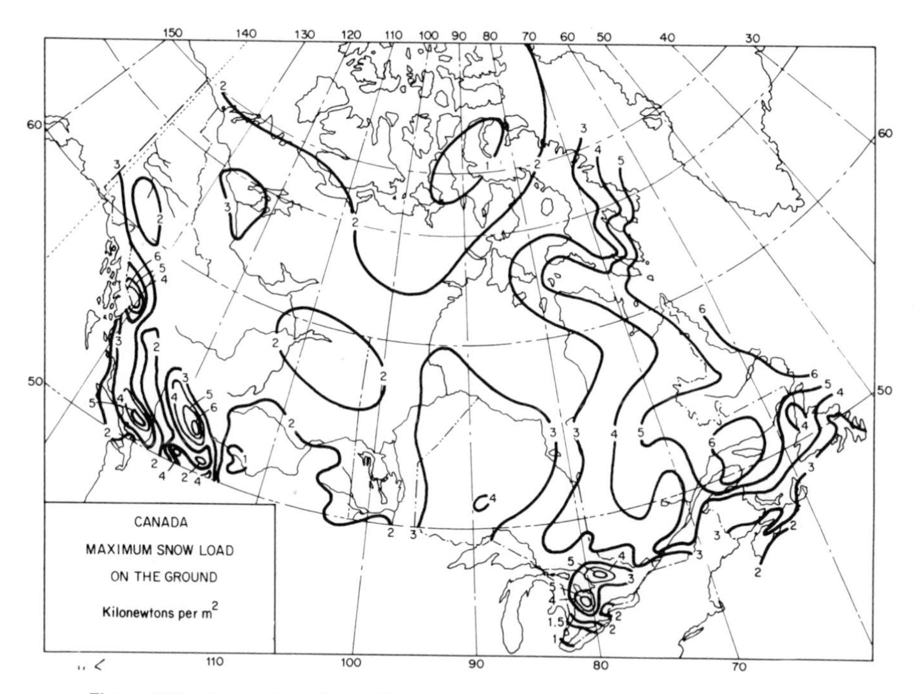

Figure 16.5 Generalized distribution of maximum snow load on the ground in Canada. Values incorporate maximum expected snow weight and the weight of maximum expected one-day fall of rain. (Data from Canadian National Research Council, National Building Code.)

of rain. A steeply pitched roof will permit wet snow to slide off, easing the stress on the roof but creating a hazard beneath the eaves. Gables and cornices tend to collect drifts on the leeward, producing an uneven weight distribution. Gentler slopes hold "dry" snow, which has valuable insulating properties in extreme cold.

To obtain benefit from a cover of snow, the roof itself must be well insulated so that the snow is not melted by heat from the house. Otherwise, ice will form, reducing the insulating capacity. Moreover, melt water which flows down the roof is likely to form icicles along the eaves; adequate drainage of melt water is thus impeded and the danger of falling icicles is introduced (see Fig. 14.9).

Hail damage is usually greatest to roofs and windows. Roofing materials should be resistant to the pounding of hailstones. Shutters, or preferably heavy screens, afford direct protection to windows, although overhangs usually can prevent hail from striking windows beneath. Skylights are inadvisable in areas subject to hailstorms.

The influence of fog on design is closely associated with sunshine. In addition, frequent fogs enhance the danger of moisture damage to building materials and paint. Where air pollution is common, its chemical effects are greater when associated with fog, dew, or frost. Streaking of walls often results from precipitation of dirty fog particles. The choice of paint should be made with these factors in mind.

Lightning occurs primarily with thunderstorms; where they are common, lightning protection should be an integral part of house design. A stroke of lightning incorporates both an electrical discharge and intense heat that may start a fire. The principle of protection involves provision of a direct and easy passage of the current to earth. Metal lightning rods attached to the highest parts of the roof and heavy cables to conduct the current to the ground serve this purpose. Utility wires, antennas, and other wiring attached to the building should be grounded, preferably on the exterior. Building codes usually include regulations for lightning protection.

AIR CONDITIONING

Rarely can site selection, orientation, materials, and design create the desired indoor climate at all times. Air conditioning is the last resort in the overall attempt to provide a suitable indoor climate. In popular usage, the term has sometimes been restricted to the artificial cooling of the interiors of buildings. In the broader sense used here, it includes all attempts to modify indoor temperature, humidity, air movement, and composition of the air by artificial means. The demands placed upon air conditioning in the control of these elements depend on the outdoor climate, building design and its related factors, and the kind of indoor climate which is wanted. Though many of the principles remain the same, air conditioning of hospital operating rooms is quite a different problem from refrigerating a cold-storage locker. In residential buildings, the functions of different rooms influence requirements for comfort. Most important, individuals differ widely in their perception of comfort.

Under certain circumstances, an air-conditioning system may be called on to cleanse the air of pollutants. Dust, soot, pollens, and other solid materials are removed by means of filters, by "washing" the air in a spray chamber, or by precipitating the particles on electrically charged screens. Objectionable gases are much more difficult to remove, although some are soluble in water and can be partially controlled in a spray chamber. Chemical modification of gases is practical only under extreme conditions and should not be necessary in a planned residential settlement.

HEATING

In the past much more attention has been given to heating than to cooling of interiors, for technology has flourished in the middle latitudes, where cold winters have been a greater problem than hot summers. Heating systems that depend on forced circulation of air must be appraised in terms of effective temperature rather than the dry-bulb temperature alone (see p. 357). Excessive air movement reduces the effective temperature by convective cooling. Cool air introduced into a room undergoes a decrease in relative humidity upon being heated. Fortunately, the range of comfortable relative humidity is rather wide—between 20 and 70 percent—so that there may be no resulting discomfort unless the cold outside air has an extremely low specific humidity. When the indoor relative humidity is too low water can be evaporated into the air, preferably in conjunction with the heating or ventilating system.

Stoves and so-called radiators actually distribute heat in a room both by radiation and by convection of air that is warmed along their surfaces. As the proportion transferred by convection increases, a slightly higher temperature at the source is required to maintain a comfortable effective temperature. Panel heating, also known as radiant heating, employs the concept of mean radiant temperature to warm people and objects in a room mainly by direct radiation from large surfaces (such as walls and floors) having comparatively low temperatures of 30° to 55°C. Panel heating has the advantage that convection is held to a minimum while the room is heated more uniformly than when smaller, high-temperature heat sources are used.

Direct solar energy is an obvious source for heating by radiation. The principles which can reduce excessive solar heating in a hot climate can also be applied beneficially in a cold climate. Appropriate siting, orientation, design, and construction not only permit simple "passive" heating during periods when sunlight is available but also enhance storage. Additional adjustments are necessary to provide adequate collection, transmission, and storage in an "active" solar heating system (see Fig. 16.6).

The gradient between temperatures indoors and those of the outside air determines, to a large extent, what amount of heating is required, for even the best-designed house exchanges heat with the exterior by conduction. Evaporation, the

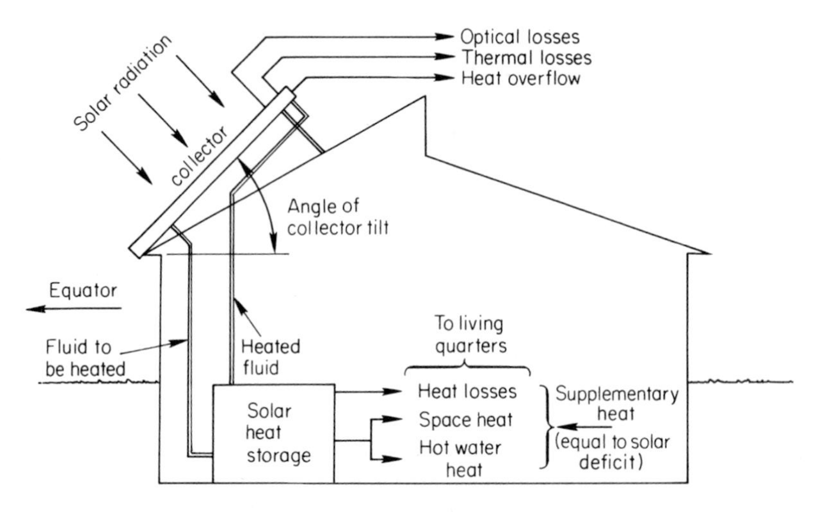

Figure 16.6 Principal features of a solar heating system. (After Tybout.)

escape of heated air, and convective cooling by wind also are processes that need consideration in planning a heating system, however. Wind promotes both convective and evaporational cooling. Direction as well as speed is significant, since a house may be more vulnerable to these types of cooling on one side than on another. A house loses part of its heat by radiation to cold ground surfaces, other buildings, and the atmosphere. In the latter instance, clear skies enhance radiative cooling, especially at night and in winter. Cooling is increased by low relative humidity as well as wind. Escape of heated air is controlled mainly by features of house design. Faulty ventilating systems often contribute to this loss. Ordinary fireplaces create a convection system that draws cool air into a room and carries warm air up the chimney.

Records of local climatic data are useful for determining the amount of heating required to overcome these cooling processes. Microclimatic observations in the immediate vicinity of a building are better than data from a weather station, where site characteristics may be quite different. Nevertheless, regional temperature averages, ranges, frequencies of extremes, and heating degree-days are employed with some success in estimating the energy requirements for a specific building. The *heating degree-day,* widely used by heating engineers, is defined as a day on which the mean daily temperature is one degree below an acceptable base value. In the United States and Canada the base temperature in general use is 18°C. Thus, a mean daily temperature of 10°C for a given day would yield 8 heating degree-days. Total heating degree-days for a month, season, or year are found by adding the accumulated degree-days for the period. This index of heating requirements does not take into account cooling by radiation, wind, or evaporation. For accurate computations, research is needed to develop an index which combines all cooling factors.

Questions involving the type of heating system, kind of fuel or power, operating costs, and efficiency are primarily problems for the heating engineer, but they have climatic implications. The design and capacity of a heating system should meet requirements imposed by weather changes and extremes. Ideally, automatic control mechanisms should have sensors exposed to outdoor as well as indoor conditions in order to avoid lag in adjustment of the indoor climate. The amount of energy needed to operate a system is estimated in terms of heating requirements, which, in turn, are closely correlated with weather conditions.

COOLING

Under summer or tropical conditions, the main objective of air conditioning is to reduce the effective temperature. The practical methods include lowering air temperature, lowering relative humidity, and increasing air movement. Design and orientation of a building should be exploited to the fullest to accomplish these aims at the lowest expense. If the desired conditions cannot be obtained, as for example in interior rooms or under extreme temperatures, some form of cooling mechanism may be necessary. Fans speed air movement and promote cooling by increasing evaporation of skin moisture. At night, cooler outside air can be drawn into a room by exhaust fans placed near outlets at high levels. Adaptations of refrigeration units placed at air intakes actually cool the air and are most effective if combined with fans and used in a room that is closed except for intake and exhaust openings. Comparatively efficient units of this type have been developed which are no more expensive than heating systems. If heat is combined with high relative humidity, the effective temperature may be lowered by passing the air through an apparatus which condenses a part of the moisture on cold coils or otherwise dehumidifies it. Air that is too dry can cause skin discomfort and irritate respiratory passages. In hot, dry climates air can be passed over water or through spray to increase its relative humidity and, at the same time, cool it by evaporation.

Cooling degree-days may be used as a basis for determining the amount of energy necessary to reduce the effective temperature of warm air. Although a cooling degree-day may be considered as a day on which the temperature is one degree above a desired base temperature, an expression incorporating a humidity value provides a more satisfactory index of the amount of cooling that will create comfortable conditions.

QUESTIONS AND PROBLEMS FOR CHAPTER 16

1. Prepare a flow chart showing the sequence of decisions that lead to a comfortable, energy-efficient home. Explain how each decision is related to climate.
2. How can heating degree-days data be used by a homeowner? By a utility company?

3. What kinds of climatic data should be considered in the planning and construction of a mountain cabin in the mid-latitudes? Where might the required data be obtained?
4. Compare the practicable methods of cooling a house interior in the humid subtropics with methods that are suited to a tropical arid climate.
5. What climatic factors should be taken into account in planning a residential community? Cite examples to illustrate.
6. Why are soil moisture and temperature important factors in building climatology?
7. What precautions can a homeowner take to reduce damage by a hurricane? A tornado? A blizzard? What are the hazards during a Santa Ana?

17

Modification of Weather and Climate

In view of the impact of weather and climate on human activities, it is logical to inquire whether manipulation of the climate system might produce benefits or disasters for humanity. Chapter 10 reviewed some of the evidence that climates have been changed by natural causes. The possibility of duplicating those causes suggests modification on a grand scale. Accounts of attempts at rainmaking persist throughout human history, from ritual rain dances, through unmitigated hoaxes and trial-and-error experimentation, to modern theoretical approaches. Many people continue to equate weather modification with rainmaking, but its scope is much broader, encompassing any inadvertent or deliberate change in atmospheric processes; because climate is an aggregate of weather such changes imply climatic effects to some degree. As in the case of natural climatic fluctuations, it is important to consider the scales of time and space. This final chapter examines theoretical bases of atmospheric modification, experimental methods, and attendant problems. It will be evident that these topics entail a review of physical principles that govern exchanges of heat, moisture, and momentum in the climate system.

MODIFYING MICROCLIMATES

A fundamental consideration in weather modification is that of scale. Changes on a small scale are relatively easy to initiate and control, yet their cumulative effects locally and their possible extension to areas of larger scale are difficult to assess. Human beings have been modifying their microclimatic environment ever since they first sought the shade of a tree, built a fire, or fashioned a shelter. We have the capacity to manage microclimates for the benefit of ourselves, our animals, and cultivated plants, but we also alter microclimates to the detriment of plant and animal life and sometimes to the distress of our neighbors. Removal of vegetation changes the albedo of the land surface, thereby affecting heat and moisture exchanges. Cutting of forests reduces the amount of rain or snow intercepted above ground level and alters the processes of evapotranspiration and runoff as well as the flow of air. Irrigation, frost prevention techniques, and the creation of windbreaks are representative agricultural practices that influence microclimates (see Chapter 13).

Buildings that are designed for control of their internal climates also affect the microclimates in their vicinities by influencing air movement, heat exchange, and moisture flux. Anyone who has contrasted the microenvironment over a hot pavement with that over a green lawn has a practical appreciation of human influence. Roads are commonly drier than adjacent land, have a lower albedo, and are likely to be travelled by sources of atmospheric pollutants. The construction of an embankment across a small valley can disrupt air flow and create a distribution of temperature that is reflected in the composition of plant communities. The effects of factories, parks, hedges, swamp drainage, and umbrellas are additional random examples that illustrate the ways in which we can modify climate on a small scale. Any change in albedo, water capacity and retention, evaporation, transpiration, or surface roughness may produce a change in climate, but the results are complex, difficult to measure, and not easily predicted.

CITY CLIMATES

Cities concentrate people and their activities in small areas, thereby providing excellent opportunities to examine cultural modifications of climate. Although atmospheric pollution, which normally attains greater densities over cities, has received a large share of popular attention in recent years, it is not the only factor influencing city climates. Urban areas also differ from their rural counterparts in surface materials, surface shapes, and heat and moisture sources. In turn these affect radiation, visibility, temperature, wind, humidity, cloudiness, and precipitation. Table 17.1 summarizes the influence of urban environments on a number of climatic elements.

Concentrations of pollutants in the air above a city create an *urban aerosol,* which attenuates insolation, especially when the sun angle is low as is the case at high latitudes and in winter. The aerosol is best developed (that is, at its worst) during conditions of stable air and calms or light winds (see the section "Atmospheric Pollution" in Chapter 14). In comparison with open rural areas, the annual total direct solar radiation in the heart of large industrial cities may be decreased by 15 to 30 percent. Insolation has been observed to vary during the week, being greatest on Sundays, when industrial activity and traffic are at a minimum. The urban aerosol is somewhat selective, for it reduces the proportion of ultraviolet radiation more

TABLE 17.1

Approximate Average Effects of Urbanization on Climatic Elements *

Elements	*Comparison with rural environment*
Pollutants	
Solid particles	10 times more
Gases	5 to 25 times more
Cloud cover	5 to 10 percent greater
Fog, winter	100 percent more
Fog, summer	30 percent more
Precipitation	5 to 10 percent more
Snowfall	5 percent less
Rain days with less than 5 mm	10 percent more
Relative humidity, winter	2 percent less
Relative humidity, summer	8 percent less
Radiation	15 to 20 percent less
Ultraviolet radiation, winter	30 percent less
Ultraviolet radiation, summer	5 percent less
Duration of sunshine	5 to 15 percent less
Annual mean temperature	0.5° to 1.0C° higher
Heating degree days	10 percent fewer
Annual mean windspeed	20 to 30 percent less
Calms	5 to 20 percent more

*Adapted from Helmut E. Landsberg, "Climate and Urban Planning," in *Urban Climates*, Technical Note No. 108 (Geneva: WMO, 1970), p. 372.

than the longer wave lengths. It reduces the number of bright sunshine hours as well as the horizontal visibility.

In spite of the diminished insolation, the center of the typical metropolis constitutes a "heat island" that has a shape and size related to urban morphology, buildings, and industries and that results largely from urban heat generation and storage. Temperatures normally are highest near the city center and decline gradually toward the suburbs, beyond which there is a steep downward temperature gradient at the rural margin (see Fig. 17.1). The differences are greater at night than by day. Figure 17.2 illustrates typical daily temperature curves for a mid-latitude city and suburb. Although heat islands tend to be larger and more intense over large urban areas, the relation is not direct. Spacing of buildings and both kind and amount of activity influence heat island development.

Owing to the blanketing effect of pollutants on the radiation budget, diurnal ranges of temperature are less in urban areas than over the countryside. In view of the importance of vertical temperature lapse rates to atmospheric stability it is significant that nighttime inversions tend to be weaker over cities, where the heat island generates modest convection. To the lee of cities, an urban heat plume at several meters above the surface may intensify rural inversions.

The roughness of the city surface increases frictional drag and turbulence. Gustiness and erratic flow of wind through the maze of urban canyons are well

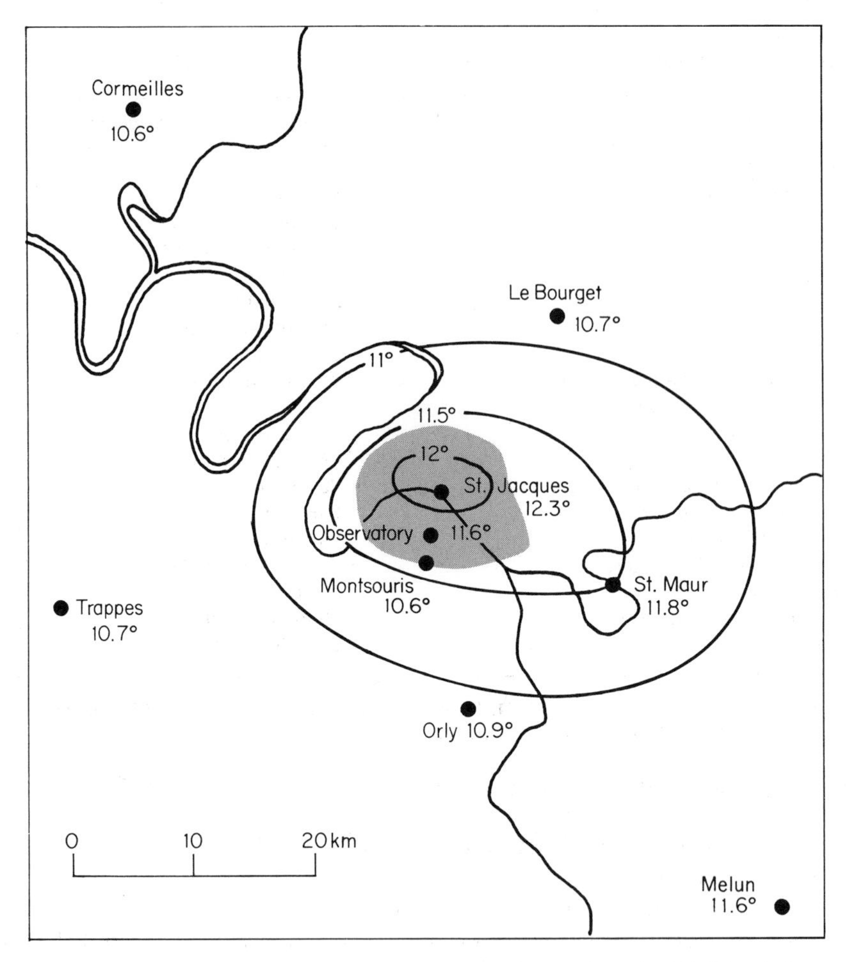

Figure 17.1 Mean annual surface temperatures of Paris and vicinity, °C. (After Dettwiller.)

known to the city dweller, although gusts are more likely to reach their maximum speeds in the open countryside. Except under conditions of low regional wind speeds the mean wind speed within the city is lower than in the surrounding rural environment. When nighttime winds are light, the speeds in the central city tend to be higher than in the country. Under nocturnal inversions the stable rural air inhibits surface flow, and calms are more frequent, whereas the relative instability of city air promotes turbulence, and stronger winds from above reach the surface more often. (Compare with the section "Diurnal Variation of Wind Speed," Chapter 4.)

A strong heat island generates its own circulation system (Fig. 17.3). The inflow of cooler rural air toward the rising air over the city is generally weaker than

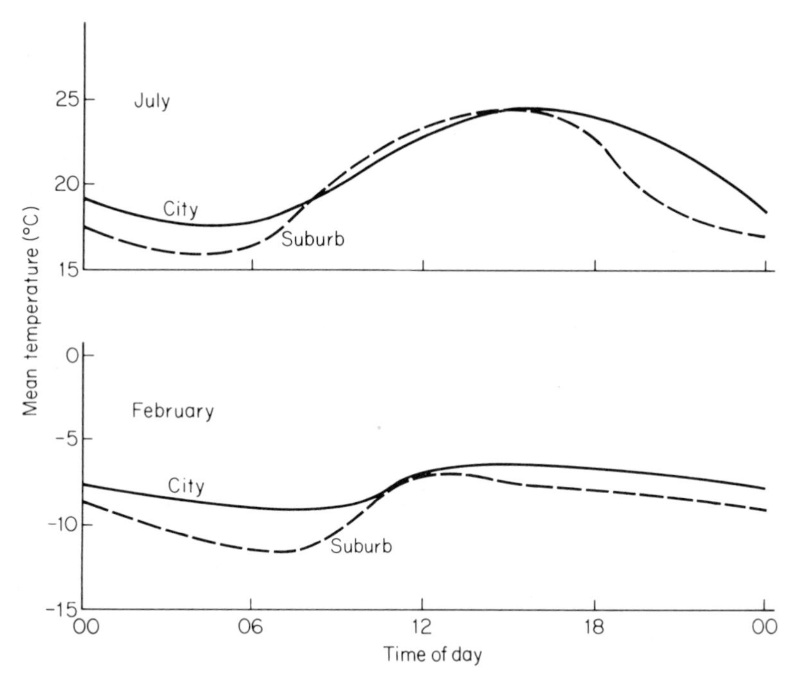

Figure 17.2 Diurnal temperature variations in a mid-latitude city and suburb. Data are for Schottenstift in urban Vienna (solid curve) and suburban Höhe Warte (dashed curve), 1956. (After Mitchell.)

might be expected, however, and is best developed on relatively calm, clear nights. Smoke plumes of residential areas have been observed to point toward a city center. Instrumental observations from towers and balloons have confirmed upward movement of air above urban heat islands.

The tendency of air to rise above the heat island is a possible explanation of greater cloudiness over cities and may account in part for greater precipitation, for

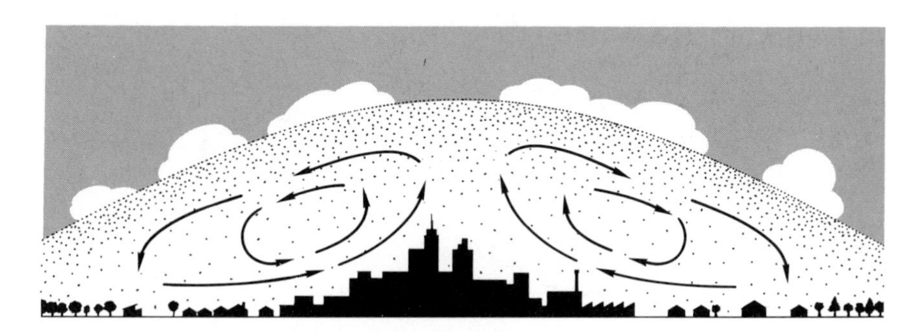

Figure 17.3 Schematic cross section of circulation generated by the heat island in an urban aerosol. (After Lowry.)

cities are good sources of condensation and ice nuclei (see Fig. 17.4). One of the problems in urban precipitation studies is the difficulty of finding unpolluted rural areas with which to make experimental comparisons. A small number of ice nuclei injected into supercooled clouds might enhance rainfall, but a massive addition of condensation nuclei to air before it reaches its saturation temperature and produces warm clouds should lead to formation of many small droplets and inhibit rainfall. The net effect of the urban aerosol on precipitation over the city and to the leeward is not clearly understood. It depends on the kind and number of nuclei emitted by urban sources, natural nuclei, relative humidity, thermal lapse rates, and mechanical turbulence. Although climatological evidence is sparse, it may be that the incidence of hail is relatively greater over cities owing to convective activity. The proportion of precipitation in the form of snow appears to be less over urban centers, presumably because of higher temperatures associated with the heat island, and when it falls its effect on the surface albedo is rapidly modified by snow removal operations and urban dust.

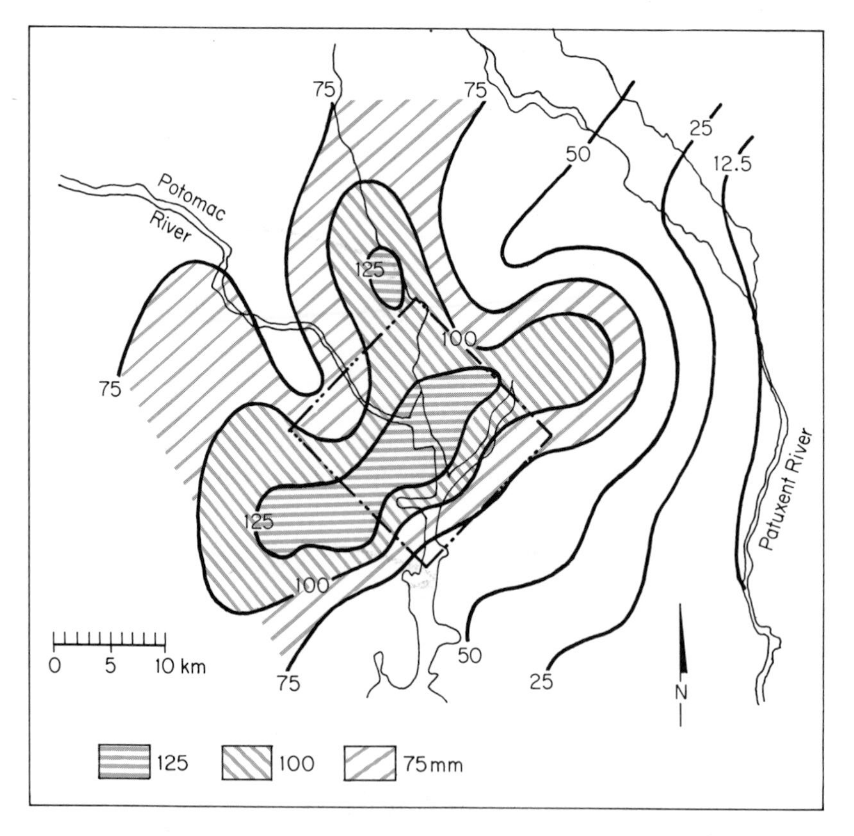

Figure 17.4 Distribution of rainfall during a storm over the Washington, D. C., area on 9 July 1970. (After Landsberg from W. N. Hess, ed., *Weather and Climate Modification.* New York: John Wiley & Sons, Inc., 1974, p. 754.)

The potential effects of the urban aerosol and heat island suggest processes leading to increased precipitation downwind from a city. Although conclusive results for a large number of cities under different conditions are lacking, specific studies appear to support a modification hypothesis. The Metropolitan Meteorological Experiment (METROMEX) was initiated at St. Louis, Missouri, in 1971 as a cooperative project of several research organizations to investigate urban influences. Its findings include: greater convective activity, more thunderstorms, denser concentrations of condensation nuclei, 10 to 30 percent more precipitation, and a greater incidence of hail along storm paths to the lee of the urban area than over adjacent rural land.

The mean relative humidity in city air is usually a few percent lower than in the surrounding country, especially at night and in summer when the heat island is well developed. Rural–urban differences in specific humidity are much more complex. Urban surfaces promote rapid runoff of precipitation, whereas vegetation and soil in the country retain moisture for evaporation over a longer period. On the other hand, the difference in availability of water vapor is offset to some extent by the many combustion sources in the city. During periods of light winds, tall buildings may inhibit the flow of air at ground level and thus reduce the upward diffusion of moist air. The net effect of all factors in a given city is to produce specific humidity values that reflect influences of urban form and function.

INADVERTENT MODIFICATION OF MACROCLIMATES

Suggestions that human activities may unintentionally alter world climates carry awesome implications that frequently are exaggerated beyond the limit of present evidence. Most of the linkages are theoretical; some are purely speculative; others undoubtedly remain to be discovered. They range through three categories: changes in the chemical or physical composition of the atmosphere, changes in direct heating, and changes in surface characteristics.

It has been fairly well established that water vapor, carbon dioxide, and ozone absorb thermal radiation selectively, producing the so-called greenhouse effect. The burning of fossil fuels as well as forests and other biomass has delivered carbon dioxide to the atmosphere at an increasing rate. Gilbert Plass and others advanced the carbon dioxide theory to account for a slight global temperature rise between 1885 and 1940. A cooling trend after 1940 suggested additional influences, however. Much remains to be learned about the effects of oceans and vegetation, both of which store carbon dioxide, and other factors in the carbon dioxide balance of the climate system. A variety of trace gases that enter the atmosphere from sources in agriculture, industries, and transportation may alter radiative exchange either directly or through complex chemical transformations. For example, gaseous sulfur compounds, notably SO_2 and H_2S, undergo chemical changes to form sulfate salts that act as aerosols. Nitrogen oxides and certain hydrocarbons react with ozone, which controls transmission of ultraviolet radiation to the earth's surface. There also has been widespread concern about the potential effects on the ozone layer of chlorofluorocarbons (used in spray cans and refrigeration) and nitrogen compound emissions from jet engines and agricultural fertilizers. The role of these and other gases in atmospheric chemistry and the global radiation budget is a leading aspect of research on inadvertent climatic change.

Atmospheric aerosols are another possible link between human beings and climatic change. Basic theories concerning the effects of dust aerosols have been developed largely from observations following volcanic eruptions. Solid particles produced by human activities do not normally reach to such heights, except in the cases of certain nuclear explosions and emissions from high-altitude rockets and aircraft. Within the troposphere the rate of return by gravity and precipitation is relatively rapid, and where the materials fall on ice or snow they may reduce the albedo and speed melting. Visual evidence and measurements indicate a worldwide increase in atmospheric dust, which reduces direct insolation and increases radiation from the sky (see Fig. 17.5). The net effect of dust layers in the upper atmosphere is likely to be an increase in the planetary albedo and a consequent cooling of the lower atmosphere. Tropospheric dust of the types introduced by industrial and land-use practices, on the other hand, is more likely to absorb heat and create upper-air inversions, leading to subsidence, warming, and a drier surface climate. The relative magnitude of climatic effects due to man-made dust and volcanic eruptions remains an important subject of research.

Among the various kinds of solids emitted from surface sources into the atmosphere, some appear to be more effective than others as nuclei for condensation or ice crystal formation. Lead particles in automobile exhausts, for example, become efficient ice nuclei when they combine with iodine or bromine. Too many

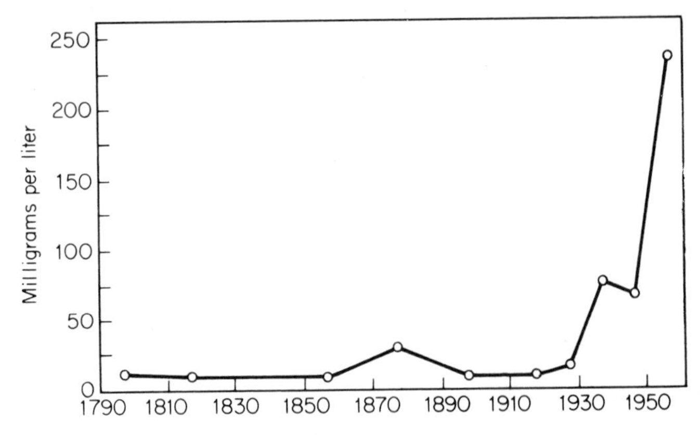

Figure 17.5 Amounts of dust measured in ice layers on the Caucasus Mountains, U. S. S. R. The small circles represent 10-year running means calculated from available samples. Explanations of the variations in dust accumulation include both local and global disturbances. (Adapted from F. F. Davitaya, "Atmospheric Dust Content as a Factor Affecting Glaciation and Climatic Change," *Ann. Assoc. Am. Geogrs.*, 59, 3 (1969), pp. 552–60.)

hygroscopic particles, such as sea salt, tend to produce a large number of small cloud droplets that do not coalesce easily. The result could be a decrease in precipitation.

Emissions by high-flying aircraft and rockets of carbon dioxide, water vapor, and solid particles that act as freezing or condensation nuclei are possible causes of unintentional atmospheric modification. The amounts of these contaminants are relatively insignificant when viewed in terms of the total atmospheric volume and compared with more abundant surface emissions, but if they accumulate in the stratosphere they could alter the radiation balance. The effect of water vapor raises unique questions. It may be that condensation trails cause cooling by increasing the planetary albedo, although sky coverage of these artificial cirrus clouds is small and ephemeral, and high-level clouds should reduce radiative loss to space. It has also been suggested that cirriform jet trails precipitate ice crystals that could modify lower clouds, but the confirming evidence is sparse. Even less is known about the potential depletion of the ozone layer by emissions from aircraft in the stratosphere or about the influence of sonic waves on cloud dynamics.

Exhausts from large rocket engines introduce water vapor well into the thermosphere in quantities sufficient to disturb the radiation budget of the upper atmosphere. The special significance of contaminants in the stratosphere and above is that they remain for months or years before being lost to space or returned to earth, whereas most particles in the troposphere settle or are washed out by precipitation within a few days (Fig. 17.6). Nuclear explosions have injected radioactive debris into the stratosphere, where isotopes such as krypton-85 may affect ionization and atmospheric electricity, in turn influencing coalescence of cloud droplets and thunderstorm activity. The specific effects on climate of high-energy and high-frequency radiation from nuclear plants, electric power lines, microwave transmitters, radio towers, or other sources are unknown. Some forms of radiation could disrupt the earth's magnetic field and climate-related processes in the outer atmosphere.

A form of pollution which would appear to have direct effects on climate is excess heat. The heat released by combustion, consumption of electrical power, mechanical friction, and even human bodies clearly influences microclimates. This raises the question of whether sustained thermal pollution from myriad sources might affect regional climates, either directly or through a triggering action. Although the annual global production of such energy equals only about one twenty-fifth of 1 percent of the total net solar radiation at the earth's surface, it is increasing at an increasing rate that warns of a future problem, especially in highly urbanized areas.

Any alteration of surface characteristics that disrupts the heat or moisture budgets or the movement of air masses is a potential cause of climatic change. Human attacks on land forms have had little evident effect on macroclimates to date. More important are changes in surface albedo and the water budget resulting from removal of vegetation, cultivation, urban expansion, and deposition of solid particles on ice and snow. One hypothesis for *desertification* suggests that overgraz-

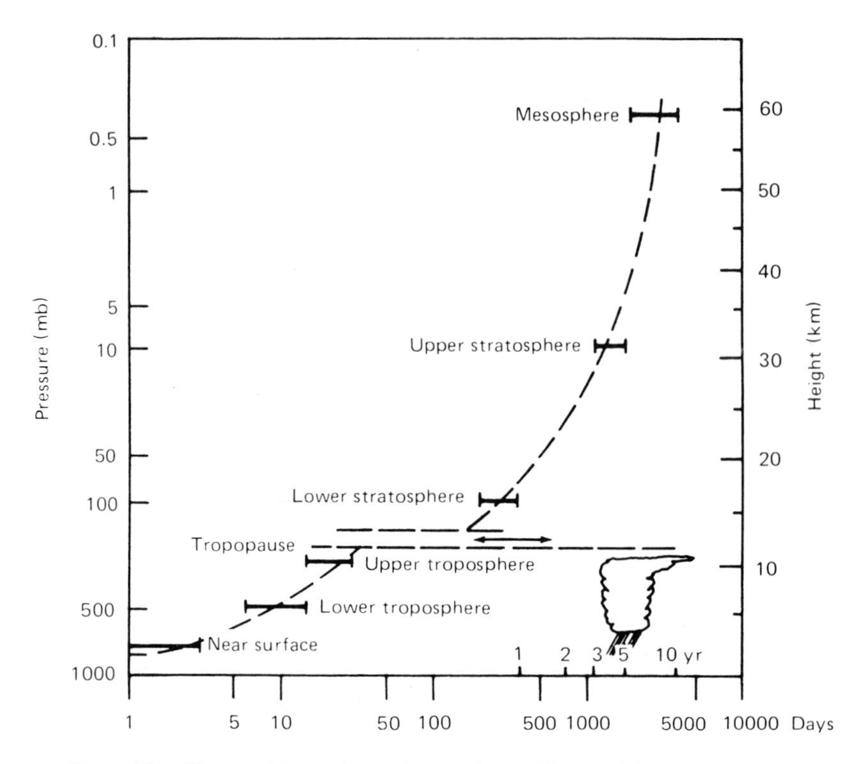

Figure 17.6 Mean residence times of aerosols at different altitudes. The double-headed arrow indicates uncertainty in the zone of fluctuating tropopause heights. (After Flohn and Bach, from Herman and Goldberg, *Sun, Weather, and Climate,* NASA, 1978.)

ing and soil erosion in semiarid and subhumid climates may increase the surface albedo, thereby reducing convection and rainfall. Accompanying increases in airborne dust presumably enhance stability in the troposphere by absorbing a larger share of radiation at upper levels. In contrast, increased dustfall on ice and snowfields should lower the albedo, speed melting, and alter the regional, if not the global, water budget.

Widespread cutting of forests, especially in the tropics, has already disrupted the hydrology of major drainage basins. Among the world's vegetation formations a forest cover is the most effective in increasing infiltration of precipitated water into the ground, and its foliage affords the greatest surface area for interception and evaporation of water. Deforestation therefore leads to increased runoff and possible floods. Because forests have a lower albedo than other types of land use (see Table 12.1), their removal also alters the surface radiation budget. Burning of vegetation adds dust and carbon dioxide to the atmosphere, while simultaneously reducing a major carbon dioxide sink.

Whether redistribution of surface waters has widespread climatic effects is a question with broad implications. It is known that large reservoirs have influences

TABLE 17.2

Annual Heat and Water Budgets of Oases and Adjacent Land in Tunisia*

	Area (km^2)	Estimated albedo (%)	Net radiation (watts/m^2)	Sensible heat flux (watts/m^2)	Latent heat flux (watts/m^2)	Evapotrans-piration (mm/yr)	Precipi-tation (mm/yr)
Oases (avg.)	150	15	100	-36	136	1,680	150
Semidesert	35,000	20	80	+67	12	150	150

*After H. Flohn adapted from SMIC, *Inadvertent Climate Modification* (Cambridge, Mass.: MIT Press, 1971), p. 174.

on the adjacent climate, reducing temperatures, increasing humidity, and sometimes generating local breezes. What is less certain is the extent to which regional precipitation is affected. Present knowledge does not lend strong support to claims that impounded water and irrigation projects lead to substantial increases in rainfall. Apparent evidence in the climatic record of enhanced seasonal or annual precipitation in years immediately following reclamation developments may be compensated by a reduction in later years. In any case, the presence of more evaporated moisture in the air is not likely to result in more rainfall unless the prevailing motion systems initiate the processes of convergence and convection. When these conditions are met in humid climates, augmented rainfall is a reasonable expectation, but in arid and semiarid areas general atmospheric subsidence inhibits the necessary lifting (see p. 403).

Irrigated vegetation has a lower albedo than typical arid land surfaces, and although local temperatures are reduced by evaporation of irrigation water, the net global effect may be a temperature increase owing to increased absorption of insolation. Table 17.2 summarizes some mesoscale influences of irrigation on surface heat and water budgets in Tunisian oases. Over an oasis part of the energy used to promote evapotranspiration is transferred from the air, which therefore is cooled. The oasis vegetation and wet soil have a lower albedo than the adjacent semiarid land surface, however. Whereas irrigation cools the local overlying air, its ultimate effect is to raise the regional temperature slightly because of the decreased reflection of incoming solar radiation. M. I. Budyko has calculated that the worldwide result of current irrigation is an increase of the mean surface temperature by about 0.07C°.

The drainage of lakes, swamps, and wet lands reduces evapotranspiration as well as the water vapor content of the air. Evidence of a reduction in local precipitation is inconclusive, at least in part because of the small areas involved. Insofar as surface albedo is increased, large drainage projects would tend to lower regional temperatures, thus offsetting the effects of irrigation elsewhere. Since drained areas are mainly in humid climates, where atmospheric processes obviously favor precipitation, their influence on global heat and water budgets is probably negligible.

PLANNED MODIFICATION

The hypothetical possibilities for intentional modification of weather and climate include any alteration of the heat and moisture budgets or motion in the climate system. Action against one necessarily affects the others. Discussion of the wisdom of attempting modification will be deferred to the end of the chapter.

To examine theoretical approaches to modification of the atmosphere's energy budget, we need review only the basic principles treated in Chapter 2 and the theories of natural climatic change in Chapter 10. At the outset we can consider the sun as the source of earth heat and reject for the forseeable future any thoughts of changing the solar constant by altering such factors as the solar fire, the shape of the earth's orbit, inclination of the polar axis, or the rate of earth rotation. Partial eclipses of the sun or diversion of insolation to desired latitudes or to the night hemisphere might be achieved by means of satellites or clouds of reflective particles, but the procedures would strain present technology.

The potential effects of changing the types and amounts of dust and gases in the atmosphere have been discussed in the previous section. Materials that affect the atmospheric albedo and transmissivity of radiation at different wave lengths would undoubtedly alter the energy exchange. Rockets and satellites could inject water, smoke, metallic particles, carbon dioxide, and other substances into the atmosphere at reasonably controlled heights, but, as we have seen, the effects are uncertain.

More is known about the processes of heat exchange at and near the earth's surface, and the prospects are brighter for modification on limited scales with less risk of disaster. Speculative proposals to alter the surface albedo range from dusting snow and ice with carbon black to enhance melting to the application of white powders for the purpose of reducing temperature and evaporation. It has been suggested that global changes in climate might be initiated by dusting the arctic ice pack with dark powder for a number of years. An ice-free Arctic Ocean would produce a vastly different regional exchange of heat and moisture. Proponents of the scheme claim that it would ameliorate the polar climates of the Northern Hemisphere, but it might generate greater storminess. Application of the dusting technique to polar ice has also been proposed as a method of forestalling an "ice age."

On a smaller scale, the darkening of surfaces in arid areas would increase thermal convection and, according to theory, stimulate cloud formation and precipitation. Applied in the path of moist winds, this procedure would create a "thermal mountain" and simulate orographic lifting. The most obvious approach to heat budget regulation is heating or cooling at the surface. Although heat losses from buildings, factories, compost heaps, orchard heaters, and other sources cause local temperature increases, the amount of energy required to effect regional or global heating by direct means far exceeds technological capabilities. The problems of regional cooling are at least as great. Only if a local triggering action were to induce processes on a larger scale could we expect such methods to be fruitful.

Precipitation, evapotranspiration, and runoff are the phases of the hydrologic system that offer possibilities for "control" of the moisture budget. An incomplete understanding of the processes of condensation, sublimation, coalescence of water droplets, and growth of ice crystals is the rather shaky foundation for modification of clouds, precipitation, and fog. Experimental tests and attendant theories will be treated separately in the following section. Meanwhile, consider the process of evaporation. One way to increase evaporation is to heat water. Among the suggestions has been the use of nuclear power to heat known sources of atmospheric moisture; this presumably would lead to greater precipitation to the leeward. Another approach is the creation of large water bodies by means of dams or engineered stream diversions. Whereas artificial lakes clearly increase local evaporation, it is not equally evident that there is a consequent increase in precipitation either locally or to the leeward on a regional scale. Nor is the effect of evapotranspiration from irrigated lands and vegetated areas upon precipitation adequately understood. Proposals to plant deep-rooting phreatophytes in semiarid areas in order to bring ground water into the atmospheric phase of the hydrologic system have not been tested sufficiently, but there is scant evidence of their potential success in modifying climate.

The origin of water that is subsequently precipitated over land is still an unresolved question. Considering the restless nature of the troposphere and the vigor of winds, a molecule of water that leaves the surface in one area is not likely to be precipitated again until it has been transported hundreds or thousands of kilometers. Even conceding that humid climates tend to perpetuate themselves in part by the regional recycling of water, the enhancement of rainfall in arid climates by introducing reservoirs, irrigation, and vegetation has doubtful potential. The partial return of evaporated surface water as precipitation in a humid climate probably occurs for the simple reason that atmospheric conditions favor precipitation there. Arid climates are dry because the requisite dynamic processes are lacking. The example of the Red Sea reveals no direct cause and effect relationship between large water bodies and precipitation. Yet there have been numerous plans to transform the climate of the Sahara by creating a huge lake in its basin areas. The reservoir behind Aswan Dam in Egypt has not produced appreciable increments of rainfall in the downwind region of the central Sahara. These and other examples of deserts adjacent to water expanses leave little doubt that air mass subsidence in the general circulation is a more potent factor than evaporation. In regions of greater atmospheric instability, extensive evaporating surfaces afford better possibilities for increasing precipitation by economically significant amounts.

Where water is scarce, it may be more profitable to conserve it by reducing rather than promoting evapotranspiration. On a small scale, this can be achieved by weed control and the use of mulches and by strategically placing windbreaks so as to reduce air flow across moisture sources. A technique of potential importance on small reservoirs and lakes is the use of substances which form monomolecular layers on the surface and inhibit the escape of water molecules. Nontoxic alkanols such as hexadecanol and octadecanol have been employed for this purpose. The objections to oil or other polluting substances are obvious. In addition to retarding evaporation the surface film, or monolayer, suppresses tiny waves or ripples generated by light wind shear and thereby reduces upward diffusion of water vapor. Under conditions of moderate to high wind speeds, however, breaks in the monolayer render it ineffective.

Runoff control by means of engineering projects and the management of forests, grasslands, and agriculture needs no elaboration here, but the control of snowmelt deserves special mention. The application of carbon black or other dark materials to speed melting of ice and snow has already been suggested in connection with modification of the energy balance. A method that would retard melting could yield economic benefits by extending the availability of water into dry periods. Although no feasible way is known to reduce melting on a large scale, limited control might result from the spreading of reflective substances on dirty snowfields.

The simplest approaches to direct interference with air movement employ barriers and mechanical forcing. A redistribution of mountains would surely affect climate. Windbreaks and fans accomplish modest results in various agricultural practices. Increasing friction over large areas would decrease the surface winds of the same areas, although turbulence might be increased to an unacceptable extent. Schemes such as those to plant extensive shelterbelts on the North American Great Plains in the late nineteenth century and again in the 1930s might reduce surface winds, but the number of trees required for a regional amelioration of wind is so great as to require an overall change in climate. Proponents of shelterbelts have contended that trees would in fact alter the moisture and energy balances as well as winds.

The close relationship between thermal energy and motion is seen in the use of orchard heaters to induce convective activity in the surface inversion layer, in the circulation of the urban heat island, or in the altered wind patterns around large reservoirs. Proposals to assault major storms or segments of the general circulation are much more ambitious and in most cases highly speculative. Application of direct counterforces to divert violent storms or major wind systems is futile. Table 17.3 shows the immense amounts of energy developed in several types of atmospheric motion systems. Clearly, it would not be feasible to concentrate the energy required to overpower these phenomena. Large-scale manipulation of the energy exchange and initiation of triggering effects in conditions of natural instability offer greater promise.

A characteristic of the general circulation as we presently understand it is the irregularity of its perturbations. Consequently, conditions that are quite similar initially may develop into very different physical states owing to slight nuances in physical processes. This suggests the possibility of triggering natural instabilities to effect changes in weather and, by continuing the triggering action, to alter climate. Because energy, moisture, and motion systems do not operate separately in the atmosphere, however, a relatively small disturbance of one may induce a chain reaction of complex processes that are difficult to predict and impossible to control.

TABLE 17.3

Estimated Kinetic Energy in Atmospheric Motion Systems and Time Required for All United States Generating Facilities to Produce Electrical Equivalent *

System	*Approximate Kinetic Energy (ergs)*	*Time Equivalent U.S. Electric Generation*
Tornado (life)	10^{21}	30 seconds
Small thunderstorm (life)	10^{22}	5 minutes
Large thunderstorm (life)	10^{23}	several hours
Hurricane (instantaneous)	10^{25}	several days
Extratropical cyclone	10^{26}	5 to 6 weeks
General circulation of Northern Hemisphere	5×10^{27}	6 years

**Weather and Climate Modification,* Report of the Special Commission on Weather Modification (Washington, D.C.: National Science Foundation, 1966), pp. 35–36.

Insofar as ocean currents and drifts influence climates, alteration of their speed or direction would initiate climatic changes. Mechanical mixing to produce artificial upwelling of cold water not only would affect heat and moisture exchanges at the sea surface but also would cause changes in the horizontal circulation. Thermal mixing could be achieved by using deep sea water as a coolant in nuclear and other thermal power generators and reintroducing it to seek higher levels in the ocean. The results of artificial mixing might be felt in the climates of areas far removed from the operation, however, and the atmospheric effects might be difficult to reverse.

One of the most persistent suggestions for climatic modification on a large scale is that to dam Bering Strait between Siberia and Alaska. If water were then pumped out of the Arctic Ocean into the North Pacific warmer water from the North Atlantic would be drawn into the arctic basin to melt the ice. This proposal, like that to melt the ice pack by reducing its albedo, is supposed to produce milder polar and tundra climates in the Northern Hemisphere. It is conceivable that it might raise temperatures, increase evaporation from the open ocean, promote greater cloudiness and snowfall, and trigger another ice age. It might also create stronger winter winds from the cold anticyclones over the continents, thus enhancing storminess. The relevant question appears to be not whether the climate would be altered but rather *how* it would be altered. Other theoretical possibilities for regulating oceanic circulation include damming the Florida Strait to modify the Gulf Stream and engineering the exchange of water through the Strait of Gibraltar. Accelerated breakup of the coastal ice around Antarctica would affect ocean currents at least in the Southern Hemisphere. Proposals to tow ice rafts to arid coasts to augment water supplies also entail heat transport. Removal of sufficient ice would increase the area of exposed water along the antarctic perimeter, probably creating climatic effects. Like many other suggestions for global engineering these ideas pose economic as well as scientific problems.

CLOUD MODIFICATION

The principle of a triggering action to initiate weather changes has had its most widespread applications in cloud modification experiments that seek to duplicate natural processes. The critical processes are condensation, sublimation, ice crystal formation, and coalescence. In view of the large amounts of heat released by condensation it might be inferred that addition of nuclei would speed cloud development, especially in supersaturated air. Natural condensation nuclei abound in the atmosphere, however, and supersaturation in clouds is rarely greater than about one percent. Thus, attempts to stimulate cloud formation by forcing condensation are not likely to be productive. On the other hand, extensive supercooling of liquid droplets is common in natural clouds. Freezing releases latent heat of fusion and may disrupt the energy system of the cloud or even an entire storm. Growth of ice crystals and coalescence of liquid droplets are the stages immediately preceding precipitation and are more realistic considerations in cloud modification.

Most experimental work on artificial stimulation of precipitation has involved clouds made up of supercooled droplets. Below a critical temperature of −38°C freezing of small droplets is spontaneous, but at higher temperatures freezing takes place only when ice nuclei are present. During laboratory experiments in 1946, Vincent J. Schaefer found dry ice to be effective in cooling droplets to produce ice crystals. He later sprinkled 3 kg of dry ice into a cloud from an airplane at 4,300 m above western Massachusetts. Snow fell from the cloud, although it did not reach the ground. This pioneer experiment led to widespread seeding with dry ice in cloud modification. The results have not been uniformly predictable, but observers have reported changes in cloud structure and falls of rain or snow in numerous experiments (see Fig. 17.7).

A related technique of cloud seeding is artificial nucleation, the purpose of which is to provide sufficient ice nuclei to trigger precipitation or, by "overseeding," to create so many ice crystals that few become large enough to fall from the cloud. In either case artificially induced ice crystals are presumed to grow at the expense of water droplets and fall as precipitation (see Chapter 3, "Precipitation: Causes, Forms, Processes and Types"). Natural nuclei initiate the process in supercooled clouds at temperatures between −40° and −15°C; certain dust particles from the North America Great Plains are effective at temperatures as high as −7°C; crystals of silver iodide start ice crystal formation at temperatures between approximately −13° and −4°C. Ice appears to form more readily on silver iodide crystals because the crystalline structures of both are hexagonal and they are of comparable size. Cloud seeding experiments have introduced silver iodide from aircraft or rockets and from ground-based generators. Massive seeding of supercooled cumulus provides a large number of nuclei for ice crystal formation, which releases heat of fusion and promotes vertical cloud development. Although it may be intended to enhance rainfall this "dynamic seeding" also could result in an excessive number of tiny crystals and inhibit the precipitation process.

Clouds having temperatures above 0°C throughout present special problems

Figure 17.7 Hole formed in supercooled clouds following dry ice seeding near Goose Bay, Labrador. (Air Force Cambridge Research Laboratories.)

in cloud seeding. Low stratus, for example, is thermodynamically stable, and ice nucleation and release of latent heat of fusion are less significant than coalescence of droplets. Where droplets of varying sizes exist together their different rates of movement may lead to collision and coalescence. Large droplets may divide upon collision, creating several drops which set off a chain reaction of raindrop growth. Freezing of drops by means of dry ice might facilitate collision. Introduction of hygroscopic materials or water spray into a cloud could provide the necessary large drops. Warm clouds have received less attention than supercooled types, and practical means of modification are lacking.

The effectiveness of cloud seeding is difficult to appraise. It is practically impossible to arrange a scientifically controlled experiment in which two or more identical clouds are available for different treatment and observation of results. The alternative is random selection from among a number of similar clouds and statistical comparisons of many experiments with each other and with climatic probabilities. Claims of changes in precipitation by cloud seeding vary widely and include both increases and decreases. Methods of evaluation also differ, but precipitation enhancement on the order of 10 to 20 percent has been reported.

In summary, the possible results of seeding convective clouds are many and varied. A 1966 report of a special National Academy of Sciences Panel on Modification of Weather and Climate listed the following possible induced responses to seeding:

1. Increase or decrease of rain, snow, or hail
2. Increase or decrease of precipitation leeward of the seeded area
3. Increase or decrease of lightning
4. Increased buoyancy and vertical cloud development
5. Altered patterns of air movement that may affect adjacent clouds

Thus, we face two fundamental questions: What kind and amount of human intervention will set changes in motion, and what are all the possible results of such action?

In addition to cloud seeding, electricity and shock waves offer theoretical possibilities in cloud modification. Lightning produces intense heat which may trigger convective activity, and electrical charges are probable factors in the coalescence of tiny cloud droplets. These effects are plausible explanations for the gushes of precipitation which often follow a lightning discharge in cumulonimbus. Alteration of the electrical field by artificial means could affect precipitation. Thunder creates shock waves that may induce collision of droplets, and the associated cooling by expansion may create ice crystals that assist the development of precipitation. Beginning in the nineteenth century, cannonading was employed in rainmaking attempts. In more recent times experimental rocket explosions within clouds have yielded few conclusive results.

MODIFICATION OF STORMS

As the scale of atmospheric modification increases the complexity of physical processes and the uncertainties of the outcome also increase. We have seen that counterforces are impractical solutions for the abatement of major storm systems. Most experiments in the modification of storms are directed toward changing their dynamic characteristics, mainly by cloud seeding. If, as some meteorologists believe, thunderstorms and tornado vortices draw critical portions of their energy from the heat of electrical discharges, then seeding at early stages of development could disrupt the energy balance and forestall violent convection. It has even been suggested that small strips of metal dropped into thunderstorms having high concentrations of electrical charges might decrease the potential for tornado development. The great number of lightning-caused forest fires in the western United States prompted a major U.S. Forest Service research project in lightning suppression beginning in 1960. Known as Project Skyfire, its aim was the modification of the electrical field in thunderstorms by silver iodide seeding on the supposition that an abundance of ice crystals increases the discharge points in the cloud and suppresses ground strikes. The use of metallic chaff is based upon a similar hypothesis. In other experiments small rockets have been sent aloft to carry long, trailing wires to induce lightning.

Hail suppression experiments are closely related to thunderstorm modification. One approach is seeding to create more small ice particles and more but smaller hail stones. Another is to seed the supercooled layers copiously, thus reducing the number of supercooled droplets available for ice accretion. Still other techniques rely on the theory that hail formation begins on large water drops near the base of the cloud, where added quantities of condensation nuclei might produce smaller drops. Russian scientists have claimed remarkable accuracy in radar detection of cloud layers in which hailstones are being formed and into which silver iodide can be seeded by artillery shells.

Cloud seeding has also been the fundamental method in tropical cyclone modification experiments. The objective has been primarily to reduce wind speeds in the storm center on the theory that heat of fusion added as a result of the seeding would alter the pressure field and lead to enlargement of the eye. Observed wind speeds decreased following seeding experiments in Atlantic hurricanes in 1961, 1963, and 1969 under Project Stormfury, a joint effort of the United States Navy and the National Oceanic and Atmospheric Administration. Whether seeding or natural changes in the hurricanes led to the decrease is not certain. Nor is there sufficient evidence to indicate that storms could be steered by seeding. After an early seeding attempt under Project Cirrus in 1947 a hurricane suddenly changed course and moved across Savannah, Georgia, causing heavy damage. The 1971 seeding of Hurricane *Ginger* was followed by damage in North Carolina. Scientists who would modify hurricanes are understandably cautious.

Inasmuch as hurricanes derive their energy primarily from the sea surface it is logical to assume that reduction of evaporation might prevent development, or at least calm their fury. Alkanols, which may be relatively effective on the nearly calm waters of lakes and reservoirs, offer little promise for suppression of evaporation from the disturbed waters beneath a developing hurricane. Oil could be expected to form a tougher film, but it would also reduce the albedo slightly, thus partially offsetting its capacity to retard evaporation. The polluting characteristics of oil slicks might be alleviated by using biodegradable synthetic oils.

It is tempting to consider jet streams, fronts, and pressure systems as direct targets for modification. However, current knowledge of the physical processes that guide these phenomena and their irregularities is inadequate for construction of reliable models for their transformation.

FOG DISPERSAL

Because fog lies at the earth's surface both the techniques and evaluation of modification are somewhat simpler than for clouds, but fog, like clouds, occurs in both supercooled and warm forms. The most successful attempts to modify cold fogs have employed seeding procedures similar to those applied to supercooled clouds. Dry ice seeding by small aircraft, flying from fog-free air strips, has been highly if not uniformly effective in clearing fog at a number of major airports. Another cooling method uses liquid propane, which is sprayed into the fog from ground dispensers. As the propane vaporizes the gas expands and cools adiabatically, thus reducing the temperature of the supercooled droplets below the critical minimum for the liquid state.

Warm fogs present many of the same problems encountered in the modification of warm clouds. They are not only more difficult to dissipate than cold fogs, but they are also more frequent and widespread. Warm advection fogs are continually renewed over a given area such as an airport, whereas stable ground fogs tend to form in conditions of near calm. Early successes in ground fog dispersal were achieved by heating. During World War II the British developed FIDO (Fog Investigation and Dispersal Operation), which evaporated fog by burning oil dispensed from perforated pipes along runways. Military expediency justified the high cost, but the principle of fog dispersal by evaporation was physically sound. Heating and turbulence created by jet engine exhaust, simple combustion blowers, or other sources of energy have potential for removal of shallow warm fogs (see Fig. 17.8). Seeding with carbon black is intended to promote evaporation by heat gained from the absorption of radiant energy. A less direct approach to forced

Figure 17.8 Fog dissipation in the wake of a Boeing 747 aircraft landing at Seattle-Tacoma Airport. Entrainment of dry air from above enhances dissipation by heat of the jet engines. (Courtesy Stinson Aircraft, Inc.)

evaporation is the use of highly hygroscopic substances which remove water vapor from the air within the fog and thereby cause fog droplets to evaporate. Calcium chloride, in powdered form and as a spray solution, was used in early experiments, but its corrosive properties are objectionable.

Japanese scientists were among the first to test water seeding of warm fogs; occasionally they achieved temporary clearing as the result of drop collision and precipitation. Other attempts to induce coalescence employ electrical charges introduced on oil spray or by means of wires. So-called "fog brooms" consist of filaments of nylon or noncorrosive wires on frames placed so as to intercept droplets. Experiments based on this principle have been designed both to improve visibility in advection fogs and to capture water along desert coasts.

The cost of fog to commercial aviation and other forms of transportation is adequate justification for experimental investigation of fog dispersal. All known techniques are expensive and must be weighed against losses that could otherwise be expected.

IMPLICATIONS OF WEATHER AND CLIMATE MODIFICATION

Thus far we have confined ourselves largely to the theories and experimental techniques of weather modification, and have avoided the pressing question of whether we should tamper with the atmosphere. Ability to forecast natural weather is a prerequisite to altering weather predictably. So long as the effects of modification activities are uncertain with respect to both time and scale, wisdom dictates discretion. Even if forecasts of natural atmospheric events were possible and the total meteorological results of human intervention were known, there remains the serious question of effects on other elements of the natural environment and on human activities. Small changes in the heat and moisture balances could produce far-reaching results by disrupting the ecological relationships of soils, plants, or animal life, all of which depend on the atmosphere. For example, increased precipitation on a plant community might favor certain species over others, promote soil leaching and erosion, encourage new insects and diseases, and create problems of runoff control. Small but sustained temperature changes, especially in critically marginal environments of plant and animal communities, could alter biological patterns by shifting the limits of vegetation formations and agriculture and perhaps interrupting the distribution and migratory habits of animals, birds, or insects. The imagination sets few limits to the implications of ice cap melting, rises in sea level, creation of deserts, or global changes in wind patterns.

If we assume that weather might be modified under reasonably predictable conditions that permit control of duration and area affected, we face the challenge of conflicting needs. Because one person's picnic weather may be another's drought there are legal aspects of weather modification. They range through questions of cloud ownership, property damage liability, accident insurance, private versus public control and benefit, the right to experiment, interstate and international conflicts, and a host of other social, economic, and political problems. Legal cases inevitably reached the courts early in the development of cloud seeding, but a coherent body of weather modification law emerged slowly to deal with planning, execution, and after-effects.

The prospect of modifying our environment by manipulation of atmospheric processes is at once alluring and frightening. Its implications are no less important to humanity than a full understanding of the complex atmosphere.

QUESTIONS AND PROBLEMS FOR CHAPTER 17

1. Summarize the known effects of urban areas on the heat budget, the water budget, and motion in the climate system.
2. Describe the actual and potential effects of rural land use on climate.
3. Explain the special climatic significance of aerosols and gases in the stratosphere.
4. Why are attempts to increase rainfall in deserts by creating large reservoirs likely to be unsuccessful?
5. How is it possible that cloud seeding might enhance precipitation in one experiment but reduce it in another?
6. Why have fog dispersal techniques been more successful in cold fogs than in warm fogs?
7. Assess the feasibility of modifying the general circulation, a hurricane, a tornado, and a dust devil.
8. What are the prospects and dangers of climatic modification through the manipulation of ocean currents?
9. Who should decide the time, place, and methods for planned weather modification?
10. What is meant by "triggering effect"? Cite an example to illustrate how it might be applied in the modification of weather and climate.

APPENDICES

Appendix A

Conversion Equivalents

LENGTH

1 kilometer = 0.6214 statute mile = 0.5396 nautical mile
1 meter = 1.0936 yards = 3.2808 feet = 39.37 inches
1 centimeter = 0.3937 inch
1 micrometer = 1 micron = 10^{-6} meter = 10^4 angstroms
1 nanometer = 1 millimicron = 10^{-9} meter = 10 angstroms

AREA

1 hectare = 10^4 square meters = 2.471 acres
1 square meter = 10^4 square centimeters = 10.76 square feet
1 square centimeter = 0.155 square inch

VOLUME

1 cubic meter = 1000 liters = 35.31 cubic feet = 264.2 U.S. gallons
1 liter = 1000 cubic centimeters = 33.81 fluid ounces = 1.057 U.S. quarts

MASS

1 metric ton (tonne) = 1000 kilograms = 2204.6 pounds (avoirdupois)
1 kilogram = 2.2046 pounds
1 gram = 0.353 ounce = 0.0022 pound

PRESSURE

1 standard atmosphere = 1013.25 millibars = 101.32 kilopascals
= 760 millimeters Hg = 29.92 inches Hg
= 14.7 pounds per square inch
1 millibar = 1000 dynes per square centimeter = 100 newtons per square meter
= 0.75006 millimeter Hg = 0.02953 inch Hg
1 newton per square meter = 1.4504×10^{-4} pound per square inch
1 kilopascal = 10 millibars = 1000 newtons per square meter

TEMPERATURE

1 Celsius degree = 1.8 Fahrenheit degrees = 1 Kelvin (Absolute) degree
Celsius temperature = $\frac{5}{9}(F - 32)$ = Kelvin − 273.16
Fahrenheit temperature = $\frac{9}{5}C + 32$

ENERGY AND POWER

1 gram calorie (thermochemical) = 4.184 joules = 4.184×10^7 ergs
1 British thermal unit = 252 grams calories = 1054. 35 joules
1 watt = 1 joule per second = 14.34 gram calories per minute
= 0.057 Btu per minute
1 kilowatt = 1000 watts = 1.341 horsepower
1 gram calorie per minute = 0.0697 watt
1 langley = 1 gram calorie per square centimeter = 3.67 Btu per square foot
1 langley per minute = 697.33 watts per square meter

Appendix B

Psychrometric Tables

TABLE B.1

Dew Point Temperature (1,000 mb)

Dry-bulb temp. (°C)	*Saturation vapor press. (mb)*	*Wet-bulb depression ($T_d - T_w$)*																					
		1	*2*	*3*	*4*	*5*	*6*	*7*	*8*	*9*	*10*	*11*	*12*	*13*	*14*	*15*	*16*	*17*	*18*	*19*	*20*	*21*	*22*
−20	1.2540	−33																					
−18	1.4877	−28																					
−16	1.7597	−24																					
−14	2.0755	−21	−36																				
−12	2.4409	−18	−28																				
−10	2.8627	−14	−22																				
−8	3.3484	−12	−18	−29																			
−6	3.9061	−10	−14	−22																			
−4	4.5451	−7	−11	−17	−29																		
−2	5.2753	−5	−8	−13	−20																		
0	6.1078	−3	−6	−9	−15	−24																	
2	7.0547	−1	−3	−6	−11	−17																	
4	8.1294	1	−1	−4	−7	−11	−19																
6	9.3465	4	1	−1	−4	−7	−13	−21															
8	10.722	6	3	1	−2	−5	−9	−14															
10	12.272	8	6	4	1	−2	−5	−9	−14	−28													
12	14.017	10	8	6	4	1	−2	−5	−9	−16													
14	15.977	12	11	9	6	4	1	−2	−5	−10	−17												
16	18.173	14	13	11	9	7	4	1	−1	−6	−10	−17											
18	20.630	16	15	13	11	9	7	4	2	−2	−5	−10	−19										
20	23.373	19	17	15	14	12	10	7	4	2	−2	−5	−10	−19									
22	26.430	21	19	17	16	14	12	10	8	5	3	−1	−5	−10	−19								
24	29.831	23	21	20	18	16	14	12	10	8	6	2	−1	−5	−10	−18							
26	33.608	25	23	22	20	18	17	15	13	11	9	6	3	0	−4	−9	−18						
28	37.796	27	25	24	22	21	19	17	16	14	11	9	7	4	1	−3	−9	−16					
30	42.430	29	27	26	24	23	21	19	18	16	14	12	10	8	5	1	−2	−8	−15				
32	47.551	31	29	28	27	25	24	22	21	19	17	15	13	11	8	5	2	−2	−7	−14			
34	53.200	33	31	30	29	27	26	24	23	21	20	18	16	14	12	9	6	3	−1	−5	−12	−29	
36	59.422	35	33	32	31	29	28	27	25	24	22	20	19	17	15	13	10	7	4	0	−4	−10	
38	66.264	37	35	34	33	32	30	29	28	26	25	23	21	19	17	15	13	11	8	5	1	−3	−9
40	73.777	39	37	36	35	34	32	31	30	28	27	25	24	22	20	18	16	14	12	9	6	2	−2

TABLE B.2

Relative Humidity in Percent (1,000 mb)

Dry-bulb temp. (°C)	Wet-bulb depression ($T_d - T_w$)																					
	1	2	3	4	5	6	7	8	9	10	11	12	13	14	15	16	17	18	19	20	21	22
−20	28																					
−18	40																					
−16	48	0																				
−14	55	11																				
−12	61	23																				
−10	66	33	0																			
−8	71	41	13																			
−6	73	48	20	0																		
−4	77	54	32	11																		
−2	79	58	37	20	1																	
0	81	63	45	28	11																	
2	83	67	51	36	20	6																
4	85	70	56	42	27	14																
6	86	72	59	46	35	22	10	0														
8	87	74	62	51	39	28	17	6														
10	88	76	65	54	43	33	24	13	4													
12	88	78	67	57	48	38	28	19	10	2												
14	89	79	69	60	50	41	33	25	16	8	1											
16	90	80	71	62	54	45	37	29	21	14	7	1										
18	91	81	72	64	56	48	40	33	26	19	12	6	0									
20	91	82	74	66	58	51	44	36	30	23	17	11	5	0								
22	92	83	75	68	60	53	46	40	33	27	21	15	10	4	0							
24	92	84	76	69	62	55	49	42	36	30	25	20	14	9	4	0						
26	92	85	77	70	64	57	51	45	39	34	28	23	18	13	9	5						
28	93	86	78	71	65	59	53	47	42	36	31	26	21	17	12	8	4					
30	93	86	79	72	66	61	55	49	44	39	34	29	25	20	16	12	8	4				
32	93	86	80	73	68	62	56	55	46	41	36	32	27	22	19	14	11	8	4			
34	93	86	81	74	69	63	58	52	48	43	38	34	30	26	22	18	14	11	8	5		
36	94	87	81	75	69	64	59	54	50	44	40	36	32	28	24	21	17	13	10	7	4	
38	94	87	82	76	70	66	60	55	51	46	42	38	34	30	26	23	20	16	13	10	7	5
40	94	89	82	76	71	67	61	57	52	48	44	40	36	33	29	25	22	19	16	13	10	7

Appendix C

Supplementary Climatic Data

TABLE C.1

Supplementary Climatic Data

T : Temperature in degrees Celsius — Latitude to nearest whole degree
P : Precipitation in millimeters — Station height in meters

Africa		J	F	M	A	M	J	J	A	S	O	N	D	Yr
Accra, Ghana	T	27	28	28	28	27	26	25	24	25	26	27	27	26
6°N; 65 m	P	16	37	73	82	145	193	49	16	40	80	38	18	787
Addis Ababa, Ethiopia	T	17	18	19	19	19	17	15	15	16	16	17	17	17
9°N; 2,408 m	P	24	25	67	93	53	105	239	266	174	43	3	18	1,110
Algiers, Algeria	T	10	11	13	15	18	22	24	25	23	19	15	12	17
37°N; 28 m	P	116	76	57	65	36	14	2	4	27	84	93	117	691
Beira, Mozambique	T	27	28	27	26	23	21	20	21	23	25	26	27	24
20° S; 8 m	P	265	225	244	105	58	42	37	30	27	29	133	134	1,429
Bulawayo, Rhodesia	T	21	21	20	19	16	13	14	16	19	22	22	21	19
20°S; 1,344 m	P	134	111	65	21	9	3	T	1	5	25	89	125	589
Cape Town, South Africa	T	20	20	19	16	14	13	12	12	14	15	18	19	16
34°S; 44 m	P	11	15	14	53	89	84	83	73	45	31	17	11	526
Douala, Cameroon	T	27	27	27	27	27	26	25	25	25	26	26	27	26
4°N; 13 m	P	61	88	226	240	353	472	710	726	628	399	146	60	4,109
Kano, Nigeria	T	21	24	28	31	30	28	26	25	26	27	25	22	26
12°N; 476 m	P	0	T	2	8	71	119	209	311	137	14	T	0	872
Mombasa, Kenya	T	28	28	28	28	26	25	24	24	25	26	27	28	26
4°S; 55 m	P	30	14	59	192	319	100	72	69	71	86	74	76	1,163

TABLE C.1 Continued

		J	F	M	A	M	J	J	A	S	O	N	D	Yr
Tamatave, Malagasy	T	26	26	26	25	23	21	21	21	22	23	25	26	24
18°S; 5 m	P	420	442	528	403	303	300	257	208	135	91	184	259	3,530
Wadi Halfa, Sudan	T	15	16	21	26	31	32	32	32	30	28	22	17	25
22°N; 160 m	P	0	0	0	0	1	0	1	0	0	1	0	0	3
Asia														
Ankara, Turkey	T	0	1	5	11	16	20	23	23	18	13	7	2	12
40°N; 902 m	P	37	36	36	37	49	30	14	9	17	24	30	43	360
Ashkhabad, U.S.S.R.	T	2	5	9	16	23	29	31	29	24	16	8	3	16
38°N; 230 m	P	22	21	44	38	28	6	2	1	3	11	15	19	210
Baghdad, Iraq	T	10	12	16	22	28	33	35	34	31	25	17	11	23
33°N; 34 m	P	26	28	28	17	7	T	0	T	T	3	21	26	156
Bangkok, Thailand	T	26	28	29	30	30	29	28	28	28	28	27	26	28
14°N; 12 m	P	9	29	34	89	166	171	178	191	306	255	57	7	1,492
Bombay, India	T	24	25	27	29	30	29	28	27	27	28	28	26	27
19°N; 11 m	P	2	1	0	3	16	520	709	419	297	88	21	2	2,078
Inchon, Korea	T	−4	−2	3	10	15	20	24	25	21	14	7	0	11
37°N; 70 m	P	16	18	50	66	73	139	304	180	137	45	35	30	1,089
Irkutsk, U.S.S.R.	T	−21	−18	−9	2	9	15	18	15	8	1	−11	−18	−1
52°N; 485 m	P	12	8	9	15	29	83	102	99	49	20	17	15	458
Jakarta, Indonesia	T	26	26	27	27	27	27	27	27	27	27	27	27	27
6°S; 7 m	P	335	241	201	141	116	97	61	50	78	91	151	193	1,755

TABLE C.1 Continued

		J	F	M	A	M	J	J	A	S	O	N	D	Yr
Lanchow, China	T	−6	−2	5	12	17	21	23	21	16	10	2	−5	10
36° N; 1,508 m	P	1	3	8	14	34	40	66	92	55	18	4	2	338
Lhasa, Tibet (1941–48)	T	−2	1	5	8	12	17	16	16	14	9	4	0	8
30° N; 3,685 m	P	1	13	8	5	25	63	122	89	66	13	3	0	408
Manila, Philippines	T	25	26	27	29	29	28	28	27	27	27	26	25	27
15° N; 15 m	P	18	7	6	24	110	236	253	480	271	201	129	56	1,791
New Delhi, India	T	14	17	23	29	34	34	31	30	29	26	20	16	25
29° N; 216 m	P	25	22	17	7	8	65	211	173	150	31	1	5	715
Peking, China	T	−5	−2	5	14	20	25	26	25	20	13	4	−3	12
40° N; 52 m	P	4	5	8	17	35	78	243	141	58	16	11	3	623
Shanghai, China	T	4	4	9	14	19	23	28	28	23	18	12	6	16
31° N; 5 m	P	46	66	77	84	107	168	139	132	160	62	53	41	1,135
Tashkent, U.S.S.R.	T	0	3	7	14	20	25	27	25	19	13	5	1	13
41° N; 428 m	P	49	51	81	58	32	12	4	3	3	23	44	57	417
Tbilisi, U.S.S.R.	T	1	3	6	12	18	21	25	24	20	14	8	3	13
42° N; 490 m	P	20	21	36	43	87	69	50	37	42	46	37	20	508
Trivandrum, India	T	27	27	28	28	28	26	26	26	27	27	27	27	27
8° N; 64 m	P	19	21	44	122	249	331	211	164	123	271	207	73	1,835
Verkhoyansk, U.S.S.R.	T	−47	−43	−30	−14	3	13	16	11	3	−14	−36	−44	−15
68° N; 137 m	P	7	5	5	4	5	25	33	30	13	11	10	7	155

TABLE C.1 Continued

Australia and New Zealand		J	F	M	A	M	J	J	A	S	O	N	D	Yr
Alice Springs, N. Terr.	T	28	27	25	20	16	12	12	14	18	23	25	28	21
24°S; 546 m	P	27	45	18	10	18	15	14	10	6	25	23	39	250
Auckland, New Zealand	T	19	19	18	16	14	12	11	11	12	14	16	18	15
37°S; 5 m	P	85	104	71	108	123	139	139	108	98	106	89	78	1,248
Brisbane, Queensland	T	25	25	24	21	18	16	15	16	18	21	22	24	20
27°S; 41 m	P	143	183	147	78	57	56	49	30	45	77	92	136	1,092
Christchurch, New Zealand	T	16	17	15	12	9	6	5	7	9	11	14	16	11
43°S; 29 m	P	47	51	52	51	81	52	53	51	48	49	42	56	633
Hobart, Tasmania	T	16	16	15	12	11	8	8	9	11	12	14	15	12
43°S; 54 m	P	42	47	52	63	51	66	47	53	53	72	58	64	668
Invercargill, New Zealand	T	13	14	12	10	8	6	5	6	8	10	11	12	10
46°S; 1 m	P	99	94	96	98	98	104	75	71	72	80	88	88	1,063
Perth, W. Australia	T	23	24	22	19	16	14	13	14	15	16	19	22	18
32°S; 60 m	P	7	12	22	52	125	192	183	135	69	54	23	15	889
Port Hedland, W. Australia	T	30	31	30	28	23	20	19	21	24	26	29	30	26
20°S; 8 m	P	56	86	36	15	45	21	15	4	1	1	1	19	299
Sydney, New South Wales	T	22	22	21	18	16	13	12	13	15	18	19	21	18
34°S; 42 m	P	104	125	129	101	115	141	94	83	72	80	77	86	1,205
Wellington, New Zealand	T	15	16	15	13	11	9	8	8	10	11	13	14	12
41°S; 119 m	P	74	104	80	90	127	123	128	116	92	116	79	95	1,224

TABLE C.1 Continued

Europe		J	F	M	A	M	J	J	A	S	O	N	D	Yr
Athens, Greece	T	9	10	11	15	20	25	28	27	24	19	15	11	18
38° N; 107 m	P	62	36	38	23	23	14	6	7	15	51	56	71	402
Berlin, Germany	T	−1	0	4	9	14	17	19	18	14	9	4	1	9
52° N; 55 m	P	45	42	31	43	49	65	72	71	47	49	48	43	605
Edinburgh, Scotland	T	4	4	5	7	10	13	15	14	12	9	6	4	8
56° N; 76 m	P	55	42	44	39	52	53	71	78	60	67	58	57	676
Kiev, U.S.S.R.	T	−6	−5	0	8	15	19	20	19	14	8	1	−3	7
50° N; 179 m	P	43	39	35	46	56	66	70	72	47	47	53	41	615
Leningrad, U.S.S.R.	T	−8	−8	−4	3	10	15	18	17	11	5	0	−4	5
60° N; 4 m	P	36	32	25	34	41	54	69	77	58	52	45	36	559
Lisbon, Portugal	T	11	12	14	16	17	20	22	22	21	18	14	12	17
39° N; 95 m	P	111	76	109	54	44	16	3	4	33	62	93	103	708
London, England	T	4	4	7	9	12	16	18	17	15	11	7	5	10
51° N; 45 m	P	51	38	36	46	46	41	51	56	46	58	63	51	583
Madrid, Spain	T	5	6	10	13	16	21	24	24	20	14	9	6	14
40° N; 660 m	P	38	34	45	44	44	27	11	14	31	53	47	48	436
Milan, Italy	T	2	4	9	14	18	22	25	24	20	14	8	3	14
45° N; 147 m	P	60	56	66	83	101	75	51	68	72	91	98	70	891
Odessa, U.S.S.R	T	−2	−2	2	8	15	20	22	22	17	11	5	0	10
46° N; 64 m	P	28	26	20	27	34	45	34	37	29	35	43	31	389
Oslo, Norway	T	−5	−4	0	5	11	15	17	16	11	6	1	−2	6
60° N; 96 m	P	49	35	26	44	44	71	84	96	83	76	69	63	740

TABLE C.1 Continued

		J	F	M	A	M	J	J	A	S	O	N	D	Yr
Rome, Italy	T	7	8	10	14	18	22	25	24	21	17	12	8	15
42° N; 18 m	P	79	69	72	67	57	39	15	25	69	126	117	99	834
Vienna, Austria	T	−1	0	5	10	15	18	20	19	16	10	5	1	10
48° N; 203 m	P	40	43	45	45	70	67	83	72	41	56	53	45	660
Warsaw, Poland	T	−2	−3	1	7	13	17	19	18	13	8	3	0	8
52° N; 107 m	P	25	28	20	32	40	60	79	47	41	31	31	37	471
North America														
Angmagssalik, Greenland	T	−7	−7	−6	−3	2	6	7	7	4	0	−3	−5	0
66° N; 36 m	P	57	81	57	55	52	45	28	70	72	96	87	75	775
Boston, Massachusetts	T	−3	−3	2	8	14	18	22	21	17	12	6	−1	9
42° N; 198 m	P	114	95	115	102	88	95	83	103	100	95	115	101	1,207
Chicago, Illinois	T	−3	−2	2	10	16	22	24	24	19	13	4	−2	10
42° N; 187 m	P	47	41	70	77	95	103	86	80	69	71	56	48	843
Churchill, Manitoba	T	−28	−26	−20	−10	−2	6	12	12	6	−1	−12	−22	−7
59° N; 35 m	P	17	17	21	36	40	48	39	60	52	42	46	25	443
Coppermine, N.W.T.	T	−29	−30	−26	−17	−6	3	9	8	3	−7	−20	−26	−11
68° N; 9 m	P	13	8	15	14	12	20	34	44	29	27	17	13	246
Dallas, Texas	T	8	10	14	18	23	27	29	29	26	20	14	9	19
33° N; 156 m	P	63	61	84	107	114	97	71	76	69	71	69	64	946
Edmonton, Alberta	T	−14	−12	−6	4	11	14	17	16	11	5	−4	−10	3
53° N; 676 m	P	24	20	21	28	47	80	85	65	34	23	22	25	474
Gander, Newfoundland	T	−6	−6	−4	1	6	12	17	16	12	6	2	−4	4
49° N; 147 m	P	83	88	81	69	59	76	84	101	86	97	107	87	1,018

TABLE C.1 Continued

		J	F	M	A	M	J	J	A	S	O	N	D	Yr
Havana, Cuba	T	22	22	23	25	26	27	28	28	28	26	24	23	25
23° N; 24 m	P	71	46	46	58	119	165	124	135	150	173	79	58	1,224
Mazatlan, Mexico	T	20	20	20	22	24	27	28	28	28	27	24	21	24
23° N; 78 m	P	12	8	3	0	1	34	174	215	250	63	17	27	805
Mexico City, Mexico	T	13	15	17	18	19	18	17	17	17	16	14	13	16
19° N; 2,259 m	P	4	5	11	18	46	100	116	114	102	37	13	5	571
Miami, Florida	T	19	20	21	23	25	27	28	28	27	25	22	20	24
26° N; 4 m	P	52	47	58	99	164	187	171	177	241	209	72	42	1,518
Nome, Alaska	T	−15	−15	−13	−6	2	8	10	9	6	−1	−9	−14	−3
64° N; 339 m	P	26	24	22	20	18	24	58	97	68	43	29	25	454
Phoenix, Arizona	T	10	12	16	20	25	30	33	32	29	22	15	11	21
33° N; 339 m	P	19	22	17	8	3	2	20	28	19	12	12	22	183
Salt Lake City, Utah	T	−2	1	5	10	15	19	25	24	18	12	3	0	11
41° N; 1,287 m	P	34	30	40	45	36	25	15	22	13	29	33	31	353
Spokane, Washington	T	−4	−1	3	8	13	16	21	20	16	9	2	−1	8
48° N; 721 m	P	62	47	38	23	31	38	10	10	19	40	57	62	437
Topeka, Kansas	T	−2	1	5	12	18	24	27	26	21	15	6	1	13
39° N; 269 m	P	24	44	58	72	149	119	141	82	75	44	19	25	854
Yakutat, Alaska	T	−3	−2	0	3	7	10	12	12	10	6	1	−2	4
60° N; 9 m	P	276	208	221	184	203	129	214	277	420	498	407	312	3,348
Yuma, Arizona	T	13	15	19	22	26	31	35	34	31	25	18	14	24
33° N; 62 m	P	10	9	6	2	T	T	6	13	10	10	3	8	77

TABLE C.1 Continued

South America		J	F	M	A	M	J	J	A	S	O	N	D	Yr
Antofagasta, Chile (1951–60)	T	20	20	19	16	15	14	13	14	14	15	17	18	16
23°S; 122 m	P	0	0	0	0	0	0	0.3	0.1	0	0	0	0	0.4
Bogota, Colombia (1951–60)	T	13	14	14	14	14	14	13	13	13	14	14	14	14
5°N; 2,556 m	P	38	48	66	96	103	60	48	38	53	158	140	93	942
Caracas, Venezuela (1951–60)	T	19	20	21	22	22	22	21	22	22	22	21	20	21
10°N; 1,042 m	P	22	28	11	43	92	121	107	104	107	110	73	44	862
Manaus, Brazil	T	26	26	26	26	26	27	27	28	28	28	27	27	27
3°S; 44 m	P	278	278	300	287	193	99	61	41	62	112	165	220	2,095
Maracaibo, Venezuela	T	26	27	27	28	28	29	29	29	29	28	28	27	28
11°N; 40 m	P	3	2	3	27	73	43	28	42	40	99	22	5	387
Montevideo, Uruguay	Y	23	22	20	17	14	11	10	11	13	15	18	21	16
35°S; 22 m	P	83	74	104	102	91	88	73	87	84	73	79	77	1,014
Rio de Janeiro, Brazil	T	26	26	26	24	22	21	21	21	22	22	23	24	23
23°S; 27 m	P	136	137	133	116	73	44	43	43	53	74	97	127	1,074
Sao Paulo, Brazil	T	22	22	21	19	17	16	15	16	17	19	19	20	18
24°S; 795 m	P	224	209	166	69	53	42	31	42	66	113	118	155	1,287
Ushuaia, Argentina	T	9	9	8	6	3	2	2	2	4	6	7	8	6
55°S; 6 m	P	58	50	57	46	48	45	47	50	38	36	50	49	574
Valdivia, Chile (1951–60)	T	16	16	15	12	10	8	8	8	9	12	14	16	12
40°S; 13 m	P	106	50	96	178	437	371	452	318	225	98	82	72	2,486

TABLE C.1 Continued

Islands		J	F	M	A	M	J	J	A	S	O	N	D	Yr
Apia, Western Samoa	T	27	27	27	27	27	26	26	26	26	26	27	27	27
14°S, 172°W; 2 m	P	424	364	352	214	186	130	115	111	147	221	279	385	2,928
Ascension, South Atlantic	T	26	27	28	28	27	26	25	25	24	24	25	26	26
8°S, 14°W; 17 m (1951–60)	P	3	3	12	20	14	19	14	12	8	9	4	5	122
Canton, Phoenix Islands	T	28	28	28	29	29	29	29	29	29	29	29	28	29
3°S, 172°W; 3 m	P	66	54	63	92	110	67	66	64	31	28	41	65	748
Guam	T	26	26	26	27	27	27	26	26	26	26	26	26	26
14°N, 145°E; 110 m	P	118	89	67	77	106	149	228	326	339	333	261	155	2,249
Honolulu, Hawaii	T	23	22	23	23	24	26	26	26	26	26	24	23	24
19°N, 158°W; 12 m	P	96	84	73	33	25	8	11	23	25	47	55	76	556
Lae, Papua New Guinea	T	25	28	27	27	26	25	25	25	25	26	27	27	26
7°S, 147°E; 8 m	P	252	243	330	420	387	414	538	542	415	320	326	351	4,538
Laurie, South Orkneys	T	0	0	−1	−3	−7	−10	−11	−10	−7	−4	−2	−1	−5
61°S, 45°W; 4 m	P	35	39	48	41	32	26	32	32	29	29	32	27	402
Macquarie, South Pacific	T	7	7	6	5	4	3	3	3	3	4	4	6	5
54°S, 159°E; 6 m	P	99	81	104	94	81	81	68	68	68	73	64	72	952
Mauritius, Indian Ocean	T	26	26	26	24	22	21	20	20	21	22	24	26	23
20°S, 58°E; 55 m	P	207	195	218	141	97	67	57	59	37	39	51	115	1,283
Noumea, New Caledonia	T	26	26	25	23	22	21	19	20	21	22	24	25	23
22°S, 166°E; 72 m	P	117	94	175	124	93	89	84	70	53	52	47	85	1,083
Reykjavik, Iceland	T	0	0	2	3	7	10	11	11	9	5	3	1	5
64°N, 8°W; 18 m	P	90	65	65	53	42	41	48	66	72	97	85	81	805
Tristan da Cunha, South	T	17	18	17	16	14	13	12	12	12	13	14	16	15
Atlantic 37°S, 12°W; 23 m	P	103	88	88	153	144	187	147	193	129	139	117	167	1,655

Bibliography

BOOKS

ANTHES, RICHARD A., et al., *The Atmosphere* (3rd ed.). Columbus, Ohio: Charles E. Merrill Publishing Company, 1981.

BARRETT, E. C., *Climatology from Satellites*. London: Methuen & Company Ltd, 1974.

BARRY, ROGER G., *Mountain Weather and Climate*. New York: Methuen & Company Ltd., 1981.

——, and R. J. CHORLEY, *Atmosphere, Weather, and Climate* (3rd ed.). London: Methuen & Company Ltd., 1976.

——, and A. H. PERRY, *Synoptic Climatology: Methods and Applications*. London: Methuen & Company Ltd., 1973.

BATTAN, LOUIS J., *Fundamentals of Meteorology*. Englewood Cliffs, N.J.: Prentice-Hall, Inc., 1979.

BAUMGARTNER, ALBERT, and EBERHARD REICHEL, *The World Water Balance, Mean Annual Global Continental and Maritime Precipitation, Evaporation, and Runoff*. Amsterdam: Elsevier Scientific Publishing Company, 1975.

BOUCHER, KEITH, *Global Climate*. New York: John Wiley & Sons, Inc., 1975.

BREUER, GEORG, *Weather Modification: Prospects and Problems*. New York: Cambridge University Press, 1979.

BRYSON, REID A., and THOMAS J. MURRAY, *Climates of Hunger*. Madison: University of Wisconsin Press, 1977.

BUDYKO, M. I., *Climate and Life*, trans. David H. Miller. New York: Academic Press, Inc., 1974.

——, *Climatic Changes*. Washington, D.C.: American Geophysical Union, 1977.

Campbell, Ian M., *Energy and the Atmosphere.* New York: John Wiley & Sons, Inc., 1977.

Chang, Jen-Hu, *Atmospheric Circulation Systems and Climates.* Honolulu: Oriental Publishing Company, 1972.

Claiborne, Robert, *Climate, Man, and History.* New York: W. W. Norton & Company, Inc., 1970.

Cole, Franklyn W., *Introduction to Meteorology* (3rd ed.). New York: John Wiley & Sons, Inc., 1980.

Crowe, P. R., *Concepts in Climatology.* New York: St. Martin's Press, Inc., 1971.

Day, John A., and Gilbert L. Sternes, *Climate and Weather.* Reading, Mass.: Addison-Wesley Publishing Company, Inc., 1970.

Dutton, John A., *The Ceaseless Wind: An Introduction to the Theory of Atmospheric Motion.* New York: McGraw-Hill Book Company, 1976.

Eagleman, Joe R., *The Visualization of Climate.* Lexington, Mass.: D. C. Heath and Company, 1976.

——, *Meteorology: The Atmosphere in Action.* New York: D. Van Nostrand Company, 1980.

Evans, Martin, *Housing, Climate and Comfort.* London: The Architectural Press, 1980.

Ferrar, Terry A., ed., *The Urban Costs of Climate Modification.* New York: John Wiley & Sons, Inc., 1976.

Flohn, Hermann, *Climate and Weather.* New York: McGraw-Hill Book Company, 1969.

Frakes, L. A., *Climates throughout Geologic Time.* Amsterdam: Elsevier Publishing Company, 1979.

Fritts, Harold C., *Tree Rings and Climate.* New York: Academic Press, Inc., 1976.

Gaskell, T. F., and Martin Morris, *World Climate: The Weather, the Environment and Man.* New York: Thames and Hudson, 1979.

Gedzelman, Stanley David, *The Science and Wonders of the Atmosphere.* New York: John Wiley & Sons, Inc., 1980.

Geiger, Rudolf, *The Climate near the Ground* (trans. of 4th German ed.). Cambridge, Mass.: Harvard University Press, 1965.

Giese, Arthur C., *Living with Our Sun's Ultraviolet Rays.* New York: Plenum Press, 1976.

Golde, R. H., *Lightning Protection.* London: Edward Arnold (Publishers) Ltd., 1973.

Gribbin, John, ed., *Climatic Change.* New York: Cambridge University Press, 1978.

Griffiths, John F., *Applied Climatology* (2nd ed.). New York: Oxford University Press, 1976.

——, *Climate and the Environment.* Boulder, Colo.: Westview Press, 1976.

——, and Dennis M. Driscoll, *Survey of Climatology.* Columbus, Ohio: Charles E. Merrill Publishing Company, 1982.

Haurwitz, Bernhard, and James M. Austin, *Climatology.* New York: McGraw-Hill Book Company, 1944.

Herman, John R., and Richard A. Goldberg, *Sun, Weather, and Climate.* Washington, D.C.: National Aeronautics and Space Administration, 1978.

Hess, W. N., ed., *Weather and Climate Modification.* New York: John Wiley & Sons, Inc., 1974.

Hobbs, J. E., *Applied Climatology.* Boulder, Colo.: Westview Press, 1980.

JACKSON, I. J., *Climate, Water and Agriculture in the Tropics.* London: Longman Group Limited, 1977.

KENDREW, W. G., *The Climates of the Continents* (5th ed.). New York: Oxford University Press, 1961.

KONDRATYEV, K. YA., *Radiation in the Atmosphere.* New York: Academic Press, Inc., 1969.

LADURIE, EMMANUEL LE ROY, *Times of Feast, Times of Famine: A History of Climate since the Year 1000* (trans. Barbara Bray). Garden City, N.Y.: Doubleday & Company, Inc., 1971.

LAMB, H. H., *Climate: Present, Past and Future,* Vol. 1: *Fundamentals and Climate Now* (1972); Vol. 2: *Climatic History and the Future* (1977). London: Methuen & Company Ltd.

LANDSBERG, HELMUT E., *Physical Climatology* (2nd ed.). DuBois, Pa.: Gray Printing Company, Inc., 1958.

———, *The Urban Climate.* (International Geophysics Series, Vol. 28.) New York: Academic Press, Inc., 1981.

LOCKWOOD, JOHN G., *World Climatology.* New York: St. Martin's Press, Inc., 1974.

———, *Causes of Climate.* New York: Halsted Press, 1979.

LONGLEY, RICHMOND W., *Elements of Meteorology.* New York: John Wiley & Sons, Inc., 1970.

LOWRY, WILLIAM P., *Weather and Life.* New York: Academic Press, Inc., 1969.

LUTGENS, FREDERICK K., and EDWARD J. TARBUCK, *The Atmosphere: An Introduction to Meteorology* (2nd ed.). Englewood Cliffs, N.J.: Prentice-Hall, Inc., 1982.

MARKUS, T. A., and E. N. MORRIS, *Buildings, Climate and Energy.* Marshfield, Mass.: Pitman Publishing, Inc., 1980.

MASON, B. J., *Clouds, Rain and Rainmaking* (2nd ed.). New York: Cambridge University Press, 1975.

MATHER, JOHN R., *Climatology: Fundamentals and Applications.* New York: McGraw-Hill Book Company, 1974.

MCCORMAC, BILLY M., and THOMAS A. SELIGA, eds., *Solar–Terrestrial Influences on Weather and Climate.* Dordrecht, Holland: D. Reidel Publishing Company, 1979.

MILLER, ALBERT, and JACK C. THOMPSON, *Elements of Meteorology* (3rd ed.). Columbus, Ohio: Charles E. Merrill Publishing Company, 1979.

MILLER, DAVID H., *Water at the Surface of the Earth.* (International Geophysics Series, Vol. 21). New York: Academic Press, Inc., 1977.

———, *Energy at the Surface of the Earth.* (International Geophysics Series, Vol. 27.) New York: Academic Press, Inc., 1981.

MONEY, D. C., *Climate, Soils and Vegetation* (3rd ed.). Slough, U.K.: University Tutorial Press Ltd., 1978.

MONTEITH, J. L., ed., *Vegetation and the Atmosphere,* Vol. 1: *Principles* (1975); Vol. 2: *Case Studies* (1976). New York: Academic Press, Inc.

MUNN, R. E., *Biometeorological Methods.* New York: Academic Press, Inc., 1970.

NAROVLYANSKII, G. YA., *Aviation Climatology.* Jerusalem: Israel Program for Scientific Translations, 1970.

NATIONAL RESEARCH COUNCIL, *Understanding Climatic Change: A Program for Action.* Washington, D.C.: National Academy of Sciences, 1975.

———, *Climate and Food.* Washington, D.C.: National Academy of Sciences, 1976.

———, *Climate, Climatic Change, and Water Supply.* Washington, D.C.: National Academy of Sciences, 1977.

———, *Severe Storms: Prediction, Detection, and Warning.* Washington, D.C.: National Academy of Sciences, 1977.

NAVARRA, JOHN GABRIEL, *Atmosphere, Weather and Climate: An Introduction to Meteorology.* Philadelphia: W. B. Saunders Company, 1979.

NEIBURGER, MORRIS, JAMES G. EDINGER, and WILLIAM D. BONNER, *Understanding Our Atmospheric Environment* (2nd ed.). San Francisco: W. H. Freeman and Company, Publishers, 1982.

NIEUWOLT, S., *Tropical Climatology.* New York: John Wiley & Sons, Inc., 1977.

OKE, T. R., *Boundary Layer Climates.* London: Methuen & Company, Ltd, 1978.

OLIVER, JOHN E., *Climate and Man's Environment: An Introduction to Applied Climatology.* New York: John Wiley & Sons, Inc., 1979.

PEARSON, RONALD, *Climate and Evolution.* New York: Academic Press, Inc., 1979.

PERRY, A. H., and J. M. WALKER, *The Ocean–Atmosphere System.* London: Longman Group Limited, 1977.

PITTOCK, A. B., et al., eds., *Climatic Change and Variability: A Southern Perspective.* New York: Cambridge University Press, 1978.

RIEHL, HERBERT, *Introduction to the Atmosphere* (3rd ed.). New York: McGraw-Hill Book Company, 1978.

———, *Climate and Weather in the Tropics.* New York: Academic Press, Inc., 1979.

ROBERTS, WALTER ORR, and HENRY LANSFORD, *The Climate Mandate.* San Francisco: W. H. Freeman and Company, Publishers, 1979.

ROSENBERG, NORMAN J., *Microclimate: The Biological Environment.* New York: John Wiley & Sons, Inc., 1974.

RUMNEY, GEORGE R., *Climatology and the World's Climates.* New York: Macmillan Publishing Co., Inc., 1968.

SCHNEIDER, STEPHEN H., and LYNNE E. MESIROW, *The Genesis Strategy: Climate and Global Survival.* New York: Plenum Publishing Corporation, 1976.

SEEMANN, J., Y. I. CHIRKOV, J. LOMAS, and B. PRIMAULT, *Agrometeorology.* New York: Springer-Verlag Berlin Heidelberg, 1979.

SELLERS, WILLIAM D., *Physical Climatology.* Chicago: University of Chicago Press, 1965.

SHAW, ROBERT H., ed., *Ground Level Climatology,* Pub. No. 86. Washington, D.C.: American Association for the Advancement of Science, 1967.

SIMIU, EMIL, and ROBERT H. SCANLAN, *Wind Effects on Structures: An Introduction to Wind Engineering.* New York: John Wiley & Sons, Inc., 1978.

SMITH, KEITH, *Principles of Applied Climatology.* New York: Halsted Press, 1975.

SPENCE, CLARK C., *The Rainmakers: American "Pluviculture" to World War II.* Lincoln: University of Nebraska Press, 1980.

STRINGER, E. T., *Foundations of Climatology.* San Francisco: W. H. Freeman and Company, Publishers, 1972.

———, *Techniques of Climatology.* San Francisco: W. H. Freeman and Company, Publishers, 1972.

SULMAN, FELIX G., *Health, Weather and Climate.* Basel: S. Karger AG, 1976.

TICKELL, CRISPIN, *Climatic Change and World Affairs.* Elmsford, N.Y.: Pergamon Press, Inc., 1978.

TREWARTHA, GLENN T., and LYLE H. HORN, *An Introduction to Climate* (5th ed.). New York: McGraw-Hill Book Company, 1980.

———, *The Earth's Problem Climates* (2nd ed.). Madison: University of Wisconsin Press, 1981.

TROMP, S. W., *Biometeorology: The Impact of the Weather and Climate on Humans and Their Environment (Plants and Animals).* London: Heyden and Son, Ltd., 1980.

United States Department of Commerce, Weather Bureau, *World Weather Records 1941–1950.* Washington, D.C.: U.S. Government Printing Office, 1959.

———, *World Weather Records 1951–1960,* 6 vols. Washington, D.C.: U.S. Government Printing Office, 1965–68.

———, NOAA, *World Weather Records 1961–1970,* Vol. 1: *North America* (1979); Vol. 2: *Europe* (1979); Vol. 6: *Islands of the World* (1981). Asheville, N.C.: National Climatic Center.

WALKER, JAMES C. G., *Evolution of the Atmosphere.* New York: Macmillan Publishing Co., Inc., 1977.

WALLACE, JOHN M., and PETER V. HOBBS, *Atmospheric Science.* New York: Academic Press, Inc., 1977.

WILLIAMS, JILL, ed., *Carbon Dioxide, Climate and Society.* New York: Pergamon Press, Inc., 1978.

World Meteorological Organization, *International Cloud Atlas,* Vols. I and II. Geneva: WMO, 1956.

YOSHINO, MATATOSHI M., *Climate in a Small Area.* Tokyo: University of Tokyo Press, 1975.

SERIALS

Agricultural Meteorology. Amsterdam: Elsevier Scientific Publishing Company, bimonthly.

Atmosphere–Ocean, the Journal of the Canadian Meteorological and Oceanographic Society. Downsview, Ont.: University of Toronto Press, quarterly. (Published as *Atmosphere* by the Canadian Meteorological Society prior to 1978.)

Bulletin of the American Meteorological Society. Boston: American Meteorological Society, monthly.

Canadian Weather Review. Downsview, Ont.: Atmospheric Environment Service, monthly.

Climate Monitor. Norwich, U.K.: Climatic Research Unit, University of East Anglia, quarterly.

Climatic Change. Dordrecht, Holland: D. Reidel Publishing Company, quarterly.

Climatological Data, National Summary. Asheville, N.C.: NOAA, National Climatic Center, monthly and annual.

Daily Weather Maps, Weekly Series. Washington, D.C.: NOAA, Environmental Data and Information Service, weekly.

Hourly Precipitation Data. Asheville, N.C.: NOAA, National Climatic Center, monthly and annual summary.

International Journal of Biometeorology. Lisse, Holland: Swets & Zeitlinger, B.V., quarterly.

International Journal of Meteorology. Trowbridge, U.K.: Journal of Meteorology, ten issues per year.

Journal of Applied Meteorology. Boston: American Meteorological Society, bimonthly.

Journal of Atmospheric Sciences. Boston: American Meteorological Society, bimonthly.

Journal of Climatology. Chichester, U.K.: John Wiley & Sons Ltd., quarterly.

Journal of the Meteorological Society of Japan. Tokyo: Meteorological Society of Japan, bimonthly.

Local Climatological Data. Asheville, N.C.: NOAA, National Climatic Center, monthly and annual summary.

Marine Observer. London: Her Majesty's Stationery Office, quarterly.

Mariners Weather Log. Washington, D.C.: NOAA, Environmental Data and Information Service, bimonthly.

Meteorological Magazine. London: Her Majesty's Stationery Office, monthly.

Monthly Climatic Data for the World. Asheville, N.C.: NOAA, National Climatic Center, monthly.

Monthly Meteorological Summary. Downsview, Ont.: Atmospheric Environment Service, monthly and annual.

Monthly Weather Report. London: Her Majesty's Stationery Office, monthly.

Monthly Weather Review. Boston: American Meteorological Society, monthly. (Published by Department of Commerce, NOAA prior to 1974).

National Weather Digest. Marlow Heights, Md.: National Weather Association, quarterly.

Quarterly Journal of the Royal Meteorological Society. London: Royal Meteorological Society, quarterly.

Solar Radiation Data, Monthly Summary. Asheville, N.C.: NOAA, National Climatic Center, monthly.

Storm Data. Asheville, N.C.: NOAA, National Climatic Center, monthly.

Weather. London: Royal Meteorological Society, monthly.

Weatherwise. Washington, D.C.: Heldref Publications, bimonthly.

WMO Bulletin. Geneva: World Meteorological Organization, quarterly.

Index

A

B

C

D

G

H

I

J

K

L

M

N

O

P

Q

R

S

T

U

V

W

X

Y

Z